T0190338

Lecture Notes in Computer Science 12467

More information about this series at http://www.springer.com/series/7409

Samin Aref · Kalina Bontcheva ·
Marco Braghieri · Frank Dignum ·
Fosca Giannotti · Francesco Grisolia ·
Dino Pedreschi (Eds.)

Social Informatics

12th International Conference, SocInfo 2020
Pisa, Italy, October 6–9, 2020
Proceedings

Editors
Samin Aref
Max Planck Institute for Demographic Research
Rostock, Germany

Marco Braghieri
King's College London
London, UK

Fosca Giannotti
ISTI-CNR
Pisa, Italy

Dino Pedreschi
University of Pisa
Pisa, Italy

Kalina Bontcheva
University of Sheffield
Sheffield, UK

Frank Dignum
Umeå University
Umeå, Sweden

Francesco Grisolia
University of Pisa
Pisa, Italy

ISSN 0302-9743 ISSN 1611-3349 (electronic)
Lecture Notes in Computer Science
ISBN 978-3-030-60974-0 ISBN 978-3-030-60975-7 (eBook)
https://doi.org/10.1007/978-3-030-60975-7

LNCS Sublibrary: SL3 – Information Systems and Applications, incl. Internet/Web, and HCI

This Springer imprint is published by the registered company Springer Nature Switzerland AG
The registered company address is: Gewerbestrasse 11, 6330 Cham, Switzerland

Preface

This volume contains the proceedings of the 12th Conference on Social Informatics (SocInfo 2020), held in Pisa, Italy during October 6–9, 2020.

This conference series has always aimed to create a venue of discussion for researchers in computational and social sciences, acting as a bridge between these two different and often separated academic communities. SocInfo 2020 had to face challenges brought on by the COVID-19 disease, which was declared a pandemic by the World Health Organization on March 11, 2020. Thus, the conference was conducted virtually, in order to better adapt to the current travel difficulties. However, the conference's goal remained unchanged, that is providing a forum for academics to explore methodological novelties, both in the computational and social science fields, and identify shared research trajectories. Despite the difficulties brought by the pandemic, the organizers received a wide number of diverse and interdisciplinary contributions, ranging from works that ground information-system design on social concepts, to papers that analyze complex social systems using computational methods, or explore socio-technical systems using social sciences methods.

This year SocInfo received 99 submitted papers from a total of 294 distinct authors, located in 37 different countries. We were glad to have a broad and diverse committee of 27 Senior Program Committee members and 152 Program Committee members with a strong interdisciplinary background from all over the world. The Program Committee reviewed all submissions and provided the authors with in-depth feedback on how to improve their work. As was the case last year, SocInfo 2020 employed a double-blind peer-review process involving a two-tiered Program Committee. Papers received an average of three reviews by the Program Committee members, as well as a meta-review from a Senior Program Committee member. Based on their input, the Program Committee chairs selected 30 full and 3 short papers for oral presentation, and 14 submissions for poster presentations.

In addition to posters and paper presentations, SocInfo 2020 hosted four great keynotes: Dr. Leticia Bode, Provost Distinguished Associate Professor at Georgetown University, USA; Professor Virginia Dignum from the Department of Computing Science at Umeå University, Sweden; Professor Alessandro Vespignani, Director of Network Science Institute, Northeastern University, USA; Dr. Bruno Lepri, head of the Mobile and Social Computing Lab at Bruno Kessler Foundation, Italy.

Especially this year, we would like to express our deep gratitude to all the authors, and attendees, for choosing SocInfo 2020 as a forum to present and discuss their research. Moreover, we would like to sincerely thank all those that contributed to a successful conference organization, which has been more challenging than ever due to the COVID-19 pandemic. We would also like to express our appreciation for the input and support of the Steering Committee. Finally, we would like to underline our appreciation for the efforts of the Program Committee members, who played the key in assuring that submissions adhered to the highest standards of originality and scientific rigor.

We are also grateful to Rémy Cazabet and Giulio Rossetti as workshop chairs, to Luca Pappalardo and Marco De Nadai as tutorial chairs, to Vasiliki Voukelatou as poster chair, to Daniele Fadda as web chair, to Ioanna Miliou as publicity chair, to Michela Natilli as sponsorship chair, Chiara Falchi as financial chair, and to Letizia Milli and Francesca Pratesi as registration chairs. We are very thankful to our sponsors, SoBigData++, XAI, AI4EU, HumMingBird, WeVerify, NoBias, HumanAI, the Internet Festival, and Springer, for providing generous support. A special thanks to Asti for their help with logistics and planning.

Finally, we hope that these proceedings will constitute a valid input for your research interests and that you will continue the scientific exchange with the SocInfo community.

October 2020

Samin Aref
Kalina Bontcheva
Marco Braghieri
Frank Dignum
Fosca Giannotti
Francesco Grisolia
Dino Pedreschi

Organization

General Chairs

Fosca Giannotti	National Research Council, Italy
Dino Pedreschi	University of Pisa, Italy

Program Committee Chairs

Kalina Bontcheva	The University of Sheffield, UK
Frank Dignum	Umeå University, Sweden

Workshop Chairs

Remy Cazabet	Claude Bernard University Lyon 1 and French National Centre for Scientific Research, France
Giulio Rossetti	National Research Council, Italy

Proceedings Chairs

Marco Braghieri	King's College London, UK
Francesco Grisolia	University of Pisa and University of Turin, Italy

Tutorial Chairs

Marco De Nadai	Bruno Kessler Foundation, Italy
Luca Pappalardo	University of Pisa and National Research Council, Italy

Submission Chair

Samin Aref	Max Planck Institute for Demographic Research, Germany

Web Chair

Daniele Fadda	National Research Council, Italy

Publicity Chair

Ioanna Miliou	University of Pisa, Italy

Poster Chair

Vasiliki Voukelatou — Scuola Normale Superiore, Italy

Registration Chairs

Francesca Pratesi — National Research Council, Italy
Letizia Milli — National Research Council, Italy

Sponsorship Chair

Michela Natilli — National Research Council, Italy

Financial Chair

Chiara Falchi — National Research Council, Italy

Steering Committee

Karl Aberer	Swiss Federal Institute of Technology Lausanne, Switzerland
Luca Maria Aiello	Nokia Bell Labs, UK
Hsinchun Chen	University of Arizona, USA
Noshir Contractor	Northwestern University, USA
Anwitaman Datta	Nanyang Technological University, Singapore
Andreas Ernst	University of Kassel, Germany
Andreas Flache	University of Groningen, The Netherlands
Dirk Helbing	ETH Zürich, Switzerland
Irwin King	The Chinese University of Hong Kong, Hong Kong
Ee-Peng Lim	Singapore Management University, Singapore
Michael Macy	Cornell University, USA
Daniel A. McFarland	University of Stanford, USA
Sue B. Moon	Korea Advanced Institute of Science and Technology, South Korea
Katsumi Tanaka	Kyoto University, Japan
Adam Wierzbicki	Polish-Japanese Academy of Information Technology, Poland

Senior Program Committee

Luca Maria Aiello	Nokia Bell Labs, UK
Harith Alani	The Open University, UK
Samin Aref	Max Planck Institute for Demographic Research, Germany
Ciro Cattuto	Institute for Scientific Interchange Foundation, Italy
Stephen Cranefield	University of Otago, New Zealand

Kareem Darwish	Qatar Computing Research Institute, Qatar
Jana Diesner	University of Illinois at Urbana-Champaign, USA
Emilio Ferrara	University of Southern California, USA
André Grow	Max Planck Institute for Demographic Research, Germany
Renaud Lambiotte	University of Oxford, UK
Kristina Lerman	University of Southern California, USA
Elisabeth Lex	Graz University of Technology, Austria
Walid Magdy	The University of Edinburgh, UK
Naoki Masuda	State University of New York at Buffalo, USA
Yelena Mejova	Institute for Scientific Interchange Foundation, Italy
Zachary Neal	Michigan State University, USA
Laura Nelson	Northeastern University, USA
Ana Paiva	Institute for Systems and Computer Engineering, Technology and Science, Portugal
Jürgen Pfeffer	The Technical University of Munich, Germany
Jason Radford	University of Chicago, USA
Blaine Robbins	New York University Abu Dhabi, UAE
Frank Schweitzer	ETH Zürich, Switzerland
Megan Squire	Elon University, USA
Steffen Staab	University of Stuttgart, Germany, and University of Southampton, UK
Christoph Trattner	University of Bergen, Germany
Ingmar Weber	Qatar Computing Research Institute, Qatar
Taha Yasseri	University of Oxford, UK

Program Committee

Palakorn Achananuparp	Singapore Management University, Singapore
Thomas Ågotnes	University of Bergen, Germany
Wei Ai	University of Maryland, USA
Mikhail Alexandrov	Autonomous University of Barcelona, Spain
Hamed Alhoori	Northern Illinois University, USA
Kristen Altenburger	Independent
Jisun An	Hamad Bin Khalifa University, Qatar
Aris Anagnostopoulos	Sapienza University of Rome, Italy
Stuart Anderson	The University of Edinburgh, UK
Panagiotis Andriotis	University of the West of England, UK
Pablo Aragón	Pompeu Fabra University, Spain
Makan Arastuie	University of Toledo, USA
Ebrahim Bagheri	Ryerson University, Canada
Vladimir Barash	Cornell University, USA
Dominik Batorski	University of Warsaw, Poland
Christian Bauckhage	Fraunhofer, Germany
Livio Bioglio	University of Torino, Italy
Matteo Bohm	Sapienza University of Rome, Italy

Ludovico Boratto	Eurecat, Spain
Marco Brambilla	The Polytechnic University of Milan, Italy
Ulrik Brandes	ETH Zürich, Switzerland
Pavel Braslavski	Ural Federal University, Russia
Cody Buntain	New Jersey Institute of Technology, USA
Remy Cazabet	Claude Bernard University Lyon 1 and French National Centre for Scientific Research, France
Fabio Celli	Maggioli R&D, Italy
Nina Cesare	University of Washington, USA
Charalampos Chelmis	University at Albany, State University of New York, USA
Yi-Shin Chen	National Tsing Hua University, Taiwan
Dimitris Christopoulos	MODUL University Vienna, Austria; Edinburgh Business School, UK
David Corney	Full Fact, UK
Michele Coscia	IT University of Copenhagen, Denmark
Andrew Crooks	George Mason University, USA
Rubén Cuevas	University Carlos III de Madrid, Spain; UC3M-Santander Big Data Institute, Spain
Ángel Cuevas	University Carlos III de Madrid, Spain
Tiago Cunha	IT University of Copenhagen, Denmark
Marco De Nadai	Bruno Kessler Foundation, Italy
Gianluca Demartini	The University of Queensland, Australia
Djellel Difallah	New York University, USA
Ly Dinh	University of Illinois at Urbana-Champaign, USA
Meixing Dong	University of Michigan, USA
Victor M. Eguiluz	University of the Balearic Islands, Spain
Reza Farahbakhsh	Telecom SudParis, France
Rosta Farzan	University of Pittsburgh, USA
Leo Ferres	University of Desarrollo, Chile, and Telefónica R&D, Spain
Fabian Flöck	GESIS - Leibniz Institute for the Social Sciences, Germany
Gerhard Fuchs	University of Stuttgart, Germany
Sabrina Gaito	University of Milan, Italy
Floriana Gargiulo	French National Centre for Scientific Research and Paris-Sorbonne University, France
Francesca Giardini	University of Groningen, The Netherlands
Maria Glenski	Pacific Northwest National Laboratory, USA
Kwang-Il Goh	Korea University, South Korea
Genevieve Gorrell	The University of Sheffield, UK
Christophe Guéret	Accenture Labs, Ireland
Barbara Guidi	University of Pisa, Italy
Riccardo Guidotti	University of Pisa, Italy
William Hamilton	McGill University, Canada
Oul Han	University of Koblenz-Landau, Germany

Denis Helic	Graz University of Technology, Austria
Tuan-Anh Hoang	L3S Research Center and University of Hanover, Germany
Geert-Jan Houben	Delft University of Technology, The Netherlands
Gholamreza Jafari	Shahid Beheshti University, Iran; Central European University, Hungary
Alejandro Jaimes	Aicure, USA
Adam Jatowt	Kyoto University, Japan
Kazuhiro Kazama	Wakayama University, Japan
Mark Keane	University College Dublin, Ireland
Masahiro Kimura	Ryukoku University, Japan
Katharina Kinder-Kurlanda	GESIS - Leibniz Institute for the Social Sciences, Germany
Ryota Kobayashi	The University of Tokyo, Japan
Andreas Koch	University of Salzburg, Austria
Olessia Koltsova	NRU Higher School of Economics, Russia
Juhi Kulshrestha	GESIS - Leibniz Institute for the Social Sciences, Germany
Haewoon Kwak	Qatar Computing Research Institute, Qatar
Kiran Lakkaraju	Sandia National Laboratories, USA
Hemank Lamba	Carnegie Mellon University, USA
Walter Lamendola	University of Denver, USA
Georgios Lappas	University of Western Macedonia, Greece
Deok-Sun Lee	Inha University, South Korea
Florian Lemmerich	Aachen University, Germany
Bruno Lepri	Bruno Kessler Foundation, Italy
Zoran Levnajic	Faculty of Information Studies, Slovenia
Yong Li	Tsinghua University, China
Ee-Peng Lim	Singapore Management University, Singapore
Claudio Lucchese	Ca' Foscari University of Venice, Italy
Eric Malmi	Google, Switzerland
Rosario Mantegna	University of Palermo, Italy
Gianluca Manzo	French National Centre for Scientific Research and Paris-Sorbonne University, France
Peter Marbach	University of Toronto, Canada
Afra Mashhadi	University of Washington, USA
Binny Mathew	Indian Institute of Technology Kharagpur, India
Friedolin Merhout	IT University of Copenhagen, Denmark
Hisashi Miyamori	Kyoto Sangyo University, Japan
Bamshad Mobasher	DePaul University, USA
Jose Moreno	Toulouse Institute of Computer Science Research, France
Claudia Müller-Birn	Free University of Berlin, Germany
Tsuyoshi Murata	Tokyo Institute of Technology, Japan
Oleg Nagornyy	NRU Higher School of Economics, Russia
Keiichi Nakata	Henley Business School, UK

Mirco Nanni	National Research Council, Italy
Carlos Nunes Silva	University of Lisbon, Portugal
Cecilia Panigutti	Scuola Normale Superiore, Italy
Daniela Paolotti	Institute for Scientific Interchange Foundation, Italy
Symeon Papadopoulos	Information Technologies Institute, Greece
Luca Pappalardo	University of Pisa and National Research Council, Italy
Kunwoo Park	University of California, Los Angeles, USA
Sergei Pashakhin	NRU Higher School of Economics, Russia
Marco Pellegrini	National Research Council, Italy
María Pereda	Polytechnic University of Madrid, Spain
Evaggelia Pitoura	University of Ioannina, Greece
Edward Platt	University of Michigan, USA
Alexander Porshnev	NRU Higher School of Economics, Russia
Polyvios Pratikakis	Foundation for Research & Technology, Greece
Jose J. Ramasco	Spanish National Research Council and University of the Balearic Islands, Spain
Salvatore Rinzivillo	National Research Council, Italy
Giulio Rossetti	National Research Council, Italy
Luca Rossi	IT University of Copenhagen, Denmark
Giancarlo Ruffo	University of Turin, Italy
Mostafa Salehi	University of Tehran, Iran
Mattia Samory	GESIS - Leibniz Institute for the Social Sciences, Germany
Eric Sanders	Radboud University, The Netherlands
Kazutoshi Sasahara	Nagoya University, Japan
Sanja Scepanovic	Nokia Bell Labs, UK
Rajesh Sharma	University of Tartu, Estonia
Roberta Sinatra	IT University of Copenhagen, Denmark
Alina Sirbu	University of Pisa, Italy
Rok Sosic	Stanford University, USA
Konstantinos Sotiropoulos	Boston University, USA
Srinath Srinivasa	International Institute of Information Technology, India
Bogdan State	Stanford University, USA
Pål Sundsøy	Norges Bank Investment Management, Norway
Michael Szell	IT University of Copenhagen, Denmark
Rochelle Terman	University of Chicago, USA
John Ternovski	Yale University, USA
Thanassis Tiropanis	University of Southampton, UK
Michele Tizzoni	Institute for Scientific Interchange Foundation, Italy
Amalia Triantafillidou	University of Western Macedonia, Greece
Klaus G. Troitzsch	University of Koblenz-Landau, Germany (Retired in 2012)
Onur Varol	Northeastern University, USA
Vasiliki Voukelatou	Scuola Normale Superiore, Italy
Wenbo Wang	GoDaddy Inc., USA

Tharindu Cyril Weerasooriya	Rochester Institute of Technology, USA
Rui Wen	Helmholtz Center for Information Security, Germany
Mark Wilson	The University of Auckland, New Zealand
He Xinlei	Helmholtz Center for Information Security, Germany
Kevin S. Xu	University of Toledo, USA
Hirozumi Yamaguchi	Osaka University, Japan
Jie Yang	Delft University of Technology, The Netherlands
Arjumand Younus	University College Dublin, Ireland
Daria Yudenkova	HSE Laboratory for Internet Studies, Russia
Yang Zhang	Helmholtz Center for Information Security, Germany
Zhiqiang Zhong	University of Luxembourg, Luxembourg
Xingquan Zhu	Florida Atlantic University, USA
Arkaitz Zubiaga	Queen Mary University of London, UK

Additional Reviewers

Min Chen
Nighan Chen
Vera Danilova
Mithun Das
Rahul Goel
Masahito Kumano
Haiko Lietz
Yugeng Liu
Shaghayegh Najari
Jun Sun
Koosha Zarei

Sponsors

SoBigData++
XAI
AI4EU
HumMingBird
WeVerify
NoBias
HumanAI
Internet Festival
Springer

Contents

The Determinants of Social Connectedness in Europe 1
Michael Bailey, Drew Johnston, Theresa Kuchler, Dominic Russel, Bogdan State, and Johannes Stroebel

Combining Language Models and Network Features for Relevance-Based Tweet Classification 15
Mohamed Barbouch, Frank W. Takes, and Suzan Verberne

Co-spread of Misinformation and Fact-Checking Content During the Covid-19 Pandemic 28
Grégoire Burel, Tracie Farrell, Martino Mensio, Prashant Khare, and Harith Alani

Facebook Ads: Politics of Migration in Italy. 43
Arthur Capozzi, Gianmarco De Francisci Morales, Yelena Mejova, Corrado Monti, André Panisson, and Daniela Paolotti

It's Not Just About Sad Songs: The Effect of Depression on Posting Lyrics and Quotes. 58
Lucia Lushi Chen, Walid Magdy, Heather Whalley, and Maria Wolters

Understanding the *MeToo* Movement Through the Lens of the Twitter 67
Rahul Goel and Rajesh Sharma

Discovering Daily POI Exploitation Using LTE Cell Tower Access Traces in Urban Environment 81
Sumin Han, Kinam Park, and Dongman Lee

Identifying the Hierarchical Influence Structure Behind Smart Sanctions Using Network Analysis 95
Ryohei Hisano, Hiroshi Iyetomi, and Takayuki Mizuno

Coverage and Evolution of Cancer and Its Risk Factors - A Quantitative Study with Social Signals and Web-Data 108
Saransh Khandelwal and Aurobinda Routray

Detecting Engagement Bots on Social Influencer Marketing 124
Seungbae Kim and Jinyoung Han

Social Capital as Engagement and Belief Revision 137
Gaurav Koley, Jayati Deshmukh, and Srinath Srinivasa

Employee Satisfaction in Online Reviews. 152
Philipp Koncar and Denis Helic

Stable Community Structures and Social Exclusion 168
Boxuan Li, Martin Carrington, and Peter Marbach

Inside the X-Rated World of "Premium" Social Media Accounts 181
Nikolaos Lykousas, Fran Casino, and Constantinos Patsakis

Impact of Online Health Awareness Campaign: Case of National Eating Disorders Association 192
Yelena Mejova and Víctor Suarez-Lledó

Moral Framing and Ideological Bias of News 206
Negar Mokhberian, Andrés Abeliuk, Patrick Cummings, and Kristina Lerman

Malicious Bot Detection in Online Social Networks: Arming Handcrafted Features with Deep Learning 220
Guanyi Mou and Kyumin Lee

Spam Detection on Arabic Twitter 237
Hamdy Mubarak, Ahmed Abdelali, Sabit Hassan, and Kareem Darwish

Beyond Groups: Uncovering Dynamic Communities on the WhatsApp Network of Information Dissemination. 252
Gabriel Peres Nobre, Carlos Henrique Gomes Ferreira, and Jussara Marques Almeida

Structural Invariants in Individuals Language Use: The "Ego Network" of Words 267
Kilian Ollivier, Chiara Boldrini, Andrea Passarella, and Marco Conti

Dynamics of Scientific Collaboration Networks Due to Academic Migrations 283
Pavlos Paraskevopoulos, Chiara Boldrini, Andrea Passarella, and Marco Conti

Social Search and Task-Related Relevance Dimensions in Microblogging Sites 297
Divi Galih Prasetyo Putri, Marco Viviani, and Gabriella Pasi

Regional Influences on Tourists Mobility Through the Lens of Social Sensing 312
Helen Senefonte, Gabriel Frizzo, Myriam Delgado, Ricardo Lüders, Daniel Silver, and Thiago Silva

Genuine Personal Identifiers and Mutual Sureties for Sybil-Resilient Community Growth. 320
Gal Shahaf, Ehud Shapiro, and Nimrod Talmon

Political Framing: US COVID19 Blame Game 333
Chereen Shurafa, Kareem Darwish, and Wajdi Zaghouani

Measuring Adolescents' Well-Being: Correspondence of Naïve Digital Traces to Survey Data 352
Elizaveta Sivak and Ivan Smirnov

A Computational Analysis of Polarization on Indian and Pakistani Social Media 364
Aman Tyagi, Anjalie Field, Priyank Lathwal, Yulia Tsvetkov, and Kathleen M. Carley

Young Adult Unemployment Through the Lens of Social Media: Italy as a Case Study. 380
Alessandra Urbinati, Kyriaki Kalimeri, Andrea Bonanomi, Alessandro Rosina, Ciro Cattuto, and Daniela Paolotti

Women Worry About Family, Men About the Economy: Gender Differences in Emotional Responses to COVID-19. 397
Isabelle van der Vegt and Bennett Kleinberg

Jump on the Bandwagon? – Characterizing Bandwagon Phenomenon in Online NBA Fan Communities 410
Yichen Wang, Jason Shuo Zhang, Xu Han, and Qin Lv

ALONE: A Dataset for Toxic Behavior Among Adolescents on Twitter. 427
Thilini Wijesiriwardene, Hale Inan, Ugur Kursuncu, Manas Gaur, Valerie L. Shalin, Krishnaprasad Thirunarayan, Amit Sheth, and I. Budak Arpinar

Cross-Domain Classification of Facial Appearance of Leaders 440
Jeewoo Yoon, Jungseock Joo, Eunil Park, and Jinyoung Han

The Effect of Structural Affinity on the Diffusion of a Transnational Online Movement: The Case of #MeToo 447
Xinchen Yu, Shashidhar Reddy Daida, Jeremy Boy, and Lingzi Hong

Author Index 461

The Determinants of Social Connectedness in Europe

Michael Bailey[1(✉)], Drew Johnston[2], Theresa Kuchler[3], Dominic Russel[3(✉)], Bogdan State[1], and Johannes Stroebel[3]

[1] Facebook, Menlo Park, USA
{mcbailey,bogdanstate}@fb.com
[2] Harvard University, Cambridge, USA
drewjohnston@g.harvard.edu
[3] New York University, New York, USA
{tkuchler,drussel}@stern.nyu.edu, johannes.stroebel@nyu.edu

Abstract. We use de-identified and aggregated data from Facebook to study the structure of social networks across European regions. Social connectedness declines strongly in geographic distance and at country borders. Historical borders and unions—such as the Austro-Hungarian Empire, Czechoslovakia, and East/West Germany—shape present-day social connectedness over and above today's political boundaries and other controls. All else equal, social connectedness is stronger between regions with residents of similar ages and education levels, as well as between regions that share a language and religion. In contrast, region-pairs with *dissimilar* incomes tend to be more connected, likely due to increased migration from poorer to richer regions.

Keywords: Social connectedness · Homophily · Border effects

1 Introduction

Social networks shape many aspects of global society including patterns of migration and travel, social mobility, and political preferences. In turn, social networks reflect both past and present political borders and migration patterns, as well as geographic proximity, culture, and other factors. While understanding the determinants and effects of these networks across regions and countries can be informative for a wide range of questions in the social sciences, researchers have traditionally been limited by the scarcity of large-scale representative data on regional social connections.

In this paper, we investigate the spatial structure of social networks in Europe. We measure social networks using de-identified and aggregated data

S. Aref et al. (Eds.): SocInfo 2020, LNCS 12467, pp. 1–14, 2020.
https://doi.org/10.1007/978-3-030-60975-7_1

from Facebook, a global online social network.[1] We construct a measure of social connectedness across European NUTS2 regions—regions with between 800,000 and 3 million inhabitants—which captures the probability that Facebook users located in these regions are Facebook friends with each other. Europe consists of a number of proximate nations, has a relatively high population density, and includes a diversity of areas with distinct cultural and linguistic identities. Each of these factors differentiates Europe from the U.S., which has been the primary focus of prior research on social connectedness. This paper documents the important role that these and other factors play in shaping social connections, and thereby advances our understanding of the determinants of social networks.

We begin by discussing a number of case studies that show the relationship of European social connections with patterns of migration, past and present political borders, geographic distance, language, and other demographic characteristics. We then explore the association between social connectedness and these factors more formally. We find that social connectedness strongly declines in geographic distance: a 10% increase in distance is associated with a 13% decline in social connectedness. Social connectedness also drops off sharply at country borders. Controlling for geographic distance, the probability of friendship between two individuals living in the same country is five to eighteen times as large as the probability for two individuals living in different countries. Furthermore, using a number of 20th century European border changes, we find that this relationship between political borders and connectedness can persist decades after boundaries change. For example, we find higher social connectedness across regions that were originally part of the Austro-Hungarian empire, even after controlling for distance, current country borders, and a number of other relevant factors.

In addition to distance and political borders, we find that regions more similar along demographic measures such as language, religion, education, and age are more socially connected. In particular, social connectedness between two regions with the same most common language is about 4.5 times larger than for two regions without a common language, again controlling for same and border country effects, distance, and other factors. In contrast, we see that pairs of regions with *dissimilar* incomes are more connected. This finding may be explained by patterns of migration from regions with low incomes to regions with high income. This finding in Europe contrasts with prior research that finds a positive relationship between connectedness and income similarity across U.S. counties and New York zip codes [3,5].

2 Data

Our measures of social connectedness across locations builds on de-identified administrative data from Facebook, a global online social networking service.

[1] The European social connectedness data that we compile and use in this project is accessible at http://data.humdata.org/dataset/social-connectedness-index. See [3] for information on county-level U.S. social network data and [5] for zip code-level data in the New York Combined Statistical area. Additional results using the European data are available in our Online Appendix, located at: http://arxiv.org/abs/2007.12177.

Facebook was created in 2004 and, by the fourth quarter of 2019, had about 2.5 billion monthly active users globally, including 394 million in Europe.

While Facebook users are unlikely to be entirely representative of the populations we study, it has a wide user base. One independent resource estimates 80% of European social media site visits from September 2018 to September 2019 were to Facebook [21]. A separate study found that the number of active accounts on the most used social network in each country, as a share of population, was 66% in Northern Europe, 56% in Southern Europe, 54% in Western Europe, and 45% in Eastern Europe [24]. Another 2018 survey found that the share of adults who used any social networking site in 10 European countries was between 40% and 67% [19].

A related question evolves around the extent to which friendship links on Facebook correspond to real world friendship links. We believe that this is likely. Establishing a Facebook friendship link requires the consent of both individuals, and the total number of friends for a person is limited to 5,000. As a result, networks formed on Facebook more closely resemble real-world social networks than those on other online platforms, such as Twitter, where uni-directional links to non-acquaintances, such as celebrities, are common.

We observed a de-identified snapshot of all active Facebook users from July 2019. We focus on those users who reside in one of 37 European countries and who had interacted with Facebook over the 30 days prior to the date of the snapshot. The 37 countries are the members of the European Union and European Free Trade Association, as well as European Union candidate countries as of 2016; these countries were selected because they have standardized administrative boundaries at the NUTS2 (Nomenclature of Territorial Units for Statistics level 2) level.[2] NUTS2 regions contain between 800,000 and 3 million people, and are generally based on existing sub-national administrative borders. For example, NUTS2 corresponds to 21 "regions" in Italy, 12 "provinces" in the Netherlands, and a single unit for all of Latvia.

To measure social connections between NUTS2 regions, we follow [3] and construct our measure of $SocialConnectedness_{ij}$ as follows:

$$SocialConnectedness_{ij} = \frac{FB\_Connections_{ij}}{FB\_Users_i * FB\_Users_j} \tag{1}$$

Here, $FB\_Connections_{ij}$ is the total number of connections between individuals living in NUTS2 region i and individuals living in NUTS2 region j. FB_Users_i and FB_Users_j are the number of eligible Facebook users in each region. Dividing by the product of regional Facebook users allows us to take into account the fact that we will see more friendship links between regions with more Facebook

[2] Specifically the list of countries is: Albania, Austria, Belgium, Bulgaria, Croatia, Cyprus, Czech Republic, Denmark, Estonia, Germany, Greece, Finland, France, Hungary, Iceland, Ireland, Italy, Latvia, Lichtenstein, Lithuania, Luxembourg, Malta, Montenegro, the Netherlands, North Macedonia, Norway, Poland, Portugal, Romania, Serbia, Spain, Slovakia, Slovenia, Sweden, Switzerland, Turkey, and the United Kingdom.

users. This measure captures the probability that two arbitrary Facebook users across the two countries are friends with each other: if $SocialConnectedness_{ij}$ is twice as large, a Facebook user in region i is about twice as likely to be connected with a given Facebook user in region j.

We have shown in previous work that this measure of social connectedness is useful for describing real-world social networks. We also documented that it predicts a large number of important economic and social interactions. For example, social connectedness as measured through Facebook friendship links is strongly related to patterns of sub-national and international trade [6], patent citations [3], travel flows [5], investment decisions [13] and the spread of COVID-19 [14]. More generally, we have found that information on individuals' Facebook friendship links can help understand their product adoption decisions [7] and their housing and mortgage choices [2,4].

3 Determinants of European Social Connectedness

To illustrate the data and explore the factors that shape social connections within Europe, we first highlight the geographic structure of social connections of a few European regions. We provide additional cases studies in the Online Appendix.

Figure 1 maps the social network of South-West Oltenia in Romania in Panel A and the Samsun Subregion in Turkey in Panel B; darker shading indicates greater connectedness. In both examples, the strongest social connections are to nearby regions in the same country. Residents of South-West Oltenia have relatively strong social connections throughout Europe, especially to Italy, Spain, Germany, and the United Kingdom. This is likely related to patterns of migration. Romania became a member of the European Union in 2007, which entitled its citizens to certain freedoms to travel and work in other EU member states. According to a report by the World Bank, between 3 and 5 million Romanians currently live and work abroad, representing around a fifth of the country's population. The top destination countries in 2017 were Italy, Spain, Germany, the United States, and the United Kingdom [25]. By contrast, Panel B shows that the connections between the Samsun Subregion in Turkey, which is not an EU member state, and other European regions are much weaker. The strongest connections between the Samsun Subregion and other countries are concentrated in western Germany and Berlin, with substantially weaker connections in eastern Germany (former German Democratic Republic). These connections likely reflect the lasting impacts of the West Germany's 1961–1973 labor recruitment agreement *Anwerbeabkommen* with Turkey, which resulted in many Turkish workers re-settling in West Germany (see the discussion in [1]).

Assessing Potential Determinants of Social Connectedness. We next assess the role of the determinants of European social connectedness in a regression framework. To estimate the relationship between various factors and social connectedness between European regions, we estimate the following equation:

$$\log(SocialConnectedness_{ij}) = \beta_0 + \beta_1 \log(d_{ij}) + X_{ij} + \psi_i + \psi_j + \epsilon_{ij} \quad (2)$$

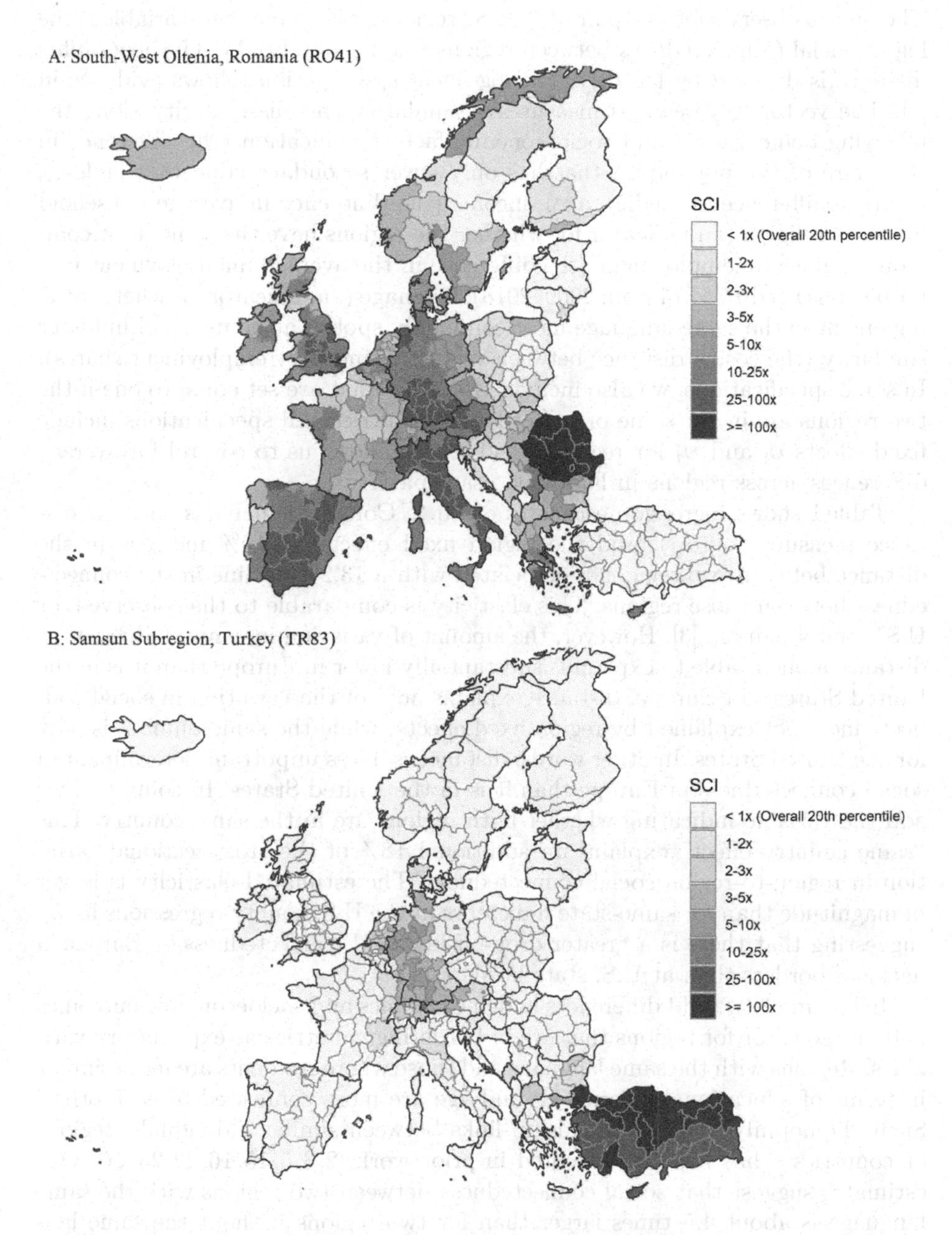

Fig. 1. Social Network Distributions in Romania and Turkey (*Note:* Figure shows the relative probability of connection, measured by $SocialConnectedness_{ij}$, of all European regions j with two regions i: South-West Oltenia, RO (Panel A) and Samsun Subregion, TR (Panel B). The measures are scaled from the 20th percentile of all i, j pairs in Europe. Darker regions have a higher probability of connection).

The unit of observation is a pair of NUTS2 regions. The dependent variable is the log of Social Connectedness between regions i and j (see Eq. 1). The geographic distance is denoted by $\log(d_{ij})$. The log-linear specification follows evidence in [3]. The vector X_{ij} includes measures of similarity and dissimilarity along the following demographic and socioeconomic factors: education (the difference in the share of the population that has only lower secondary education or less), age (the difference in median age), income (the difference in average household income), religion (an indicator for whether the regions have the same most common religion), unemployment (the difference in the average unemployment rate for persons aged 15 to 74 from 2009–2018), language (an indicator for whether the regions have the same language most commonly spoken at home), and industry similarity (the cosine distance between vectors of industry employment shares). In some specifications, we also include indicators that are set equal to one if the two regions are in the same or in bordering countries. All specifications include fixed effects ψ_i and ψ_j for regions i and j; this allows us to control for average differences across regions in Facebook usage patterns.

Table 1 shows regression estimates of Eq. 2. Column 1 includes only the distance measure, $\log(d_{ij})$, and the region fixed effects. A 10% increase in the distance between two regions is associated with a 13.2% decline in the connectedness between those regions. This elasticity is comparable to that observed for U.S. county pairs in [3]. However, the amount of variation in connectedness that distance alone is able to explain is substantially lower in Europe than it is in the United States—in Europe, distance explains 36% of the variation in social connectedness not explained by region fixed effects, while the same number is 65% for the United States. In other words, distance is a less important determinant of social connectedness in Europe than it is in the United States. In column 2, we add the variable indicating whether both regions are in the same country. This "same country effect" explains an additional 18% of the cross-sectional variation in region-to-region social connectedness. The estimated elasticity is larger in magnitude than for same-state indicators in the U.S. county regressions in [3], suggesting that there is a greater drop-off in social connectedness at European national borders than at U.S. state borders.

In column 3, we add differences in demographics and socioeconomic outcomes and an indicator for regions that are in bordering countries as explanatory variables. Regions with the same language and those where residents are more similar in terms of educational attainment and age are more connected to each other. Such "homophily" – more friendship links between similar individuals, regions or countries – has been documented in prior work [2,3,5,15,16,22,23,27]. Our estimates suggest that social connectedness between two regions with the same language is about 4.5 times larger than for two regions without the same language, even after controlling for same country and border country effects, geographic distance, and other demographic and socioeconomic factors. When we include language and demographic factors, the estimated effect of being in the same country falls (from a coefficient estimate of 2.9 to 1.6) suggesting that

Table 1. Determinants of social connectedness across region Pairs

	Dependent variable: log (SocialConnectedness)						
	(1)	(2)	(3)	(4)	(5)	(6)	(7)
log(Distance in KM)	−1.318***	−0.558***	−0.582***	−0.572***	−0.737***	−1.177***	−0.591***
	(0.046)	(0.053)	(0.041)	(0.038)	(0.027)	(0.032)	(0.031)
Same Country		2.896***	1.651***				
		(0.077)	(0.124)				
Border Country			0.285***	0.340***			
			(0.044)	(0.046)			
Δ Share Pop Low Edu (%)			−0.013***	−0.012***	−0.002	−0.007**	−0.000
			(0.002)	(0.002)	(0.001)	(0.003)	(0.001)
Δ Median Age			−0.017***	−0.021***	0.000	−0.014***	0.001
			(0.004)	(0.004)	(0.003)	(0.005)	(0.002)
Δ Avg Income (k €)			0.053***	0.055***	0.015***	0.025***	0.012***
			(0.003)	(0.003)	(0.002)	(0.006)	(0.002)
Δ Unemployment (%)			−0.000	0.004	0.006*	0.021**	0.007*
			(0.005)	(0.005)	(0.003)	(0.010)	(0.004)
Same Religion			0.027	0.049*	0.044***	0.127***	0.029**
			(0.031)	(0.025)	(0.013)	(0.040)	(0.013)
Same Language			1.493***	1.548***	1.529***	2.279***	1.909***
			(0.097)	(0.120)	(0.216)	(0.133)	(0.107)
Industry Similarity			0.128	0.044	0.528***	0.242	0.633***
			(0.169)	(0.158)	(0.107)	(0.199)	(0.109)
NUTS2 FEs	Y	Y	Y	Y	Y	Y	Y
Indiv. Same Country FEs				Y			
All Country Pair FEs					Y	Y	Y
Sample						Same country	Diff. country
R^2	0.490	0.669	0.745	0.775	0.906	0.927	0.839
Number of Observations	75,900	75,900	75,900	75,900	75,900	5,266	70,634

Note: Table shows results from Regression 2. The unit of observation is a NUTS2 region pair. The dependent variable in all columns is the log of $SocialConnectedness_{ij}$. Column 1 includes the log of distance and region fixed effects. Column 2 adds a control for regions in the same country. Column 3 incorporates demographic and socioeconomic similarity measures, as well as a control for regions in countries that border. Column 4 adds fixed effect for each same-country pair. Column 5 adds fixed effects for each country pair. Columns 6 and 7 limit the observations to pairs in the same country and pairs in different countries, respectively. Standard errors are double clustered by each region i and region j in a region pair. Significance levels: *(p<0.10), **(p<0.05), ***(p<0.01).

some—but not all—of the higher in intra-country connectedness is due to common language and other demographic similarities.

Somewhat surprisingly, we see higher connectedness between regions with larger differences in income, even after controlling for country-pair fixed effects, and both limiting to regions within the same country and limiting to regions in different countries. In some of these specifications, we also see a positive relationship between connectedness and differences in unemployment. These relationships run contrary to findings from prior research that finds positive relationships between connectedness and income similarity across U.S. counties and New York zip codes [3,5]. A possible explanation is related to the migration patterns suggested by our case studies: migrants are particularly likely to move from regions with low income (or higher unemployment) to regions with higher income (or lower unemployment) and comparatively less likely to move to other low or middle income regions. Hence, we see more migration and more connections between regions with large differences in income versus those with more similar levels of income or unemployment. This finding is particularly interesting in light of a recent and substantial literature on intra-U.S. migration that documents a general decline in moves over the past three decades and the importance of opportunistic moves for the U.S. labor market (for example, [11,12,17,26]). By contrast, much less is known about regional migration flows within Europe, largely due to a lack of comprehensive data. The existing prior research has focused on country-to-country flows [10], the intensity of within country migration [8,9], or regional net-migration [20]. Our unique data set on connectedness provides insights into region-to-region migration patterns throughout the continent. For example, existing data show that within country moves in Europe are generally less common than in the United States; however, the positive relationship we observe between income dissimilarity and connectedness, compared to the negative relationship observed in the U.S., suggest that there may be higher rates of migration in Europe from less prosperous to more prosperous regions. These are exactly the opportunistic moves that increase labor market dynamism.

Column 4 adds fixed effects for each same-country pair, and column 5 adds fixed effects for every country pair. The magnitude of the coefficient on income dissimilarity falls, consistent with country-level migration flows explaining some of the connectedness between regions with dissimilar incomes; however, even holding average connectedness across country pairs fixed, social connectedness is stronger between regions with more different incomes. Columns 6 and 7 limit to pairs of regions in the same and in different countries, respectively. Social connectedness declines in geographic distance more within countries than across countries: a 10% greater geographic distance between regions within the same country implies a 11.7% decrease in social connectedness, whereas a 10% greater geographic distance between regions in different countries implies only a 5.9% decrease in connectedness (conditional on the other controls).

Strength of Within-Country Connectedness. So far, we have shown that, on average, regions in the same country are more connected than regions in different countries that are similarly far apart. We next explore the extent of heterogeneity in this within-country effect on connectedness. We do so by comparing the coefficients on the individual same-country effects estimated in column 4 of Table 1, which capture the additional connectedness associated with two regions being part of the same country. Figure 2 shows these coefficients plotted for all countries with two or more NUTS2 regions. Higher values are indicative of stronger within-country social connectedness. Within-country connectedness is generally stronger for countries with smaller populations, such as Slovenia and Croatia, than for countries with larger populations, such as the United Kingdom and Germany. There are also noticeable differences between countries of similar sizes. For example, the United Kingdom and France have roughly equal populations, yet two regions in France are on average 18 times more connected than two similarly situated regions in Europe, whereas two regions in the United Kingdom are only 1.8 times more connected. There are several possible reasons for such differences, such as historical patterns (e.g., did the nations unite at different times?), geography (e.g., are there physical barriers that separate parts of the nations?), or modern government structures (e.g., do sub-regional governments have greater autonomy in some countries than in others?). Determining the relative importance of these factors is an exciting avenue for future research.

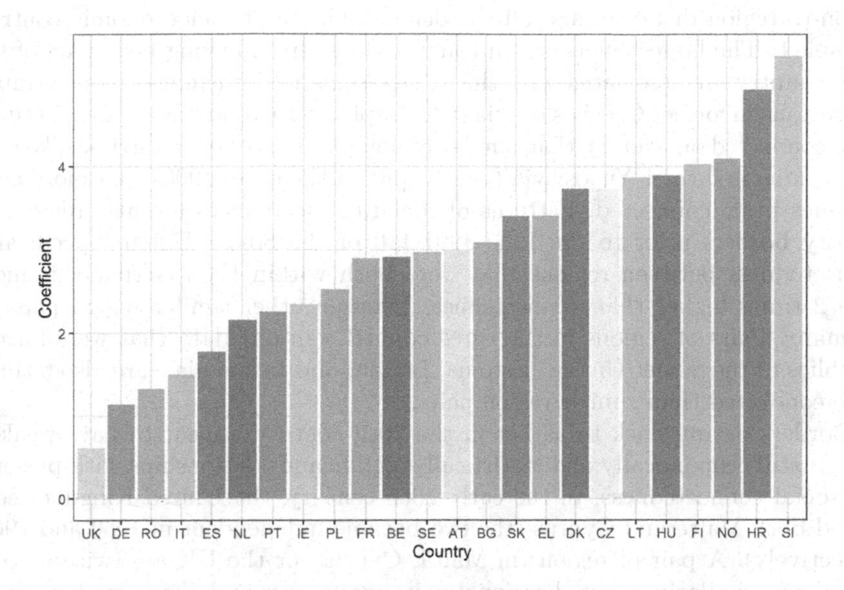

Fig. 2. Connectedness within European Countries (*Note:* Figure shows coefficients of the individual same-country effects from the regression reported in column 4 of Table 1. The coefficients are roughly the additional connectedness that is associated with two regions being part of the same country, for each country. Higher values are indicative of stronger within-country connectedness (after controlling for certain demographic and socioeconomic effects). The labels on the x-axis are the two-letter prefix of each country's NUTS codes).

Relationship Between Historical Borders and Connectedness. Next, we take a more detailed look at the relationship between historical political boundaries and today's social connectedness. Information on national borders in 1900, 1930, 1960, and 1990 comes from [18].[3] Table 2 adds additional variables based on these historical borders to the analysis in Table 1. Column 1 uses all of the same controls as column 4 in Table 1 except log(Distance); throughout this table, we instead use 100 dummy variables representing percentiles of the distribution of distance to avoid picking up non-linearities in the relationship between geographic distance and historical borders.[4] Columns 2 to 5 add indicators based on national borders at the start of 1990, 1960, 1930, and 1900, respectively.

[3] In cases when a modern NUTS2 region spans two historical countries, we classify the region as part of the country for which it had a greater land area overlap.

[4] In general, the historical coefficients in Table 2 do not change, or become slightly larger, when using log(Distance).

We look at several major European border changes dating back to the early 20th century, showing that present-day connectedness is higher between regions that have been part of the same country in the past. This result is *in addition* to the effects of being in the same country today, being in bordering countries today, region-to-region distance, and all the demographic and socioeconomic controls in Table 1. The largest increases in connectedness from having been part of the same country are associated with the most recent border changes. For example, two regions in former Czechoslovakia (which split in 1993) are more than 19 times more connected on average than similar region pairs in other countries. Likewise, two regions in former Yugoslavia (which split in the early 1990s) are more than 13 times more connected. Patterns of social connectedness are also related to country borders prior to the 1990–1991 fall of the Soviet Union. Specifically, connectedness between regions that were both within East Germany is more than 2 times higher than connectedness between other similar region pairs in Germany. Pairs of regions in the three countries in our data that were former republics of the Soviet Union—Estonia, Latvia, and Lithuania—are also 6 times more connected than similar region pairs.

Borders dating back to earlier in the 20th century appear to have weaker, though still economically and statistically significant relationships with present-day social connectedness. In the early 20th century, the United Kingdom controlled both Malta and Cyprus (the two became independent in 1960 and 1964, respectively). A pair of regions in Malta, Cyprus, or the UK are twice as connected as a similarly situated regional pair, again, over and above modern country borders. The borders of Germany in 1930 were also different than today: the country included the Liege region in modern Belgium and a number of regions in modern Poland; on the other hand, it did not include the Saarland—a formerly independent nation within modern Germany. We find a 46% increase in connectedness between regions that were part of 1930 Germany (but are not part of the same country today).

Finally, we look at three national borders that changed before or shortly after the first World War: the Austro-Hungarian Empire, the German Empire, and the United Kingdoms of Sweden and Norway. In 1900, the Austro-Hungarian Empire stretched across much of central and eastern Europe, encompassing part or all of modern Austria, Hungary, Czech Republic, Slovakia, Slovenia, Croatia, Romania, Poland, and Italy. After adding our present-day controls, we find that two regions within this empire are more than 90% more connected than a pair of otherwise similar regions. Compared to modern Germany, the German Empire in 1900 controlled large parts of modern Poland (even more so than 1930 Germany) and the Alsace region of France. We find that having been part of the German Empire in 1900 is associated with a 50% increase in present-day social connectedness, again controlling for both the effects of the modern German borders and 1930 German borders. It is interesting that the regression primarily loads on the older 1900 borders, while the coefficient for the 1930 borders decreases. One possible explanation is the period of time the borders were in effect: whereas the 1930 German borders were effective only in the 20 year

Table 2. Historical determinants of social connectedness

	Dependent variable: log(SocialConnectedness)				
	(1)	(2)	(3)	(4)	(5)
		1990	1960	1930	1900
Border Country	0.418***	0.399***	0.392***	0.372***	0.310***
	(0.045)	(0.045)	(0.045)	(0.045)	(0.043)
Both Czechoslovakia		3.525***	3.529***	3.541***	2.945***
		(0.217)	(0.217)	(0.216)	(0.217)
Both Yugoslavia		3.108***	3.110***	3.123***	2.616***
		(0.105)	(0.105)	(0.105)	(0.114)
Both West Germany		0.006	0.005	0.015	−0.005
		(0.046)	(0.046)	(0.044)	(0.043)
Both East Germany		1.088***	1.092***	1.072***	1.124***
		(0.053)	(0.053)	(0.055)	(0.050)
Both Soviet Union		1.884***	1.874***	1.882***	2.052***
		(0.080)	(0.081)	(0.081)	(0.077)
Both United Kingdom 1960			1.015***	1.016***	0.998***
			(0.155)	(0.156)	(0.157)
Both Germany 1930				0.465***	0.159**
				(0.104)	(0.063)
Both Austro-Hungarian Empire 1900					0.920***
					(0.111)
Both German Empire 1900					0.492***
					(0.074)
Both United Sweden-Norway					2.057***
					(0.123)
Distance Controls	Y	Y	Y	Y	Y
Table 1 Controls	Y	Y	Y	Y	Y
NUTS2 FEs	Y	Y	Y	Y	Y
Indiv. Same Country FEs	Y	Y	Y	Y	Y
R^2	0.784	0.790	0.791	0.792	0.801
Number of Observations	75,900	75,900	75,900	75,900	75,900

Note: Table shows results from Regression 2 with added historical country borders controls X_{ij}. The unit of observation is a NUTS2 region pair. The dependent variable in all columns is the log of $SocialConnectedness_{ij}$. Every column includes controls for same country, region i, and region j effects. Column 1 is the same as column 4 of Table 1, except with 100 dummy variables representing percentiles of distance instead of log(distance). Columns 2, 3, 4, and 5 add controls for certain historical borders in 1990, 1960, 1930, and 1900, respectively. Coefficients for the demographic and socioeconomic controls in Table 1 are excluded for brevity. Standard errors are double clustered by each region i and region j in a region-pair. Significance levels: *(p<0.10), **(p<0.05), ***(p<0.01).

interwar period (and indeed changed even during that period), the 1900 borders essentially remained unchanged for nearly 50 years between 1871 to 1918. Lastly, from 1814 to 1905 the lands of present-day Sweden and Norway were united under a common monarch as the United Kingdoms of Sweden and Norway. A pair of regions within this union are more than 7 times more connected today than similarly situated regions in otherwise similar country-pairs. As with all of our analyses, the historical patterns we observe are correlations rather than necessarily causal and may also capture the effect of other factors that relate to historical borders that we do not explicitly control for.

4 Conclusion

We use de-identified and aggregated data from Facebook to better understand social connections in Europe. We find that social connectedness declines substantially in geographic distance and at country borders. Using a number of 20th century border changes (such as the breakups of the Austro-Hungarian Empire and Czechoslovakia), we find that the relationship between political borders and social connectedness can persist decades after boundaries change. We also find evidence of homophily in Europe, as connections are stronger between regions with residents of similar ages and education levels, as well as between those that share a language and religion. However, region pairs with *dissimilar* incomes are more connected, likely due to migration from poorer to richer regions.

In our Online Appendix, we explore a number of *effects* of social connections across countries. We first look at the relationship between social connectedness and travel flows. We find that a 10% increase in social connectedness between two regions is associated with a 12% to 17% increase in the number of passengers that travel between the regions by train. This result persists even after controlling for geographic distance and travel time, by train and car, between the central points of the regions. We highlight that this result provides empirical support for a number of theoretical models suggesting social networks play an important role in individuals' travel decisions. It also provides strong evidence that the patterns of social connectedness correspond to real-world social connections.

In the Online Appendix, we also study how variation in the degree of connectedness of European regions to other countries is reflected in political outcomes. We first document substantial variation across European regions in the share of friendship links that are to individuals living in other European countries: at the 10th percentile of the distribution, less than 4.1% of connections are to individuals in a different country, compared to over 19.7% at the 90th percentile. We then explore the relationship between this variation and the share of a region's residents that hold Eurosceptic beliefs or that vote for Eurosceptic political parties. According to both measures, we find that Euroscepticism decreases with the share of a region's connections that to regions in a different European country. Specifically, a 1% point increase in the share of a region's connections that are to individuals outside of their home country is associated with a 0.5% point increase in the share of residents who trust the E.U. and a 0.76% point decrease in the share that voted for an anti-E.U. political party. These results persist, but become weaker (0.25 and −0.54% points, respectively), after adding controls for the share of residents living in the region who are born in other European countries as well as the regional average income and unemployment rate, and the shares of employment in manufacturing, construction, and professional sectors. While causality behind this result is hard to establish, it is consistent with the theory that exposure to other European countries increases pro-European views, a narrative that lies behind the creation of programs such as the Erasmus European student exchange.

Our Online Appendix & Additional Results are available at: http://arxiv.org/abs/2007.12177

References

1. Aydin, Y.: The Germany-Turkey migration corridor: Refitting policies for atransnational age. Report, Migration Policy Institute (2016)
2. Bailey, M., Cao, R., Kuchler, T., Stroebel, J.: The economic effects of social networks: evidence from the housing market. J. Polit. Econ. **126**(6), 2224–2276 (2018)
3. Bailey, M., Cao, R., Kuchler, T., Stroebel, J., Wong, A.: Social connectedness: measurements, determinants, and effects. J. Econ. Perspect. **32**(3), 259–80 (2018)
4. Bailey, M., Dávila, E., Kuchler, T., Stroebel, J.: House price beliefs and mortgage leverage choice. Rev. Econ. Stud. **86**(6), 2403–2452 (2018)
5. Bailey, M., Farrell, P., Kuchler, T., Stroebel, J.: Social connectedness in urban areas. J. Urban Econ. **188**, 103264 (2020)
6. Bailey, M., Gupta, A., Hillenbrand, S., Kuchler, T., Richmond, R., Stroebel, J.: International trade and social connectedness. Working Paper 26960, National Bureau of Economic Research (2020)
7. Bailey, M., Johnston, D.M., Kuchler, T., Stroebel, J., Wong, A.: Peer effects in product adoption. Working Paper 25843, National Bureau of Economic Research (2019)
8. Bell, M., Charles-Edwards, E., Ueffing, P., Stillwell, J., Kupiszewski, M., Kupiszewska, D.: Internal migration and development: comparing migration intensities around the world. Popul. Dev. Rev. **41**(1), 33–58 (2015)
9. Esipova, N., Pugliese, A., Ray, J.: The demographics of global internal migration. Migrat. Policy Pract. **3**(2), 3–5 (2013)
10. World Migration Report 2018, International Organization for Migration (2017)
11. Kaplan, G., Schulhofer-Wohl, S.: Understanding the long-run decline in interstate migration. Working Paper 18507, National Bureau of Economic Research, November 2012
12. Karahan, F., Li, D.: What caused the decline in interstate migration in the united states? In: Liberty Street Economics (blog). Federal Reserve Bank of New York (2016)
13. Kuchler, T., Peng, L., Stroebel, J., Li, Y., Zhou, D.: Social proximity to capital: implications for investors and firms. Working Paper 27299, National Bureau of Economic Research (2020)
14. Kuchler, T., Russel, D., Stroebel, J.: The geographic spread of COVID-19 correlates with structure of social networks as measured by Facebook. Working Paper 26990, National Bureau of Economic Research (2020)
15. Lazarsfeld, P., Merton, R.K.: Friendship as a social process: a substantive and methodological analysis. In: Berger, M., Abel, T., Page, C.H. (eds.) Freedom and Control in Modern Society, pp. 18–66 (1954)
16. Marmaros, D., Sacerdote, B.: How do friendships form? Q. J. Econ. **121**(1), 79–119 (2006)
17. Molloy, R., Smith, C.L., Wozniak, A.: Job changing and the decline in long-distance migration in the United States. Demography **54**(2), 631–653 (2017)
18. Max Planck Institute for Demographic Research and Chair for Geodesy and Geoinformatics, University of Rostock. MPIDR Population History GIS Collection - Europe (partly based on CEuroGeographics for the administrative boundaries) (2013)

19. Poushter, J., Bishop, C., Chwe, H.: Social media use continues to rise indeveloping countries but plateaus across developed ones. Report, Pew Research Center (2018)
20. Sardadvar, S., Rocha-Akis, S.: Interregional migration within the european union in the aftermath of the eastern enlargements: a spatial approach. Rev. Regional Res. **36**(1), 51–79 (2016)
21. StatCounter: Social media stats Europe (2019). https://gs.statcounter.com/social-media-stats/all/europe
22. State, B., Park, P., Weber, I., Macy, M.: The mesh of civilizations in the global network of digital communication. PLoS One **10**(5), e0122543 (2015)
23. Verbugge, L.M.: A research note on adult friendship contact: a dyadic perspective. Soc. Forces **62**(1), 78–83 (1983)
24. Digital in 2018. Report (2018). https://wearesocial.com/blog/2018/01/global-digital-report-2018
25. Romania systematic country diagnostic: Migration background note. Technical report, World Bank Group (2018)
26. Yagan, D.: Moving to opportunity? Migratory insurance over the great recession. Job market paper (2014)
27. Zipf, G.K.: Human Behavior and the Principle of Least Effort: An Introduction to Human Ecology. Addison-Wesley Press, Boston (1949)

Combining Language Models and Network Features for Relevance-Based Tweet Classification

Mohamed Barbouch(✉), Frank W. Takes, and Suzan Verberne

Leiden University (LIACS), Leiden, The Netherlands
m.barbouch@umail.leidenuniv.nl,
{f.w.takes,s.verberne}@liacs.leidenuniv.nl

Abstract. In this paper we present methods for categorizing Twitter data from eight natural disasters into topical classes. Automatically categorizing social media content is of great importance to crisis management organizations that must quickly identify relevant information during situations such as floodings and earthquakes. Unique to our approach is that we leverage both the content of the tweets and the influence of the users producing this content. We compare the effectiveness of traditional text classifiers to a transfer learning method with a large pre-trained language model (BERT). To understand user influence, the rank of the user in the underlying Twitter mention network is included in the classification process. The final approach consists of an ensemble of the best content-based model as well as various user rank features. We find that BERT outperforms traditional text classifiers, in particular for the larger categories. In addition, we find that the influence of a user based on his or her social position, is of high relevance in some particular tweet categories. The proposed approach may prove useful in the automated real-time detection of relevant Twitter content in crisis situations.

1 Introduction

Crisis situations, such as natural disasters, political conflicts and virus outbreaks require significant human effort to control. A crisis typically indicates an urgent malfunction in the standard order, where the day-to-day behavior of larger groups of people can be severely disrupted. In these types of situations, crisis management organisations are often confronted with a large information overload originating from social media channels. Therefore, social media analytics [13] is considered of great value in this context.

There is a large body of research addressing social media analytics in crisis situations, leveraging theories and methods from both the social sciences as well as computer science, in particular natural language processing (NLP), machine learning and network analysis. Reuter et al. [9] presents an overview of this research. A number of open problems and possibly fruitful areas of further research include, according to [9], the early detection of false information

S. Aref et al. (Eds.): SocInfo 2020, LNCS 12467, pp. 15–27, 2020.
https://doi.org/10.1007/978-3-030-60975-7_2

and rumours, determining relevance and trustworthiness of information and data sources, and ultimately providing real-time analytics for decision makers.

In this paper, we address the classification of social media messages during crisis situations into relevant categories. In our approach we take both the content as well as the source of this content into account. Apart from analyzing the content of the information using text classification models, we explicitly incorporate the influence of the user (sender of the message) into the classification model. This enables insights in the extent to which trustworthiness of the source of the content is useful for finding relevant information.

We use Twitter data from eight different natural disasters, obtained from Imran et al. [3]. We train traditional text classification models based on word features, as well as a state-of-the-art neural language model [2] to classify the tweets into the relevance categories defined by Imran et al. [3].

Our first contribution is that we analyze to what extent user influence features can contribute to the effectiveness of the classification. We address this question by enriching the model with information on the position of the user in the underlying mention network, utilizing well-known node ranking methods from the field of social network analysis [10]. Furthermore, we assess which types of crisis information are classified best using which type of approach, enabling a better understanding of how automated approaches can be employed in the real-time understanding of social media data in crisis situations.

The remainder of this paper is organized as follows. Related work is presented in Sect. 2, followed by a description of the data in Sect. 3. Our proposed approach as well as the incorporated methodological building blocks, is outlined in Sect. 4. Results of applying this approach to eight large Twitter datasets are presented and discussed in Sect. 5. Finally, Sect. 6 concludes the paper and provides suggestions for future work.

2 Related Work

In this section we briefly survey related work on social media analysis and crisis events. The need for an understanding of reliability of sources and content on social media during crisis events was approached in [7] using the concept of trust. A theoretical framework for information on social media in times of crisis was proposed, leaving an implementation and evaluation for future work. The work presented in [3] does make an empirical evaluation step and uses Twitter data of 19 different crisis events, collected between 2013 and 2015. The evaluation was done on eight labeled datasets (which are also used in this work), by classifying the messages into relevant and not relevant. The content was enrichted using out-of-vocabulary word normalization, since the language used on Twitter is informal and may contain numerous errors. For feature selection, unigrams and bi-grams were used. Moreover, a word2vec word embeddings model was trained on the whole corpus and is made available for usage.

A deep learning framework for social media posts classification during two natural disasters is presented in [1], using adversarial training and a graph based semi-supervised learning to leverage unlabeled data, given the scarcity of labeled data. The aim is to perform binary classification between relevant and non-relevant tweets by deriving distinctive features from the older events. Their graph-based approach improved the baseline's F1-score by up to 13%.

In [9], an overview of research on social media in conflicts and crises is provided. Three methodological approaches are in general deemed useful: text mining, text classification and network analysis. Our research uses each of these three aspects, but crucially considers a network built using user mentions in tweets, rather than a network based on textual similarity between tweets, as done in [1]. In doing so, we aim to incorporate the trust of the information source, i.e., the user's influence, in the classification process.

3 Data

In this section, we describe our data collection, enrichment and cleaning steps, ending with how the mention network was constructed from this data.

Data Preparation. Table 1 provides details of the eight natural disaster datasets used in this work, listing for each dataset its number of tweets, number of labeled tweets, country and year in which the data was gathered. Note that reported numbers may slightly deviate from what is reported in [3], as the data contained duplicate tweets. Furthermore, the data contains both unlabeled tweets and tweets labeled according to a classification scheme used by the United Nations Office for the Coordination of Humanitarian Affairs (UN OCHA). This scheme consists of nine classes: Injured or dead people; Missing, trapped, or found people; Displaced people and evacuations; Infrastructure and utilities damage; Donation needs, offers or volunteering services; Caution and advice; Sympathy and emotional support; Other useful information; Not related or irrelevant.

Tweet Retrieval. The subset of the original data that was labeled, contained tweet IDs, tweets and labels, whereas the unlabeled part originally only contained tweet IDs and user IDs. We used the Twitter API to obtain the content of these

Table 1. The eight natural disaster datasets used in this work

Name	Unlabeled		Labeled			
	Tweets	Retrieved	Tweets	Retrieved	Country	Year
Nepal earthquake	4,223,516	2,845,416	3,003	2,107	Nepal	2015
Chile earthquake	317,879	208,698	1,932	1,251	Chile	2014
California earthquake	187,963	122,896	1,701	1,106	USA	2014
Pakistan earthquake	98,022	65,612	1,881	1,262	Pakistan	2013
Cyclone PAM	490,394	305,569	2,004	1,391	Vanuatu	2015
Typhoon Hagupit	625,974	362,898	2,010	1,271	Phillippines	2014
Pakistan floods	789,536	492,666	1,769	1,179	Pakistan	2014
India floods	5,259,637	2,786,805	1,820	1,185	India	2014

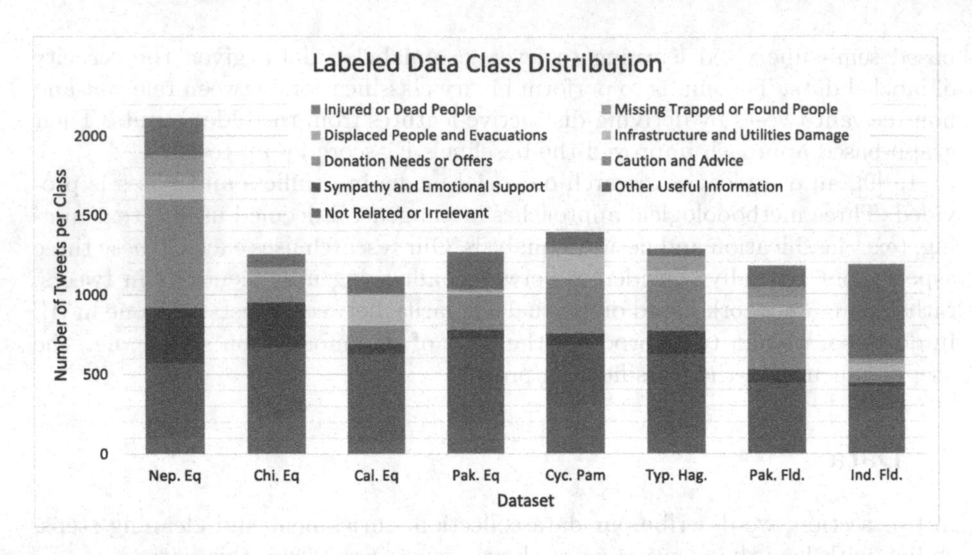

Fig. 1. Class distribution in retrieved labeled dataset (see Table 1).

tweets. As some tweets were no longer available, this resulted in fewer tweets that could be analyzed in total, as listed for both the unlabeled and labeled data in the "Retrieved" column of Table 1; this number was between 30% and 50% on average.[1]

Cleaning Tweets. To prepare the tweet content for text classification, steps similar to what was done in [3] were performed. We applied lower casing and removed punctuation, stopwords, urls and mentions. For stopwords the English list from the NLTK package[2] is used. URLs and usernames are removed using regular expressions. We applied stemming using Lovins stemmer,[3] following the approach in Imran et al. [3]. However, it failed to handle a few cases, for which we used the Porter stemmer in NLTK.

Class Distribution. The data has a large class imbalance within and across different datasets (Fig. 1). For example, the distribution of "Missing Trapped or Found People" class is extremely underrepresented in six datasets. The same applies to "Displaced People and Evacuation" in four cases. On the other hand, "Other Useful Information" is rather dominant in all sets; this can be seen as a container class in which all other relevant tweets are put. In between, other classes such as "Injured or Dead People" and "Donation Needs or Offers" differ from being the largest in one set to one of the smallest class in another dataset.

Mention Network Construction. The unlabeled dataset was used to build a mention network. With a "mention" we refer to the fact that a user explicitly names

[1] The final dataset used in this research can be found at https://github.com/mbarbouch/BERT_Net.
[2] https://www.nltk.org/.
[3] https://github.com/Rafiot/Stemming-Python.

Table 2. Descriptive statistics of the extracted mention network.

	Tweets	Mentions	Tweets with mentions	Nodes	Edges
Nep. Eq	2,845,416	2,607,972	2,024,837	830,591	2,004,708
Chi. Eq	208,698	159,802	133,920	97,541	137,349
Cal. Eq	122,896	86,091	68,203	51,163	75,007
Pak. Eq	65,612	41,572	31,989	16,399	31,239
Cyc. PAM	305,569	213,797	161,524	97,149	167,652
Typ. Hag.	362,898	310,610	246,827	109,230	229,233
Pak. Fld.	492,666	338,656	268,740	154,162	279,719
Ind. Fld.	2,786,805	1,737,763	1,360,050	1,236,762	1,452,705

another user in a tweet, indicating a form of direct communication. This process resulted in a directed graph in which there is an edge from user A to user B if user A mentioned user B in a tweet.

Descriptive statistics of the networks for each dataset are given in Table 2, listing (again for reference) the number of tweets, the number of mentions over all of these tweets, and the number of unique tweets with at least one mention. The number of unique users makes up column "Nodes" and the number of mentions between these users is shown in the column "Edges". Mentions of users with the @null tag, meaning the account did not exist on Twitter anymore upon retrieval of the tweet, resulted in between 3% and 20% of the mentions originally present in these tweets being left out. Furthermore, self-mentions, so users mentioning themselves, were left out given their limited value in assessing user influence.

4 Methods

We train models that classify tweets from the datasets given in Table 1 automatically into the classes listed in Sect. 3. This is a multi-class text classification task, where the tweets are assigned to one of the classes.

4.1 Text Classification

Traditional Text Classification Models. As traditional text classification models we use the set of learning algorithms listed in by Imran et al. [3, p. 2], i.e.: Support Vector Machines (SVM), Random Forest (RF), and Naive Bayes (NB). These are commonly used algorithms in text classification [4, p. 28–31, 32–33, and 25–27]. After the cleaning steps described in Sect. 3, the data is transformed to a sparse bag-of-words representation with TF-IDF (term frequency–inverse document frequency) feature weights using scikit-learn.[4] We extract two types

[4] https://scikit-learn.org/stable/modules/generated/sklearn.feature_extraction.text.TfidfVectorizer.html.

of sparse features: words (unigrams and bigrams) and characters (unigrams and bigrams). These were combined in one feature matrix. We used the top-1000 most frequent features to train our models.

Transfer Learning with BERT. In addition to the previous method we also use BERT (Bidirectional Encoder Representations for Transformers) [2], the current state-of-the-art language modelling paradigm. The idea of pre-trained language models is that they learn the meaning of words on a large (unlabeled) corpus, e.g. the English Wikipedia. The language knowledge about words can then be transferred to downstream tasks, such text classification, where the words are represented by dense vector representations. This has shown to improve effectiveness on many tasks. We expect for social media data that BERT can outperform the traditional classifiers because tweets are short texts for which the traditional bag-of-words features are extremely sparse. In addition, BERT should be able to deal with class imbalance without additional steps to explicitly balance the data [6]. We assume this will adequately address the underrepresented classes described in Sect. 3 .

BERT is a well-established model from which other variants have emerged. In this work we use RoBERTa (A Robustly Optimized BERT Pretraining Approach) [5], a variant that was published by Facebook in 2019. This improves on BERT by using more data, training the model longer and using bigger batches. We use RoBERTa$_{BASE}$ model from HuggingFace's Transformers library [12]. In the remainder of this paper we use the term 'BERT' for simplicity.

4.2 Network Features

To capture user influence in our classification model, we include features based on the mention network constructed in Sect. 3. Here we expect that the rank of a user in the network has an impact on the classification effectiveness, hypothesizing that a user with a higher rank posts more relevant tweets than a user with lower rank. We select four well-known centrality metrics [10] commonly used in the field of social network analysis to capture this for each node (user):

- Indegree centrality: the number of times that the user is directly mentioned by other users (number of incoming edges in the mention graph);
- Closeness centrality: the extent to which the node is easy to reach in the full mention graph (inverse of average shortest path length from all nodes);
- Betweenness centrality: the relative number of times that a user is involved in efficiently passing on information (involved in realizing a shortest path);
- Eigenvector centrality: user importance depends on importance of other users that mention this user (higher value when incoming links have higher values).

Computation was done using NetworkX.[5] The rank of a user is then determined by sorting the users in descending order by their centrality value.

[5] https://networkx.github.io.

4.3 Ensemble Model

We aim to create a model that combines the user rank features and the text classifier in one classifier. Of course, merely adding the user rank features to the (high-dimensional) content-features matrix will hardly have any effect. A better representation to balance the contribution of all features is to use a low-dimensional feature matrix. For this, we use Ensembling by Stacking [14, p. 83]. Stacking is an ensemble technique where the combination of multiple first-level models is learned by a second meta-classifier. The first-level output is fed to the top classifier in a low-dimensional feature matrix. This enables us to combine user rank features with content features in a proportional way.

Our feature matrix consists of one feature containing the predictions of the best text classification model, and four features containing the user ranks. As meta-classifier of the ensemble model we use Random Forest.

4.4 Evaluation

We classify the data for each disaster tweet collection according to the 9 predefined classes listed in Sect. 3. We evaluate our models using 10-fold cross-validation for each set, with stratified sampling of the categories. We use Precision, Recall, and weighted F1-score as evaluation metrics. For the contribution of each feature to the final score in the ensemble model, we use feature importances of the RandomForest classifier from Scikit-learn[6].

To assess the informativeness of user ranks with respect to tweet relevance, we rely on lift charts. This is a technique that is for example used in marketing campaigns to estimate the performance of a prediction model in terms of profitability, given a ranked list of records [8]. The lift is based on the cumulative gain, which measures the accumulated relevance when using user ranks against the accumulated relevance when we do not respect the ranks [11, p. 4]. In our case, we consider all tweets in the class "Not Related or Irrelevant" as *irrelevant*, and tweets in all other classes as *relevant*. We then go through the ranked list of users and select their tweets, while keeping track of counters for their relevant and irrelevant tweets. We do the same with a randomized list of users. In the end, we plot the cumulative counts to show the difference in relevance between ranked and random selections. To then better express the effect of each rank on (ir)relevance, we use lift charts. Lift can be calculated by taking the ratio between cumulative gain of ranked users and cumulative gain of random users.

Furthermore, to quantify the rank informativeness in one value, we calculate the correlation between user rank and tweet relevance. First we divide the cumulative values of relevance by those of irrelevance and cancel out the weight of randomness by further division by the corresponding random values.

[6] https://scikit-learn.org/stable/modules/generated/sklearn.ensemble.RandomForestClassifier.html.

Table 3. Weighted F1 results of content-based multi-class tweet classification using SVM and BERT, in addition to Random Forest for BERT-UserRank ensemble model. Classes: CaA = Caution and Advice | DPaE = Displaced People and Evacuations | DNoO = Donation Needs or Offers | IaUD = Infrastructure and Utilities Damage | IoDP = Injured or Dead People | MToFP = Missing Trapped or Found People | SaES = Sympathy and Emotional Support | OUI = Other Useful Information | NRoI = Not Related or Irrelevant. Colors: Best F1-score in dataset **Best class F1-score** 0 F1-score.

Dataset	Class.	**CaA**	**DPaE**	**DNoO**	**IaUD**	**IoDP**	**MToFP**	**SaES**	**OUI**	**NRoI**	Avg / Total
Nepal Eq	SVM	0.000	0.309	0.767	0.446	0.795	0.588	0.732	0.496	0.210	0.629
	BERT	0.000	**0.446**	**0.832**	**0.554**	**0.859**	**0.720**	**0.822**	**0.590**	**0.370**	**0.717**
	Ens.	**0.051**	0.311	0.730	0.401	0.803	0.640	0.733	0.456	0.308	0.619
	Size	30	60	653	88	226	133	346	462	109	2107
Chile Eq	SVM	0.567	**0.520**	0.000	**0.317**	0.804	0.000	0.826	0.778	0.832	0.756
	BERT	**0.608**	0.512	0.000	0.000	0.846	0.000	**0.884**	**0.807**	**0.892**	**0.790**
	Ens.	0.593	0.504	0.000	0.000	**0.855**	0.000	0.844	0.750	0.832	0.751
	Size	155	39	7	19	80	5	277	439	230	1251
Calif. Eq	SVM	**0.327**	0.000	0.702	0.664	**0.906**	0.000	0.659	0.766	**0.029**	0.692
	BERT	0.155	0.000	**0.827**	**0.722**	0.894	0.000	0.802	**0.815**	0.000	**0.727**
	Ens.	0.051	0.000	0.748	0.630	0.889	0.000	**0.845**	0.740	0.015	0.668
	Size	48	4	63	192	105	4	64	559	67	1106
Pak. Eq	SVM	0.000	**0.167**	0.765	**0.408**	0.820	**0.200**	0.494	0.822	0.562	0.721
	BERT	0.000	0.000	**0.826**	0.000	**0.854**	0.000	**0.688**	**0.833**	**0.644**	**0.752**
	Ens.	**0.062**	0.100	0.767	0.000	0.836	0.000	0.660	0.779	0.527	0.703
	Size	45	10	177	18	231	4	57	530	190	1262
Cyc. PAM	SVM	**0.256**	**0.300**	0.753	0.547	0.853	0.000	0.375	0.608	0.746	0.639
	BERT	0.000	0.217	**0.811**	**0.645**	**0.869**	0.000	**0.510**	**0.683**	**0.838**	**0.697**
	Ens.	0.176	0.187	0.731	0.538	0.842	0.000	0.442	0.595	0.780	0.633
	Size	62	36	284	154	88	13	72	374	308	1391
Typ. Hag.	SVM	0.450	**0.889**	0.554	0.315	0.781	0.000	0.728	0.677	0.392	0.611
	BERT	**0.576**	0.864	**0.699**	**0.471**	0.845	0.000	**0.789**	**0.721**	**0.592**	**0.694**
	Ens.	0.445	0.861	0.565	0.400	**0.862**	0.000	0.718	0.627	0.465	0.605
	Size	223	90	91	60	41	4	143	497	122	1271
Pak. Fld.	SVM	**0.417**	**0.350**	0.682	**0.284**	0.804	**0.499**	0.477	0.682	0.000	0.624
	BERT	0.175	0.333	**0.804**	0.270	**0.848**	0.488	**0.605**	**0.765**	0.000	**0.688**
	Ens.	0.372	0.294	0.723	0.167	0.839	0.375	0.555	0.672	0.000	0.622
	Size	39	64	289	60	131	68	79	436	13	1179
Ind. Fld.	SVM	**0.497**	**0.235**	0.465	0.468	0.938	**0.157**	0.000	0.579	0.869	0.797
	BERT	0.029	0.117	0.450	**0.472**	**0.960**	0.000	0.000	**0.649**	**0.925**	**0.812**
	Ens.	0.133	0.036	**0.453**	0.249	0.947	0.000	0.000	0.514	0.884	0.773
	Size	33	22	32	49	593	11	22	150	273	1185

5 Results

In this section we present the results of applying the models proposed in Sect. 4 to the data discussed in Sect. 3, evaluating the results using the measures discussed in Sect. 4.4.

5.1 Content-Based Tweet Classification

Table 3 shows for each dataset the F1-scores for the content-based models discussed in Sect. 4.1 (first two rows of each block in the table). Of the traditional text classifiers with tf-idf feature weights, SVM obtained the best results on the initial labeled dataset: F-scores between 0.61 and 0.80 with standard deviations between 0.018 and 0.050. Overall, BERT outperforms SVM, although SVM also achieved satisfying results. Averaged over categories, BERT scores better in all datasets: F-scores between 0.69 and 0.82 with standard deviations between 0.027 and 0.064. This confirms that pre-trained language models are effective for short documents such as tweets. However, if we zoom in on individual categories, we see some differences. It appears that BERT often performs better on medium to large categories (e.g. *Donation Needs or Offers* in Nepal Eq, *Other Useful Information* in Chile Eq). On the smallest categories, on the other hand, BERT performs poorly: some of the smallest categories are never identified by BERT, resulting in an F1 score of 0.0, while SVM was able to identify the category and thus obtained a non-zero score (e.g.*Infrastructure and Utilities Damage* in Chile Eq, and *Displaced People and Evacuations* and *Infrastructure and Utilities Damage* in Pakistan Eq).

The highest F1-scores were obtained for the category *Injured or Dead People*, with scores ranging between 84.8% and 96.0%. One could argue that this is good news given that it is a critical situation in natural disasters.

For comparison, the weighted F1-score achieved with BERT on Nepal Earthquake is 6.6% points higher than in the paper by Alam et al. on the same reference dataset [1].

5.2 Results Ensemble Model

Table 3 shows the results in terms of F1 values for the Ensemble model from Sect. 4.3 (3rd row per block in the table, 'Ens.'). As our aim is to see the effect of user ranks on classification, we present the results of a stacked model of BERT predictions with user ranks.

For none of the datasets does the ensemble model outperform BERT in terms of overall performance. However, results are slightly better on a number of classes. On *Caution and Advice* for Nepal Eq and Pakistan Eq, F1-scores of 0.051 and 0.062 were found, where BERT reported 0.0 in both cases. The model managed to prevent the zero score for this relatively small class in all datasets. *Missing Trapped or Found People* appears again in most cases not possible to classify due to its small size.

We also looked at feature importance to see to what extent each feature has contributed to the final predictions; the result of this analysis is in Table 4. In all datasets, BERT was most informative in all cases. Pakistan Floods has achieved the highest user rank feature importance of 0.131, based on betweenness centrality. Here we discover an interesting finding. Compared to BERT, the ensemble model got the largest increase in F1-score for *Caution and Advice* class; the score went from 0.175 to 0.372. When we look at the corresponding lift chart, Fig. 2, we see that *Caution and Advice* also has the highest relevance value among all classes. This information was probably relevant enough to improve the score of BERT on this class.

Table 4. Feature importance of ensemble model.

Dataset	BERT	Indeg.	Clos.	Betw.	Eig.
Nep. Eq	0.514	0.122	0.121	0.120	0.122
Chi. Eq	0.667	0.083	0.083	0.086	0.081
Cal. Eq	0.549	0.113	0.114	0.114	0.109
Pak. Eq	0.578	0.112	0.103	0.104	0.103
Cyc. PAM	0.549	0.111	0.111	0.116	0.112
Typ. Hag.	0.492	0.125	0.126	0.129	0.129
Pak. Fld.	0.508	0.125	0.119	**0.131**	0.118
Ind. Fld.	0.729	0.069	0.068	0.069	0.065

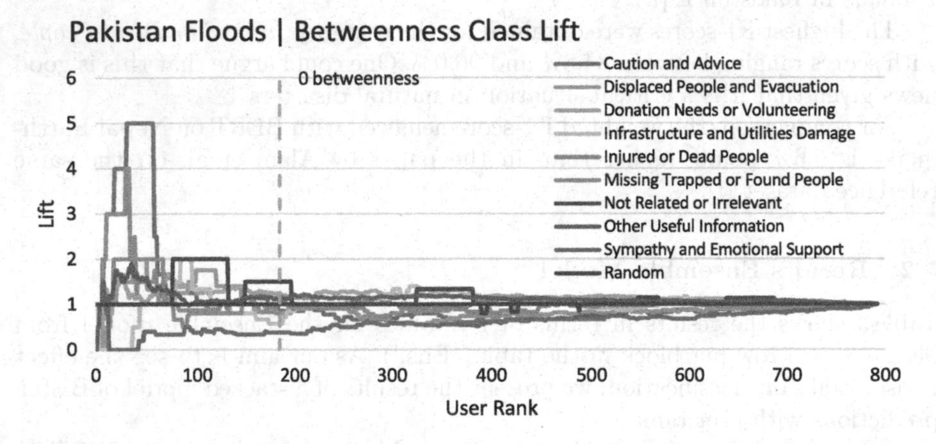

Fig. 2. Class lift in Pakistan Floods based on betweenness rank.

On the contrary, India Floods and Chile Eq have the lowest user rank feature importance values. As a difference with Pakistan Floods, we see in their eigenvector-relevance lift charts (Fig. 3) that the relevance curves are fluctuating over user ranks[7]. Presumably any dominance of relevance by top users is taken over by irrelevance curves of lower ranked users. So, any information about (ir)relevance is then almost completely eliminated at the end.

[7] Lift relevance includes all first 8 relevant classes (Sect. 3).

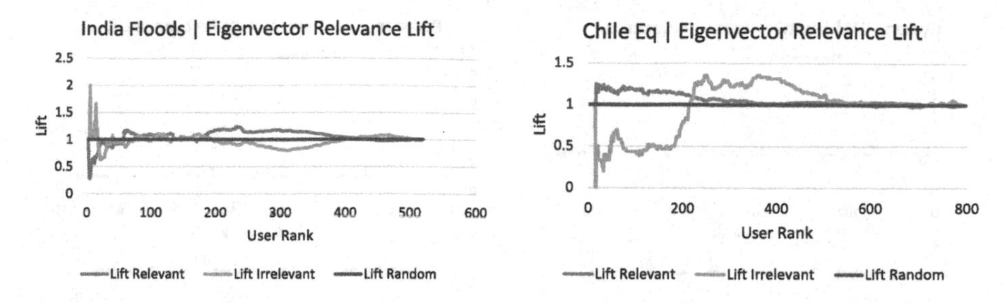

Fig. 3. Relevance lift in Chile Earthquake and India Floods by eigenvector rank.

5.3 Analysis of User Rank Contribution to Tweet Relevance

Apart from feature importance, we quantify the contribution of each type of user rank to tweet relevance by computing correlations between lift curve and user rank (between -1 and 1), as explained in Sect. 4.4. The values are given in Table 5. The more negative the value for ratios based on ranked users, the better the relevance is; the more positive the worse. A favorable scenario is when top k users post more relevant- and fewer (preferably none) irrelevant tweets. Hence a negative value (last column), in which both ranked and random correlations are incorporated, represents an overall score of better relevance than random.

Table 5. Computed relevance correlations for each centrality metric. * Compared to random. Colors: **Best correlated feature for dataset**, Worst feature for dataset, Best found correlation across all datasets , Worst found correlation .

	Random	Indegree	Closeness	Betweenness	Eigenvector	Relevance*
Nep. Eq	−0.303	0.510	0.472	**0.377**	0.399	−
Chi. Eq	−0.278	−0.544	−0.595	−0.325	**−0.619**	+
Cal. Eq	−0.245	**−0.267**	0.126	0.171	0.164	+
Pak. Eq	−0.110	**0.511**	0.650	0.939	0.558	−
Cyc. PAM	0.264	−0.479	−0.586	**−0.696**	−0.535	+
Typ. Hag.	0.091	−0.282	−0.295	**−0.431**	−0.280	+
Pak. Fld.	0.150	**−0.351**	−0.338	0.008	−0.315	+
Ind. Fld.	−0.374	−0.114	**−0.155**	0.241	0.067	−

We found in 5 cases better relevance quantities with user ranks, i.e, in Chile Eq, California Eq, Cyclone PAM, Typhoon Hagupit and Pakistan Floods. The global best and worst correlations were shown by betweenness on Cyclone PAM and Pakistan Eq, with values of −0.696 and −0.939, respectively. It is striking that betweenness did not get the highest or the lowest feature importance in Table 4 for these two datasets. In fact, feature importance is also influenced by irrelevant tweets in "Not Related or Irrelevant" class. So in these cases, having a high level of relevance will be balanced by a low(er) level of irrelevance in the

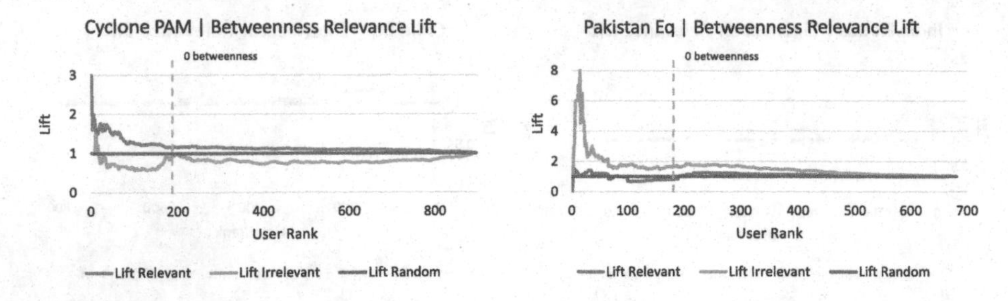

Fig. 4. Relevance lift in Cyclone PAM and Pakistan Eq by betweenness rank.

feature importances, and vice versa. See Fig. 4 for the tweet relevance in the two datasets based on user rank by betweenness centrality.

The correlations given in Table 5 are taken between the number of tweets and all ranked users. Since we find only top k users interesting, one would argue to take the correlation only for a predefined value of k. This argument would be further strengthened, given that a number of centralities becomes 0 at some point. Although this is all reasonable, our goal is – when detecting relevance – to extend textual features with centrality values to improve the classification. For this we therefore need to compute the correlation with all users. However, the way in which relevance w.r.t. user rank has been evaluated has shown a representative picture in this work for quantifying model's effectiveness.

Furthermore, although user ranks were not able to improve the overall classification, the analysis carried out with regard to the relevance of network features has shown useful patterns that could be exploited the other way around. Credibility of users, for instance, can be assessed by the observed relevance patterns.

6 Conclusions

This research was motivated by the necessity to filter relevant content in big data streams during natural disasters. We used several millions of tweets from 8 different disasters and did tweet classification based on content and user ranks. Our final model is a stacked ensemble consisting of BERT predictions and user features based on node centrality in the underlying mention graph.

We find that BERT is outperforming traditional text classification models when there is sufficient data to train. *Injured or Dead People* turned out to be the easiest category to classify. Small classes such as *Missing Trapped or Found People* remain on the other hand difficult to identify. Surprisingly, SVM did slightly better on these classes when compared to the BERT model. We suspect that classes with very few tweets would not get a pre-trained language model like BERT sufficiently fine-tuned on. Sparse feature-based models trained only on the corpus itself, like SVM, would learn the class specificities better.

Using user ranks based on centrality metrics in the ensemble model did not improve the overall classification scores, but did contribute at the level of individual classes. At the class level we have seen improvements when a rank strongly correlates with a corresponding class. A reasonable explanation is that a ranked user does not necessarily tweet about just one relevant topic. Given a tweet, the classifier would be confused to assign the right class if historical tweets of the user cover multiple relevant classes. Likely, the last approach can be better leveraged only for binary classification between relevant and irrelevant tweets.

Further research is necessary to confirm (or disprove) the statement that when a rank strongly correlates with a class, the rank features would improve binary classification. In our approach we hand-crafted the rank features and used them in a low-dimensional feature matrix. Network relationships have however the potential to contain much more information. In future work it might be interesting to use graph embeddings rather than centrality values. Although explainability of the final model will decrease, embeddings may learn to just the right network features in each particular category, potentially creating a superior overall classification model.

References

1. Alam, F., Joty, S., Imran, M.: Domain adaptation with adversarial training and graph embeddings. arXiv preprint arXiv:1805.05151 (2018)
2. Devlin, J., Chang, M.-W., Lee, K., Toutanova, K.: BERT: pre-training of deep bidirectional transformers for language understanding. In: Proceedings of the 2019 Conference of the North American Chapter of the Association for Computational Linguistics: Human Language Technologies, Volume 1 (Long and Short Papers), pp. 4171–4186 (2019)
3. Imran, M., Mitra, P., Castillo, C.: Twitter as a lifeline: Human-annotated twitter corpora for NLP of crisis-related messages. arXiv preprint arXiv:1605.05894 (2016)
4. Kowsari, K., Jafari Meimandi, K., Heidarysafa, M., Mendu, S., Barnes, L., Brown, D.: Text classification algorithms: a survey. Information **10**(4), 150 (2019)
5. Liu, Y., et al.: RoBERTa: a robustly optimized BERT pretraining approach. arXiv preprint arXiv:1907.11692 (2019)
6. Madabushi, H.T., Kochkina, E., Castelle, M.: Cost-sensitive BERT for generalisable sentence classification with imbalanced data. arXiv preprint arXiv:2003.11563 (2020)
7. Mehta, A.M., Bruns, A., Newton, J.: Trust, but verify: social media models for disaster management. Disasters **41**(3), 549–565 (2017)
8. Piatetsky-Shapiro, G., Masand, B.: Estimating campaign benefits and modeling lift. In: Proceedings of the Fifth ACM SIGKDD International Conference on Knowledge Discovery and Data Mining, pp. 185–193 (1999)
9. Reuter, C., Stieglitz, S., Imran, M.: Social media in conflicts and crises. Behav. Inf. Technol. 1–11 (2019)
10. Scott, J.: Social Network Analysis, 4th edn. SAGE, New York (2017)
11. Shmueli, G.: Lift up and act! Classifier performance in resource-constrained applications. arXiv preprint arXiv:1906.03374 (2019)
12. Wolf, T., et al.: Huggingface's transformers: state-of-the-art natural language processing. ArXiv, abs/1910.03771 (2019)
13. Zeng, D., Chen, H., Lusch, R., Li, S.-H.: Social media analytics and intelligence. IEEE Intell. Syst. **25**(6), 13–16 (2010)
14. Zhou, Z.-H.: Ensemble Methods: Foundations and Algorithms. CRC Press, Boca Raton (2012)

Co-spread of Misinformation and Fact-Checking Content During the Covid-19 Pandemic

Grégoire Burel(✉), Tracie Farrell, Martino Mensio, Prashant Khare, and Harith Alani

Knowledge Media Institute, The Open University, Milton Keynes, UK
{g.burel,t.farrell,m.mensio,p.khare,h.alani}@open.ac.uk

Abstract. In the context of the Covid-19 pandemic, the consequences of misinformation are a matter of life and death. Correcting misconceptions and false beliefs are important for injecting reliable information about the outbreak. Fact-checking organisations produce content with the aim of reducing misinformation spread, but our knowledge of its impact on misinformation is limited. In this paper, we explore the relation between misinformation and fact-checking spread during the Covid-19 pandemic. We specifically follow misinformation and fact-checks emerging from December 2019 to early May 2020. Through a combination of spread variance analysis, impulse response modelling and causal analysis, we show similarities in how misinformation and fact-checking information spread and that fact-checking information has a positive impact in reducing misinformation. However, we observe that its efficacy can be reduced, due to the general amount of online misinformation and the short-term spread of fact-checking information compared to misinformation.

Keywords: Covid-19 · Misinformation · Fact-checking · Social media

1 Introduction

Recent research indicates that misinformation spreads much faster than true information by exploiting emotions [32]. At the moment, public attention to danger is heightened and fear may influence behaviour [9]. Misinformation about Covid-19 has been rampant on social media [5,6], with some tragic results[1,2]. Studying the spread of misinformation about Covid-19 helps us to understand what information correction the public needs during a health crisis. It also helps to distinguish patterns and timings that are significant in the spread of misinformation.

[1] BBC – The cost of virus misinformation, https://www.bbc.co.uk/news/stories-52731624.

[2] The Guardian – UN warns of deadly effect of Covid-19 misinformation in Pacific, https://www.theguardian.com/world/2020/apr/17/un-warns-of-deadly-effect-of-covid-19-misinformation-in-pacific.

S. Aref et al. (Eds.): SocInfo 2020, LNCS 12467, pp. 28–42, 2020.
https://doi.org/10.1007/978-3-030-60975-7_3

We compare the diffusion of 2,830 misinformation and 734 fact-checking URLs about Covid-19 on Twitter,[3] from early December 2019 to early May 2020 in order to understand the spread of misinformation and fact-checks over time.

First, we analyse how spread differs during the pandemic by observing changes between the initial pandemic onset, the ramping up phase and late pandemic period (Covid-19 level). Second, we study the relative misinformation and fact-checking diffusion patterns by aligning individual URL spreads and analysing how individual misinformation spread after their initial appearance (relative level).

We address the following research questions: Are misinformation and fact-checking information shared similarly? How do misinformation and fact-checking spread patterns vary at the pandemic level and relative level? and How does fact-checking spread affect the diffusion of misinformation about Covid-19?

2 Related Work

In this section, we discuss some of the propositions that researchers have made regarding the spread of misinformation and fact-checks on social networks. We highlight the complexity of establishing the impact of fact-checks on misinformation sharing, to which our study contributes.

2.1 Misinformation Spread Analysis

Most studies of misinformation spread tend to be focused on early intervention and removal [3]. In this context, many works have focused on the application and extensions of epidemiological models [3,12,13] with additional features like weighted values for particular users [29] or, more recently, information about debunkers and the dynamics of opinion evolution [23]. Notably, Saxena *et al.* [23] demonstrated that identifying influential nodes may also help exploit the spread of fact-checked information and impact user opinion over time.

Several works investigate the role of different topological features in misinformation spread [2,8,30,35], finding that some topic/audience interdependencies may increase the spread of misinformation, perhaps related to cultural norms, experiences or values [7]. Likewise, chains or groups of nodes may accelerate the spread of misinformation [22] and, as Xian *et al.* demonstrate, individuals can be exposed to and share misinformation across platforms [34]. In the context of the current crisis, Cinelli *et al.* [6] analysed spread patterns of different Covid-19 related misinformation across several platforms. The authors noted different diffusion patterns for different types of misinformation on each platform. Researchers are beginning to explore the role of social media *hype* in accelerating both panic and therefore uptake of misinformation about Covid-19 and other viral pandemics on social media [10].

---

[3] Twitter, http://twitter.com.

Most researchers agree that the biggest impacts of misinformation happen within a short time span from the initial circulation [26]. Misinformation spikes are prevalent during times of conflict and war [16], and during politic events [15]. Misinformation often accompanies breaking news developments, when people are looking for more details, as well as during disasters, when they might desperately need information about where to go or what to do next [26].

Existing research has mostly focused on analysing misinformation spread alone without much focus on whether fact-checking information impacts the spread of misinformation. Although topological and social features are important for characterising misinformation spread, we decide to leave these features in our study and focus on the co-spread of misinformation and fact-checks with a particular focus on different time periods. We leave additional topological and user analysis as future work.

2.2 Fact-Checking Information Spread

Fact-checks are a type of information that is distinct from just "true", or "false" information. Researchers already showed that true information and misinformation spread differently [32]. Fact-checks assess claims for accuracy [31], hence representing a new category of information [11]. At the time of writing, we could not identify work looking computationally at the diffusion of fact-checking in a network, in particular to assess causal relationships to misinformation.

Tambuscio *et al.* [28] used agent-based simulations to develop a two parts epidemiological model for defining the "minimal reaction" necessary to get rid of a viral hoax, but this was not transferred to a real dataset. Later work by Kim *et al.* [14] used real-world datasets from Twitter and Weibo[4] to model how the network could be mobilised to spread corrective information effectively. Still, these models are meant to predict how future fact-checks may diffuse and not to estimate existing causal relationships.

Researchers have looked at the usefulness of fact-checking from a variety of perspectives. Nyhan and Reifler [21] found that attitudes toward fact-checking in the USA were generally supportive. However, they noted that scepticism toward fact-checking may stem from a lack of trust in fact-checking entities. More recently, Barrera *et al.* [4] found that fact-checking did improve voters' knowledge, but did not appear to impact policy, or support for individual candidates. This phenomenon was also reported by researchers in the context of the 2016 USA presidential election [27]. In their exploration of the Australian presidential election in 2017, Aird *et al.* found that the number of corrections must outweigh the number of affirmations of the misinforming claim, in order to have a stronger impact on belief and behaviour [1]. Similar findings were echoed in [26].

A recent extensive review of fact-checking literature performed by Nieminen *et al.* showed that the corrective potential of fact-checking was a dominate subject of research, but that subjectivity in fact-checking assessment, the overem-

[4] Weibo, http://weibo.com.

phasis of fact-checking in the US, and a lack of clarity around correcting beliefs were continued challenges [20]. Assessing the impact of fact-checking from the perspective of individuals consuming fact-checks is difficult to do in laboratory settings. In our work, we focus on assessing the presence and diffusion of both misinformation and fact-checks on Twitter, to explore temporal patterns and evidence of causal relationship.

3 Co-spread of Misinformation and Corrective Information During the Covid-19 Pandemic

The review of existing work investigating misinformation spread shows a gap in understanding the relation and interaction between the corrective information propagated by fact-checking and misinformation spread. We conduct an analysis on the co-spread of fact-checking information and misinformation on Twitter based on the sharing of misinforming URLs that were collected from claim reviews collected from fact-checking websites. The data was collected as part of the misinfo.me platform [17] up to the 4th May 2020.

In our approach, first, we collect Twitter data by looking for the appearance of misinforming URLs that we have collected. Second, misinformation and fact-checks spread is aggregated for three different time periods at two different granularity: 1) From the Covid-19 worldwide spread perspective (Covid-19 level analysis), and; 2) From the initial emergence of a misinforming URL (relative level analysis). This allows for a better understanding of spread at different levels. Third, we perform multiple analyses to investigate how fact-checks and misinformation spread behaviour differs. This analysis allows the identification of significant relations between misinformation spread and fact-checking information, which can be used for designing better methods for spreading fact-checking information on social media. Finally, weak causation and impulse response analyses are performed between fact-checks and misinformation in order to identify if fact-checking information diffusion impacts misinformation spread.

3.1 Dataset

For our analysis, we need to create a dataset that contains both misinformation and fact-checking information. We focus our work on Twitter due to its popularity and its accessibility. We rely on COVID19-related reports from fact-checking websites that identify misinforming content by their URLs, and search the occurrences of these URLs in user posts on Twitter.

Fact-Checker URLs Dataset. The dataset of fact-checks comes from the misinfo.me tool [17], that collects URLs that have been fact-checked, labelled and provided with a fact-checker review. The reviews are published by multiple fact-checking websites belonging to the International Fact-Checking Network[5] (IFCN) using the standard `ClaimReview` schema[6], which was defined appositely

[5] IFCN, https://ifcncodeofprinciples.poynter.org/signatories.

[6] ClaimReview Schema, https://schema.org/ClaimReview.

for the purpose of annotating reviews of claims. The data collection is primarily based on the Data Commons `ClaimReview` public feed.[7] From this public feed, ratings are extracted and normalised between $[-1; +1]$ depending on their credibility [18]. Using these ratings, we only select misinforming URLs (ratings ≤ 0). Although, different levels of misinformation exist (e.g., manipulation, misleading information, forgery, etc.), we focus our investigation on any type of misinformation in order to simplify the analysis. We also keep the URLs of the original fact-checking articles and then filter all the URLs to only get the Covid-19 fact-checks by using a set of relevant keywords[8] based the title and content of the fact-checks. The final URL dataset includes fact-checks published until the 4th of May 2020, with a total of 2,830 distinct misinforming URLs and 734 fact-checking URLs.

Twitter Dataset. Using the misinforming and fact-checking URLs, we create the Twitter dataset by searching their occurrences on Twitter by adapting an existing Twitter Hashtag crawler that collect posts using Twitter's mobile interface.[9] Out of all the seed URLs, we find posts for only 1,190 distinct URLs for a total of 21,394 posts from 16,308 different users. On average, there are 17.54 posts for each URL ($\sigma = 28.35$, $min = 1$, $max = 232$).

Figure 1 shows the cumulative spread of misinforming and fact-checking information URLs shared over time in our dataset. The figure also shows the amount of Covid-19 casualties and cases over the same period as well as the Covid-19 *initial*, *early* and *late* periods (vertical dashed lines).

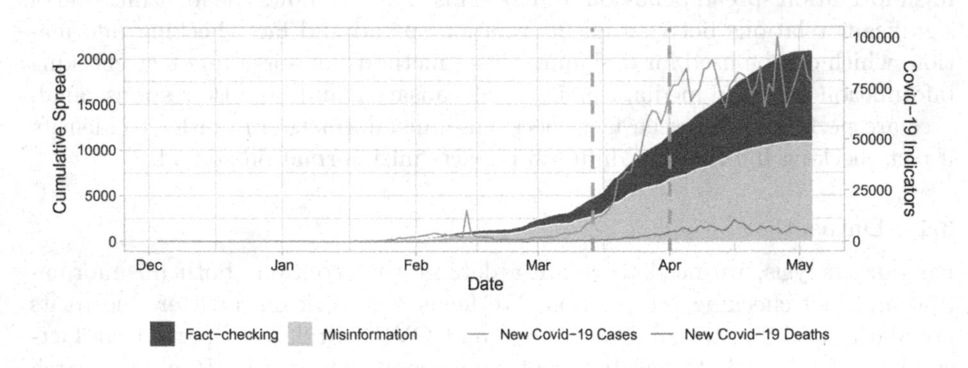

Fig. 1. Stacked cumulative spread of misinforming and corrective information URLs over time and amount of Covid-19 casualties and cases over the same time period.

[7] ClaimReview Public Feed, https://www.datacommons.org/factcheck/download.
[8] Twitter Covid-19 keywords, https://developer.twitter.com/en/docs/labs/covid19-stream/filtering-rules.
[9] Twitter Scrapper, https://github.com/amitupreti/Hands-on-WebScraping.

Covid-19 Cases Dataset. To generate the different time periods at the Covid-19 pandemic granularity, we use the data produced by the European Center for Disease Prevention and Control (ECDPC).[10] The ECDPC collects daily statistics about the number of Covid-19 cases and causalities worldwide for multiple countries. Although, the data is continuously updated, in our work we focus on the 1st Dec. 2019 to 4th May 2020 period since it matches the data we collected on Twitter. The beginning date is selected as the 1st Dec. 2019 since this date tends to be associated with the first traceable case of the pandemic [33].

3.2 Analysed Periods Generation

We analyse the behaviour of fact-checking information and misinformation at two different granularity levels. First, at the pandemic level (Covid-19 level analysis), we are interested in understanding if spread behaviour varies during three different time period within the pandemic based on the the amount of worldwide cases. Second, at the URL level (relative analysis), we are interested in understanding how behaviour differs based on the number of days since the first occurrence of a misinformation-related URL (i.e., a particular misinforming content or its associated fact-checking information). This would show how misinformation and fact-checks spread over time independently from when they were posted.

Covid-19 Periods. We generate three *initial*, *early* and *late* time periods to analyse fact-checking information and misinformation spreads at the level of the Covid-19 pandemic. We fit multiple linear regression models for the daily worldwide Covid-19 cases curve in order to identify inflection points in the amount of Covid-19 cases [19].

Looking for two inflection points in the curve, the *initial* time period is specified as any tweet posted before Saturday, Mar 14, 2020. *Early* period corresponds to any tweet between Saturday, Mar 14, 2020 and Thursday, Apr 2, 2020. *Late* period is for any posts after Thursday, Apr 2, 2020.

Relative Periods. To understand sharing behaviour independently from when each URL has been initially shared, we align the initial sharing of each URL so that all the URLs shared in the dataset always start from the same initial time (i.e., we normalise the dates for each analysed URL). We identify the first occurrence of each URL and then obtain the number of times it has been shared per day for each day following its initial appearance.

Following the same approach outlined in the previous section, we use the daily aggregated curve containing all the shared URLs (i.e., misinforming and fact-checking URLs) for identifying the *initial*, *early* and *late* relative time periods by obtaining the inflection point in the daily shared URLs. Using the aforementioned method, the *initial* time period is specified as any URL shares happening

[10] ECDC Covid-19 Data, https://opendata.ecdc.europa.eu/covid19/casedistribution/csv.

within the first 2 days after its first occurrence. The *early* period correspond to share between day 2 and day 14. Finally the *late* period is for any shares happening after 14 days.

4 Multivariate Spread Variance Analysis

The first part of the analysis is to identify the different patterns of appearance of misinformation and fact-check URLs over varying periods of time. In order to perform such analysis, we use the one way Multivariate ANalysis Of VAriance (MANOVA) and the one way ANalysis Of VAriance (ANOVA) methods. This approach allows us to determine if there are significant differences in information spread between the fact-checking information and misinformation groups in each *initial*, *early* and *late* periods.

4.1 Experimental Setup

MANOVA and ANOVA rely on the definition of independent variables and dependent variables. For our analysis, the amount of information spread associated with each information type is our dependent variable whereas each information type (i.e., misinformation and corrective information) is an independent variable.

Since our data does not follow all the assumptions required for the standard ANOVA and MANOVA methods (i.e., multicollinearity, normality and homogeneity), we use non-parametric versions of MANOVA and ANOVA for the analysis, using F-approximations permutation tests. The F-approximation of ANOVA's test, as well as Wilks' Lambda Type Statistic are obtained with their p-value and the associated permutation test p-value.

Our analysis is divided into two different parts for the Covid-19 and relative level analyses: 1) A Non-parametric MANOVA analysis is performed for identifying if there are differences in spread between the different periods and information types, then; 2) Non-parametric ANOVA analysis is then performed if the MANOVA results are significant for each individual time period for determining in which sub-period (i.e., *initial*, *early* and *late*) the pattern differs.

For the non-parametric ANOVA analysis, the Kruskal-Wallis test is used and the p values are adjusted using Bonferroni correction (since multiple dependent variables are analysed). Significant results mean that the behaviour of corrective information and misinformation are significantly different whereas a non-significant result means that the distribution of spread for each time period is non-significant.

4.2 Results

In the following section we only report significant results for brevity.

Covid-19 Period Analysis. The one way MANOVA analysis comparison at the Covid-19 level URL shares for misinforming URLs and fact-checking URLs shows a significant permuted p-value of 0.01. This means that at the Covid-19 pandemic level, there are significant differences in how misinforming URLs and fact-checking URLs spread and that the type of shared URLs has an effect on the amount of shared URLs during the pandemic. Following the significant result of the MANOVA analysis, a one way ANOVA analysis is performed for each Covid-19 time period. The Bonferroni adjusted Kruskall-Wallis tests are only significant for the *initial* ($p = 0.00558$) and *late* ($p = 0.0234$) periods. This result means that sharing behaviour does not differ fundamentally during the *early* Covid-19 period ($p = 1$) whereas sharing behaviour differs in the other periods.

Looking at the individual URLs shares for each time periods, we observe higher deviations in sharing behaviour for misinformation ($\sigma \in \{20.5, 24.9, 26.3\}$) compared to fact-checking information ($\sigma \in \{7.52, 6.94, 11.3\}$). It also appears that fact-checked information is shared less often than the corresponding misinforming URLs in term of means with lower means for all the time periods ($2.42 < 5.88$, $3.60 < 8.73$ and $6.34 < 10$). This suggests that perhaps the types of users that share misinformation is more varied than the types of users that share fact-checks. Similarly, there may be a variation in what misinforming topic attracts the most shares compared to the fact-checking content.

Relative Period Analysis. The one way MANOVA analysis comparison at the relative URL shares level for misinforming URLs and fact-checking URLs shows a significant permuted p-value of 0. This means that at the relative URL level, there are significant differences in how misinforming URLs and fact-checking URLs spread and that the type of shared URLs has an effect on the amount of spread at different relative time periods.

Following the significant result of the MANOVA analysis, a one way ANOVA analysis is performed for each relative time period. The Bonferroni adjusted Kruskall-Wallis tests are only significant for the *early* ($p = 4.74 \times 10^{-4}$) and *late* ($p = 1.338 \times 10^{-5}$) periods. This means that sharing behaviour during the *initial* ($p = 0.522$) relative period does not differs during that period whereas differences exists when looking at the *early* and *late* periods.

The individual distribution of misinforming and fact-checking URLs for each time period show that the amount of shares tends to be similar across the URL types with a slightly higher spread for the misinforming URLs in general. Interestingly, the highest difference in term of mean and standard deviation between the different URL types appears to be mostly during the initial phase with a more important standard deviation for the misinforming URLs ($\sigma = 12.6$ for misinforming content and $\sigma = 3.31$ for fact-checks).

5 Fact-Checking Misinformation Impact Analysis

In this section we investigate how the two types of information (fact-checking URLs and misinforming URLs) impact each other. In particular, we are interested in understanding if the spread of fact-checking information has a beneficial impact in reducing the diffusion of misinformation. For this analysis, we focus on modelling the spread of URLs as a Vector AutoRegression (VAR) model using the misinformation and fact-checking URLs as endogenous variables. We perform this analysis at the relative level (i.e., the relative number of days since the first appearance of a URL related to a particular misinformation) and determines if weak causation relations between each information types exists.

5.1 Experimental Setup

Although it is not simple to identify causation relations between each information types, it is possible to estimates if the spread of a given information type can be used to predict the spread of another information type using a Granger causality test. In order to compute the Granger causality test we first build a Vector AutoRegression (VAR) model using the combined misinformation spread and fact-checking information for the analysed period. However, since our data is non-stationary, we first integrate each analysed information types so that the spread amount for each day is represented as the difference between the current day value and the previous day value.

A 14 days order value is used for the VAR model based on Akaike's information criterion. Using the VAR(14) model, we perform a bootstrapped Granger causality test for determining if misinformation spread can be associated with fact-checking URL spread or if fact-checking spread can be inferred from misinformation spread.

In order to understand the dynamics that relate fact-checking and misinformation, impulse response analysis is performed as well as Forecast Error Variance Decomposition (FEVD). For the impulse response analysis, we use orthogonal impulse responses in order to evaluate the spread response of the different types of URLs for 14 days steps. This approach allow us to determines how a particular sharing behaviour may affect other types of URLs shares in future. We are particularly interested in determining if an increase in fact-checking information shares trigger a reduction in misinformation diffusion. We run the FEVD with the same 14 days periods in order to obtain the contribution importance of each information types on both misinforming URLs and fact-checking URLs spread.

5.2 Results

Using the VAR(14) model, we observe a Granger causality relation showing that fact-checking spread has predictive causality over misinformation spread ($p = 0.02$). This observation is not found in the opposite direction ($p = 0.93$). This result suggests that at the relative-level, change in fact-checking information spread may cause a change in misinformation spread and therefore fact-checking articles have an impact on misinformation spread. Surprisingly, the opposite result shows that fact-checking spread may not be influenced by misinformation.

The impulse response for the orthogonal shock in the amount of shared fact-checking URLs (Fig. 2) shows an initial drop in misinformation shares (first day) but mixed results afterwards. Despite this observation, a general downward misinformation spread trend can be observed. This suggests that fact-checking tend to have a short significant impact on the spread of misinformation. A shock in misinformation leads to a sharp drop in misinformation spread. This confirms our previous observation that misinformation spreads tend to occurs mostly after its initial spread and decrease quickly in the following days.

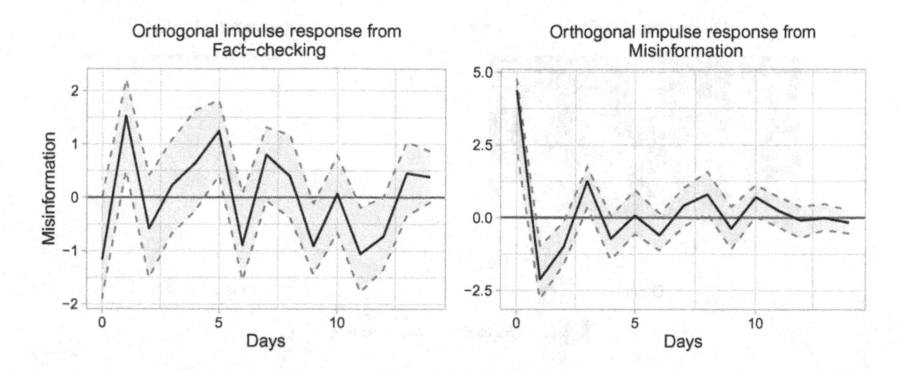

Fig. 2. Bootstrapped relative-level orthogonal impulse response from fact-checking shock (95% confidence interval).

The impulse response for the orthogonal shock in the amount of shared misinforming URLs (Fig. 3) shows a delayed fact-checking increase two days after the initial misinformation spread. This result suggests that fact-checking spread tend to follow misinformation spread despite a lack of causal relation (i.e., fact-checking articles are created as a response to misinformation). As with the misinformation sharing behaviour, we observe a sharp decrease in fact-checking sharing behaviour after the initial shock as initial sharing behaviour reduces.

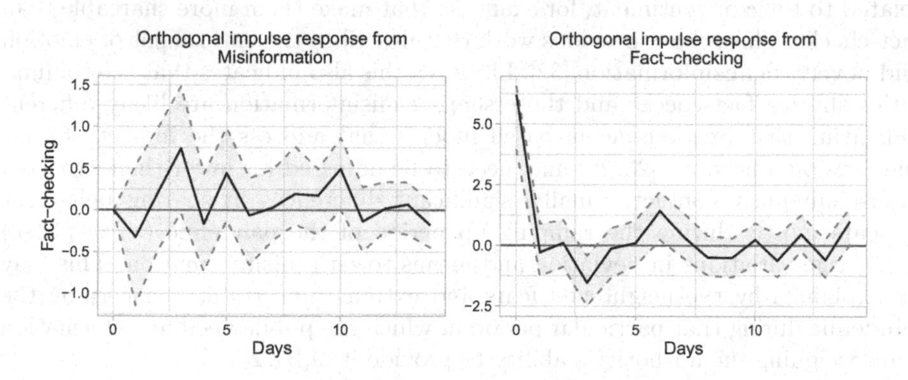

Fig. 3. Bootstrapped relative-level orthogonal impulse response from misinformation shock (95% confidence interval).

The FEVD results displayed in Fig. 4 show that misinformation spread predictions are directly affected by the spread of fact-checking information with misinformation prediction getting more affected by fact-checking spread as time goes by whereas fact-checking spread appears to be unaffected by past misinformation spread. This result adds to our previous causality observation between fact-checking information and misinformation spread.

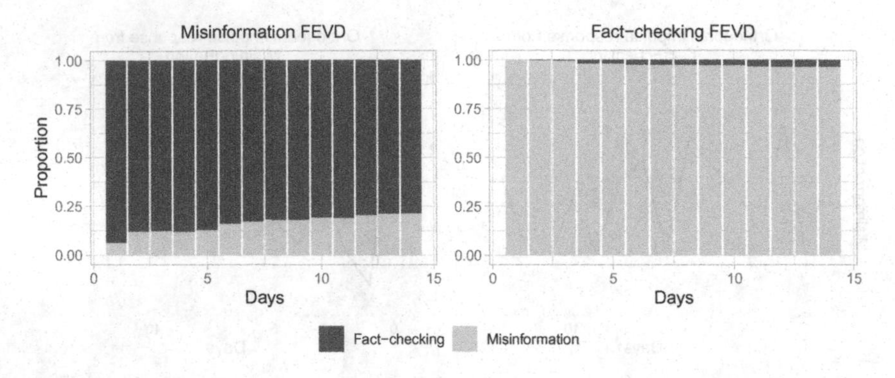

Fig. 4. Forecast Error Variance Decomposition (FEVD) for the relative-level misinformation and fact-checking spread.

6 Discussion

Our results show that at the Covid-19 level, fact-checked URLs are less shared (compared to misinformation in term of mean) during all the periods and that the standard deviation and mean are much higher for misinforming URLs. This indicates that there may be some intrinsic features of misinforming URLs, potentially related to topic or sentiment, for example, that make them more shareable than fact-checks. This echoes previous work that describes the enticement of emotion and novelty in misinformation [32]. Likewise, this also indicates that the communities sharing fact-checks and those sharing misinformation are likely different indicating that previous agent-based models that address the impact of fact-checkers on a network [23,28] may need to be adjusted for lower-than-expected inter-community contact. Finally, significant differences in sharing behaviour appears mostly during the ramping up period of the pandemic (*early* phase) with large variations in deviation and means toward misinformation. This may be explained by the heightened fears and extreme uncertainty concerning the pandemic during that particular period in which the public need for information is outweighing the authority's ability to provide it [9,24,25].

At the relative level, we confirm previous findings showing the initial stage of circulation is associated with highest information spread in general [26]. The absence of significance in general behaviour during the initial period and the observed high difference in standard deviation during that period shows that most difference in spread behaviour happens in the later periods and may be associated with the virality of misinforming content and its ability to spread deeper compared to fact-checks [32]. This result also highlights that the difference in spread may be highly related to the initial amount of shares of a given URL and to external contextual factors rather than the intrinsic properties of the shared URLs (e.g, the relation between the pandemic state and the misinforming URLs topics rather than simply the misinforming URLs topics).

Causality analysis confirms that misinformation spread can be predicted from fact-checking spread. This relation is also confirmed by the FEVD analysis (Fig. 4). However, the opposite relation is not observed meaning that fact-checking spread behaviour is not causally related to misinforming behaviour even though impulse analysis show that to some extent misinformation spread shocks tend to lead to an initial increase in fact-checking spread.

Although the previous observation is encouraging, our results (Sect. 5) show that the reduction in misinformation spread associated with an increase in fact-checking information is mostly temporary. This indicates that the misinformation reduction power of fact-checking is impeded by its apparent inability to be shared over long periods of time. This echoes previous research that suggested that the amount of corrective information may play an essential role in reducing misinformation [1,26]. To this end, better fact-checking campaigns may be required to increase the virality of fact-checking content for increasing its shareability.

7 Limitations and Future Work

Although our approach is really accurate, since it does not depends on automatic annotations for identifying misinformation and fact-checks, our data covers only a small amount of misinformation and does not contain variations of the same posts. Similarly, the amount of collected posts is limited by the data collection method. A relatively simple approach for future work would be to use automatic misinformation detection methods coupled with semantic similarity measures to detect content that is already fact-checked but associated with different URLs. We could also combine different data collection methods for improving the fidelity of our study. As our results have shown, additional topological and community analysis is required to better characterise the deviations and mean differences observed in the multivariate spread analysis (Sect. 6). We plan to increase the granularity of our analysis by obtaining more fine grained information about the users (e.g., demographics) that share misinformation as well as intrinsic misinformation and fact-checking content features such as topical information.

8 Conclusion

We have presented an initial analysis of the relation between misinformation spread and fact-checking information during the initial period of the Covid-

19 pandemic. Although our results show that fact-checking spread has a positive impact in reducing misinformation, we have found that the impact of fact-checking is seriously impeded by three different factors: the amount of shared misinformation (which is disproportionately higher than fact-checking content), the different communities of fact-check sharers *versus* misinformation sharers, and the short period of time in which fact-checks are likely to spread. To overcome this, it will be necessary to build interaction bridges between fact-checking and misinformation spreaders, and create fact-checking content that is more appealing. This will help create a sustainable fact-checking information spread over time.

Acknowledgements. This work has received support from the European Union's Horizon 2020 research and innovation programme under grants agreement No 770302 (Co-Inform) and No 101003606 (HERoS).

References

1. Aird, M.J., Ecker, U.K., Swire, B., Berinsky, A.J., Lewandowsky, S.: Does truth matter to voters? The effects of correcting political misinformation in an australian sample. Roy. Soc. Open Sci. **5**(12), 180593 (2018)
2. Allgaier, J., Svalastog, A.L.: The communication aspects of the ebola virus disease outbreak in western africa-do we need to counter one, two, or many epidemics? Croatian Med. J. **56**(5), 496 (2015)
3. Almaliki, M.: Online misinformation spread: a systematic literature map. In: Proceedings of the 2019 3rd International Conference on Information System and Data Mining, pp. 171–178 (2019)
4. Barrera, O., Guriev, S., Henry, E., Zhuravskaya, E.: Facts, alternative facts, and fact checking in times of post-truth politics. J. Public Econ. **182**, 104123 (2020)
5. Brennen, J.S., Simon, F., Howard, P.N., Nielsen, R.K.: Types, sources, and claims of Covid-19 misinformation. Reuters Institute (2020)
6. Cinelli, M., et al.: The Covid-19 social media infodemic. arXiv preprint arXiv:2003.05004 (2020)
7. Farrell, T., Piccolo, L., Perfumi, S.C., Alani, H., Mensio, M.: Understanding the role of human values in the spread of misinformation. In: Conference for Truth and Trust Online (2019)
8. Harman, S.: The danger of stories in global health. Lancet **395**(10226), 776–777 (2020)
9. Huang, Y.L., Starbird, K., Orand, M., Stanek, S.A., Pedersen, H.T.: Connected through crisis: emotional proximity and the spread of misinformation online. In: Proceedings of the 18th ACM Conference on Computer Supported Cooperative Work & Social Computing, pp. 969–980 (2015)
10. Ippolito, G., Hui, D.S., Ntoumi, F., Maeurer, M., Zumla, A.: Toning down the 2019-nCoV media hype- and restoring hope. Lancet Respirat. Med. **8**(3), 230–231 (2020)
11. Jiang, S., Wilson, C.: Linguistic signals under misinformation and fact-checking: evidence from user comments on social media. In: Proceedings of the ACM on Human-Computer Interaction, vol. 2, no. CSCW, pp. 1–23 (2018)

12. Jin, F., Dougherty, E., Saraf, P., Cao, Y., Ramakrishnan, N.: Epidemiological modeling of news and rumors on Twitter. In: Proceedings of the 7th Workshop on Social Network Mining and Analysis, pp. 1–9 (2013)
13. Jin, F., et al.: Misinformation propagation in the age of Twitter. Computer **47**(12), 90–94 (2014)
14. Kim, J., Tabibian, B., Oh, A., Schölkopf, B., Gomez-Rodriguez, M.: Leveraging the crowd to detect and reduce the spread of fake news and misinformation. In: Proceedings of the Eleventh ACM International Conference on Web Search and Data Mining, pp. 324–332 (2018)
15. Kuklinski, J.H., Quirk, P.J., Jerit, J., Schwieder, D., Rich, R.F.: Misinformation and the currency of democratic citizenship. J. Polit. **62**(3), 790–816 (2000)
16. Lewandowsky, S., Stritzke, W.G., Freund, A.M., Oberauer, K., Krueger, J.I.: Misinformation, disinformation, and violent conflict: from iraq and the "war on terror" to future threats to peace. Am. Psychol. **68**(7), 487 (2013)
17. Mensio, M., Alani, H.: MisinfoMe: who is interacting with misinformation? In: ISWC Satellites (2019)
18. Mensio, M., Alani, H.: News source credibility in the eyes of different assessors. In: Truth and Trust Conference (2019). https://doi.org/10.36370/tto.2019.3
19. Muggeo, V.: Estimating regression models with unknown break-points. Stat. Med. **22**, 3055–3071 (2003). https://doi.org/10.1002/sim.1545
20. Nieminen, S., Rapeli, L.: Fighting misperceptions and doubting journalists' objectivity: a review of fact-checking literature. Polit. Stud. Rev. **17**(3), 296–309 (2019)
21. Nyhan, B., Reifler, J.: Estimating Fact-Checking's Effects. American Press Institute, Arlington (2015)
22. Sarkar, S., Guo, R., Shakarian, P.: Using network motifs to characterize temporal network evolution leading to diffusion inhibition. Soc. Netw. Anal. Min. **9**(1), 1–24 (2019). https://doi.org/10.1007/s13278-019-0556-z
23. Saxena, A., Hsu, W., Lee, M.L., Leong Chieu, H., Ng, L., Teow, L.N.: Mitigating misinformation in online social network with top-k debunkers and evolving user opinions. In: Companion Proceedings of the Web Conference 2020, pp. 363–370 (2020)
24. Spence, P.R., Lachlan, K., Burke, J.M., Seeger, M.W.: Media use and information needs of the disabled during a natural disaster. J. Health Care Poor Underserv. **18**(2), 394–404 (2007)
25. Spence, P.R., et al.: Proxemic effects on information seeking after the september 11 attacks. Commun. Res. Rep. **22**(1), 39–46 (2005)
26. Starbird, K., Dailey, D., Mohamed, O., Lee, G., Spiro, E.S.: Engage early, correct more: how journalists participate in false rumors online during crisis events. In: Proceedings of the 2018 CHI Conference on Human Factors in Computing Systems, pp. 1–12 (2018)
27. Swire, B., Berinsky, A.J., Lewandowsky, S., Ecker, U.K.: Processing political misinformation: comprehending the trump phenomenon. Roy. Soc. Open Sci. **4**(3), 160802 (2017)
28. Tambuscio, M., Ruffo, G., Flammini, A., Menczer, F.: Fact-checking effect on viral hoaxes: a model of misinformation spread in social networks. In: Proceedings of the 24th International Conference on World Wide Web, pp. 977–982 (2015)
29. Tong, G.A., Du, D.Z.: Beyond uniform reverse sampling: a hybrid sampling technique for misinformation prevention. In: IEEE INFOCOM 2019-IEEE Conference on Computer Communications, pp. 1711–1719. IEEE (2019)
30. Vaezi, A., Javanmard, S.H.: Infodemic and risk communication in the era of CoV-19. Adv. Biomed. Res. **9** (2020)

31. Vlachos, A., Riedel, S.: Fact checking: task definition and dataset construction. In: Proceedings of the ACL 2014 Workshop on Language Technologies and Computational Social Science, pp. 18–22 (2014)
32. Vosoughi, S., Roy, D., Aral, S.: The spread of true and false news online. Science **359**(6380), 1146–1151 (2018)
33. Wang, D., et al.: Clinical characteristics of 138 hospitalized patients with 2019 novel coronavirus-infected pneumonia in Wuhan, China. JAMA **323**(11), 1061–1069 (2020)
34. Xian, J., Yang, D., Pan, L., Wang, W., Wang, Z.: Misinformation spreading on correlated multiplex networks. Chaos: Interdisc. J. Nonlinear Sci. **29**(11), 113123 (2019)
35. Xie, B., et al.: Global health crises are also information crises: a call to action. J. Assoc. Inf. Sci. Technol. (2020)

Facebook Ads: Politics of Migration in Italy

Arthur Capozzi, Gianmarco De Francisci Morales, Yelena Mejova(✉), Corrado Monti, André Panisson, and Daniela Paolotti

ISI Foundation, Turin, Italy
yelena.mejova@gmail.com

Abstract. Targeted online advertising is on the forefront of political communication, allowing hyper-local advertising campaigns around elections and issues. In this study, we employ a new resource for political ad monitoring – Facebook Ads Library – to examine advertising concerning the issue of immigration in Italy. A crucial topic in Italian politics, it has recently been a focus of several populist movements, some of which have adopted social media as a powerful tool for voter engagement. Indeed, we find evidence of targeting by the parties both in terms of geography and demographics (age and gender). For instance, Five Star Movement reaches a younger audience when advertising about immigration, while other parties' ads have a more male audience when advertising on this issue. We also notice a marked rise in advertising volume around elections, as well as a shift to more general audience. Thus, we illustrate political advertising targeting that likely has an impact on public opinion on a topic involving potentially vulnerable populations, and urge the research community to include online advertising in the monitoring of public discourse.

Keywords: Targeted advertising · Social media · Politics · Immigration

1 Introduction

According to the International Organization for Migration, almost 130k migrants have arrived in Europe in 2019.[1] Global migration is a systemic challenge for Europe [18], and for Italy in particular being on the Mediterranean route [12]. Therefore, it is not surprising that migration is a central issue in European and Italian politics [1].

At the same time, Europe and the world in general have seen a resurgence of nationalism with a populist derive, such as what seen in USA, Brazil, Philippines, Turkey, UK, Hungary, and Italy to name a few [28]. Nationalist parties often espouse nativist positions, denouncing the negative effects of migration, and

After first author, the author names are in alphabetical order.
[1] https://migration.iom.int/europe?type=arrivals.

S. Aref et al. (Eds.): SocInfo 2020, LNCS 12467, pp. 43–57, 2020.
https://doi.org/10.1007/978-3-030-60975-7_4

emphasizing the loss of cultural identity [10]. Indeed, migration is a controversial issue often used as a political tool by parties across the aisle. In this work, we focus on the Italian debate around migration, which has been of paramount importance in recent history.[2]

Some of the parties have risen to success by embracing the new communication technologies available on the Web and social media [3]. The Five Star Movement (*Movimento 5 Stelle*, M5S) [23] is a notorious example of this trend, with its focus on anti-establishment rhetoric paired with an organizational focus around new communication technologies. In contrast with traditional media which focus on mass communication (TV and news), social media allows direct and personal communication. This *micro-targeting* feature is highly controversial and has caused wide backlash towards Facebook [16].

Given this socio-political context, our main research aim is to study the *political messaging around migration in Italy via Facebook ads.* In particular, we look for evidence of micro-targeting and analyze different focus of the major parties on Italian stage. We find that different parties have different demographic foci, and the target audience shifts during events such as elections. In addition, we find evidence that nationalist parties focus on more male audience for their migration ads compared to their normal targets. Finally, we discuss the advantages and limitations of this methodology, and future integration of online data in political discourse analysis.

2 Background

Political advertising. As efforts toward transparency have recently resulted in Facebook and other major platforms releasing information about political advertising, the research is just beginning to measure the extent and impact of online targeting on vulnerable groups [3]. For instance, Hegelich and Serrano [15] examined political ads made available through Facebook Ad Library API and the Google Cloud BigQuery API, finding targeting differences among major German political parties. While some parties published ads with broad appeal, others had specific targets, such as women of working age. However, no differentiation between topical focus was made in this analysis. Other sources of ad-related datasets become available as political events unfold, such as in the case of Ribeiro et al. [25] who examined a dataset of 3,517 Facebook ads that were linked to a Russian group Internet Research Agency (IRA). Targeted to U.S. audience, the ads achieved a higher click through rate (CTR) than usual Facebook ads. After running a survey on the most impactful ads, the authors found that many of these ads were severely divisive, and "generated strongly varied opinions across the two ideological groups of liberals and conservatives". In particular, these ads were targeted to those who were more likely to believe, and approve and subsequently less likely to report or identify false claims in them. However, the Facebook Ads Library API has been shown to be unstable

[2] https://www.bbc.com/news/world-europe-43167699
https://www.politico.eu/article/italy-immigration-debate-facts-dont-matter.

during the 2019 European Parliament Election [22,26] – this study is the latest attempt to negotiate the API to extract useful information about political advertising and targeting.

Italian Parties and Migration. In the time period we analyzed, March 2019 to March 2020, the main Italian parties are the Five-Star Movement (*Movimento 5 Stelle*, M5S), the League (Lega), the Democratic Party (*Partito Democratico*, PD). M5S was the largest party in Italy at the previous national elections in March 2018; it currently holds about one third of the parliament, while Lega, PD, and Forza Italia hold about one fifth each. M5S has been unanimously described as anti-establishment and populist; however, its position on the left-right spectrum has been harder to describe [21]. Emanuele et al. [9] found that "M5S voters are leftist on the economy but quite close to right-wing voters on Europe and immigration". M5S's positions on migration is wavering [24], mixing humanitarianism with defense of national borders [21] – for instance, they accused NGOs of increasing illegal immigration by rescuing migrants at sea [5]. According to Gerbaudo and Screti [13] this strategy has the goal of competing on the issue with Lega.

Lega under Salvini is in fact keeping a strong anti-immigration and nativist focus, adopting "stop the invasion" as a slogan, calling for immediate repatriations, and depicting Islam as a threat to Italian Christian identity [17]. As the interior minister in June 2018, Salvini declared Italian ports closed to NGO ships rescuing migrants [7]. Brothers of Italy (*Fratelli d'Italia*, FdI) holds similar positions, and was allied with Lega and Forza Italia in the center right coalition in 2018.

The Democratic Party, who has dominated Italian government coalitions after 2013, is the main target of such critiques from other parties. PD adopted a strategy of "setting up a decentralized system for the management of asylum requests" [8]. In their view, migration must be managed more than stopped: for instance, they aimed at reducing migrant flows to Italy [29], through agreements with Libya [4] and by negotiating the redistribution of migrants with the rest of EU. However, they kept a mostly humanitarian position in public discourse about rescuing operations at sea. During the analyzed time period, former prime minister Matteo Renzi split from PD to form a new formation, Italia Viva (IV).

3 Data

The data for this collection comes from Facebook Ads Library, which provides an API[3] for accessing advertisement shown on Facebook and Instagram platforms about "social issues, elections or politics" since March 2019. On March 30, 2020 we run a query to collect all advertisements originating from Italy containing keywords related to migrants, immigrants, and refugees (see Appendix A for the exact keywords). We use keywords in Italian language because we are interested

[3] https://www.facebook.com/ads/library/api.

in the domestic perception of the topic of migration. The keywords were compiled using an iterative approach, by starting from "migrante" (migrant) and "immigrato" (immigrant), and then adding keywords to the set using FastText embedding similarity [14]. In particular, we search for ads appearing only on Facebook (not Instagram).

The API returns ads that were published since March 2019, including those which are running at the time of the collection, and excluding those which have been deleted (mostly removed by the platform). For each ad, the API provides the ad ID, title, and body, and URL, its creation time and the time span of the campaign, the Facebook page authoring the ad, as well as a funding entity. Additional information includes a range for the cost (as well as its currency) and impressions of the ad. The API also provides a distribution over impressions broken down by gender (male, female, unknown), age (7 groups), and location (down to region in Italy). We restrict our analysis by considering only impressions localized in Italy. In further analysis of age, cost and impression information, which come in a range, we take the average of the end points of the range, and for open-ended ranges such as 65+ we take the known closed end point (in this case, 65). Note that in further analysis we focus on the expenditure and impressions, rather than the raw number of ads, as the same message can be presented in several ads, thus making a single ad not a meaningful unit of analysis. For example, to compute the cost per thousand impressions for an author, we divide their total expenditure on all their campaigns by the total number of impressions received. Upon manual examination of top authors, we exclude 7 who are not relevant for the aim of our analysis, such as brands and emigration advertising.[4] The dataset consists of 2312 ads from 733 unique pages.

To better understand the pages in our dataset, we query WikiData[5] with all n-grams of words in the name of each page. We restricted the matched entities in WikiData to a specific set of categories: parties, politicians, journalists, and NGOs. In all the cases where we had more than one match, we manually disambiguate the matched entities and select the relevant one. To further refine our data, we match the remaining pages against a list of local Italian politicians.[6] The process results in 249 pages identified as individual politicians and 53 pages identified as parties (including regional branches).

We then consider the pages of politicians affiliated with one of the four parties in Italy most present in our data – Democratic Party (PD), League (Lega), 5 Stars Movement (M5S), Brothers of Italy (Fratelli d'Italia, FdI), and Italy Alive (Italia Viva, IV) – resulting in 208 pages. For each page, we retrieve all ads without keyword constraint, 17014 ads in total, which we later use as a general political baseline.

[4] The full list of authors excluded is reported in Appendix A.

[5] https://www.wikidata.org.

[6] https://dait.interno.gov.it/elezioni/open-data/dati-amministratori-locali-carica-al-31-dicembre-2018.

4 Results

4.1 Characterizing Advertising Around Migration

We begin by considering the top spenders in our data, alongside those whose ads received the most impressions. These statistics are plotted in Fig. 1, with bottom x axis showing the impressions and top the spending, summed over all ads by the given author. The most prolific author is Matteo Salvini, the leader of the Lega party – a major right-wing nationalist party – having spent in total over 50000 Euros and attracted almost 8000000 impressions. Interestingly, we find other party leaders in the top places: the third most prolific author is Matteo Renzi (former leader of PD, current leader of IV), ahead of Giorgia Meloni (leader of FdI), despite spending less, pointing to a different cost per impression. The second most prolific author is an NGO, the Italian branch of Amnesty International, followed by others such as Save the Children Italia and Refugees Welcome Italia. As we detailed in Sect. 2, NGOs have been at the center of political controversies for their rescuing operations. Other authors include a journalist (Giulio Gambino), a petition website (petizioni.it), and an online learning website (ICOTEA).

Thus, we find that the biggest spender on ads related to migration comes from the political domain. In particular, Matteo Salvini being the largest spender on ads about immigration is not surprising: first, his party has been recognized as "one of the first political entrepreneurs of anti-immigration sentiments in the Italian arena" [4]; moreover, political commenters recognized Matteo Salvini as successful on leveraging social media to promote his views.[7] Facebook has been recognized in previous analysis [20] as an important tool also for the other party leaders we find in the top positions.

The resources used for this political advertising make for an unequal distribution of impressions. The Gini coefficient of impressions per page is 0.465, while for individual ads even larger at 0.800, thus indicating a highly skewed distribution of impressions across the field.

Considering the efficiency of the campaigns, we find that most are able to achieve better impression per euro spent ratios than the top spender. Whereas Matteo Salvini paid on average 6.8 ([3.2, 11]) Euros per thousand impressions (CPM), everybody else in the top 30 has spent 4.1 ([0.6, 10.7]) Euros per thousand impressions. The most expensive impressions were purchased by Giorgio Maria Bergesio (local politician from Lega party) at 10.3 CPM, while the cheapest by Save the Children Italia at 0.8 CPM. We address whether this difference may be linked to microtargeting in the next section.

Figure 2 shows the temporal distribution of impressions and expenditure per day. Specifically, we divide the total impressions (expenditure) uniformly over the duration of each ad campaign. The two time series are highly correlated, and

[7] See for instance The Atlantic, *"The New Populist Playbook"* (https://www.theatlantic.com/international/archive/2019/09/matteo-salvini-italy-populist-playbook/597298/) and Reuters.com, *"Chestnuts, swagger and good grammar: how Italy's 'Captain' builds his brand"* (https://www.reuters.com/article/us-italy-politics-salvini-socialmedia/chestnuts-swagger-and-good-grammar-how-italys-captain-builds-his-brand-idUSKCN1MS1S6).

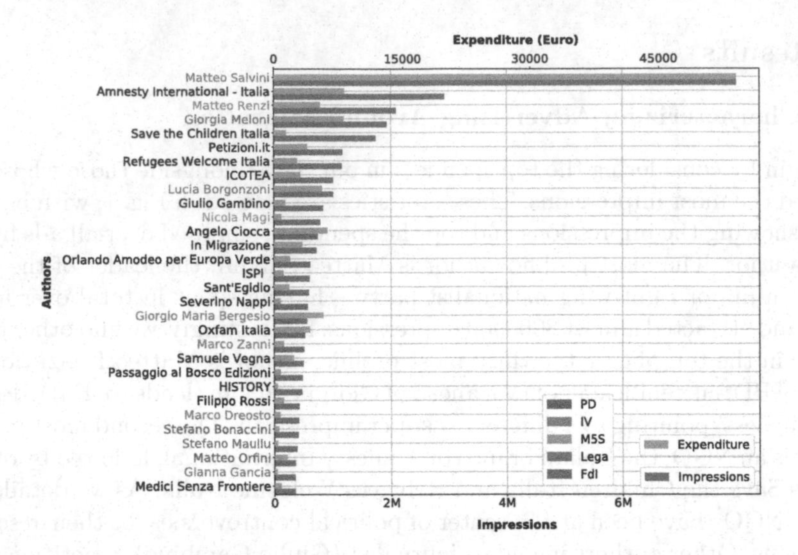

Fig. 1. Authors by impressions and expenditure per page (top 30 for impressions).

are bursty. We also show major events: all major elections (European elections, as well as regional in Piedmont, Umbria, Calabria, and Emilia-Romagna) and the government crisis. We observe that some of those events correspond to spikes in ads expenditure. During these elections, the major players were political parties, which behave differently at different elections, and we explore their behavior in further sections. The largest spike is in the weeks preceding the European elections on 23–26 May (other regional elections also take place then).

Figure 3 shows the geographical distribution over regions of the number of impressions of political ads normalized by population (left) and of the fraction of impressions of migrant ads (right). The most impressions per capita of political ads are Emilia-Romagna, (with an average of 22.2), followed by Umbria (12.4), and Calabria (11.4). These are in fact the regions that had regional elections in the reference period. The fact that these impressions per person are larger than 1 implies that an average user saw several ads in that time. Emilia-Romagna having the highest number of impressions per capita (despite having a large population) confirms the perceived political importance of their 2020 regional elections: many observers noted in fact that those local elections were particularly important for Italian politics, as they could affect the national government [2].

The map on the right in Fig. 3 shows the fraction of political ads that are related to immigration. We note how there is a geographical gradient in attention to this topic: northern regions and islands (Sardinia and Sicily) show more focus on this topic. The northern regions are typically viewed as strongholds of Lega,[8] whose political agenda is strongly focused on immigration. The prevalence of this topic in Sicily and Sardinia is more surprising. It is interesting to note that these regions emerged in a survey as having the largest fraction of the population overestimating the presence of immigrants [27]. Emilia-Romagna and

[8] https://commons.wikimedia.org/wiki/File:Camera_2018_Partiti.svg.

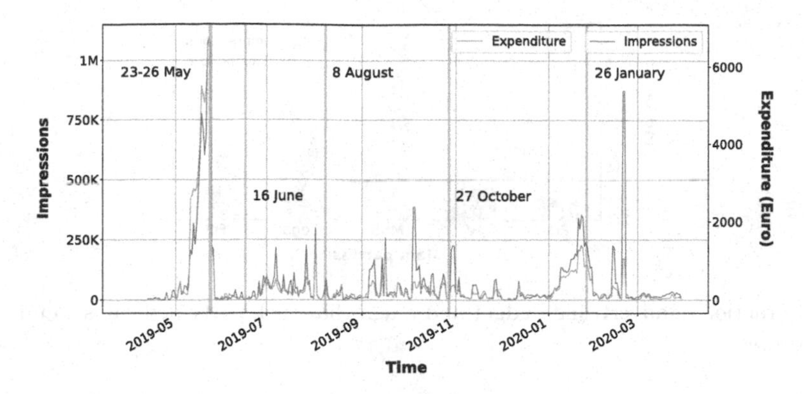

Fig. 2. Impressions and expenditure for ads related to migrations, per day. We also highlight important political events with vertical bars: 23–26 May - European and regional (Piedmont) elections; 16 June - Regional elections in Sardinia; 8 Aug - Lega leaves the government; 27 Oct - Regional election in Umbria; 26 Jan - Regional elections in Calabria and Emilia-Romagna.

Calabria show the least prevalence of the topic. In the next section, we find that these regions also display a more generalist targeting: the age and gender distribution of their ads is closer to the population average. Finally, we note that the attention to the topic in Sicily could be connected to the fact that most rescue ships operations happen near its shores. We also find that a single advertising campaign (Nicola Magi, M5S) is responsible for the large portion of the impressions in Marche (44%).

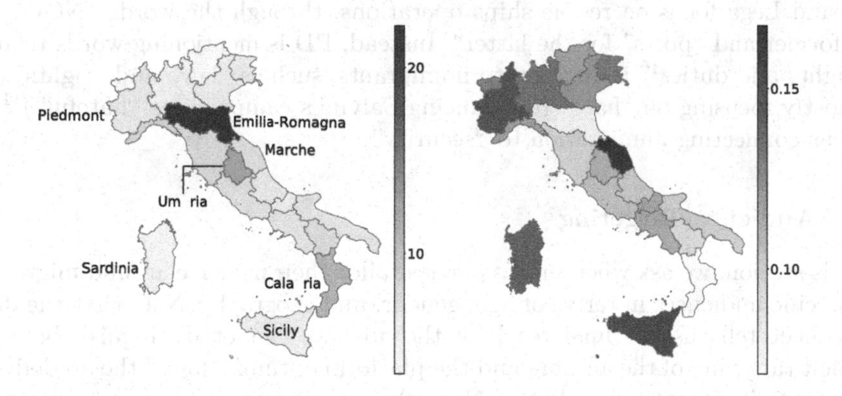

Fig. 3. Impressions per capita of all political ads (left), fraction of impressions of migrant-related ads over all political ads (right).

In Fig. 4 we put into context the expenditure committed and impressions gained by the major political parties, as a proportion of the total political advertising. We find that Lega and M5S have invested the most into migrant-related

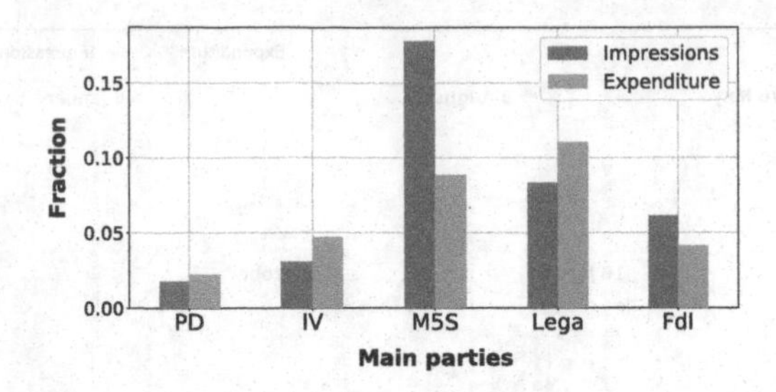

Fig. 4. Fraction of advertising budget and reach that each party dedicates to migrant-related topics.

ads as a proportion of their total budget. Among the parties, M5S attains the largest proportion of their viewership from the migrant-related ad impressions. The party having the least focus on the migrant topic is PD, both in terms of relative budget spent and impressions. Out of these, the one that has spent the most on migrant-related advertising is Lega spending in total 87314 €, followed by PD (13786 €), FdI (10963 €), IV (6023 €) and M5S (3533 €).

Finally, we look at the messages conveyed by each party, by analyzing the text of each ad. Table 1 in Appendix A shows the top terms used by each party about migration, sorted by odds ratio of frequency in migrant-related ads to all political ads. First, we find a difference in the way each party speaks about migration: Lega and FdI both focus on "clandestines" (i.e., illegal immigrants) in their ads. Second, all parties speak about themselves, but interestingly Lega and Salvini are often mentioned by their political opponents. Third, FdI, refers to law, territory, nation, and citizenship, thus emphasizing the defense of Italian sovereignty. Both FdI and Lega focus on rescue ships operations, through the words "NGO" for the former and "ports" for the latter[9]. Instead, PD is mentioning words related to rights and duties[10] in relation to immigrants, such as "law" and "rights". IV is mostly focusing on "hate" (denouncing Salvini's campaign as "hateful")[11] as well as connecting immigration to "security".

4.2 Audience Targeting

In this section, we ask whether the parties tailor their message around migration to specific audiences in terms of age, gender, and geography. Note that the data we collect tells us the final reach of the ads, we cannot distinguish between explicit targeting of the authors and the platform optimization of the ad delivery to potentially interested audience. Nevertheless, this gives us a precise measure of the final reach of the ads, and as such it is still a strong clue for investigating

[9] E.g., using the hashtag **`#portichiusi`** – "close the ports".

[10] The definition of "rights and duties" for immigrants is PD's first point of intervention for their integration (see https://www.partitodemocratico.it/archivio/il-pd-e-limmigrazione/).

[11] E.g., https://www.facebook.com/ads/library/?id=2095571910748061.

the ads' target audience. Figure 5 (left) plots a circle for each ad, located at the mean age and gender of audience reached, colored by the authoring party, and sized by impressions. For each party, we also plot the global impression-weighted average as an ellipse, such that each axis reflects one standard deviation of age and gender. Overall, we find most ads to be targeting older and more male audiences. However, there are differences among the parties, which we compare in Fig. 6 by plotting kernel density estimates of impressions for each party over age (left) and gender (right). We find M5S to have a broader age distribution, having more young audience, followed by IV. Considering gender, IV captures more female audience, while FdI is more male-oriented.

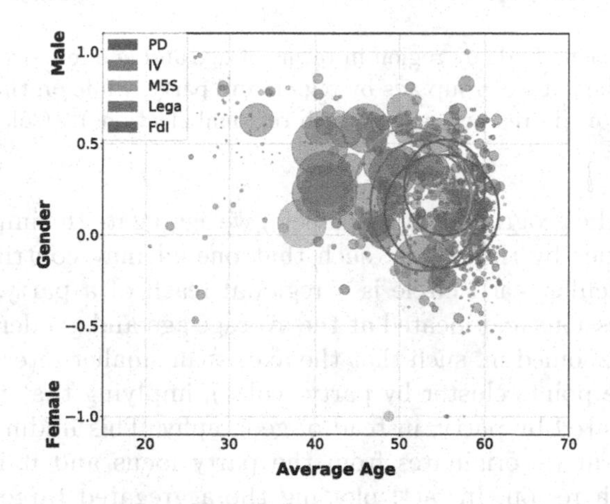

Fig. 5. Average reach of individual ads authored by political parties over age and gender, size proportional to impressions. Ellipses represent one standard deviation for each axis around the average for each party (weighted by impressions).

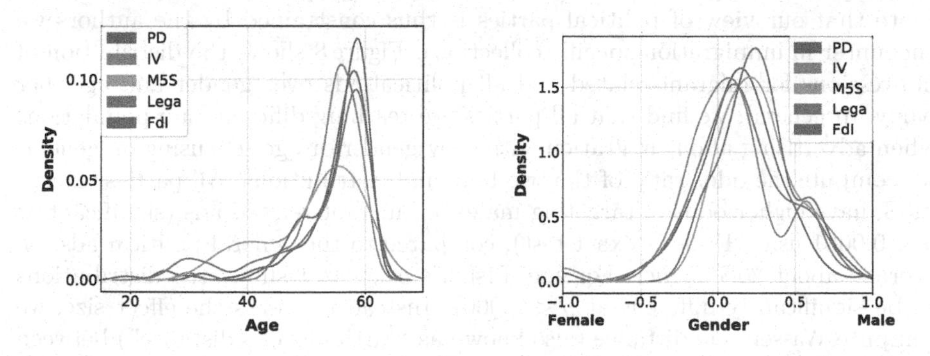

Fig. 6. Probability density functions of age (left) and gender (right) impressions by party computed by kernel density estimation.

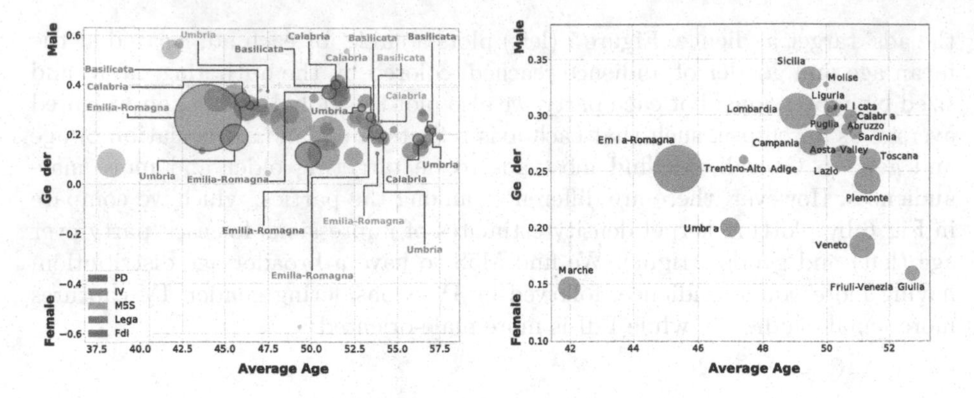

Fig. 7. Average reach of ads by region in terms of age and gender, size proportional to impressions. On the left we group ads by region and party, while on the right we group them only by region (in red are regions with regional elections). (Color figure online)

Considering the geography, in Fig. 7 (left) we aggregate the impressions of ads by different parties by the region, such that one ad may contribute to several regions. In particular, each circle is a regional reach of a party, scaled by the number of impressions and located at the average age and gender of the targets. Note the plot is zoomed in, such that the axes span smaller interval than Fig. 5. We find that the points cluster by party (color), implying that the targeting is mostly differentiated by party, instead of geography. This finding suggests that the targeting strategy originates from the party focus and it is only slightly adjusted for each region. In fact, plotting the aggregated target audiences by region in Fig. 7 (right), we find that most of the difference among regions are due to events such as elections. In fact, elections seem to make the targeting more gender-balanced and younger in age.

Next, we ask whether this targeting is due to the specific topic of immigration, as compared to overall political ads. Recall that we collect all other ads posted by the politicians present in our collection, making up a topic-neutral baseline (note that our view of political parties is thus constrained by the authors we encounter in immigration-specific collection). Figure 8 shows the distribution of impressions for migrant-related and all political ads over gender and age, per party. In general, we find that all parties are reaching different sub-populations when advertising about migration, either by gender or age. Focusing on gender, we compute the odds ratio of the two binomial distributions. All parties, except M5S, have higher odds of targeting males on migrant-related ads (significant at $p < 0.0001$ using Fisher's exact test), compared to the general political ads, on average about 20% higher. For age, Fisher's exact test shows the distributions to be significantly different at $p < 0.0001$. Instead to assess the effect size, we compute Wasserstein distance (also known as "earth mover's distance") between the two age distributions. The higher this metric, the more one needs to "move" from one distribution to achieve the other. We find M5S and PD to have the most different migrant-related ad audience, compared to that of all of their political ads, with the focus shifting drastically from older to younger segments for M5S.

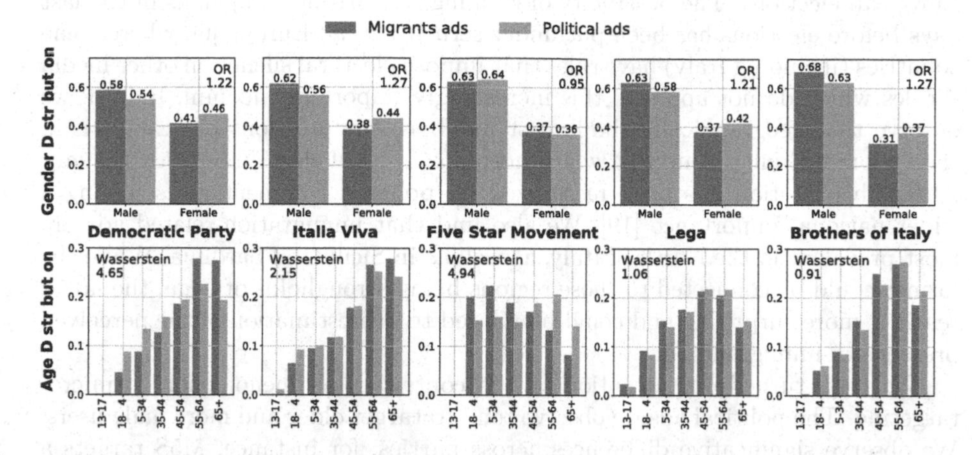

Fig. 8. Distributions of impressions for migrant-related and all political ads over gender (top) and age (bottom), per party.

Lastly, we wonder whether a higher cost per impression may be due to narrower targeting of the viewership. To test this hypothesis, we measure the extent of micro-targeting by computing entropy of the distribution of impressions over demographic buckets and over geographic regions. Correlating this measure with cost per impression, we find the correlation between cost per impression and demographics entropy to be $\rho_d = -0.112$, and correlation between cost per impression and geographic entropy $\rho_g = -0.243$, which indicates very little relationship between the narrowness of the targeting and efficiency.

5 Discussion

In this work, we have analyzed the behavior of political and social advertisers on Facebook on the topic of immigration in Italy. We believe our data set, including the ad metadata and enrichment by using external sources, to be of value to the community, and thus we make it openly available.[12]

In our analysis, we find evidence that Facebook advertising is considered important by many political actors. Among the top four authors by total number of impressions, we find that three are party leaders. In particular, Matteo Salvini (leader of the Lega party) is the top one, both by impressions and by total expenditure. This results reflects two main characteristics of his leadership: a focus on nativist discourse [4] and the usage of social media as a political platform.

[12] https://github.com/PotenteOpossum/Facebook-Ads-Politics-of-Migration-in-Italy.

We find ads impressions to be largely correlated with national and local elections, thus corroborating the hypothesis that they are viewed as an important means to attract votes. In particular, we find the largest peak to coincide with European elections. The possibility of running advertising campaigns in the last days before elections has been put under scrutiny in all Europe [6], where some countries (including Italy) have rules that impose electoral silence on other media – rules which do not apply to this increasingly important medium. Indeed, we observe that geographically the largest number of impressions per capita is in the region of Emilia-Romagna, which hosted a regional election on 26th January 2020. This election has been recognized by political commenters as having a wider national importance [19]. We also find that immigration-related ads are most prevalent in the North of Italy, as well as in Sicily and Sardinia. While the former could be connected to those regions being strongholds of Lega, the latter result is more surprising and could be related to overestimation of the perceived presence of immigrants [6].

We then turned our attention to the controversial phenomenon of micro-targeting. The political ads we observe tend to target older and more male users. We observe significative differences across parties: for instance, M5S targets a younger audience, while FdI a more male one. This is consistent with the demographic distribution of their voters, as estimated from surveys [11]. However, we find that regional elections can shift the audience towards the population average, both in terms of age and gender. Furthermore, the ads targets change with the topic: some parties appear to reach a significantly different audience when speaking about immigration. In particular, most parties tend to reach a more male-oriented audience when speaking about immigration, while in terms of age each party has a different bias compared to their general audience. We believe these results to be an indication of micro-targeting, i.e., parties trying to reach different sub-populations with different messages [25].

Limitations. We restricted our attention to a specific use case of political advertising on Facebook: the topic of immigration in Italy. Since immigration has become one of the most loaded issues in Italian party politics [4], our findings are limited to this specific context. Narrowing our scope in topic, time, and space, however, has the beneficial effect of removing potential confounding factors.

Some specific aspects of the advertising campaigns are not available to the public through the Facebook Ads Library API, thus limiting our findings. In particular, different mechanisms could be driving the micro-targeting we observed: on the one hand, a certain sub-population could be engaging more with immigration-related ads, thus increasing their impressions; on the other, they could be explicitly targeted by the parties. For instance, Hegelich et al. [15] find that "organic" interactions – those resulting from free posts that are usually disseminated to existing party followers – may attract more views than paid advertising. Without further data shared by the platform, there cannot be a conclusive answer to this important question. It has also been reported that the API appeared unstable and unreliable [22]. To test the stability of our data, we re-query both the migrant-related ads and general political ones two weeks after, and find that 1 ad is missing in the former collection and 3 in latter. Thus we conclude that, in our case, the API is fairly stable.

Despite these limitations, we believe our work brings important elements in the debate on Facebook advertising in politics, showing its importance in electoral campaigns, and bringing statistical significance to the hypothesis that different parts of the population are being targeted by different ads. More work is necessary on this topic. In order to understand the different political messages, it would be possible to use text and image mining to characterize ads, and understand which views and behaviors they promote. Furthermore, comparing the data we obtained from Facebook with other data sources – e.g., surveys, news coverage, election results – could help in understanding the complex relationship of these ads with real-world events.

A Appendix

Table 1. Top 10 words for each party by odds ratio of presence in migrant ads compared to all political ads (translated to English).

PD	IV	M5S	Lega	FdI
PD	Hate	Movement	Lega	Europe
Salvini	Italians	Italy	Salvini	Italy
Politics	September	Countries	Clandestines	Brothers
Migrants	Matteo	Conte	Italians	Law
Italy	Death	Answer	Truth	Clandestines
Party	Voice	Policies	Migrants	Territory
Europe	Mario	Lega	Matteo	NGO
Rights	Salvini	Europe	Immigrants	Activity
Law	Community	Agreement	Immigration	Nations
Women	Security	Migrants	Ports	Citizenship

List of Ad Authors Manually Removed: 'Move To Canada Today', 'Patagonia', 'VisaPlace - Niren & Associates Immigration Law Firm', 'Immigration Spot', 'Battlefield Italia-La pagina', 'Videodrome'.

Keywords Used to Retrieve Ads from the Facebook Ads Library: 'migrante', 'migranti', 'immigrato', 'immigrati', 'immigrata', 'immigrate', 'ius soli', 'ius culturae', 'sbarchi', 'sbarco', 'migrazioni', 'migrazione', 'clandestino', 'clandestini', 'clandestina', 'clandestine', 'profugo', 'profughi', 'profughe', 'profuga', 'scafisti', 'scafista', 'extracomunitario', 'extracomunitari', 'extracomunitaria', 'extracomunitarie'.

References

1. European Union, Events of 2019. Human Rights Watch (2020). https://www.hrw.org/world-report/2020/country-chapters/european-union
2. Amaro, S.: Salvini's return? A regional vote in Italy risks further chaos in Rome (2020). https://www.cnbc.com/2020/01/13/matteo-salvini-return-a-regional-vote-in-italy-risks-further-chaos-in-rome.html
3. Beer, D., Redden, J., Williamson, B., Yuill, S.: Landscape summary: online targeting: what is online targeting, what impact does it have, and how can we maximise benefits and minimise harms? Centre for Data Ethics and Innovation (2019)
4. Bulli, G., Soare, S.C.: Immigration and crisis in a new immigration country: the case of Italy. Hrvatska komparativna javna uprava: časopis za teoriju praksu javne uprave **18**(1), 127–156 (2018)
5. Coticchia, F., Vignoli, V.: Populist parties and foreign policy: the case of Italy's five star movement. Br. J. Polit. Int. Relat. 1369148120922808 (2020)
6. Council of Europe: Internet and electoral campaigns - study on the use of internet in electoral campaigns (2018). https://edoc.coe.int/en/internet/7614-internet-and-electoral-campaigns-study-on-the-use-of-internet-in-electoral-campaigns.html
7. Cusumano, E., Gombeer, K.: In deep waters: the legal, humanitarian and political implications of closing Italian ports to migrant rescuers. Mediterr. Polit. **25**(2), 245–253 (2020)
8. Diamond, P.: The Italian democratic party and social democratic parties in Europe. Italian Polit. (2019)
9. Emanuele, V., Maggini, N., Paparo, A.: The times they are a-changin': party campaign strategies in the 2018 Italian election. West Eur. Polit. **43**(3), 665–687 (2020)
10. Falk, F.: Invasion, infection, invisibility: an iconology of illegalized immigration. Technical report, Basel University Library (2010)
11. Formigoni, L., Forni, M.: Elezioni politiche (2018). www.ipsos.com/sites/default/files/ct/news/documents/2018--03/elezioni_politiche_2018_analisi_post-voto_ipsos-twig_0.pdf
12. FRONTEX: Migratory routes. European Border and Coast Guard Agency (2020). https://frontex.europa.eu/along-eu-borders/migratory-routes/central-mediterranean-route
13. Gerbaudo, P., Screti, F.: Reclaiming popular sovereignty: the vision of the state in the discourse of podemos and the movimento 5 stelle. Javnost Public **24**(4), 320–335 (2017)
14. Grave, E., Bojanowski, P., Gupta, P., Joulin, A., Mikolov, T.: Learning word vectors for 157 languages. In: Proceedings of the International Conference on Language Resources and Evaluation (LREC 2018) (2018)
15. Hegelich, S., Serrano, J.C.M.: Microtargeting (2019)
16. Hern, A.: Facebook to curb microtargeting in political advertising (2019). https://www.theguardian.com/technology/2019/nov/22/facebook-to-curb-microtargeting-in-political-advertising
17. Ivaldi, G., Lanzone, M.E., Woods, D.: Varieties of populism across a left-right spectrum: the case of the front national, the northern league, podemos and five star movement. Swiss Polit. Sci. Rev. **23**(4), 354–376 (2017)
18. Joppke, C., et al.: Challenge to the Nation-State: Immigration in Western Europe and the United States. Oxford University Press on Demand, Oxford (1998)
19. Frosini, J.O., Jones, E., Pasquino, G.: Emilia-Romagna: a setback for Salvini or a comeback for the left? (2020). https://www.bipr.eu/PROFILESUMMARIES/20200130.pdf

20. Mazzoleni, G., Bracciale, R.: Socially mediated populism: the communicative strategies of political leaders on Facebook. Palgrave Commun. **4**(1), 1–10 (2018)
21. Mosca, L., Tronconi, F.: Beyond left and right: the eclectic populism of the five star movement. West Eur. Polit. **42**(6), 1258–1283 (2019)
22. Mozilla: Data Collection Log – EU Ad Transparency Report (2019). https://adtransparency.mozilla.org/eu/log/
23. Natale, S., Ballatore, A.: The web will kill them all: new media, digital utopia, and political struggle in the Italian 5-star movement. Media Cult. Soc. **36**(1), 105–121 (2014)
24. Pasini, N., Regalia, M.: The 2018 Italian general elections: focus on immigration. The Twenty-fourth Italian Report on Migrations 2018 (2019)
25. Ribeiro, F.N., et al.: On microtargeting socially divisive ads: a case study of Russia-linked ad campaigns on Facebook. In: Proceedings of the Conference on Fairness, Accountability, and Transparency, pp. 140–149 (2019)
26. Rosenberg, M.: Ad tool Facebook built to fight disinformation doesn't work as advertised. The New York Times, July 2019. https://www.nytimes.com/2019/07/25/technology/facebook-ad-library.html
27. Saso, R.: L'immigrazione in Italia: tra dati reali, (dis)informazione e percezione. l'Eurispes (2019). https://www.leurispes.it/limmigrazione-in-italia-tra-dati-reali-disinformazione-e-percezione/
28. Smith, A.: Nations and Nationalism in a Global Era. Wiley, Hoboken (2013)
29. Wintour, P.: Italian minister defends methods that led to 87% drop in migrants from Libya, September 2017. https://www.theguardian.com/world/2017/sep/07/italian-minister-migrants-libya-marco-minniti

It's Not Just About Sad Songs: The Effect of Depression on Posting Lyrics and Quotes

Lucia Lushi Chen[1(✉)], Walid Magdy[1], Heather Whalley[2], and Maria Wolters[1]

[1] School of Informatics, University of Edinburgh, Edinburgh, UK
{lushi.chen,wmagdy,maria.wolters}@ed.ac.uk
[2] Centre for Clinical Brain Sciences, University of Edinburgh, Edinburgh, UK
heather.whalley@ed.ac.uk

Abstract. When studying how mental illness may be reflected in people's social media use, content not written by the users is often ignored, because it might not reflect their own emotions. In this paper, we examine whether the mood of quotes posted on Facebook is affected by underlying symptoms of depression. We extracted quotes and song lyrics from the feeds of 781 Facebook users from the MyPersonality database who had also completed the CES-D depression scale. We found that participants with elevated depressive symptoms tend to post more song lyrics, especially lyrics with neutral or mixed sentiment. By analysing the topics of those lyrics, we found they center around overwhelming emotions, self-empowerment and retrospection of romantic relationships. Our findings suggest removing quotes, especially lyrics, might eliminate content that reflects users' mental health conditions.

Keywords: Social media · Quotes · Lyrics · Depression

1 Introduction

Social media records provide psychologists with a novel way of examining mental illness symptoms [4,5]. Existing studies often focus on the emotional content written by the social media users themselves, which we refer to as "self-created Content" (SC) [15]. Non-self created content, such as reposts, music and videos, is seen as reflecting indirect emotions of the user and thus receives less attention in the analysis [18].

There are two types of non-self created content: *repost* (e.g. shares and retweets), and *copy-and-paste quotes* (e.g. song lyrics, religious verses, and famous quotes). A repost is easy to identify, since it is a functionality on social media platforms to share a post from another user. However, quotes are more complicated to identify, since usually there are no quote marks or references to the source. Furthermore, sometimes quotes and lyrics are posted from memory, which can introduce distortions. There are only a few studies that examine reposts (e.g., [17]), while, to our knowledge, quotes have not been studied yet.

S. Aref et al. (Eds.): SocInfo 2020, LNCS 12467, pp. 58–66, 2020.
https://doi.org/10.1007/978-3-030-60975-7_5

There is extensive work on how affective disorders, especially depression [4, 5, 15, 17], are reflected in social media data. However, existing studies in this line of research focus on self-created content only. To the best of our knowledge, this study is the first that examines whether quotes and lyrics in social media are associated with users' emotional state. This is surprising, because music is associated with mood regulation [6, 9, 14].

Our research questions are:

1. Is posting lyrics and quotes associated with levels of depression symptoms?
2. What are the themes and emotions conveyed in the lyrics and quotes posted by people with high symptoms of depression, and how might they relate to symptoms?

We analysed a set of 93,378 posts from 781 Facebook users who consented to take part in the myPersonality study [1] and who completed a measure of depressive symptom levels (CES-D) in addition to the personality scales. Potential quotes and song lyrics were detected using an automatic classifier, and logistic regression was used to examine links between the emotions expressed in quotes and lyrics and depressive symptom levels. Topic modelling was used to identify the themes of lyrics that are posted by people with high versus low depressive symptoms.

We found that quotes account for more than 10% of the content in 12.6% of participants[1]. Users with higher depressive symptom levels tend to post more lyrics with neutral sentiment. Our findings suggest that lyrics are used as an agent for users to communicate their emotions indirectly. Therefore, not all the non self-created content should be regarded as noise.

2 Background

Social Media Behavior and Depressive Symptom Level. Experiencing negative emotions can increase the amount of social interaction and sharing of emotions, both of which are part of the mood regulation process that leads to mood improvement [7]. Posting patterns on social media and the mood of posts may reflect symptoms of depression [4, 15, 17]. To study the emotions expressed in text, researchers often use sentiment analysis that categorizes affect or opinions expressed in the text. Several studies have shown that users with depressive symptoms use more negative affective words (e.g., sad, cry, hate) in their text than those who do not [5, 17].

Effects of Lyrics and Quotes on Depression. A recent study by Chen et al. [3] on a subset of the current data set found a potential association between posting quote on Facebook and users' depressive symptom levels, but they did not examine the sentiment expressed in the quotes. Surprisingly, the link between the content of quotes and symptoms of depression has not been examined in the existing literature, even though music is strongly linked to emotions.

[1] For distribution details, see Fig. 1 (Appendix).

People often choose music that is in congruence with their mood. Listening to songs centered around hurt, pain, and grief is part of the mood regulation process for coping with adverse life events [6]. Retrieving nostalgic memories from music may enhance the mood, especially when these memories are related to meaningful moments in life [13]. Listeners may find some consolation in lyrics when they realize they are not alone in dealing with the painful situations [6,14]. Quotes, especially song lyrics, may induce congruent emotions [6,9,14].

3 Data Collection and Preparation

3.1 myPersonality Dataset

We used the myPersonality data set [1]. myPersonality collected Facebook posts from 180,000 participants from 2010 to 2012, with the consent from Facebook users. The data collection process complied with Facebook's terms of service, and we obtained the required permission to use the data. Ethical approval for the secondary data analysis was obtained from the Ethics Committee of the School of Informatics, University of Edinburgh.

781 participants over the age of 18 also completed the Center for Epidemiologic Studies Depression Scale (CES-D). We extracted all posts that these participants posted during the year before completing the CES-D ($N = 93,378$). On average, participants posted 120 posts during this time window. Most participants are young ($M = 26$, $SD = 11.7$) female ($N = 448$, 57%) and White American ($N = 309$, 39%). Detailed demographics are provided in Table 5.

Depressive Symptom Screening Test. The 20-item CES-D scale is a widely used tool that measures the presence of depressive symptoms in the general population [12]. It has high internal consistency, test-retest reliability [11,12], and validity [11]. Scores range between 0 and 60. Following common practice, we adopted 22 as a cutoff point to divide participants in our dataset into high symptom (P_{HS}, $N = 478$) and low symptom groups (P_{LS}, $N = 303$) [5,17].

3.2 Identifying Quotes in User Timelines

For each post, we retrieved the first page of search results via the Google search API, which included the link, title, and snippet. Since we observed that quotes often contain misspellings or small variations, we created a rule-based classifier outlined in Algorithm 1, which assigns each post to one of three classes: 1) Lyric, 2) non-lyric quote (NL-quote), and 3) self-created content (SC).

In order to calculate the cosine similarity between post and snippet, we created document embeddings for each by converting each word to word embedding using the pre-trained word embeddings from Python Package Spacy [8], and summing the word embeddings into a single document embeddings. The first author (LC) annotated a subset of 750 posts for the quote classifier to determine the values of the thresholds in Algorithm 1 $\{X, Y, Z, N, N_l\}$ using grid search, and to test the performance on a separate test set. Of those 750 posts, 523 were

Algorithm 1. Quotes identification algorithm. $\cos(\theta)$ is the max cosine similarity between post and each of the retrieved snippets. C_q and C_l are the counts of search results that contain the word "quote" or "lyric" respectively

```
cos(θ) = argmax(cosine(post, snippet))
C = C_q + C_l
if (cos(θ) > X) OR (X > cos θ > Y and C > 0) OR (Y > cos θ > Z and C > N) then
   label ← Quote
   if (C_l > N_l) then
      label ← Lyric
   end if
else
   label ← Self-create
end if
```

Table 1. Result of quotes and lyrics classifier shown in Algorithm 1

	Validation			Test		
	F1	Recall	Precision	F1	Recall	Precision
NL-quote	87.8	88.3	87.2	89.1	85.3	94.0
Lyrics	76.6	78.0	80.3	79.6	80.0	82.2

used as validation for threshold optimisation and 227 were used for testing. The final values for the threshold are: $X = 0.998$, $Y = 0.975$, $Z = 0.85$, $N = 3$, and $N_l = 2$. Table 1 shows the classifier performance on the validation and test sets for detecting quotes and lyrics.

Of the 93,378 Facebook status updates in our data set, 3,722 were p classified as quotes or lyrics. They were posted by 305 (39%) of our 781 participants. 1,488 (40%) of the 3722 posts were song lyrics, posted by 102 (13%) of the participants.

Figure 1 (Appendix) shows the percentage of posts which were quotes or lyrics for those 305 participants. For roughly a third ($N = 99$), 10–40% of their posts were quotes.

4 Quotes/Lyrics and Depressive Symptom Levels

4.1 Frequency and Sentiment of Quotes and Lyrics

When considering all 781 participants, 198 (41%) of the P_{HS} (high symptoms) share quotes compared to 107 (35%) of P_{LS} (low symptoms). (P_{HS}) also share significantly more quotes and lyrics on their timeline ($M = 5.55, SD = 14.07$) than those with low symptoms (P_{LS}, $M = 3.52$, $SD = 9.12$) (t-test, $t(2.4) = 778.61$, $p = 0.015$).

Table 2. Sentiment distribution in identified quotes for P_{HS} and P_{LS}

	High symptoms (P_{HS})				Low symptoms (P_{LS})			
	total	pos	neg	neut/mix	total	pos	neg	neut/mix
Lyrics	1056	422 (40%)	253 (24%)	381 (36%)	432	172 (40%)	117 (27%)	143 (33%)
NL-Quotes	1597	655 (41%)	399 (25%)	543 (34%)	637	261 (41%)	140 (22%)	236 (37%)
Total	2653	1061 (40%)	663 (25%)	929 (35%)	1069	428 (40%)	256 (24%)	385 (36%)

Notes: NL-quote: non-lyrics quotes

The sentiment of quotes was analyzed based on the sentiment scores calculated by SentiStrength, which has been validated and adopted by researchers [16]. Posts that are dominated by a polarised sentiment score were labeled accordingly as Positive or Negative, while posts with no dominant polarised sentiment (neutral or equivalent magnitude of positive and negative words) are labeled as Neutral/Mixed. Table 2 shows the full distribution of sentiment in the quote posts shared by participants for each P_{HS} and P_{LS}.

4.2 Sentiment of Quotes

Since P_{HS} are more likely to post quotes than P_{LS}, we now examine the relationship between the sentiment of those quotes and level of depressive symptoms more closely. We use logistic regression, with symptom group (low versus high) as the dependent variable, frequency of lyrics and quotes (expressed as ratios), and sentiment of lyrics and quotes as independent variables.

We derived 13 relevant metrics of lyric and quote sentiment, which are listed in Table 3. Most variables are weakly correlated ($r < 0.25$) with each other, and all the correlations are significant ($p < 0.001$; see also Fig. 1). However, the magnitude of sentiment variables are moderately correlated with ratio of lyrics and quotes ($r > 0.40$).

We constructed two general linear models, one with all 13 metrisc as independent variables (Model 1), and one excluding variables that have high collinearity with others (Model 2).

Model 1 (AIC = 1028) and Model 2 (AIC = 1021) are not significantly different from each other (ANOVA, $F(2,6) = -4.86$, $p > 0.05$). Both models are shown in Table 3. We see the strongest association between symptom level and the ratio of lyrics to total post count. People with higher symptom levels are more likely to post lyrics, and when they post non-self created content, lyrics are somewhat more likely to be of mixed or neutral content. Non-lyric quotes may also be somewhat more likely to carry a negative sentiment.

Table 3. Logistic Regression Models. NL-quote: non-lyrics quotes, B: Beta coefficient, SE: standard error of the coefficient, *: $p < 0.05$

Variables	Model 1		Model 2	
	B	SE	B	SE
Ratio of positive lyrics to quote	0.52	1.02	0.40	0.79
Ratio of negative lyrics to quote	−0.59	0.85	−0.14	0.69
Ratio of neutral or mixed lyrics to quote	1.82	0.85*	1.89	0.84*
Ratio of lyrics to total post count	−8.75	4.15*	−6.92	3.76*
Ratio of positive NL-quote to quote	−0.26	0.61	−0.23	0.40
Ratio of negative NL-quote to quote	0.56	0.77	1.15	0.58*
Ratio of neutral or mixed NL-quote to quote	−0.22	0.41	−0.10	0.41
Averaged sentiment magnitude of positive lyrics	−0.06	0.19		
Averaged sentiment magnitude of negative lyrics	−0.18	0.18		
Averaged sentiment magnitude of positive NL-quote	−0.02	0.19		
Averaged sentiment magnitude of negative NL-quote	−0.17	0.18		
Ratio of mixed/neutral posts to total post count	0.95	0.70		
Ratio of negative posts to total post count	0.90	0.73		

4.3 Themes in Quotes

We have seen that people with higher levels of depressive symptoms are more likely to post quotes, in particular lyrics. Now, we investigate whether there are differences in content between quotes and lyrics posted by P_{HS} versus P_{LS} We used LDA topic modelling [2, 10] to extract common themes in quotes from both groups, using Verbs, nouns, and adjectives as input. Each input word was tagged with its source (from a post by P_{HS} or P_{LS}). The best performing model yielded 15 topics.

Most of the topics in the lyrics reflect hurt and grief in romantic love. Among the five most prevalent topics in lyrics, nearly all of them mainly comprise of words from P_{HS}. Table 4 (Appendix) shows the 10 most frequent keywords and themes of the seven most prevalent topics. Three of the most common topics of lyrics deal with empowerment, in particular self-empowerment. Topic 0, with keywords such as love, want, feel, need, think, is highly emotionally charged, and mainly comprised of words from P_{HS}, while Topic 7 comprises of lyrics that indicate introspection (e.g. feel, know). Topics from non-lyrics quotes are less varied. Most of the non-lyrics posts are dominated by two topics, which centre around life, love, and feelings towards various entities. The dominant words from topic 13 are mainly from high symptom individuals, whereas those in topic 9 are from low symptom individuals.

5 Discussion and Conclusion

In this study, we showed that people with high levels of depressive symptoms are more likely to post quotes and lyrics on Facebook. However, there is no strong association between negative sentiment of lyrics and quotes and depressive symptom levels. Instead, people with more depressive symptoms are more likely to post lyrics, and the sentiment of those lyrics tends to be mixed or neutral.

Most of the lyrics centered around hurt, pain, and grief in a romantic relationship, which may indicate a mood regulation process [6]. Some of the lyrics reflect introspection and the desire of self-empowerment, which is part of the coping process. Therefore, we argue that lyrics and quotes should not be excluded from studies of the ways in which people with depression use social media—they may hold important clues to coping strategies.

A limitation of our study is that this sample is predominantly White American female and was collected in the early 2010's. Future work should focus on digital footprints in other platforms and seek to replicate these findings on today's Facebook, where usage patterns may have evolved in the past decade.

Appendix

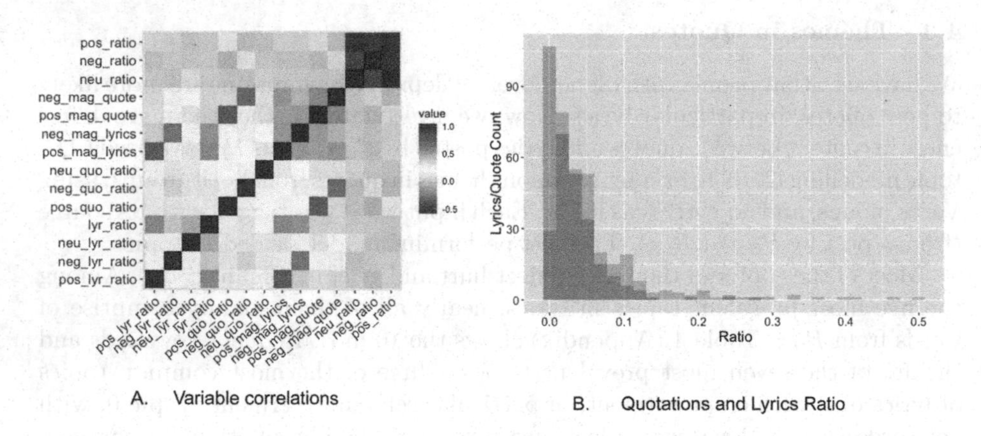

Fig. 1. Variable statistics. graph A: $p < 0.001$ for all correlations, graph B: blue: NL-quotes; red: lyrics; ratio: lyrics or quotation ratio to all post count. (Color figure online)

Table 4. Quotes Topics, H, L: high or low symptom users

#	Theme	Docs	Top 10 keywords	Example
Lyrics				
0	Overwhelming	245	$love_H$, got_H, $want_H$, $thing_H$, $feel_H$, $need_H$, $make_H$, $come_H$, $think_H$, say_H	Example 1: You traded in your wings for everything freedom brings. You never left me
3	Self-empowerment	305	$love_L$, see_L, $know_L$, $make_L$, $feel_L$, let_L, $time_L$, go_L, got_L, $life_L$	Example 1: If you're trying to turn me into someone else. Its easy to see I'm not down with that. I'm not nobody's fool
5	Self-empowerment	129	$know_H$, $hold_H$, $wait_H$, $want_H$, $tell_H$, day_H, $love_H$, $heart_H$, $dark_H$, $live_H$	Example 1: So, so you think you can tell heaven from Hell blue skies from pain, can you tell a green field
7	Introspection	130	$take_H$, say_H, $good_H$, $feel_H$, got_H, $time_H$, $sleep_H$, see_H, $know_H$, $change_H$	Example 1: I feel angry. I feel helpless. Want to change the world. I feel violent. I feel alone. Don't try to change my mind
12	Self-empowerment	264	go_H, $know_H$, let_H, $love_H$, $time_H$, day_H, $come_H$, $fall_H$, $make_H$, gon_H	Example 1: we all got holes to fill, them holes are all that's real some fall on you like a storm, sometimes you dig your own, the choice is yours to make, time is yours to take
Non-lyrics quotes				
13		1432	$love_H$, $life_H$, go_H, $know_H$, day_H, $make_H$, $thing_H$, $time_H$, $feel_H$, $people_H$	Example 1: Gratitude unlocks the fullness of life. It turns what we have into enough, and more Example 2: As a girl you see the world as a giant candy store filled with sweet candy and such. But one day you look around and you see a prison and you're on death row
9		343	$life_L$, $love_L$, $thing_L$, day_L, go_L, $time_L$, $come_L$, see_L, $think_L$, $want_L$	Example 1: If you can't make it good, at least make it look good Example 2: Would you dare? Would you dare to believe that you still have a reason to sing? Cause the pain that you've been feeling can't compare to the joy that's coming. So hold on

Table 5. Demographics

Ethnicity	No.	%
White	511	65.3
Asian	110	14.1
Black	38	4.3
Native American	13	1.6
Middle Eastern	13	1.7
Not Specified	96	12.2
Marital Status	**No.**	**%**
Single	574	73.8
Divorced	28	3.5
Married	27	3.4
Married with Children	38	4
Not specified	36	4.5

References

1. Bachrach, Y., Kosinski, M., Graepel, T., Kohli, P., Stillwell, D.: Personality and patterns of Facebook usage. In: Proceedings of the 4th Annual ACM Web Science Conference, pp. 24–32. ACM (2012)
2. Blei, D.M., Ng, A.Y., Jordan, M.I.: Latent dirichlet allocation. J. Mach. Learn. Res. **3**(Jan), 993–1022 (2003)

3. Chen, L.L., Magdy, W., Wolters, M.K.: The effect of user psychology on the content of social media posts: originality and transitions matter. Front. Psychol. **11**, 526 (2020)
4. Chen, L.L., Magdy, W., Whalley, H., Wolters, M.A.: Examining the role of mood patterns in predicting self-reported depressive symptoms. In: WebSci2020, vol. 2020, pp. 1–8 (2020)
5. De Choudhury, M., Gamon, M., Counts, S., Horvitz, E.: Predicting depression via social media. In: Seventh International AAAI Conference on Weblogs and Social Media (2013)
6. Gladding, S.T., Newsome, D., Binkley, E., Henderson, D.A.: The lyrics of hurting and healing: finding words that are revealing. J. Creativity Mental Health **3**(3), 212–219 (2008)
7. Hill, C.A.: Seeking emotional support: the influence of affiliative need and partner warmth. J. Pers. Soc. Psychol. **60**(1), 112 (1991)
8. Honnibal, M., Johnson, M.: An improved non-monotonic transition system for dependency parsing. In: Proceedings of the 2015 Conference on Empirical Methods in Natural Language Processing, pp. 1373–1378. Association for Computational Linguistics, Lisbon, Portugal, September 2015. https://aclweb.org/anthology/D/D15/D15-1162
9. Hunter, P.G., Schellenberg, E.G., Griffith, A.T.: Misery loves company: mood-congruent emotional responding to music. Emotion **11**(5), 1068 (2011)
10. Newman, D., Chemudugunta, C., Smyth, P.: Statistical entity-topic models. In: Proceedings of the 12th ACM SIGKDD International Conference on Knowledge Discovery and Data Mining, pp. 680–686 (2006)
11. Orme, J.G., Reis, J., Herz, E.J.: Factorial and discriminant validity of the center for epidemiological studies depression (CES-D) scale. J. Clin. Psychol. **42**(1), 28–33 (1986)
12. Radloff, L.S.: The CES-D scale: a self-report depression scale for research in the general population. Appl. Psychol. Meas. **1**(3), 385–401 (1977)
13. Routledge, C., Wildschut, T., Sedikides, C., Juhl, J., Arndt, J.: The power of the past: nostalgia as a meaning-making resource. Memory **20**(5), 452–460 (2012)
14. Sachs, M.E., Damasio, A., Habibi, A.: The pleasures of sad music: a systematic review. Front. Hum. Neurosci. **9**, 404 (2015)
15. Seabrook, E.M., Kern, M.L., Fulcher, B.D., Rickard, N.S.: Predicting depression from language-based emotion dynamics: longitudinal analysis of Facebook and Twitter status updates. J. Med. Internet Res. **20**(5), e168 (2018)
16. Thelwall, M., Buckley, K., Paltoglou, G., Cai, D., Kappas, A.: Sentiment strength detection in short informal text. J. Am. Soc. Inform. Sci. Technol. **61**(12), 2544–2558 (2010)
17. Tsugawa, S., Kikuchi, Y., Kishino, F., Nakajima, K., Itoh, Y., Ohsaki, H.: Recognizing depression from Twitter activity. In: Proceedings of the 33rd Annual ACM Conference on Human Factors in Computing Systems, pp. 3187–3196. ACM (2015)
18. Wang, B., Zhuang, J.: Crisis information distribution on Twitter: a content analysis of tweets during hurricane sandy. Nat. Hazards **89**(1), 161–181 (2017)

Understanding the *MeToo* Movement Through the Lens of the Twitter

Rahul Goel(✉) and Rajesh Sharma

Institute of Computer Science, University of Tartu, Tartu, Estonia
{rahul.goel,rajesh.sharma}@ut.ee

Abstract. In recent years, social media has provided platforms for raising the voice against sexual harassment (SH). The *MeToo* movement is one such online movement that aims to show the magnitude of this stigmatized issue in society. In particular, on *Twitter*, which is the focus of this study, has attracted a large number of tweets from all over the world regarding the *MeToo* movement. The studies of the *MeToo* movement focus on the SH and sexual assault (SA) incidents but fails to analyze its other hidden facets. In this work, we perform micro-analysis of the *MeToo* movement using tweets and present a descriptive analysis coupled with macro level tweets analysis in order to reveal and understand the diverse subtopics of the *MeToo* movement. In addition, we also identify and characterize varied user-groups derived through social network analysis. We find that users discussing a similar facet forms a strong community. Some of the facets out of many being discovered are as follows (1) SH incidents reporting is high for people of color (color means other than white); (2) discussion over color often leads to the use of hate and offensive vocabulary; and (3) along with workplaces, domestic SH cases are higher.

Keywords: MeToo Movement · Twitter · LDA · NLP · WordCloud · Social network analysis

1 Introduction

Social media platforms, particularly Twitter, is often used for online social movements [30]. Several studies [1,14,19,27,33] have explored tweets data to understand various social movement such as anti-discrimination, awareness, political, and women's rights, to name a few.

In this study, we focus on a particular online movement termed as *MeToo*. The phrase "*MeToo*" was initially used on myspace, an online social media platform, in 2006 by sexual harassment (SH) survivor and activist Tarana Burke [24]. However, the phrase attracted a lot more attention when Milano tweeted it around noon on October 15, 2017. It was used more than 200,000 times by the end of the day [31], and was tweeted more than 500,000 times by October 16, 2017 [13].

S. Aref et al. (Eds.): SocInfo 2020, LNCS 12467, pp. 67–80, 2020.
https://doi.org/10.1007/978-3-030-60975-7_6

In this work, we analyze the 3.5 million *#MeToo* tweets of 1.03 million users, collected using Twitter API over a period of ~5 months to cover the following research directions:

1. **Revealing the hidden facets:** In each of the online social movements, some facets are more evident than others. We termed these not so evident facets as hidden facets, which are often not easy to detect as they are the topics of discussion of a smaller group, which gets buried under the heap of more popular topics. Our findings show that the *MeToo* movement covers many hard-to-see facets apart from Sexual Harassment (SH) and Sexual Assault (SA) such as (1) SH incidents reporting is high for people of color[1]; (2) discussion over color often leads to the use of **hate and offensive vocabulary**; and (3) majority of SH incidents belong to workplaces or are domestic, compared to public places (Sect. 4).
2. **Topic aligned communities:** To understand user interactions with respect to various subtopics in the online *MeToo* movement, we employ social network analysis techniques. We observe that users tend to create communities based on topics of their interest. We also identified leaders of these communities. By leaders, we mean users who are retweeted a relatively higher number of times. We also observe that the content from activists, journalists, and media profiles gets more acknowledged (Sect. 5).

Various researchers have worked on the topic of *MeToo* movement either by using qualitative methodologies such as surveys and questionnaires or using a small amount of data and focusing on one particular facet such as feminism [15], men's psychology [26,32], SA places [18] and the role of social platforms [22]. To the best of our knowledge this is the first work which has analysed such a large datasets and revealed hidden facets along with the intuitively evident facets of SH and SA in the *Metoo* movement.

The rest of the paper is organized as follows. Next, we discuss related works. We then describe the dataset in Sect. 3. In Sect. 4, we reveal the hidden facets of the *MeToo* movement. Section 5 studies the topic aligned communities and leaders. We conclude with a discussion of future directions in Sect. 6.

2 Related Work

In this section, we discuss relevant literature with respect to the *MeToo* movement, which involves two different lines of work. First involving qualitative analysis and second using quantitative analysis. Qualitative analysis works are based on surveys, interviews with survivors or public opinion, whereas data from online social platforms or records of SH and SA incidents are used for quantitative analysis.

Qualitative Analysis. In [15], the authors explain that the *MeToo* movement is making a change in the era of feminism by helping women to share their

[1] Color means other than white.

anger and stories, which is challenging to do otherwise. They also state that under *#MeToo*, a few men have also shared their stories. [32] and [26] study the psychology of men and masculinity in the *MeToo* movement. They highlights that in response to the *MeToo* movement, some men in positions of power are afraid to participate in mentoring relationships with women. In [17], authors exclaim that being afraid to mentor women is not simply about fearing false accusations of sexual misconduct rather it is about discrediting women who speak out against sexual assault and harassment.

In other work [16], authors discuss the SH in medicine. This highlights that the SH and SA incidents exist among educated and esteem professions as well. Despite this, the reporting of SH incidents is limited and the main reason behind hiding these incidents is that people tend to avoid or are afraid to talk about SA [29]. The issue of SA and SH are spread in defense forces as well. For example, there are several studies to support and protect victims and to hold culprit liable in the army [11].

In a different work [25], authors focus on highlighting the critical points to explain why the experiences of women of color are ignored in the *MeToo* movement. They argue that the presence of racial biases in the movement shows that the SH doctrine must need some reasonable improvements.

Quantitative Analysis. In another line of work [22], authors present how individuals on different platforms (Twitter and Reddit) share their own experiences and respond to the experiences shared by the others. Their research shows that Reddit allows individuals to share their personal experiences in-depth while individuals on Twitter prefer to follow the *MeToo* movement with others. In [18], the authors try to identify the risk factors associated with SA places. Deep learning-based lexical approaches are used in their study to identify SA in terms of locations and offenders.

This work is different from the existing studies as the focus of previous works were mostly on analyzing a single facet of the *MeToo* movement considering some prior knowledge or assumptions. However, in this work, we extract the hidden facets apart from SH and SA under the *#MeToo*. These subtopics are extracted directly from the collected tweets' data without any prior assumptions.

3 Dataset Description

This section provides information about the usability of Twitter, an online social media platform for understanding social movements. The procedure of collecting and pre-processing the Twitter data is addressed in the following subsections.

3.1 Twitter and Social Movements

Twitter[2] is an online social media platform classified as a microblog with which users can share messages, links, images, or videos with other users. There are

[2] http://twitter.com.

other microblogging platforms such as Tumblr[3], FourSquare[4], Google+[5], and LinkedIn[6] of which Twitter is the most popular microblog launched in 2006 and since then has attracted a large number of users. As of 2019, Twitter has more than 326 million monthly active users[7].

In recent years, *Twitter* has provided a platform in facilitating discussions regarding stigmatized issues in society through online activism [10]. An online activism is considered as an effective way to broadcast a specific information to large and targeted audiences on social media [8,23]. One way to achieve this is by using hashtags, also known as "Hashtag activism" [6,8,21,23]. This is the act of showing support for a cause through a like, share, comment, etc on any social platform. The hashtag activism on Twitter covers multi-dimensional real-world issues like **Human rights** (#BlackLivesMatter, #IStandWithAhmed, #YessAllWoman, etc), **Political** (#ArabSpring, #NotOneMore, #NODPL, #UmbrellaRevolution, etc), **Trends** (#icebucketchallenge, #ALS, #Hallyu), **Awareness** (#FakeNews, #AmINext, #HimToo, #MeToo, etc) and others.

3.2 Data Collection and Preprocessing

This work utilizes the dataset which has been collected using Twitter's Streaming Application Programming Interface (API) and Python (*tweepy* package) for 5 months (Oct 2018 to Feb 2019) of tweets containing either "metoo" as hashtag or keyword. The size of the collected dataset is 6.5 GB, containing 5.1 million tweets.

Table 1. Statistics of the dataset.

Parameter	Value
Time period	30-09-2018 to 18-02-2019
Total number of tweets	3,529,607
Number of original tweets	870,516
Number of unique users	1,034,831
Number of unique features	91
Tweets with username	1,335,885
Tweets with hyperlink	2,306,485
Tweets with one hashtag	1,935,752
Tweets with two or more hashtags	1,593,855

[3] https://www.tumblr.com/.
[4] https://foursquare.com/.
[5] http://plus.google.com.
[6] http://linkedin.com/.
[7] https://eu.usatoday.com/story/tech/news/2017/10/26/twitter-overcounted-active-users-since-2014-shares-surge/801968001/.

Table 2. Topic modelling using LDA with gibbs sampling.

S. no.	Words	Topic (Proposed)
1	white, sex, black, brown, money, hollywood, racism, rapist, free, color	People of Color
2	sexual, harassment, assault, abuse, report, violence, public, law, issues, workplace	Places of harassment
3	fake, victims, rape, start, real, victim, person, false, true	Trust deficit among users
4	shit, hard, hell, fuck, accused, feel, stop, bad, wrong, call	Hate and Offensive vocabulary
5	woman, life, girls, sexually, day, happened, girl, raped, assaulted, remember	Female sexual harassment
6	court, vote, justice, political, left, ford, evidence, democrats, guilty, innocent	Role of justice system
7	allegations, media, hollywood, accused, news, align, industry, bollywood	News and media industry
8	speak, read, survivors, hope, talk, coming, forward, powerful, speaking, voice	Public speak-outs for survivors
9	power, world, male, change, respect, society, gender, rights, fight, culture	Role of world, society, gender and culture
10	time, love, female, video, watch, hard, era, consent, ladies, lady	Feminism on Twitter

On initial investigation, we find that some tweets containing *MeToo* as part of the tweet, but not having the hashtag (*#MeToo*) is irrelevant. Therefore, we remove the tweets without the *#MeToo*. Furthermore, we observe that tweets that contains large number of hashtags are often promotional and not link to the actual *MeToo* movement. These are removed using the outlier treatment method. For outlier treatment, we follow the standard statistical method of 1.5*IQR, where we follow the IQR (Inter Quartile Range) of the number of hashtags. Based on the distribution of co-hashtags, we removed tweets containing more than 4 hashtags. Finally, after the cleaning process, our dataset contains 3.5 million tweets from 1.03 million unique users. Table 1 summarises various statistics about this dataset.

4 Hidden Facets of the *MeToo* Movement

This section's focus is to reveal the hidden facets of the *MeToo* movement along with SH and SA on Twitter. We use the topic modelling technique to extract these subtopics. In particular, we use the Latent Dirichlet Allocation (LDA) [2] method with Gibbs sampling [28] which is an unsupervised and probabilistic machine-learning topic modeling method that extracts topics from text data. The key assumption behind LDA is that each given text is a mix of multiple topics. The model also tells in what percentage each document talks about each topic. Hence, a topic is represented as a weighted list of words.

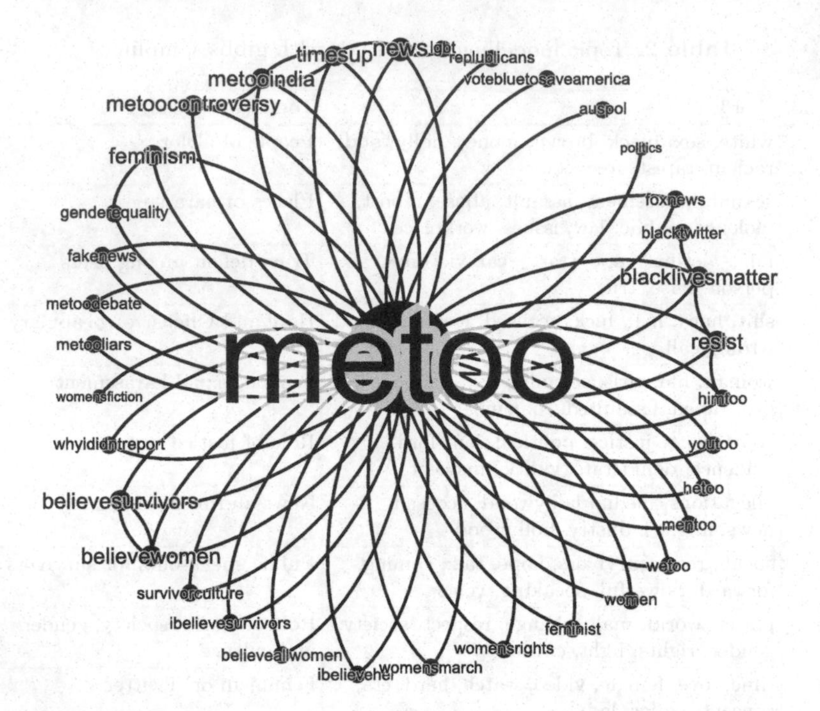

Fig. 1. Using the hashtags of the original tweets, we performed a bi-gram analysis. This identify the most frequent co-occurring hashtags. This makes it convenient to see the hashtags that belongs to each issue.

We extract ten topics from tweets' text and the LDA returns a set of ten words relating related to each identified topic (but not the title of the topic) (Table 2, Column 2). We then assign appropriate topic titles to each set of words that closely reflect the topic at an abstract level (Table 2, Column 3). To show the connectivity among various facets of discussion in the *MeToo* movement, we analyze the frequently co-appearing hashtags with *#MeToo* (see Fig. 1). Our analysis highlights the hashtags that belong to these hidden facets. For example, *#HimToo*, which represents the *trust deficit among individuals* is a movement against false rape allegations [12].

Our analysis reveals that the *MeToo* movement covers many hard-to-see facets: 1) **People of color:** Individuals are mainly discussing the color facet in SH incidents using the *#MeToo*; 2) **Places of harassment:** Mentioning the probable places of SH and SA incidents; 3) **Trust deficit among individuals:** Individuals discusses the possibility of attracting attention by using false allegations; 4) Use of **hate and offensive vocabulary** in tweets; 5) **Female sexual harassment**; 6) **Role of justice system**; 7) **News and media industry**; 8) **Public speak-outs for survivors**; 9) **Role of world society, gender and culture**; and 10) **Feminism on Twitter**. In this work, due to space limitation, we concentrate on only five important facets (S.No. 1 to 5) as shown in Table 2.

Table 3. Topic modelling for color, places of assault, hate and offensive vocabulary using LDA.

Type	Words
People of Color	black, white, sexual, racist, woman, support, brown, assault, male, hate
Hate	racist, white, stop, supremacy, human, angry, wrong, rich, color, read
Offensive	shit, fuck, fucking, ass, bitch, bullshit, guys, liar, fake, yall
Domestic	family, house, sexual, support, violence, sister, parents, abuse, harassment, domestic
Work	workplace, harassment, office, era, hire, job, sexual, light, forced, judge
Public	stop, guy, girl, stay, home, day, sexually, feel, night, public
Study	sexual, students, assault, harassment, report, abuse, faculty, victims, violence, survivors

4.1 People of Color

We investigate the color facet in the *MeToo* movement and identify that 2.73% of total tweets belong to color. Tweets containing the hashtags such as *#blacklivesmatter*, *#blacktwitter*, *#blm* etc. along with the *#MeToo* hashtag are labeled under this category.

The topic modeling of tweets listed in the color category (see Table 3, *Type* is *People of Color*) indicates that users are discussing gender and color under this in the *MeToo* movement. We understand that gender and color are sensitive topics and during our analysis, we further observe that discussions over these often lead to hate and offensive vocabulary.

Case Study of USA: For further investigation on color category, we study the USA tweets data. We start by comparing the state-wise total, white, and color population percentage [5] with the *MeToo* tweets percentage originated from that state (see Fig. 2). We observe a direct relationship between the total population of a state and tweet traffic generated from that state. In other words, higher populated states contribute more to tweeting activity. These findings are further supported by the positive correlation value and p-value (see Table 4).

The relationship between the white population and the tweet traffic generated from that state shows a negative correlation with a low p-value. On the other hand, we observe a direct relationship between the color population and tweet traffic generated from each state. This means that states with higher color population contribute more in tweets belonging to the *MeToo* movement. This is further supported by the positive correlation value. Thus, we can infer that the number of tweets generated and the color population in states of the USA are correlated.

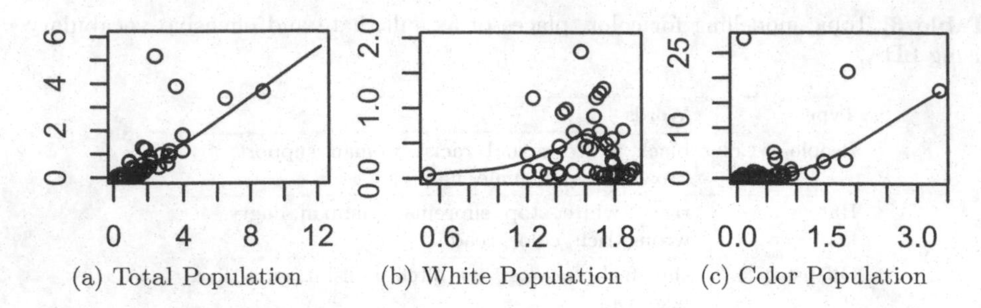

(a) Total Population (b) White Population (c) Color Population

Fig. 2. Fitting Curve: Here, x-axis represents state-wise different USA population i.e., in (a) total population, (b) white population, and (c) color population; and y-axis represents tweets traffic.

Table 4. USA: population vs tweets (statistical data)

Population type	p-value	Corr. value	Corr. type
Total population and tweets	0.00124	0.4483	Positive
White population and tweets	0.01642	−0.4312	Negative
Color population and tweets	0.00022	0.5043	Positive

The analysis of color facet of the *MeToo* movement shows that color people are tweeting higher compared to white people during this movement. We further notice that discussion over color often leads to the use of hate and offensive vocabulary which is discussed next.

4.2 Hate and Offensive Vocabulary

We identify the *MeToo* tweets that belong to either hate or offensive vocabulary using the HateSonar library from python [9] and find that 10% of the total tweets are either hate or offensive.

For better insights of each category, we perform the topic modelling (see Table 3, *Type* are *Hate* and *Offensive*). For instance, hate category tweets indicate that the hate vocabulary is connected to racism as most frequent words are *racist*, *white* and *black*. Likewise, based on offensive category tweets, we can infer that offensive vocabulary is the sign of strongly impolite and rude behavior. These conclusions are further supported by the word cloud analysis for both hate and offensive vocabulary tweets, as shown in Figs. 3b and 3c respectively. Additionally, we observe that the individuals tend to use the offensive vocabulary in case of distrust towards the incidents or survivors (see Table 3, third row).

4.3 Trust and Distrust Among Individuals

Even though *MeToo* movement helps the survivors to share their SH and SA incidents, there is also fear of false allegations among people. A poll by Leanin.org

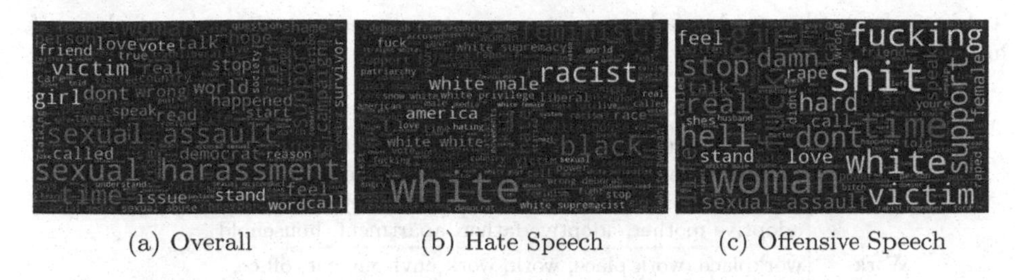

(a) Overall (b) Hate Speech (c) Offensive Speech

Fig. 3. Users' vocabulary: identifying the most frequent words under different vocabulary category.

shows that 60% of male managers don't want to mentor women out of fear of a false accusation [20]. On the other hand, according to [4], fears of false accusations are not supported by statistics. This results in some sort of both trust and distrust among people. This has also been observed from the tweets having conflicting hashtags such as *#BelieveWomen*, *#IBelieveSurvivors*, *#IBelieveHer*, or *#BelieveSurvivors* signifying trust category and tweets containing hashtags such as *#metooliars*, *#fakenews*, *#womensfiction*, or *#HimToo*, representing distrust category.

The tweets percentage identified in trust and distrust category are 5.46% and 1.27% respectively. On Twitter, we notice two separate groups of individuals in the *MeToo* movement. First, who trust the survivors; and others who show distrust towards the incidents shared by survivors.

4.4 Places of Harassment

According to the survey by [7], the majority of women (66%) had been sexually harassed in public spaces, 38% of women experienced it at the workplace, 35% had experienced it at their residence. However, workplace and domestic experiences are more likely to be assaults and the "most severe forms" of harassment [7]. Inspired by these studies, we annotate the tweets into four places of harassment categories, (1) Domestic, (2) Workplace, (3) Public place, and (4) Educational institutions. We use specific keywords to annotate these tweets into different categories. The complete list of keywords in each category is shown in Table 5. The percentage of total tweets identified in each category are as follows: *domestic* (3.23%), *work* (3.42%), *public* (0.90%) and *study* (1.40%). We can infer that under the *#MeToo*, domestic and work place SH incidents are higher compared to the educational institutions and the public places.

The topic modeling analysis of tweets for each mentioned places of harassment categories (see Table 3, the last four rows) provide collective insights. For instance, from workplace violence tweets, we can infer that workplace SH incidents often occur during recruitment or interview processes. This also indicate that disclosing SH events in the office environment may lead to the victim being judged. Similarly, from public place tweets, we may infer that women feel unsafe in public places.

Table 5. List of keywords used to annotate tweets into various categories of places of harassment.

Type	Keywords
Domestic	home, parent, parents, friend, legal guardian, domestic, family, house, residence, mother, father, single parent, lone parent, brother, sister, step brother, step sister, stepmother, stepfather, adoptive mother, adoptive father, apartment, household
Work	workplace, work place, work, work environment, office, employment, interview, employer, employee, job, business, organization, working, factories, co-worker, client, supervisor, hire, company, colleague, workmate
Public	park, bus, public place, theater, stranger, train, restaurant, bar, bus stop, public park, mall, street
Study	school, college, student, academic, educational, teacher, professor, secondary school, university, faculty, study, studies, classmate, friend

In this section, we explore the extracted hidden facets of the *MeToo* movement in detail using tweets text. Our results show that people of color are reporting higher SH incidents. We also find that discussion about color often leads to the use of hate and offensive vocabulary. This further follows by the distrust among individuals towards the incidents or survivors of SH. Next, we study the user interactions with respect to these facets in the *MeToo* movement.

5 Users' Communities and Leaders

To understand the topic distribution across various users of our dataset, we employ social network analysis technique. We build the directed "retweet" network among users: an edge ($u \longrightarrow v$) indicates that user u retweet user v (see Fig. 4). The descriptive statistics of the retweet network is provided in Table 6. The lower value of the average clustering coefficient (Average C.C.) and edge density, can be used to infer that the network is sparse. This further indicates that the network is spread out, which is further confirmed by a large diameter. From the values of weakly connected components (WCC) and the number of components, and the average path length we can conclude that there are a large number of small communities.

In Fig. 4, users are grouped into communities based on Louvain community detection algorithm [3] and for better representation each community is color-coded. We observe that users discussing the similar facets form a strong community. For example, users discussing the *color* facets are shown in *green* and they forms a strong community. The legend in Fig. 4, provide the complete list of colors assigned to individual users that belongs to various facet of the *MeToo* movement.

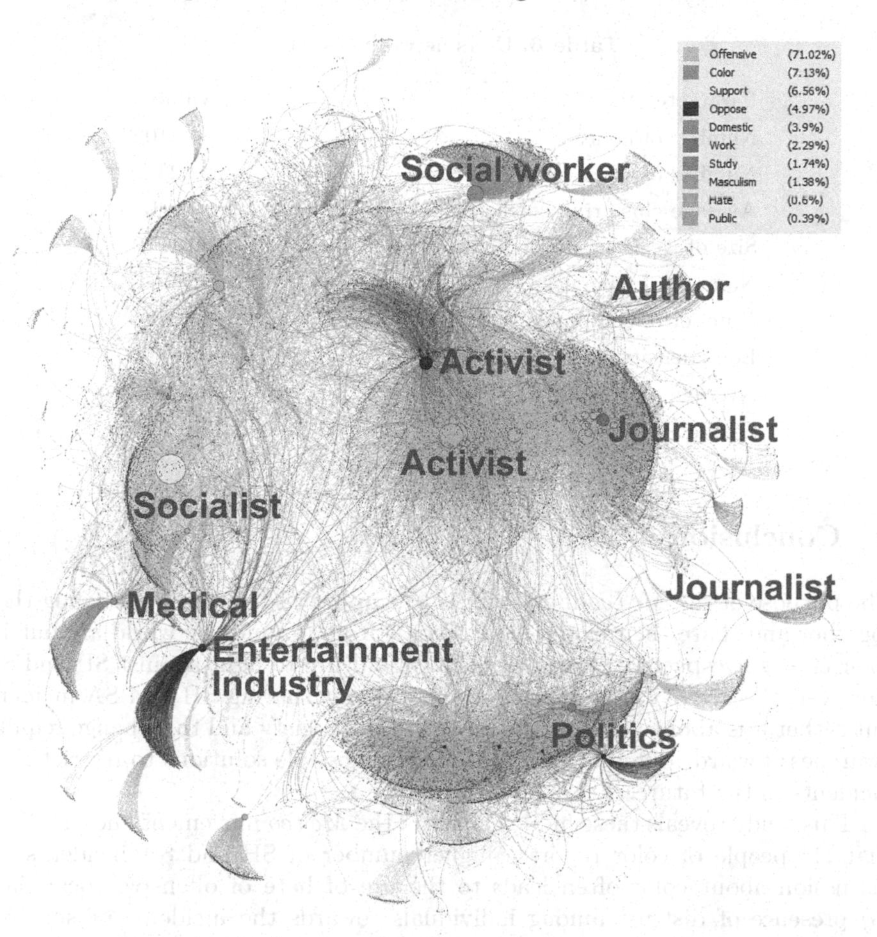

Fig. 4. Leaders' profession and user-communities.

Next, we identify the leaders of these communities. Our definition of leaders relies on the fact that they are often retweeted a relatively higher number of times. We employ the well-known degree centrality algorithm to identify them and are shown with relative bigger node size in the Fig. 4. To study leaders in more detail, we explore their professions. We check their publicly available information such as profile description, tweets' frequency and contents. We observe that the content from activists, journalists, authors, singers, and media people are highly appreciated (i.e., retweet) during the *MeToo* movement on Twitter.

Table 6. Users network statistics.

Parameter	Value
Number of nodes	846,102
Number of edges	2,223,281
Average clustering coefficient	0.0946
Size of weak and strong clustering coefficient	809,979
Number of triangles	5,149,422
Number of components	14,870
Edge density	6.2e-6
Average path length	8.51
Diameter	30

6 Conclusion

The purpose of the *MeToo* movement is to empower survivors by bringing them together and share their experiences. People from around the world are putting their efforts irrespective of gender, color, or culture to fight against SH and SA. However, the *MeToo* movement is not just about sharing SH and SA incidents but rather it is about bringing the change in the society and this change requires awareness towards different causes of SH and possible solutions to restrict these incidents in the future.

This study reveals these hidden facets of the *MeToo* movement and highlights that (1) people of color report a higher number of SH and SA incidents; (2) discussion about color often leads to the use of hate of offensive vocabulary; (3) presence of distrust among individuals towards the incidents or survivors of SH; (4) domestic and workplace SH incidents are drastically high; (5) users discussing a similar alignment forms a strong community; and (6) the content from activists, journalists, and media people are highly appreciated (i.e., retweet) during the *MeToo* movement on Twitter.

The major limitation to this work is that our analysis may not be generalized to other similar movements such as *#genderequality* which is a feminism movement.

Our future work would consider several important directions, such as to determine the effect of users' response and media (such as images and videos) in such movement. First, we plan to investigate how users respond to positive and supporting tweets compared to abusive and offensive tweets. For instance, abusive and offensive tweets may cause users to leave the discussion otherwise, they wouldn't have. On the other hand, positive and supporting tweets may encourage more survivors to share their incident(s), which leads to a healthy discussion. Also, nowadays, images and videos are a significant portion of the data generated on social media sites. Hence, we plan to analyze them in-depth for the understanding of such movements.

Acknowledgments. This research was funded by ERDF via the IT Academy Research Programme and H2020 project, SoBigData++. We would like to thanks Nishkal Prakash for the data collection.

References

1. Ahmed, W.: Public health implications of# shoutyourabortion. Public Health **163**, 35–41 (2018)
2. Blei, D.M., Ng, A.Y., Jordan, M.I.: Latent dirichlet allocation. J. Mach. Learn. Res. **3**(Jan), 993–1022 (2003)
3. Blondel, V.D., Guillaume, J.L., Lambiotte, R., Lefebvre, E.: Fast unfolding of communities in large networks. J. Stat. Mech: Theory Exp. **2008**(10), P10008 (2008)
4. Brett, M.: Many in media distort the framing of #metoo (2019)
5. Bureau, U.C.: Annual estimates of the resident population by sex, race, and hispanic origin for the united states: April 1, 2000 to July 1, 2008 (NC-EST2008-03) (2010)
6. Carr, D.: Hashtag activism, and its limits. New York Times, vol. 25 (2012)
7. Chatterjee, R.: A new survey finds 81 percent of women have experienced sexual harassment. National Public Radio (2018)
8. Dadas, C.: Hashtag activism: the promise and risk of "attention". In: Social Writing/Social Media: Publics, Presentations, Pedagogies, pp. 17–36 (2014)
9. Davidson, T., Warmsley, D., Macy, M., Weber, I.: Automated hate speech detection and the problem of offensive language. In: Eleventh International AAAI Conference on Web and Social Media (2017)
10. Edwards, F., Howard, P.N., Joyce, M.: Digital activism and non-violentconflict. Available at SSRN 2595115 (2013)
11. Elliman, T.D., Shannahoff, M.E., Metzler, J.N., Toblin, R.L.: Prevalence of bystander intervention opportunities and behaviors among us army soldiers. Health Educ. Behav. **45**(5), 741–747 (2018)
12. Flynn, M.: 'this is my son': navy vet horrified as mom's tweet miscasts him as# himtoo poster boy-and goes viral. The Washington Post (2018)
13. France, L.R.: # metoo: Social media flooded with personal stories of assault. CNN.com, vol. 16 (2017)
14. Grinberg, N., Joseph, K., Friedland, L., Swire-Thompson, B., Lazer, D.: Fake news on twitter during the 2016 us presidential election. Science **363**(6425), 374–378 (2019)
15. Jaffe, S.: The collective power of# metoo. Dissent **65**(2), 80–87 (2018)
16. Jagsi, R.: Sexual harassment in medicine-# metoo. N. Engl. J. Med. **378**(3), 209–211 (2018)
17. Kelly, L.: The (in) credible words of women: false allegations in European rape research. Violence Against Women **16**(12), 1345–1355 (2010)
18. Khatua, A., Cambria, E., Khatua, A.: Sounds of silence breakers: exploring sexual violence on Twitter. In: 2018 IEEE/ACM International Conference on Advances in Social Networks Analysis and Mining (ASONAM), pp. 397–400. IEEE (2018)
19. Killham, J.E., Chandler, P.: From tweets to telegrams: using social media to promote historical thinking. Soc. Educ. **80**(2), 118–122 (2016)
20. Leanin.org: Working relationships in the #metoo era (2019)
21. Loken, M.: # bringbackourgirls and the invisibility of imperialism. Feminist Media Stud. **14**(6), 1100–1101 (2014)

22. Manikonda, L., Beigi, G., Liu, H., Kambhampati, S.: Twitter for sparking a movement, reddit for sharing the moment:# metoo through the lens of social media. arXiv preprint arXiv:1803.08022 (2018)
23. Mutsvairo, B. (ed.): Digital Activism in the Social Media Era. Springer, Cham (2016). https://doi.org/10.1007/978-3-319-40949-8
24. Ohlheiser, A.: The woman behind 'me too' knew the power of the phrase when she created it—10 years ago. The Washington Post, vol. 19 (2017)
25. Onwuachi-Willig, A.: What about# ustoo: the invisibility of race in the# metoo movement. Yale LJF **128**, 105 (2018)
26. PettyJohn, M.E., Muzzey, F.K., Maas, M.K., McCauley, H.L.: # howiwillchange: Engaging men and boys in the# metoo movement. Psychology of Men & Masculinity (2018)
27. Phillips, W., Milner, R.M.: Decoding memes: barthes' punctum, feminist standpoint theory, and the political significance of #YesAllWomen. In: Harrington, S. (ed.) Entertainment Values. PEI, pp. 195–211. Palgrave Macmillan UK, London (2017). https://doi.org/10.1057/978-1-137-47290-8_13
28. Porteous, I., Newman, D., Ihler, A., Asuncion, A., Smyth, P., Welling, M.: Fast collapsed gibbs sampling for latent dirichlet allocation. In: Proceedings of the 14th ACM SIGKDD International Conference on Knowledge Discovery and Data Mining, pp. 569–577. ACM (2008)
29. Ram, Y., Tribe, J., Biran, A.: Sexual harassment: overlooked and under-researched. Int. J. Contemp. Hosp. Manag. **28**(10), 2110–2131 (2016)
30. Scott, J., Marshall, G.: A Dictionary of Sociology. OUP, Oxford (2009)
31. Sini, R.: How 'metoo' is exposing the scale of sexual abuse. BBC.com (2017)
32. Soklaridis, S., Zahn, C., Kuper, A., Gillis, D., Taylor, V.H., Whitehead, C.: Men's fear of mentoring in the# metoo era-what's at stake for academic medicine? (2018)
33. Waseem, Z., Hovy, D.: Hateful symbols or hateful people? predictive features for hate speech detection on Twitter. In: Proceedings of the NAACL Student Research Workshop, pp. 88–93 (2016)

Discovering Daily POI Exploitation Using LTE Cell Tower Access Traces in Urban Environment

Sumin Han, Kinam Park, and Dongman Lee(✉)

School of Computing, KAIST, Daejeon, South Korea
{suminhan,parkinam321,dlee}@kaist.ac.kr

Abstract. Point of interest (POI) in an urban space represents the perception of city dwellers and visitors of a certain place. LTE cell tower access trace data is one of the promising data sources which has the potential to show real-time POI exploitation analysis. However, there is not much discussion on how it is correlated to diachronic POIs and their exploitation pattern. In this paper, we first show that the access trace pattern from the LTE cell tower can be used to discover which types of POIs exist in a certain area. Then, we propose a daily POI exploitation discovery scheme which can extract patterns of how POIs are daily used. Our analysis can provide a good insight into future urban space-based services such as urban planning and tourism.

Keywords: LTE cell tower access trace · Place of interest · POI exploitation · Classification · Regression

1 Introduction

Understanding how urban spaces are exploited by people can change the way we perceive and conceptualize the city. Service providers can use it to provide citizens with modern urban services such as tourism or personalized advertisement. We can discover such place exploitation based not only on the physical form of a place but also on the projection of human perception. The former is composed of a space and its buildings, while the latter is built from the social and physical perceptions of dwellers accumulated over a long period [1]. Among various forms of data sources to understand the place exploitation [2–6], LTE cell tower access trace data has recently attracted the attention of urban researchers as a promising data source. It can capture the mobility patterns of urban users, which may represent fundamental aspects of social and economic interactions. This advantage with the widespread adoption of mobile devices enables urban researchers to conduct various spatio-temporal analysis without extensive fieldwork. Discovering the correlation between the LTE cell tower access traces and the urban circumstances can broaden the understanding of the placeness of an area, and have the potential to make the real-time place analysis possible as they can provide more instant observation about an urban area.

S. Aref et al. (Eds.): SocInfo 2020, LNCS 12467, pp. 81–94, 2020.
https://doi.org/10.1007/978-3-030-60975-7_7

Liu et al. [6] cluster the moving patterns of citizens read from the mobile traffic data to discover different types of urban functional areas. Zinman et al. [8] identify the land use of urban areas using a classification method on cellular communication usage patterns. Xu et al. [10] cluster the mobile traffic patterns from LTE cell towers to identify each cluster's urban function recognized by the distribution of POIs. Cici et al. [5] use a clustering approach on mobile traffic data to discover the geographical relationship between the clusters and the POIs in the urban area. Although these clustering and classification methods are informative to a macroscopic urban design, they are limited to discover microscopic exploitation of places and their daily changes, which can provide more temporal information for urban services such as urban tourism and popular place discovery.

In this paper, we propose a POI exploitation discovery scheme that infers which POIs diachronically exist as well as how they are daily exploited in a given area, using LTE cell tower access traces. To analyze spatio-temporal characteristics of urban places, we apply time decomposition to LTE access trace collected from 4,096 LTE cell towers and extract 7-day (168-h) seasonal components. We leverage several classification models that can identify urban space POIs, and regression models for discovering daily POI exploitation, respectively, on the data. For ground truth for POI identification, we annotate each LTE cell using the places and their types from the Google Place API. We also annotate the daily exploitation patterns of each POI in an LTE cell using Google Place popular-times data. Based on these two datasets, our scheme can discover the types of diachronic POIs and the daily exploitation patterns of POIs of an LTE cell area. Evaluation results show that the proposed scheme performs well on the POI identification and can discover explainable changes in POI exploitation.

2 Related Works

Liu et al. [6] use CDR (Call Daily Records) data to discover the fabric of urban city. They extract snapshot map to represent each region and leverage embedded features from the convolutional auto encoder trained by their dataset. Although their evaluation was limited due to the lack of data size, each cluster was shown to include same type of urban functional areas.

Zinman et al. [8] use CDR data to identify the land use of an area which also includes road information such as highway and street. They extract several different features and measure the variable importance (VI) of each feature to make better input feature for the model. They apply random forest classification and their result shows that their analysis of the feature improved accuracy.

Xu et al. [10] extract and model the traffic patterns of large scale cell towers placed in an urban area. They apply hierarchical clustering to seasonal components of mobile LTE traffic data and find a correlation between the cluster and urban ecology. Besides, through traffic spectrum analysis in the frequency domain using DFT, they find a linear combination that could express human activity behavior and indirectly capture the movement of people over time.

Zhang et al. [9] aim to understand the patterns of mobile traffic data and to find correlations between mobile traffic patterns and human activities in urban environments. They separate traffic data into three components: seasonal, trend, and residual. The seasonal component can identify the relationship with the POI of the place through clustering, and the trend component can identify some POI through the pattern difference between weekday and weekend. Residual components show to enable the capture of unexpected behaviors.

Cici et al. [5] aim to infer the characteristics of urban ecology (social and economic activities, and social interaction) from the patterns of cell phone data. They divide the cell phone traffic data in Milan into seasonal and residual patterns using Fourier time decomposition. They find the seasonal data characterize the patterns related to socio-economic activity in the region using agglomerative hierarchical clustering, while residual data make it possible to analyze how irregularity from the new events affect in other regions.

Shafiq et al. [7] aim to identify application usage patterns in each region so that network operators can easily optimize the distribution of cellular data. They collect traffic volume in terms of byte, packet, flow count, and unique user count, which can be extracted from 3G cellular data, and clustered for each term to specify the characteristics of each region (Downtown, University, Suburb, etc). The network operators can optimize the distribution of cellular data that fits clusters.

3 Dataset

In this section, we describe how we construct our dataset in three steps. We first explain how we preprocess LTE signal patterns. Then we illustrate how we annotate each LTE cell with POI type label, and collect POI exploitation patterns for the experiment.

3.1 Step1: Extracting a 7-Day Seasonal Component from an LTE Cell Tower Access Trace

Our dataset consists of anonymized LTE cell tower access traces on 4,096 LTE cell towers, collected by an internet service provider (ISP), Korea Telecom (KT) from March 1st, 2018 to February 28th, 2019. These LTE cell towers are located across three most popular commercial areas in Seoul as shown in Fig. 1. Each cell tower covers a $50\,\mathrm{m} \times 50\,\mathrm{m}$ area and access traces represent the number of people connected to each cell per hour. To analyze embedded patterns from each LTE cell tower access trace, we leverage the STL time decomposition method [22] to extract a 7-day (168-h) seasonal components. Then we conduct min-max scaling to normalize each pattern.

3.2 Step 2: Associating Diachronic POI Type Labels with Each Cell

To understand which types of diachronic POIs exist in each LTE cell, we establish a ground-truth of (LTE access pattern, POI labels) pairs. First of all, we crawl

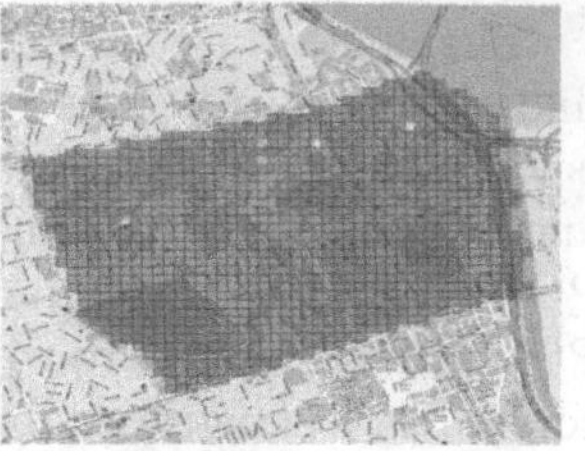

Fig. 1. We use LTE cell tower access traces from 50 m × 50 m LTE cells in the described area. Hongik-University (Area 1, left-most: N37.551°, E126.925°), Yeoksam-station (Area 2, center: N37.501°, E127.036°), and Samseong-station (Area 3, right-most: N37.514°, E127.052°).

a list of places in all the LTE cells using Google Place API[1] by querying any places within a radius of 36 m from each centroid of the LTE cells to be able to cover each cell area. Then we use the place type tags from Google query result to annotate LTE cell whether it contains certain types of POI. To prevent the associations of the POI labels with the LTE cell tower access patterns from being diversified too much, we choose 8 most frequent place types (out of 100) found by the Google Place API in our three target areas: *restaurant, store, cafe, health*[2], *finance, bar, clothing_store, and lodging.* Table 1 shows the number of LTE cells which contains places of corresponding POI types. Figure 2 shows the median of the extracted 7-day seasonal component for each type. We can see a clear difference between weekdays and weekends. Weekdays have regular patterns, while weekends have irregular patterns.

Table 1. The number of LTE cells containing each POI type.

	restaurant	store	cafe	health	finance	bar	clothing_store	lodging
Area 1	1,953	1,503	913	378	126	449	346	337
Area 2	1,317	1,009	417	651	418	218	85	172
Area 3	623	683	181	277	173	102	154	53
Total	3,893	3,195	1,511	1,306	717	769	585	562

3.3 Step 3: Annotating Daily POI Exploitation Using Google Popular-Times

To understand how the exploitation patterns can be inferred from LTE access pattern, we establish a ground-truth of (LTE access pattern, a daily POI

[1] https://developers.google.com/places/web-service/search.
[2] *This category is mostly comprised of medical clinics.*

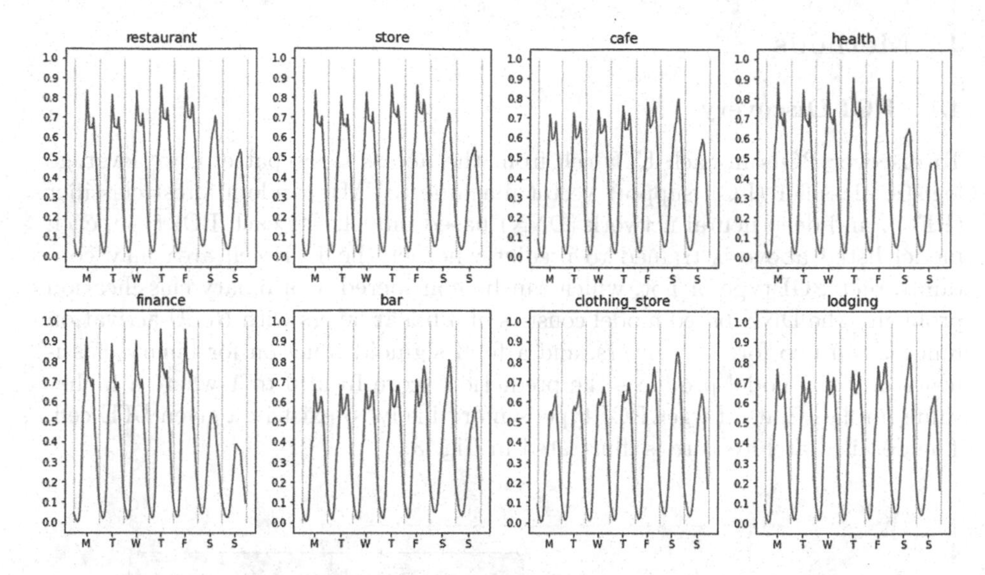

Fig. 2. The median of min-max scaled weakly seasonal patterns from each POI type. Every 24 h represents the day of the week starting from Monday to Sunday.

exploitation pattern) pairs. This dataset requires POI exploitation patterns on a daily basis to figure out how POIs in a given cell are exploited during a week. To construct such a dataset, we use Google popular-times[3] for each POI. Since official Google Place API does not provide popular-times, we leverage the third-party crawler[4] that provides a 7-day exploitation pattern normalized from 0 to 100 for a given place. We simply scale the pattern by 1/100 to have it represented by values between 0 and 1. Table 2 shows the number of collected POI exploitation data in each area.

Table 2. The number of POIs contain popular-times information.

	restaurant	store	cafe	health	finance	bar	clothing_store	lodging
Area 1	541	193	260	2	19	108	19	0
Area 2	323	105	65	6	26	38	13	3
Area 3	133	71	29	1	14	16	13	2
Total	997	369	354	9	59	162	45	5

[3] https://support.google.com/business/answer/6263531?hl=en.
[4] https://github.com/m-wrzr/populartimes.

4 Methods

4.1 POI Discovery

To discover POIs in each LTE cell from the access trace pattern, we leverage logistic classifier (LC), support vector machine (SVM), random forest classifier (RFC), and deep neural network (DNN) based model. For each POI type, each model listed above is trained to predict whether the LTE cell area may contain target POI type or not, which can be considered as a binary classification problem. The DNN based model consists of 3 linear layers with ReLU activation function on two foremost layers, and a final sigmoid function for binary classification. This model produces the prediction score from 0 to 1 where a higher value implies that a target POI type is more likely to exist in a given LTE cell. The detailed architecture is illustrated in Fig. 3.

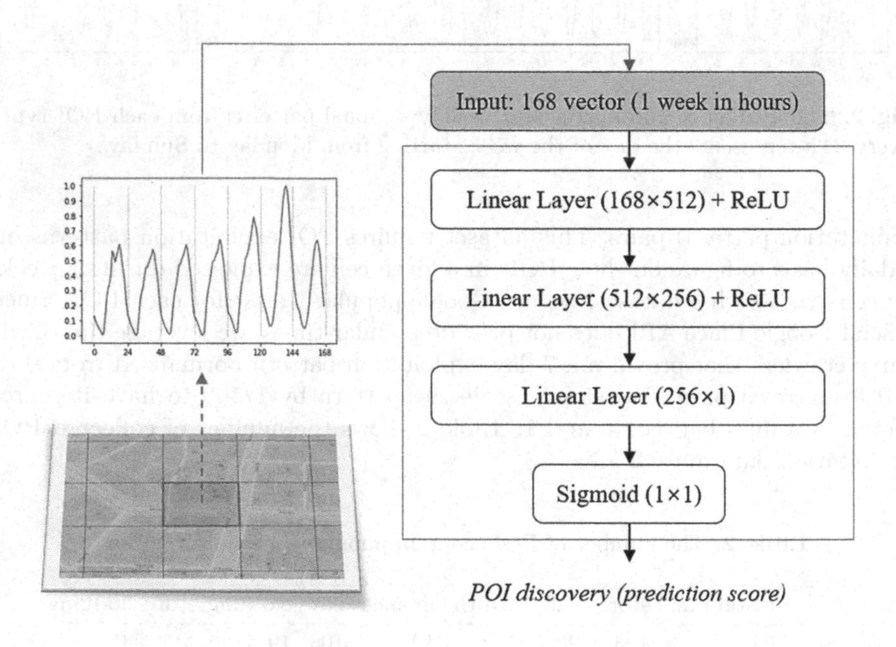

Fig. 3. The DNN based POI discovery model architecture.

4.2 Daily POI Exploitation Pattern Discovery

To discover the daily POI exploitation pattern from an LTE access pattern, we leverage a random forest regression (RFR) model which can be used for time-series pattern prediction [20,21]. Figure 4 illustrates our RFR model and its training process. First of all, find POIs with exploitation patterns for each LTE cell. Next, pair each POI exploitation pattern with corresponding LTE access pattern. During this process, pairs of an LTE access pattern and a POI exploitation pattern will be separated into different datasets each belongs to the same POI type. Then, we train a RFR model to predict the exploitation pattern for each POI type. Finally, this model can be used to predict unknown POI exploitation pattern using corresponding LTE access pattern in a given area.

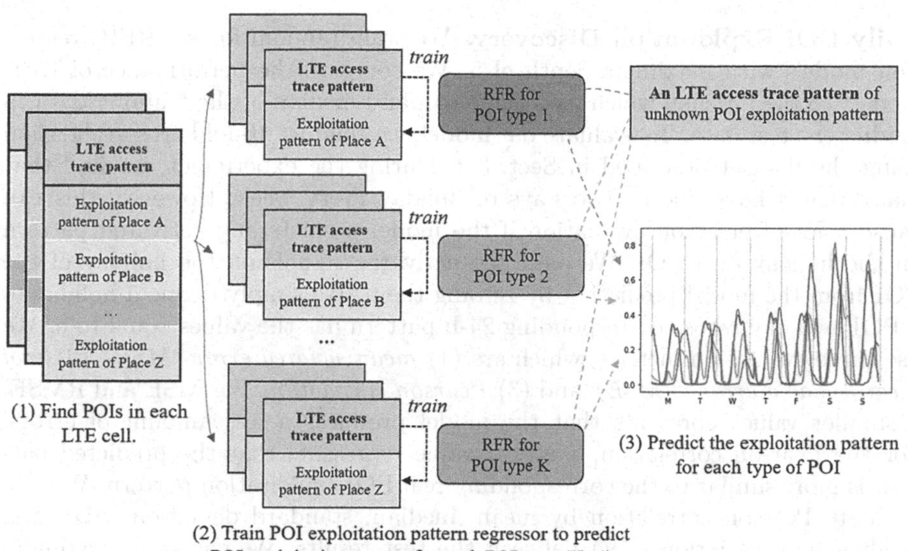

Fig. 4. Daily POI exploitation discovery scheme on an LTE access trace pattern.

5 Evaluation

5.1 Evaluation Setup

POI Discovery. We train binary classification models (LC, SVM, RFC, DNN model) for each POI type to predict whether the cell contains such POI or not. When we train the models, they may suffer from the unbalanced amount of positive and negative labels of the dataset. To prevent this, we use a random undersampling technique that removes either positive or negative samples randomly to have an equal amount to each other. We conduct 10-fold cross-validation using the dataset described in Sect. 3.2. To analyze the performance of the model, we use five standard metrics: (1) *Accuracy*, (2) *Precision*, (3) *Recall*, (4) *F1-macro*, (5) *ROC-AUC*. When we evaluate the first four metrics above, we use the threshold value as 0.5 to make a positive prediction. For each metric in this experiment, a higher value represents that a model shows better performance. The parameters used in each model are as follows:

- LC: L2 norm loss, tolerance = 0.1, LBFGS solver
- SVM: radial basis function (RBF) kernel, regularization parameter = 1
- RFC: the maximum depth of 5, the number of estimators = 100

Table 3 shows the mean value for each metric after cross validation with respect to each POI type.

Daily POI Exploitation Discovery. We train random forest (RFR) regression models with maximum-depth of 5. We compare the performance of RFR models to the baseline which uses the mean and median of the training data to predict the test data. To evaluate our model, we conduct 10-fold cross-validation using the dataset described in Sect. 3.3. During the experiment, we find that many places have one or two days of holiday every week. However, this can cause a lot of error on evaluation if the model predicts any activated pattern on the holiday of a POI. We omit the activated signals on the holiday of the POI from the model prediction by zeroing them. We simply define a holiday of a POI as a day whose corresponding 24-h pattern has the values equal to 0. We use three evaluation metrics, which are (1) *mean squared error (MSE)*, (2) *root mean squared error (RMSE)*, and (3) *Pearson correlation*. For MSE and RMSE, a smaller value represents that the model produces a less amount of errors. For the Pearson correlation, a higher value represents that the predicted pattern is more similar to the corresponding real POI exploitation pattern. We also evaluate Pearson correlation by mean, median, standard deviation (SD), and coefficient of variation (CV) between the test results. We choose a maximum depth of 5 for an RFR model.

5.2 Validation of POI Discovery

Table 3 shows the performance results of each model for each POI type. When we compare the overall performance across the categories, the models trained for bar category show the highest accuracy while that for the health category show the lowest. Although the accuracy of a model for bar might be seemed marginally better than others, the precision is higher than any other category. This is because the signal patterns of LTE cells encompassing bars are quite distinguishable from others as bars mainly open during the night and their existence are easy to be discovered from the patterns. On the other hand, POIs in the health category are difficult to be predicted as its corresponding places do have a small number of visitors, compared with other categories. This makes it difficult to find a common pattern from the LTE access pattern belonging to this category. The table shows that RFC and DNN model can be considered as a model which produces a fine prediction in terms of accuracy, as they show good performance on all the categories except for clothing_store and lodging. LC and SVM also performs pretty well and worth to be applied when the researcher needs an prototype result. However, evaluating the models only on accuracy can be dangerous as we conduct our experiment on a small amount of dataset after

Table 3. The evaluated performance measures for diachronic POI discovery.

POI	Model	Accuracy	Precision	Recall	F1-macro	AUC-ROC
restaurant	LC	0.6881	0.6789	0.7107	0.6940	0.7567
	SVM	0.7072	0.6969	0.7301	0.7127	0.7675
	RFC	0.6858	0.6704	0.7331	0.6990	0.7643
	DNN	**0.7093**	0.7076	0.7442	0.7159	**0.7828**
store	LC	0.6566	0.6447	0.6992	0.6697	0.7316
	SVM	0.6771	0.6643	0.7204	0.6896	0.7447
	RFC	**0.6987**	0.6917	0.7150	0.7025	**0.7730**
	DNN	0.6911	0.6972	0.7565	0.7062	0.7614
cafe	LC	0.6902	0.6958	0.6774	0.6849	0.6903
	SVM	0.6952	0.7053	0.6724	0.6867	0.6946
	RFC	0.6957	0.6925	0.6996	0.6948	**0.7682**
	DNN	**0.6979**	0.6938	0.7243	0.6994	0.7648
health	LC	0.6181	0.5998	0.7159	0.6507	0.6457
	SVM	0.6160	0.5946	0.7244	0.6520	0.6587
	RFC	**0.6368**	0.6237	0.6910	0.6548	**0.6844**
	DNN	0.6271	0.6185	0.7121	0.6555	0.6685
finance	LC	0.6617	0.6463	0.7186	0.6787	0.6603
	SVM	0.6878	0.6684	0.7360	0.6992	0.6852
	RFC	0.6992	0.6960	0.7044	0.6990	0.7459
	DNN	**0.7013**	0.7027	0.7420	0.7077	**0.7477**
bar	LC	0.7204	0.7264	0.7052	0.7138	0.7219
	SVM	0.7320	0.7594	0.6728	0.7119	0.7335
	RFC	0.7278	0.7271	0.7185	0.7212	0.7998
	DNN	**0.7384**	0.7404	0.7417	0.7304	**0.8068**
clothing_store	LC	0.6677	0.6799	0.6330	0.6533	0.6680
	SVM	**0.6895**	0.7272	0.6127	0.6612	0.6896
	RFC	0.6664	0.6855	0.6249	0.6502	0.7272
	DNN	0.6581	0.6705	0.6607	0.6582	**0.7396**
lodging	LC	0.6830	0.6764	0.7009	0.6857	0.6822
	SVM	**0.6925**	0.7020	0.6656	0.6815	0.6922
	RFC	0.6845	0.6818	0.6909	0.6849	0.7246
	DNN	0.6736	0.6662	0.7290	0.6870	**0.7314**

the under-sampling technique. So, we also evaluate the models based on AUC-ROC scores, which can take into account of sensitivity and specificity. In terms of AUC-ROC scores, DNN based model shows better performance overall. We assume that this is because the DNN architecture captures a feature that most commonly appears in each POI type, resulting in a more consistent model.

Table 4. The performance measures of POI exploitation pattern prediction.

POI	Model	MSE	RMSE	Pearson correlation			
				Mean	Median	SD	CV
restaurant	Mean	0.0313	0.1769	0.6979	0.7264	0.1636	0.2346
	Median	0.0336	0.1832	0.6832	0.7156	0.1793	0.2627
	RFR	**0.0274**	**0.1655**	**0.7459**	**0.7827**	0.1702	**0.2284**
store	Mean	0.0336	0.1831	0.7486	0.8137	0.2092	0.2809
	Median	0.0353	0.1877	0.7414	0.8207	0.2286	0.3101
	RFR	**0.0310**	**0.1761**	**0.7685**	**0.8303**	0.2000	**0.2615**
cafe	Mean	0.0286	0.1690	0.8000	0.8450	0.1482	0.1857
	Median	0.0303	0.1739	0.7922	0.8437	0.1591	0.2013
	RFR	**0.0252**	**0.1585**	**0.8264**	**0.8699**	0.1412	**0.1713**
bar	Mean	0.0250	0.1580	0.7569	0.8353	0.2229	0.2967
	Median	0.0267	0.1632	0.7496	0.8336	0.2334	0.3137
	RFR	**0.0231**	**0.1519**	**0.7731**	**0.8579**	0.2222	**0.2892**

5.3 Validation of Daily POI Exploitation Discovery

We conduct our experiment on *restaurant*, *store*, *cafe*, and *bar* since these categories have enough data to evaluate our experiment while others suffer from data sparsity. Table 4 shows the evaluated performance measures of daily POI exploitation pattern prediction. The results show that RFR-based models produce best prediction patterns than baseline mean and median models as they shows less MSE and RMSE. The models for bar category shows lowest MSE and RMSE as the exploitation patterns of this category are more distinguishable than others. The higher mean value of Pearson correlations means that RFR models predict rather similar exploitation patterns to the real value than the baseline models. Plus, a small coefficient of variance means that the level of dispersion of the Pearson Correlation is smaller around the mean. Therefore, we can conclude that RFR models produce a similar exploitation pattern to the corresponding real value more frequently for all types of POI. In addition, a gap between the mean of the Pearson correlation from the best RFR model and that from the baseline mean model is biggest in the restaurant category and smallest in the bar category as shown in Table 4. Considering that the amount of training data is biggest at the restaurant category and smallest at the bar category, this observation implies that the current RFR model can be reinforced as more training data is provided.

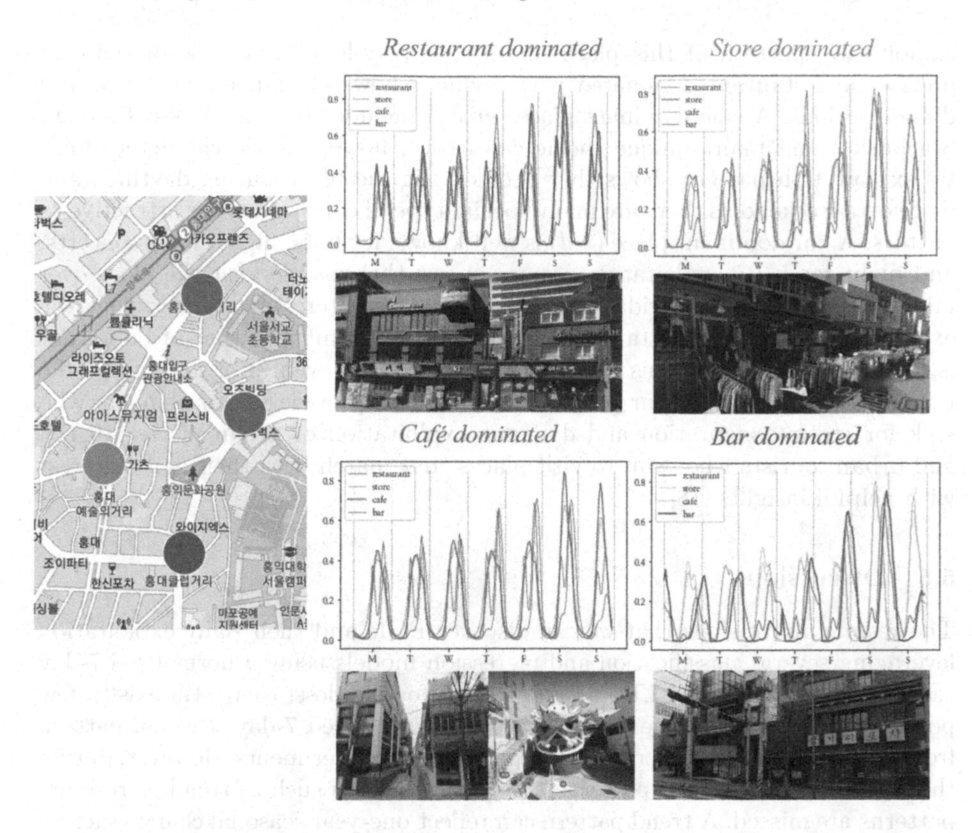

Fig. 5. Different daily changes of POI exploitation of 4 places with the same POI set derived from the LTE access pattern in Area 1. The colored circle on the map represents each area: *restaurant (blue)*, *store (orange)*, *cafe (green)*, and *bar (red)*. We emphasis the dominated POI pattern for each POI exploitation graph by thickening the line width. (Color figure online)

5.4 Finding POI Exploitation Variances Across Places with the Same POI Types

By conducting a cross-analysis of the daily POI exploitation changes, it can be found that urban places with the same POI types can be exploited differently by dwellers and tourists from the ne-grained perspective. Figure 5 shows the daily changes of POI exploitation of four different places identified by the same diachronic POIs (i.e., having similar urban compositions), while their dominant POI types are different. By using the Google street view, we nd that exploitation patterns are different, depending on the urban setting of each place. A restaurant dominant place (*blue*) is located near a main street where people gather and socialize while having lunch and dinner. Its exploitation pattern shows that this area is highly visited at every lunch and dinner periods, and even higher at the dinner period on Friday and Saturday. A store dominant place (*orange*) is a popular shopping street mainly for buying clothes and accessories. The exploitation pattern of this place shows that people visit during all the day, unlike the restaurant dominated place, which shows clear peaks in lunch and dinner periods. A cafe dominant place (*green*) includes big franchise cafes (e.g., Starbucks) and many unique theme cafes (pet, flower, movie character, etc.). Its exploitation pattern shows that people visit the place during daytime, and more on a weekend as there are many special type of cafes which are attractive to visitors. A bar dominant place (*red*) is well-known for bars, night clubs, concerts, and night food. Its exploitation pattern shows that people visit the place over midnight, and more on Friday and Saturday. It is interesting to note that the exploitation pattern of restaurants in this place is similar to that of bars, that is, many restaurants in this area are also not closed even after midnight. The ndings above show that our proposed scheme can provide urban planners who seek for static conguration and dynamic exploitation of urban places [16,17], and urban tourists who want to visit places that match with their daily interest with helpful insights [18].

5.5 Discussion

The results above show that we can discover POIs and their daily exploitation leveraging several classification and regression models using a normalized 7-day seasonal pattern from an LTE access traces. Nevertheless, there still exist a few points to investigate further. First, we use a normalized 7-day seasonal pattern from each LTE access pattern as our input for both experiments. However, during the extraction of seasonal patterns, many information such as trend or residual patterns are missed. A trend pattern can reflect one-year seasonal change such as the weather or public traveling behavior, and it can provide more insight on the long-term exploitation pattern change. A residual pattern can reflect unexpected signal away from seasonal and trend, so it can show the frequency of unexpected event in a given area. So, the combined analysis of trend and residual as well as seasonal patterns would provide another dimension to analyze the LTE access pattern. Another way to read social information from LTE access patterns is to analyze in terms of age and gender. This will enable us to understand the demographics of the area and help to figure out what age and gender groups facilitate the POI (or a combination of POIs) of a given place - that is, understand what placeness is leveraged not only when but also by whom.

6 Conclusion

In this paper, we leverage LTE cell tower access traces and develop a decent scheme to identify what kinds of points of interest (POIs) exhibit in a given space and how these POIs are exploited in a daily manner. For the former, we leverage several binary classification models such as LC, SVM, RFC, and DNN. For the latter, we leverage the random forest regression model for discovering daily changes in POI exploitation. Evaluation results show that our approach performs well on each task, and can discover the poi exploitation variance of the different urban environment. Our analysis based on LTE cell tower access data will provide a good insight into future urban space-based services such as travel and tourism.

Acknowledgment. This work was supported by Institute of Information & communications Technology Planning & Evaluation (IITP) grant funded by the Korea government (MSIT) (No.2019-0-01126, Self-learning based Autonomic IoT Edge Computing). We appreciate Korea Telelcom(KT) for providing valuable LTE access data.

References

1. Loukaitou-Sideris, A., Banerjee, T.: Urban Design Downtown: Poetics and Politics of Form. University of California Press, Berkeley (1998)
2. Ferrari, L., et al.: Extracting urban patterns from location-based social networks. In: Proceedings of the 3rd ACM SIGSPATIAL International Workshop on Location-Based Social Networks (2011)
3. Salamon, J., Bello, J.P.: Deep convolutional neural networks and data augmentation for environmental sound classification. IEEE Signal Process. Lett. **24**(3), 279–283 (2017)
4. Ferreira, N., et al.: Visual exploration of big spatio-temporal urban data: a study of New York city taxi trips. IEEE Trans. Vis. Comput. Graph. **19**(12), 2149–2158 (2013)
5. Cici, B., et al.: On the decomposition of cell phone activity patterns and their connection with urban ecology. In: Proceedings of the 16th ACM International Symposium on Mobile Ad Hoc Networking and Computing (2015)
6. Liu, C., et al.: Clustering analysis of urban fabric detection based on mobile traffic data. In: Journal of Physics: Conference Series, vol. 1453 (2020)
7. Shafiq, M.Z., et al.: Characterizing geospatial dynamics of application usage in a 3G cellular data network. In: 2012 Proceedings IEEE INFOCOM. IEEE (2012)
8. Zinman, O., Lerner, B.: Utilizing digital traces of mobile phones for understanding social dynamics in urban areas. Pers. Ubiquit. Comput. **24**(4), 535–549 (2019). https://doi.org/10.1007/s00779-019-01318-w
9. Zhang, M., Xu, F., Li, Y.: Mobile traffic data decomposition for understanding human urban activities. In: 2016 IEEE 13th International Conference on Mobile Ad Hoc and Sensor Systems (MASS). IEEE (2016)
10. Xu, F., et al.: Understanding mobile traffic patterns of large scale cellular towers in urban environment. IEEE/ACM Trans. Netw. **25**(2), 1147–1161 (2016)
11. Han, D., Kim, J., Kim, J.: Deep pyramidal residual networks. In: Proceedings of the IEEE Conference on Computer Vision and Pattern Recognition (2017)
12. Shoji, Y., Takahashi, K., Dürst, M.J., Yamamoto, Y., Ohshima, H.: Location2Vec: generating distributed representation of location by using geo-tagged microblog posts. In: Staab, S., Koltsova, O., Ignatov, D.I. (eds.) SocInfo 2018. LNCS, vol. 11186, pp. 261–270. Springer, Cham (2018). https://doi.org/10.1007/978-3-030-01159-8_25
13. Breiman, L.: Random forests. Mach. Learn. **45**(1), 5–32 (2001)
14. Mitchell, S.: Evaluating impacts and defining public perceptions of police body-worn cameras (BWCs). Diss. Kent State University (2019)
15. Hair, J.F., Ringle, C.M., Sarstedt, M.: Partial least squares structural equation modeling: rigorous applications, better results and higher acceptance. Long Range Plan. **46**(1–2), 1–12 (2013)

16. Frias-Martinez, V., Frias-Martinez, E.: Spectral clustering for sensing urban land use using Twitter activity. Eng. Appl. Artif. Intell. **35**, 237–245 (2014)
17. Southworth, M., et al.: People in the design of urban places, pp. 461–465 (2012)
18. Karduni, A., et al.: Urban space explorer: a visual analytics system for urban planning. IEEE Comput. Graph. Appl. **37**(5), 50–60 (2017)
19. Abbar, S., Zanouda, T., Al-Emadi, N., Zegour, R.: City of the people, for the people: sensing urban dynamics via social media interactions. In: Staab, S., Koltsova, O., Ignatov, D.I. (eds.) SocInfo 2018. LNCS, vol. 11186, pp. 3–14. Springer, Cham (2018). https://doi.org/10.1007/978-3-030-01159-8_1
20. Wu, H., et al.: Time series analysis of weekly influenza-like illness rate using a one-year period of factors in random forest regression. Bioscience Trends (2017)
21. Shirmohammadi-Khorram, N., et al.: A comparison of three data mining time series models in prediction of monthly brucellosis surveillance data. Zoonoses Public Health **66**(7), 759–772 (2019)
22. Cleveland, R.B., et al.: STL: a seasonal-trend decomposition. J. Official Stat. **6**(1), 3–73 (1990)

Identifying the Hierarchical Influence Structure Behind Smart Sanctions Using Network Analysis

Ryohei Hisano[1,2(✉)], Hiroshi Iyetomi[2,3], and Takayuki Mizuno[2,4]

[1] Graduate School of Information Science and Technology, The University of Tokyo, Tokyo, Japan
em072010@yahoo.co.jp

[2] The Canon Institute for Global Studies, Tokyo, Japan

[3] Niigata University, Niigata, Japan

[4] National Institute of Informatics, Tokyo, Japan

Abstract. Smart sanctions are an increasingly popular tool in foreign policy. Countries and international institutions worldwide issue such lists to sanction targeted entities through financial asset freezing, embargoes, and travel restrictions. The relationships between the issuer and the targeted entities in such lists reflect what kind of entities the issuer intends to be against. Thus, analyzing the similarities of sets of targeted entities created by several issuers might pave the way toward understanding the foreign political power structure that influences institutions to take similar actions. In the current paper, by analyzing the smart sanctions lists issued by major countries and international institutions worldwide (a total of 73 countries, 12 international organizations, and 1,700 lists), we identify the hierarchical structure of influence among these institutions that encourages them to take such actions. The Helmholtz–Hodge decomposition is a method that decomposes network flow into a hierarchical gradient component and a loop component and is especially suited for this task. Hence, by performing a Helmholtz–Hodge decomposition of the influence network of these institutions, as constructed from the smart sanctions lists they have issued, we show that meaningful insights about the hierarchical influence structure behind smart sanctions can be obtained.

Keywords: Political networks · Smart sanctions · Helmholtz–Hodge decomposition

1 Introduction

As globalization progresses, economic trade among nations has been growing at an increasing rate [13]. There is no doubt that this global expansion of economic trade contributes to the growth of prosperity in the global economy. Still, the rising geographical and cultural distances among the participants make it challenging to avoid trading with entities (such as companies and individuals) that

S. Aref et al. (Eds.): SocInfo 2020, LNCS 12467, pp. 95–107, 2020.
https://doi.org/10.1007/978-3-030-60975-7_8

are involved with illegal criminal activities such as money laundering, terrorism, drug cartels, and human trafficking. To tackle such challenges, governments and international organizations around the world are increasingly interested in issuing smart sanctions lists, which contain the names of entities involved in such criminal acts [4,7,11,15].

However, these lists vary quite substantially, both in the set of entities being banned and the timing of their inclusion, even when the target category that they aim to ban is the same. This difference stems from several sources. For instance, some institutions might be vigorous in collecting intelligence that influences others to follow, whereas other lists might be ignored because of a different understanding of the problem. Some authoritative institutions might have a higher standard for banning an entity, while others might take actions without much undeniable evidence. It is also possible that an institution is merely copying the entities added to another, prestigious list to project a sense of international cooperation without actively collecting intelligence themselves [6]. Thus, by analyzing the similarities and differences among such lists, we can shed some light on the influence network that governs the institutions issuing these lists. The understanding of such an influence network could enrich our knowledge of the global political structure.

In this paper, we describe the application of network analysis to the influence network that governs the major institutions issuing smart sanctions lists. We use a dataset that includes over 1,700 smart sanctions lists, mainly focused on banning global criminal activities, and perform a Helmholtz–Hodge decomposition on the network constructed from this dataset. We show that this simple analysis readily provides meaningful results, which enables a better understanding of the global political process behind smart sanctions lists.

Many empirical studies measuring the impact of economic sanctions have restricted their scope to state-level effects, focusing on the period in which a comprehensive sanction was enacted at the state level [5]. Meanwhile, studies focused on economic sanctions targeting specific entities such as firms, individuals, and vessels (that is, smart sanctions) are relatively new and less well understood, and studies using firm-level data are scarce [1]. Of the few such studies to date, [14] undertook an event study and measured the impact of sanctions-related news items on the stock market. To the best of our knowledge, no study has yet focused on analyzing the influence structure among institutions issuing smart sanctions lists by constructing an influence network from the smart sanctions lists they have released. The current paper opens the door to applying and developing a new network algorithm to study the foreign power structure in the global society, which would significantly affect political sciences.

The rest of the paper is organized as follows. In the next section, we review the dataset used in this paper. We describe how we construct the influence network from a set of smart sanctions lists. In Sect. 3, we briefly explain the Helmholtz–Hodge decomposition technique used in this paper. In Sect. 4, we summarize our results. The final section concludes the paper.

Table 1. Ranking of institutions by how many smart sanctions lists they have issued.

Rank	Institution	Count
1	European Union	925
2	United States	160
3	Japan	151
4	Canada	53
5	United Nations	39
6	Switzerland	39
7	United Kingdom	35
8	Australia	19
9	Singapore	17
10	Brazil	15

2 Data

Smart sanctions lists are available online for public use. However, named entity recognition of banned targets, such as companies and individuals, can be a difficult task because of the variety of ways each institution handles their lists. Thus, we resort to information provided from professional sources.[1] We use the smart sanctions list data included in the Dow Jones Adverse Media Entities, State-Owned Companies, and the Watchlist datasets. The Dow Jones datasets contain approximately 1,700 smart sanctions lists from 2001 to the present. The purpose behind these smart sanctions are to curb illegal activities such as money laundering, drug use, fraud, organized crime, human trafficking, banned air carriers, and terrorist activity.

A total of 85 institutions (such as countries and international organizations) have issued sanctions lists in our dataset. The top ten institutions, in terms of the total number of sanctions lists issued, are provided in Table 1. We can see that the majority are from countries worldwide, but international organizations such as the EU and the United Nations are also included. The number of smart sanctions lists issued by each institution varies quite substantially. The average number of lists issued is 16.7, and the standard deviation is 92.7, which confirms this insight.

Each entry in a sanctions list comprises the name of the entity sanctioned and the date of their inclusion, and we can build two types of influence networks from this dataset. One is the influence network at the smart sanctions list level, which treats each smart sanctions list as a node. The network is constructed as follows. For each pair of smart sanctions lists, if list B includes an entity that is the same as an entity added earlier to list A, we add a weight (that is, 1) to the edge from A to B. A pair of smart sanctions lists have no connecting edges if there are no common target entities in the lists. We ignore cases in which two

[1] Currently, we are recollecting the data for open use.

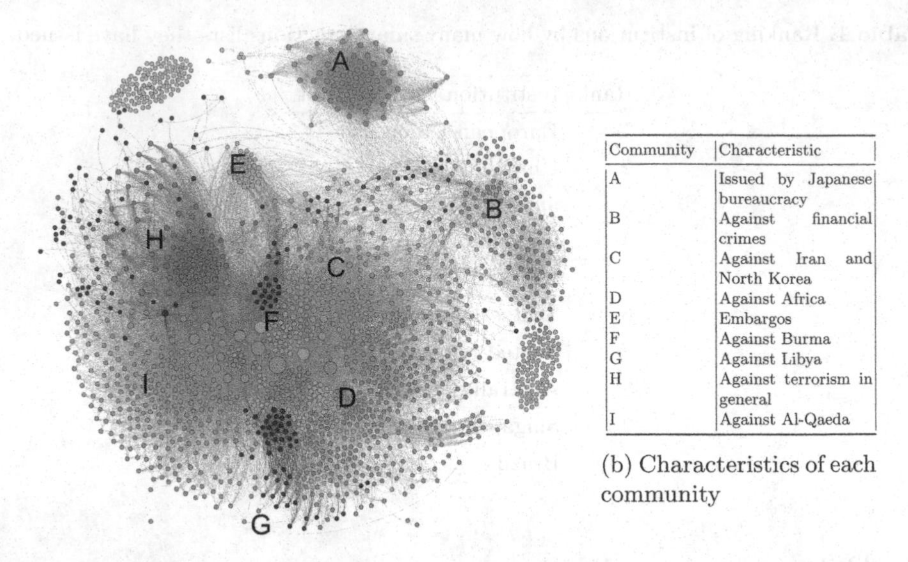

Community	Characteristic
A	Issued by Japanese bureaucracy
B	Against financial crimes
C	Against Iran and North Korea
D	Against Africa
E	Embargos
F	Against Burma
G	Against Libya
H	Against terrorism in general
I	Against Al-Qaeda

(b) Characteristics of each community

Fig. 1. Analysis of the community structure of the influence network at the level of smart sanctions lists

lists include entities precisely the same date because the direction of influence is not clear.

We show the result in Fig. 1a. The color of the nodes indicates the major communities found by standard modularity minimizing algorithms [2,10]. We see many isolated nodes (that is, lists) located at the top of community H and below community B. These lists are mainly domestic wanted lists issued by countries that are of little interest to other institutions. Excluding these lists, the algorithms identified nine major communities. Figure 1b summarizes the characteristics of each community. We observe that smart sanctions lists targeting terrorism are generally located on the left (in community H) close to community I, which targets Al-Qaeda. On the bottom of the network, there is a community that targets Libya (community G), and right next to it, we have a community that deals with Africa in general. Smart sanctions lists dealing with domestic Japanese issues have, in many cases, nothing in common with foreign lists, which explains why community A is isolated. However, we can also see several edges between communities A and B, where B deals with financial crimes. We use these distinctive communities (the categories of smart sanctions lists) in the following sections.

We can also create an influence network at the country level, and this network is used to identify the hierarchical influence structure among the institutions behind the smart sanctions. The steps taken are quite similar to those used to create the influence network at the smart sanctions list level. We first treat each institution that has issued a list as a node. For each pair of institutions, if institution B included the same entity on their list at a later time than institution A, we add a weight (that is, 1) to the edges between A and B. A pair of institutions have no edges if there are no common entities on the lists that they have issued. We ignore cases when two institutions add an entity on precisely the same date, as the direction of influence is not clear. This procedure produces a weighted directed network of institutions, as shown in Fig. 2. The network can be decomposed into two communities, distinguished by two colors, using standard modularity maximization techniques [2,10]. Note that the position on the y-axis reflects the hierarchical position (that is, the Helmholtz–Hodge potential) defined in the next section.

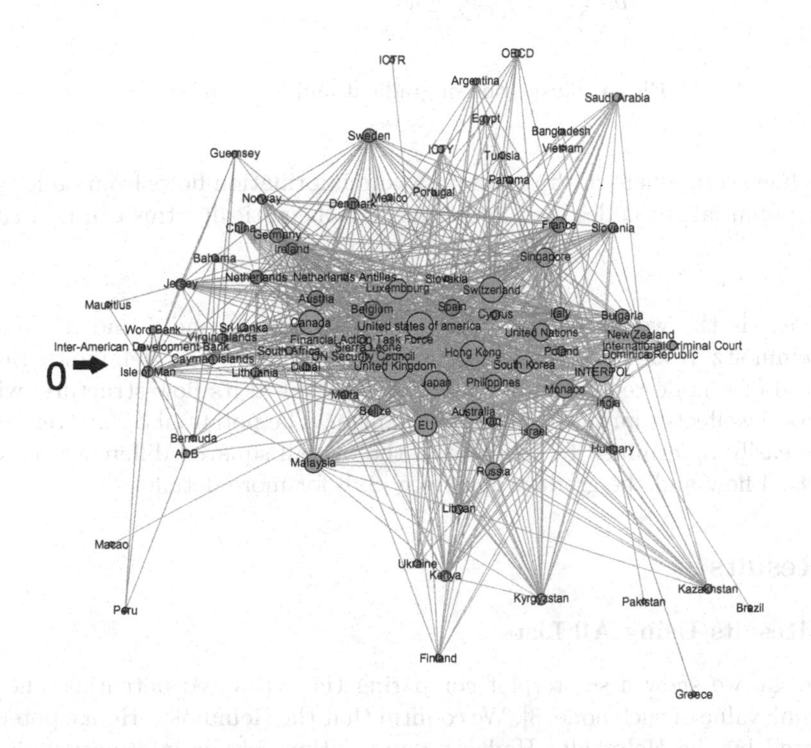

Fig. 2. Influence network at institution level using all smart sanctions lists depicted in Fig. 1a.

3 Helmholtz–Hodge Decomposition

The flow F_{ij} running from node i to node j in a directed network can be decomposed into

$$F_{ij} = F_{ij}^{p} + F_{ij}^{c}, \tag{1}$$

where F_{ij}^{p} denotes the gradient flow and F_{ij}^{c} denotes the circular flow (see Fig. 3 for a visual illustration). Circular flow F_{ij}^{c} corresponds to the feedback loops that are inherent in such networks. Gradient flow F_{ij}^{p} can be understood as the

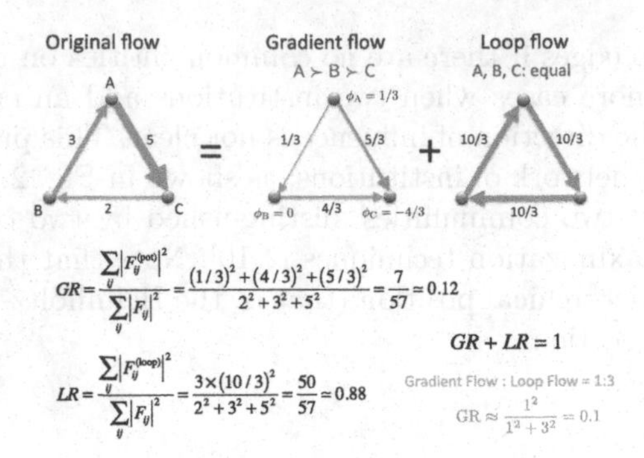

Fig. 3. Illustration of gradient and loop ratios.

hierarchical component of the network, where information flows from nodes with higher potentials to nodes with lower ones. Mathematically, this can be written as

$$F^{p}_{ij} = w_{ij}(\phi_i - \phi_j), \tag{2}$$

where w_{ij} is the weight of the edges between nodes i, and j and ϕ_i denotes the Helmholtz–Hodge potential associated with node i. The Helmholtz–Hodge potential of a node reflects its hierarchical position in its flow structure, which neglects the effect from the feedback mechanism. The potential ϕ_i for every node can be easily determined by minimizing the overall squared difference between the actual flow and the gradient flow (see [8,9] for more details).

4 Results

4.1 Results Using All Lists

In Fig. 4a, we show a scatterplot comparing the estimated potentials and the page rank value of each node [3]. We confirm that the Helmholtz–Hodge potential estimated by the Helmholtz–Hodge decomposition reveals information that is independent of page rank value.

Figure 4b shows a subset of the estimated Helmholtz–Hodge potential for the 85 institutions analyzed in this paper (the potential of the rest of the institutions is indicated by their position on the y-axis in Fig. 2[2]). The arrow shows the location of where the y-axis being 0. The G7 countries (Canada, France, Germany, Italy, Japan, the United Kingdom, and the United States of America) are shown in bold font. We can see that the OECD and the International Criminal Tribunal for Rwanda (ICTR), which are primarily focused on specific issues, top the list, which shows that they are less influenced by the sanctions lists issued by other institutions. This observation is quite intuitive because the smart sanctions issued by these institutions are restricted to focusing on specific issues. However,

[2] The actual position were slightly adjusted so that the nodes do not overlap.

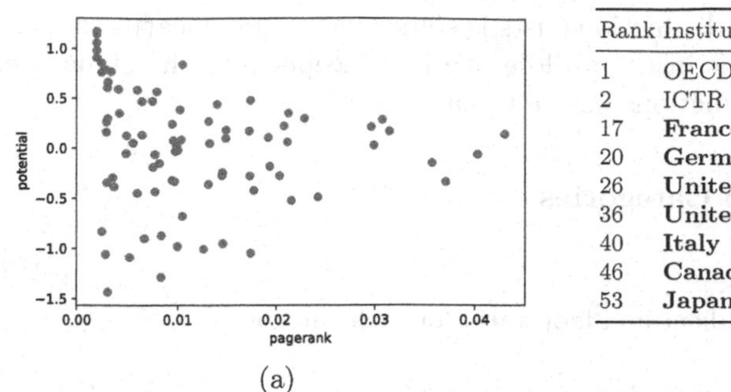

(a)

Rank	Institution	Potential
1	OECD	1.176
2	ICTR	1.148
17	**France**	0.482
20	**Germany**	0.441
26	**United Kingdom**	0.284
36	**United States of America**	0.148
40	**Italy**	0.101
46	**Canada**	0.029
53	**Japan**	-0.145

(b)

Fig. 4. (a) Scatterplot comparing the page rank value and Helmholtz–Hodge potential of each node. (b) Estimated Helmholtz–Hodge potentials. G7 countries are denoted in bold font.

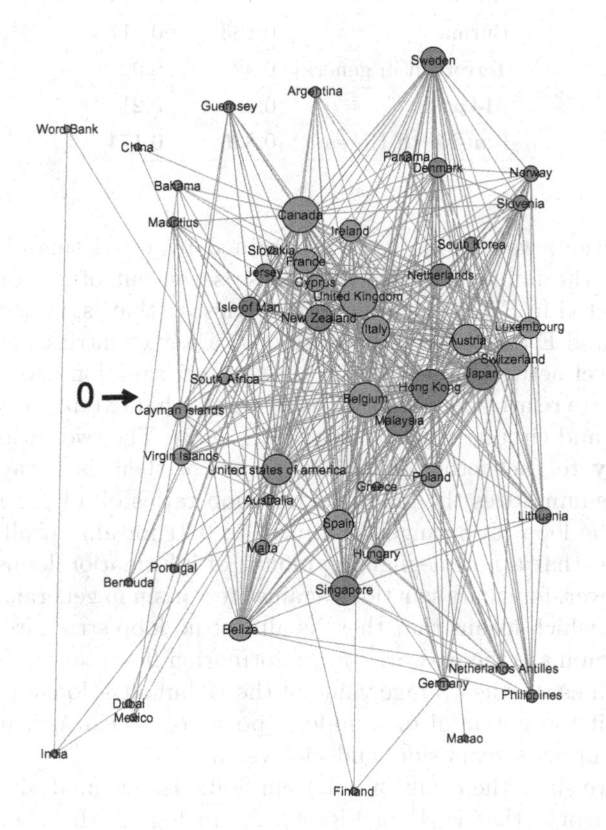

Fig. 5. Influence network at institution level using the smart sanctions lists grouped as financial crimes in Fig. 1a.

an analysis using all the smart sanctions lists is somewhat vague, locating a large proportion of the institutions in the middle. To look deeper into the structure, we must divide the smart sanctions lists into categories.

4.2 Restricting Lists to Categories

Table 2. Gradient and loop ratio for each category.

Category	Gradient	Loop
Financial crimes	0.55	0.45
Libya	0.865	0.135
Africa	0.886	0.114
Burma	0.883	0.117
Terrorism in general	0.98	0.02
Al-Qaeda	0.79	0.21
Iran-North Korea	0.826	0.174

In this section, we provide results obtained by restricting the set of lists used to derive the influence network. We use seven out of the nine categories found, as depicted in Fig. 1, excluding categories A (that is, issued by Japanese bureaucracy) and E (embargoes) because only a few countries were involved in the country-level network created from these smart sanction lists. The gradient and loop ratio are relative measures quantifying to what extent the flow structure is hierarchical and circular respectively (i.e., Fig. 3). The two ratios are thereby complementary to each other so that the sum of them is always normalized to 1. Table 2 summarizes the gradient and loop ratios of all the networks. We observe that the loop ratio ranges from 0.11 to 0.21 for almost all the network, which indicates that a non–negligible amount of information loops through the network. However, for smart sanctions against terrorism in general, the loop ratio is close to 0.0, which means that there is almost no loop structure. Hierarchical components dominate the network, and information flows to a hub, as shown in Fig. 6(b). In all cases, the average value of the Helmholtz-Hodge potential is set to 0.0. Thus, if the potential of a node is positive, we can conclude that it is located in the upper stream side, and vice versa.

In Fig. 5, we show the result of our Helmholtz–Hodge analysis for the financial crime category (that is, B in Fig. 1). As in Fig. 2, the y-axis position is determined using the Helmholtz–Hodge potential and the x-position is determined using the method of [12]. In this influence network, we see that there are many players involved, which creates four different communities (North America, Europe, island nations in the Indian Ocean, and other significant countries) where each community focuses on different sets of targeted entities. Moreover, the North American community and the island nation's community are located

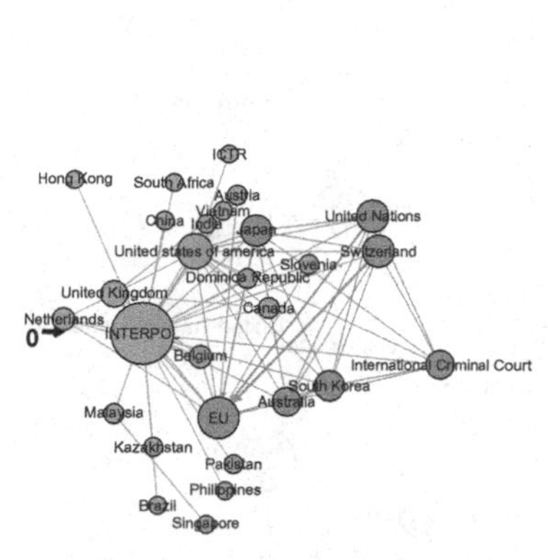

(a) Influence network at the institution level using smart sanctions lists grouped as against Libya in Fig. 1a.

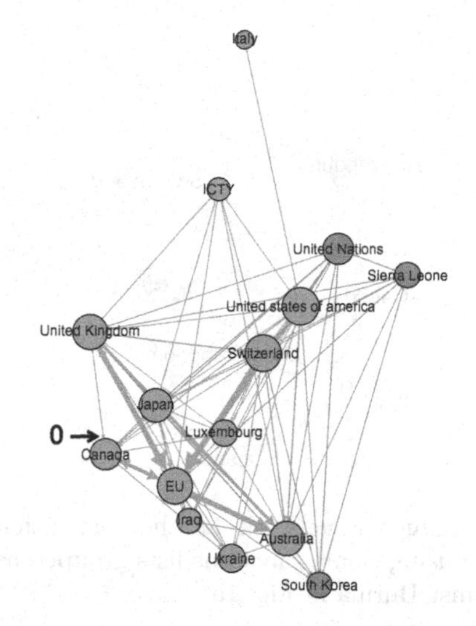

(b) Influence network at the institution level using smart sanctions lists grouped as against Africa in Fig. 1a.

Fig. 6. Analysis of the community structure of the influence network at the level of smart sanctions lists

in distinct positions on the left, which reflects the similarity between these communities.

In Fig. 6a, we show the influence network for the Libya category (G in Fig. 1). In this example, we find two distinct communities. One reflects countries and institutions that heavily influence the European Union (such as the United Nations, Switzerland, and Japan) and the other reflecting institutions that influence INTERPOL. Figure 6b shows the result for sanctions against Africa (D in Fig. 1). Italy is at the far top of the hierarchy, but there is only one edge pointing to the United States. Switzerland and the United Kingdom are key players that strongly influence the European Union, whereas Australia seems to be a follower of the European Union decisions.

Figures 7a and 7b showcases in which there is only one community in the network. Figure 7a corresponds to smart sanctions against Burma (F in Fig. 1), and Fig. 7b corresponds to terrorism in general. In both cases, Switzerland is the key player that strongly affects the European Union's decision. Luxembourg is another key player in the Burma case, but it is quite interesting that Switzerland and Luxembourg do not have much influence on each other.

Figure 8a shows the result for the Al-Qaeda case (I in Fig. 1). This is a quite complex case in which there are three communities and many edges among the institutions, which makes the network quite dense. We can see that the United Nations and the European Union are both working as hubs, gathering information from various countries. Still, the United Nations seems to be leading

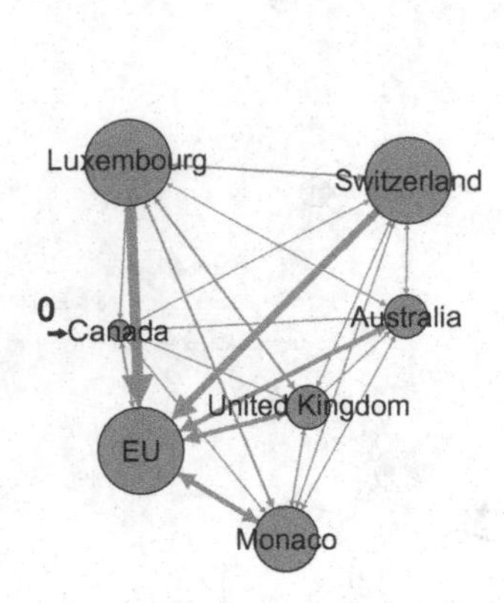

(a) Influence network at the institution level using smart sanctions lists grouped as against Burma in Fig. 1a.

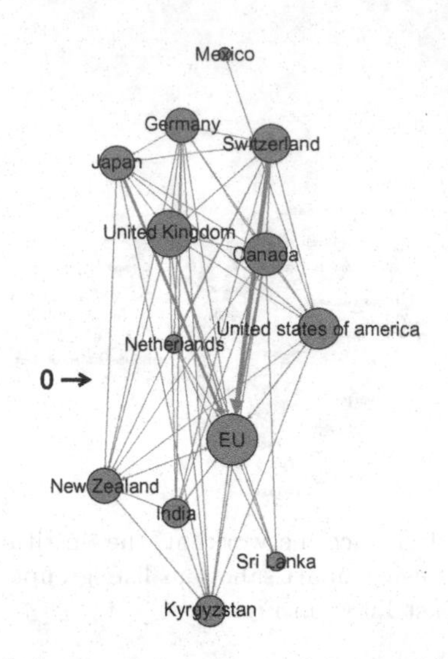

(b) Influence network at the institution level using smart sanctions lists grouped as against terrorism in general in Fig. 1a.

Fig. 7. Analysis of community structure of the influence network at the level of smart sanctions lists

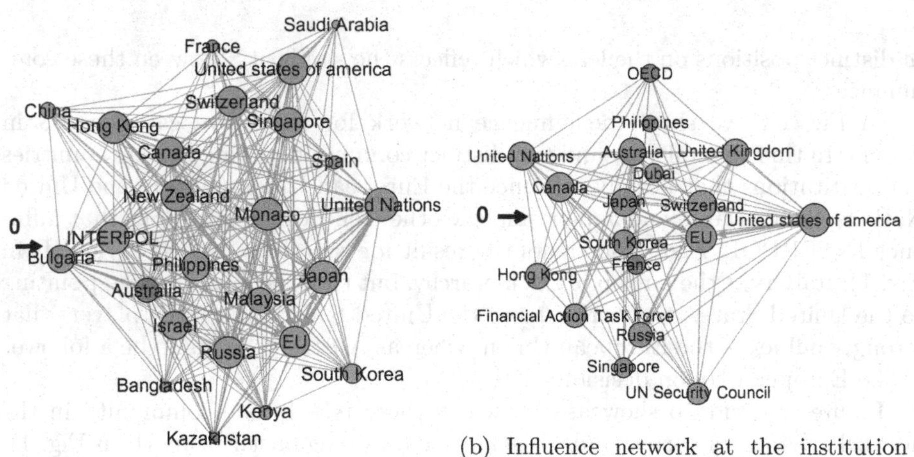

(a) Influence network at the institution level using smart sanctions lists grouped as against Al-Qaeda in Fig. 1a.

(b) Influence network at the institution level using smart sanctions lists grouped as against Iran and North Korea in Fig. 1a.

Fig. 8. Analysis of community structure of the influence network at the level of smart sanctions lists

the European Union, as could be suspected by its position. In Fig. 8b, we show the case concerning the sanctions against Iran and North Korea. In this case, we also see that the European Union acts as a hub, gathering information worldwide. It is also noteworthy that the United Nations Security Council is apparently at the bottom of the hierarchy, which indicates the complex approval mechanism of this council [6].

Table 3. Helmholtz-Hodge potential for selected countries.

Country	Financial crimes	Libya	Africa	Burma	Terrorism	Al-Qaeda	Iran-North Korea
Japan	0.073	0.256	0.053	-	0.485	−0.202	0.05
United States	−0.132	0.239	0.208	-	0.225	0.627	−0.032
China	0.507	0.597	-	-	-	0.486	-
Australia	−0.16	−0.193	−0.542	0.039	-	−0.323	0.42
United Kingdom	0.158	0.031	0.207	−0.046	0.446	-	0.379
Germany	−0.392	-	-	-	0.576	-	-
France	0.198	-	-	-	-	0.883	-0.21
Italy	0.114	-	1.208	-	-	-	
Canada	0.327	0.01	−0.105	0.0	0.356	0.243	0.122
Switzerland	0.087	0.186	0.191	0.294	0.54	0.525	0.037

Table 3 summarizes the Helmholtz–Hodge potential for selected countries. Several things are worth mentioning here. Switzerland appears in almost all categories, with positive values for each category. This observation indicates that Switzerland is quite an active player when it comes to smart sanctions. Australia, however, does appear in many categories but instead has a negative overall position (except for Iran-North Korea), which indicates that it is more of a follower. The United States is notably in a high position for Al-Qaeda, Libya, Africa, and terrorism in general, but is located in the lower part for financial crimes. Japan, conversely, is located in the lower part for Al-Qaeda but is an active player when it comes to terrorism in general. To summarize, our analysis provides meaningful insights into the hierarchical influences underlying smart sanctions lists.

5 Conclusion

This paper makes two contributions. The first is that we showed how to construct an influence network governing smart sanctions at the country level from the smart sanctions lists each country has issued. We then showed that by performing a Helmholtz–Hodge decomposition of the influence network, we could shed some light on the influence network that governs the institutions issuing such lists.

Specifically, we derived three interesting observations from our Helmholtz–Hodge analysis. First, we found that for smart sanctions lists against Iran and

North Korea, while the United Kingdom, United Nations, and the United States are at the top of the hierarchy of influencing other major countries and international institutions, the United Nations Security Council is clearly at the bottom of the hierarchy, which indicates its complex approval mechanism. Second, for the smart sanctions against Libya, the United Nations, ICTR, United States, and Japan are the key influencers in the upper stream of the hierarchy, and INTERPOL and the European Union are located in the middle of the hierarchy, acting as hubs aggregating information. Another intriguing example is the smart sanctions concerning financial crimes. In this example, there are four different communities (North America, Europe, island nations, and other countries), and each community focuses on different sets of targeted entities.

Furthermore, for some of the smart sanctions list categories, the influence network is governed almost entirely by a hierarchical structure. Still, for others, there is a significant amount of loop flow, which indicates that countries and international institutions actively influence each other. Our simple and effective analysis enables a better understanding of the hidden global political structure behind smart sanctions, possibly opening the door to understanding nations' hegemony in an increasingly complex world.

References

1. Ahn, D.P., Ludema, R.D.: The sword and the shield: the economics of targeted sanctions. Technical report (2019)
2. Blondel, V.D., Guillaume, J.L., Lambiotte, R., Lefebvre, E.: Fast unfolding of communities in large networks. J. Stat. Mech.: Theory Exp. **2008**(10), P10008 (2008). http://stacks.iop.org/1742-5468/2008/i=10/a=P10008
3. Brin, S., Page, L.: The anatomy of a large-scale hypertextual web search engine. In: Seventh International World-Wide Web Conference (WWW 1998) (1998). http://ilpubs.stanford.edu:8090/361/
4. Cortright, D., et al.: Smart Sanctions: Targeting Economic Statecraft. Rowman & Littlefield (2002). https://books.google.co.jp/books?id=tuz-eZUQapQC
5. Dreger, C., Fidrmuc, J., Kholodilin, K., Ulbricht, D.: The ruble between the hammer and the anvil: oil prices and economic sanctions. Discussion Papers of DIW Berlin 1488, DIW Berlin, German Institute for Economic Research (2015). https://EconPapers.repec.org/RePEc:diw:diwwpp:dp1488
6. Furukawa, K.: Kitacyosen Kaku no Shikingen Kokuren Sousa no Hiroku [Funding-Source of North Korea: A Note on United Nation's Investigation] Fundingsource. Tokyo Shincyosya, Tokyo, Japan (2017)
7. Gordon, J.: Smart sanctions revisited. Ethics Int. Affairs **25**, 315–335 (2011). https://doi.org/10.1017/S0892679411000323
8. Jiang, X., Lim, L.H., Yao, Y., Ye, Y.: Statistical ranking and combinatorial hodge theory. Math. Program. **127**(1), 203–244 (2011). https://doi.org/10.1007/s10107-010-0419-x
9. Kichikawa, Y., Iino, T., Iyetomi, H., Inoue, H.: Hierarchical and circular flow structure of the transaction network in japan. RIETI Discussion Paper Series 19e063, August 2019
10. Lambiotte, R., Delvenne, J.C., Barahona, M.: Laplacian Dynamics and Multiscale Modular Structure in Networks. arXiv e-prints arXiv:0812.1770, December 2008

11. Nephew, R.: The Art of Sanctions: A View from the Field. Center on Global Energy Policy Series, Columbia University Press (2017). https://books.google.co.jp/books?id=HIg_DwAAQBAJ
12. Noack, A.: Unified quality measures for clusterings, layouts, and orderings of graphs, and their application as software design criteria (2007)
13. Steger, M.: Globalization: A Very Short Introduction. Very Short Introductions, OUP Oxford (2017). https://books.google.co.jp/books?id=R1-1DgAAQBAJ
14. Stone, M.: The Response of Russian Security Prices to Economic Sanctions: Policy Effectiveness and Transmission. Technical report (2019)
15. Zarate, J.: Treasury's War: The Unleashing of a New Era of Financial Warfare. PublicAffairs (2013). https://books.google.co.jp/books?id=QFre4GHv79cC

Coverage and Evolution of Cancer and Its Risk Factors - A Quantitative Study with Social Signals and Web-Data

Saransh Khandelwal(✉) and Aurobinda Routray

Indian Institute of Technology Kharagpur, Kharagpur, India
saranshk2@gmail.com, aurobinda.routray@gmail.com

Abstract. In today's world of digitization, information is increasing at an exponential rate due to the ubiquitous nature of social media and web sources. With the advancement in Social Signal Processing, Social Informatics & Computing, and big data mining techniques, we can use this information in the health domain to determine the trend in the spread of symptoms and risk factors of diseases temporally. In this paper, our primary focus is to study coverage and evolution of risk factors of 19 most common Cancers in the last ten years (2009–2018) from the pieces of evidence collected from Tweets (Social Signal) and Wikipedia (Web Data) and demonstrate the relevant findings. Moreover, temporal variations are also shown in coverage of each cancer and risk factors to determine their evolution over time. We try to find the answers of questions like, Do different information sources show different risk factors for the same cancer ? or Is there any visible and distinctive trend in the temporal variation of any specific cancer or its risk factors? Each of the analysis unfolds various exciting observations. The most important among these is that some risk factors have the most adverse effects according to the one information source, but they are not at all harmful according to the other. Finally, the correlation of information sources, i.e., social signals and web data, is shown concerning variation in coverage of cancers and risk factors.

Keywords: Cancer trend · Social signals · Temporal analysis · Wikipedia · Twitter

1 Introduction

The severity of cancer can be understood from the fact that it is the second most common cause of death in the world after cardiovascular disease. According to the report published by the Indian Council of Medical Research[1] (ICMR) in 2016, the expected number of cancer patients in 2020 will be increased by 30 percent. Compared to the 2013 estimate of 1.32 million new cases annually by

[1] http://www.icmr.nic.in/.

S. Aref et al. (Eds.): SocInfo 2020, LNCS 12467, pp. 108–123, 2020.
https://doi.org/10.1007/978-3-030-60975-7_9

2020, the current estimate in this report is a number of 1.73 million new cases in India. The report shows a significant rise in the incidence of colon, lung, prostate, and rectum cancers among men. Among women, lungs, breast, ovary, and corpus uterus cancers are becoming common. The burden of cancer is not only limited to disease burden and mortality but also contributes to an economic and capital loss for the nation. We have done an exploratory study of cancer and its risk factors by analyzing an extensive data-set of social signal and web data. We collected pieces of evidence from Social Signals [10,17,22] using Twitter and Web data employing Wikipedia to validate the risk factors of cancer. Social Signals on Cancer can be treated as proxy information for what people think, convey, and spread about cancer. Web data is used to verify this buzz among people about cancer with the actual information.

In the following, we summarize our key contributions:

1. Developing a vast data set of tweets and Wikipedia articles which have the **cancer** keyword.
2. Quantitative coverage analysis of the 19 most common cancers and eight most harmful risk factors of Cancer (See Sect. 3) through evidence found in Twitter and Wikipedia
3. Determination of the temporal trend (from 2009 to 2018), i.e., variation through time, coverage of cancers, and risk factors on Twitter

The rest of the paper is organized as follows: Sect. 2 throws light on related works. Section 3 comprises our proposed methodology and data collection. Section 4 describes the experimental framework and results. Discussions and Conclusions are presented in Sect. 5.

2 Related Works

The widespread adoption of social media for communication creates unprecedented opportunities to monitor the opinions of large numbers of individuals. On social media, people share their experiences about symptoms, side-effects, treatments, opinions about diseases like Cancer, Influenza, Flu, and Intestinal flu. This opens up several opportunities in the healthcare sector on how to use this enormous data of health and disease-related tweets for real-time monitoring [6,7,9,15,24] and surveillance of diseases and provide an algorithmic framework to acquire relevant information. In [6], authors developed an influenza infection detection algorithm that automatically distinguishes relevant tweets from other tweets. In [15], authors did three kinds of analysis, temporal analysis to monitor disease daily activity, symptoms and treatment timeline, geographical analysis to track the spread of disease in the US, and text analysis to investigate the coverage of disease types, symptoms and treatment. There have also been several studies in determining the coverage [4,8] and trend of a particular disease in Twitter, Facebook, youtube [4], Wikipedia [13], Web data, and so on. In [4], authors monitor the Influenza trends in Web data and social media by text mining and graph-based structural mining techniques for 24 months. In [18], authors studied the coverage of Cancer in Wikipedia and the National Cancer Institute database to prove their hypothesis that content is less accurate and complete in Wikipedia than that of peer-reviewed website. Last but not least, social media is proving to be a useful tool in not only spreading awareness about symptoms, treatment, and prevention [21] but also morally support [2] the people even in the remote location [1].

As per our knowledge, there have been very few studies that combine both web sources like Wikipedia and social signals like Twitter to gather shreds of evidence about coverage of cancer and its risk factors. We try to investigate these two sources to find the coverage and temporal trend trends on a very large time-scale of 10 years.

3 Methodology in a Nutshell

Our objective is to find out pieces of evidence of occurrences of cancers and their risk factors in social and web signals. Our proposed methodology (Fig. 1) is divided into four sub-tasks:

3.1 Identification of Major Cancers and Their Risk Factors

We have implemented LDA topic modeling [5] for more than 20 research papers and Medical Reports, on cancer to cluster out the primary Cancers and Risk factors [3,11,12,14,16,19,20,23]. The 19 most common cancers are Breast, Brain, Oral, Lung, Cervical, Liver, Stomach, Liver, Stomach, Prostate, Ovarian, Bladder, Kidney, Throat, Lymphoma, Leukemia, Esophagus, Thyroid, Pancreas, Uterine, and Testicular. All the risk factors can be grouped into eight groups, which are Lifestyle, Genetic, Tobacco, Radiations, Micro-Organisms, Alcohol, Physical, and Chemicals. These are shown in Fig. 2 as a fishbone diagram.

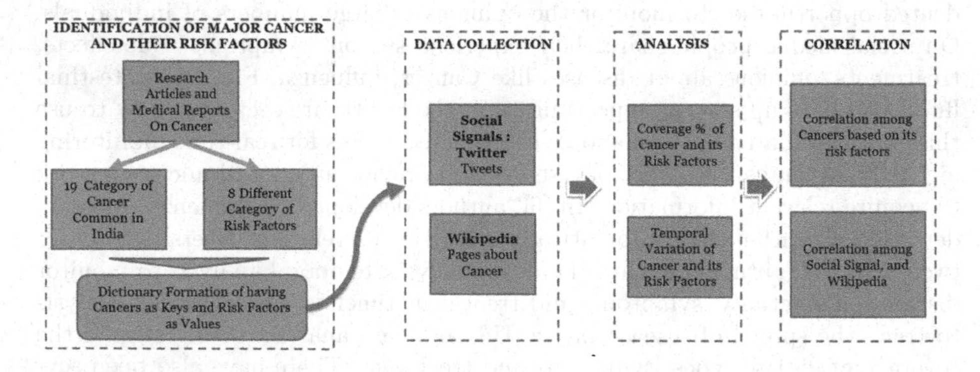

Fig. 1. Methodology

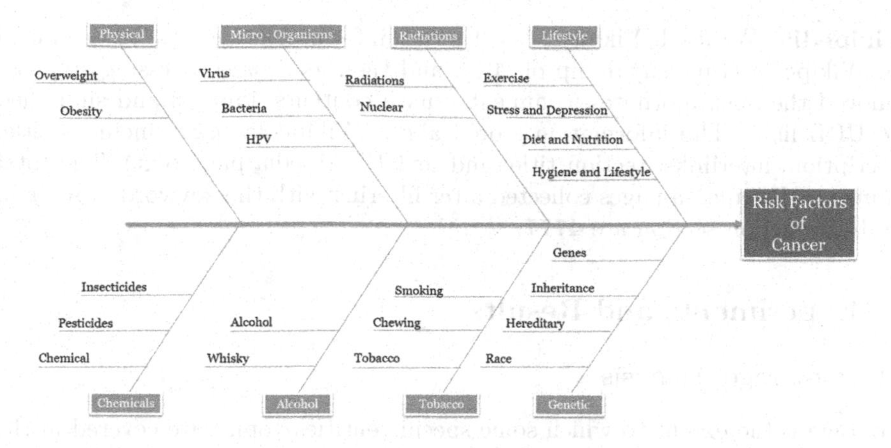

Fig. 2. Fish bone diagram for risk factors of cancer

3.2 Analysis

After getting risk factors from topic modeling, we have used all the synonyms of risk factors from **wordnet**[2] to collect pieces of evidence from the two sources. The analysis part can be classified into four essential sections: General Coverage Analysis, Specific Coverage Analysis, Temporal Variations of 19 Cancers and their corresponding eight classes of risk factors, and the correlation between sources. All of these analyses are explained in the Experiments section.

3.3 Correlation

One of the crucial aspects of this study is to determine the correlation among 19 Cancers based on their risk factors and to calculate the correlation between the sources on the coverage of both cancer and risk factors. The overall picture of correlation is discussed in the Experiments section.

3.4 Data Collection

We have collected data from two sources: Wikipedia and Twitter. All the data retrieved are filtered by **cancer** keyword.

Twitter. To investigate the extent of buzz that a piece of Cancer-related information gets on social media and to identify if there are correlations between this buzz and the evidence collected from Wikipedia, we crawled Twitter with the help of *GetOldTweets* library to get historical tweets having the keyword **cancer** for the period 2009–2018. Twitter API filters out some tweets from its internal algorithm, so this is not considered a complete dataset of tweets, but still, these tweets will give a general idea of the spread of cancer and its risk factors.

[2] https://wordnet.princeton.edu/.

Wikipedia. We used Wikipedia as the Web information source. We download the Wikipedia (English) dump of 2017, and then as a pre-processing step, we removed the tags, footnotes, comments, math contents, images, and simplified the URL links. The information stored about Wikipedia pages includes title, description, interlinks, section title, and text (Wikipedia page text). The total number of Wikipedia pages collected after filtering with the keyword **cancer** in the Introduction section are **4777**.

4 Experiments and Results

4.1 Coverage Analysis

Coverage is the extent to which some specific entities/topics are covered in the articles by a specific source. Coverage may originate from the length of articles (coverage per word published) where a specific topic is covered or from the number of articles (coverage per story) covering the specific topic. In this paper, we have done general coverage analysis and specific coverage analysis following the second criterion. In Wikipedia, the search query is performed on the description of the page, while in Twitter it is performed on the tweet.

Coverage of Cancer. Coverage of particular cancer in the specified source means how many documents contain that specific cancer keyword. For instance, in the case of Twitter, coverage of **Breast Cancer** signifies the total number of tweets that have the string **Breast Cancer** divided by the total number of the tweets having cancer keyword. Similarly, in Wikipedia, coverage means, total pages having the string **Breast Cancer** divided by total number of cancer related pages. It is clear from Fig. 17 and Fig. 18 (See Appendix) that Breast cancer is most common, followed by Brain Cancer, Lung Cancer, Prostate Cancer, and Oral cancer. The order is different in each source, which means that social signals and web data are not secondary sources.

General Coverage of Risk Factors. Similar to coverage of cancer, coverage of risk factor is defined as the number of documents in which any word related (i.e., synonyms) to that risk factor appears. For instance, the coverage of **Genetic** risk factor means all the documents containing any word from *genes, genetic, hereditary* divided by the total number of relevant documents. From Fig. 19 (See Appendix), it is evident that a spoiled lifestyle is a significant risk factor of cancer, followed by genetic and Tobacco according to Twitter. According to Wikipedia (Fig. 20), genetic factors are the significant risk factors, followed by Lifestyle and Micro-organisms (Viruses and Bacteria).

Specific Cancer - Risk Factors Analysis. The above two analyses are the general analysis of coverage of risk factors and Cancers. We have also analyzed each cancer individually for both of the information sources based on the pieces of evidence. By evidence, we mean co-occurrence of cancer and its risk factor (or synonyms as shown in Fish Bone diagram Fig. 2) at-least one time in the tweet and at least two times in the Wikipedia article. As evident from below plotted figures (Fig. 3, 4, 5, 6, 7, 8, 9, 10, 11, 12, 13 and 14), Twitter does not have the continuous distribution of risk factors; instead, it is confined to a maximum of 3-4 risk factors. One inference from this information is that either people are peculiar in tweeting about cancer or do not mention risk factors at all. In Wikipedia, for most of the Cancers, **Genetic Reasons** is the most significant risk factor.

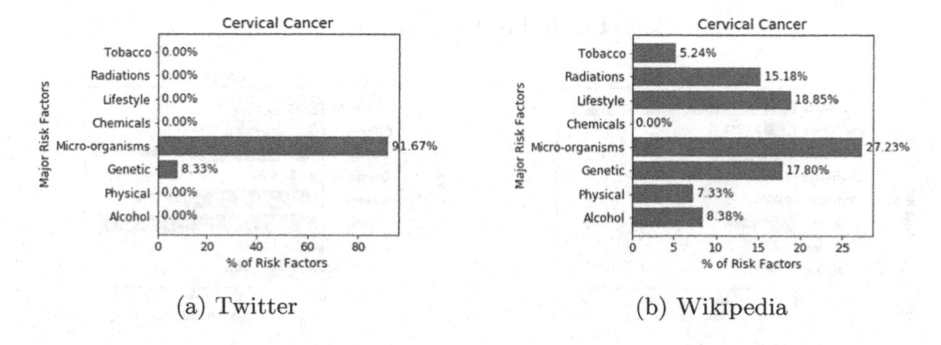

(a) Twitter (b) Wikipedia

Fig. 3. Cervical Cancer

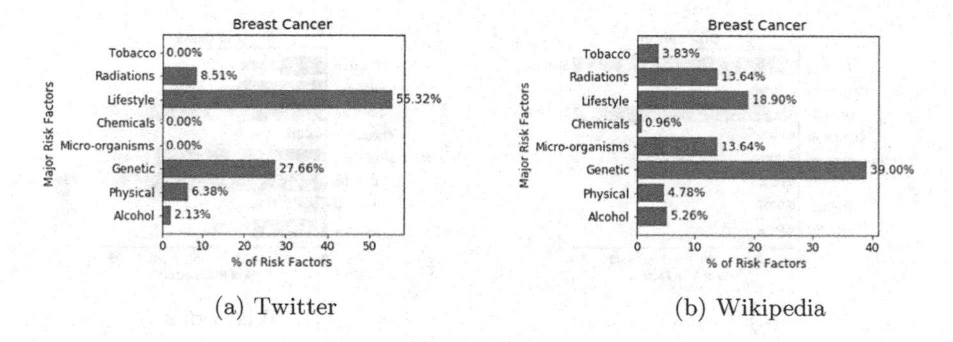

(a) Twitter (b) Wikipedia

Fig. 4. Breast Cancer

The concrete findings from this analysis are as follows:

- In Cervical Cancer (Fig. 3), both of the sources synergies well by agreeing that *Micro-organism* such as bacteria and HPV as the major risk factor.
- In Breast Cancer (Fig. 4) and Prostate Cancer (Fig. 5), according to Social signal major risk factor is *Lifestyle*, while according to Wikipedia, it is *Genetic*.

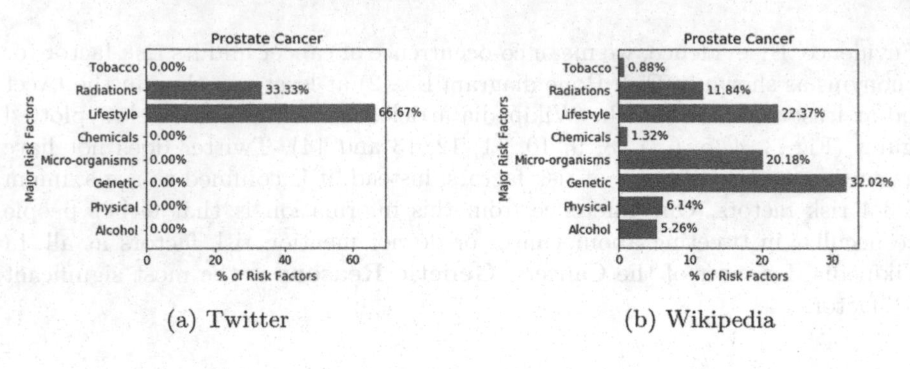

(a) Twitter (b) Wikipedia

Fig. 5. Prostate Cancer

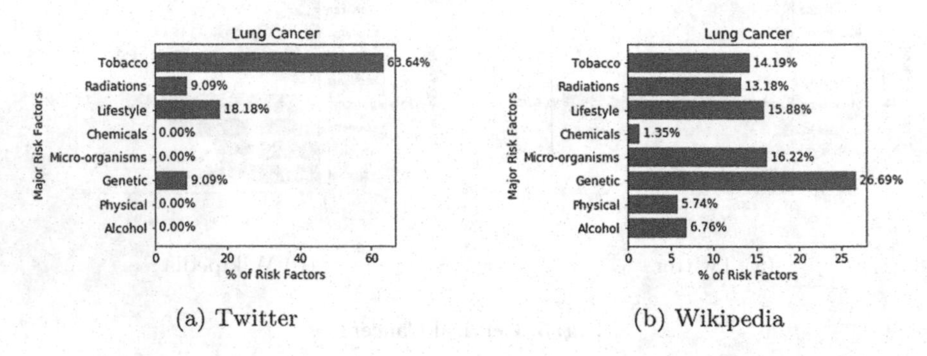

(a) Twitter (b) Wikipedia

Fig. 6. Lung Cancer

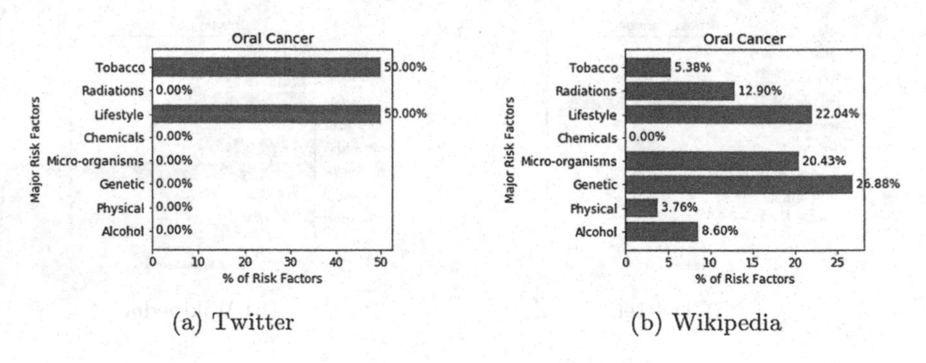

(a) Twitter (b) Wikipedia

Fig. 7. Oral Cancer

- In Lung Cancer (Fig. 6), Oral Cancer (Fig. 7), Bladder Cancer (Fig. 12) and Kidney Cancer (Fig. 13), Twitter agrees to *Tobacco* and *Lifestyle* while Wikipedia shows *Genetic* and *Lifestyle* as major risk factors.
- Similar to Cervical Cancer, both of the sources. Twitter and Wikipedia agree for Stomach Cancer (Fig. 8) in determining *Lifestyle* and *Genetic* as most severe risk factors.

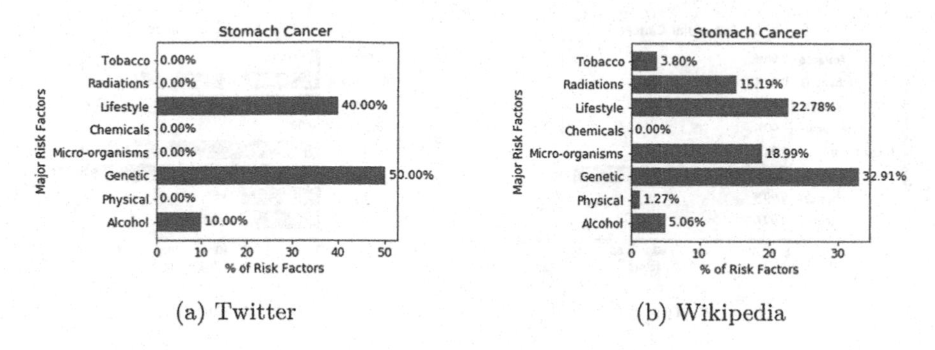

(a) Twitter (b) Wikipedia

Fig. 8. Stomach Cancer

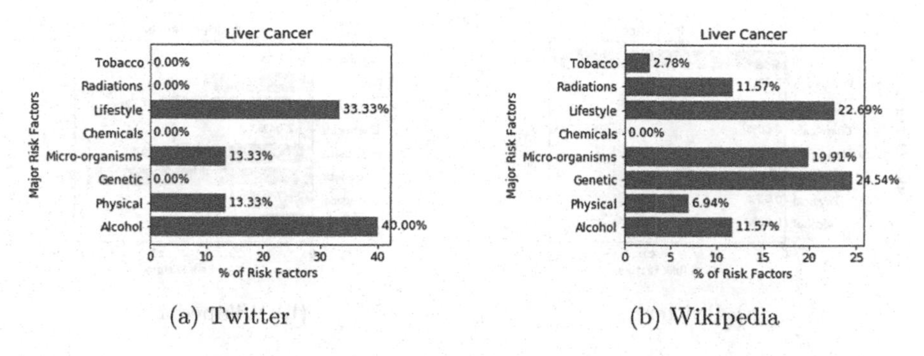

(a) Twitter (b) Wikipedia

Fig. 9. Liver Cancer

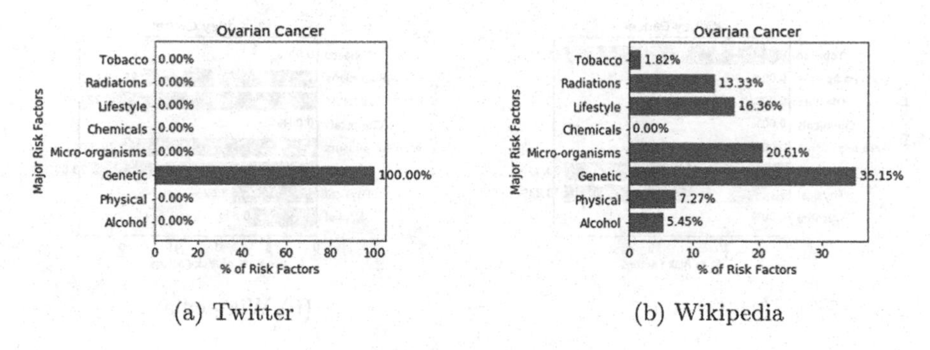

(a) Twitter (b) Wikipedia

Fig. 10. Ovarian Cancer

- *Genetic* reasons are major risk factors for Ovarian (Fig. 10) and Testicular (Fig. 11) according to both of the sources.
- Twitter do not show a uniform distribution of risk factors like Wikipedia, instead, for most of the cases it is quite specific like for Ovarian Cancer, it determines *Genetic* as a single risk factor, for Oral Cancer it depicts *Tobacco* and *Lifestyle* as key risk factors.

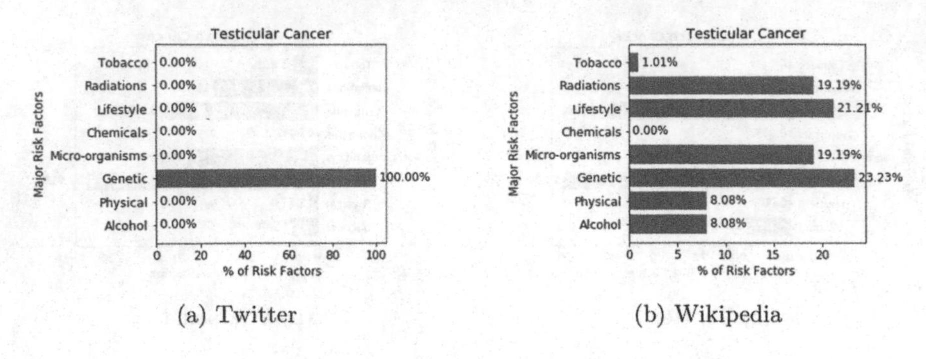

(a) Twitter (b) Wikipedia

Fig. 11. Testicular Cancer

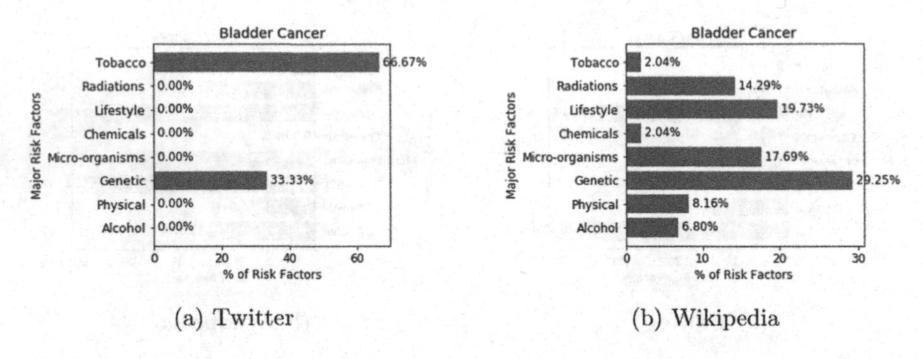

(a) Twitter (b) Wikipedia

Fig. 12. Bladder Cancer

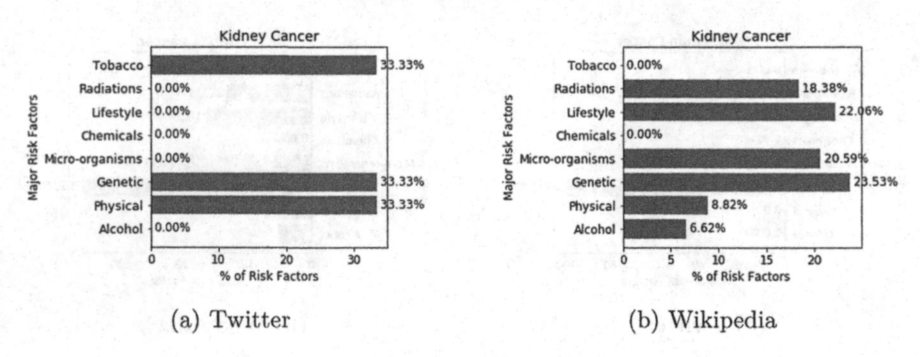

(a) Twitter (b) Wikipedia

Fig. 13. Kidney Cancer

4.2 Temporal Variation of Risk Factors over 10 Years

Temporal variation means how do these risk factors evolve with time. This study is conducted only for Social Signals i.e. Twitter. For this experiment, if one or more of the risk factors related keywords (including synonyms) is present in the content of a document, we tag that document with that risk factor. In Fig. 15, the percentage of tweets having the mention of that risk factor over the years 2009–2018 is plotted. It is quite evident from the plots that the effect of

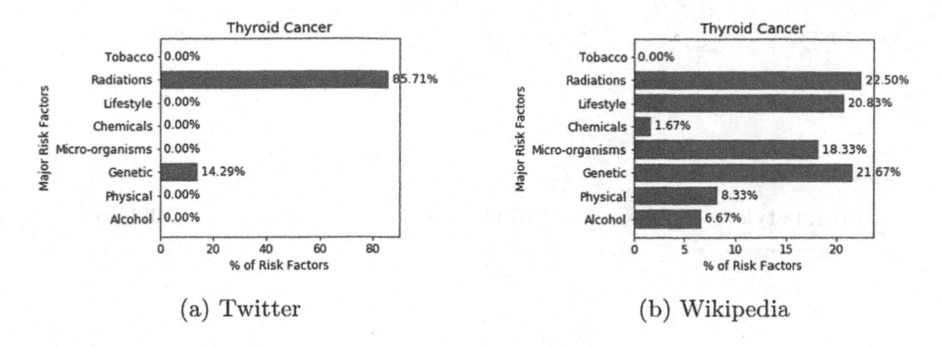

(a) Twitter (b) Wikipedia

Fig. 14. Thyroid Cancer

chemicals such as Pesticides and Insecticides, and Tobacco as a cause of cancer are increasing year by year. Although the effect of Lifestyle and Genetic risk factors are decreasing over the years, they have the highest coverage percentage (more than 15%).

4.3 Temporal Variation of Cancer Coverage over 10 Years

We have also plotted temporal coverage variation of cancers in the same manner as of risk factors. From Fig. 16, it is clear that cases of Oral, Ovarian, Kidney, Lymphoma, and Esophagus cancer are significantly increased in 10 Years. The instances of Breast and Cervical Cancer are relatively decreased, but their relative coverage percentage is still high.

4.4 Correlation

We have plotted heat-map of Spearman correlation (See Appendix Fig. 21) among cancers, with a p-value < 0.05, based on risk factors affecting them. This analysis is only done for Wikipedia since it has a more uniform distribution. The risk factors of Leukemia Cancer and Thyroid Cancer are least correlated with other cancer's risk factors. Moreover, we also calculate the Spearman rank correlation coefficient between Wikipedia and Social Signals based on the statistics of specific coverage of each cancer and its risk factors. The average correlation between Wikipedia and Social Signals is 0.68 (p-value < 0.05), which is relatively low, which necessarily means information retrieved from social signals can not be perceived as absolute truth, and both sources are independent.

5 Conclusions

5.1 Discussions

This study aims to compare and measure people's awareness of cancer risk factors through Twitter and the medical community's knowledge of risk factors through

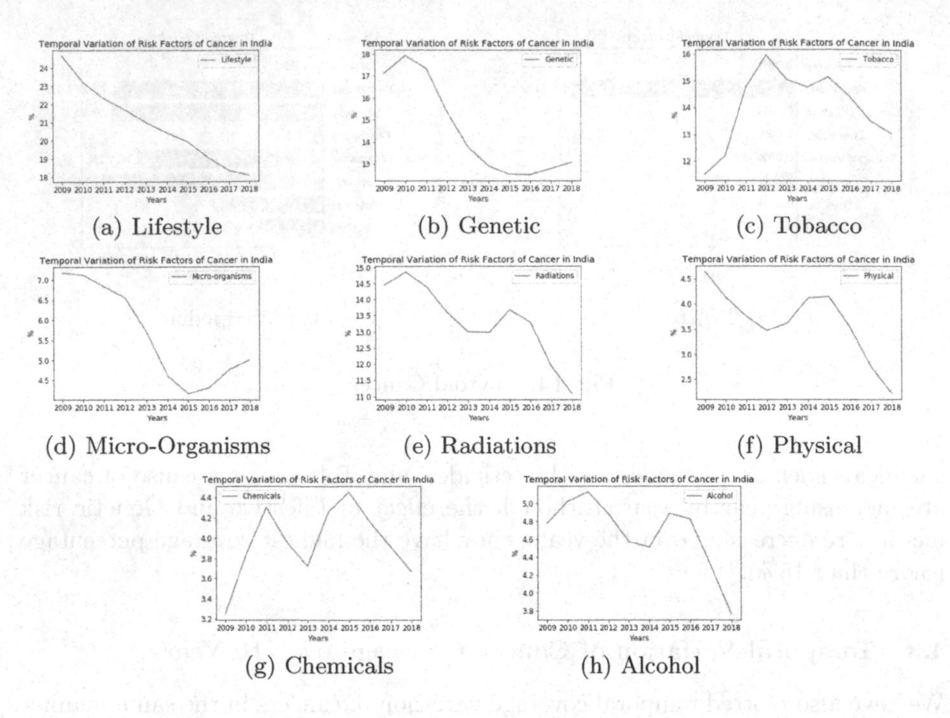

(a) Lifestyle (b) Genetic (c) Tobacco

(d) Micro-Organisms (e) Radiations (f) Physical

(g) Chemicals (h) Alcohol

Fig. 15. Temporal variations of risk factors of cancers

Wikipedia. One important thing that must be noted here is tweets are just proxy for people's opinions, they do not intend to show their medical knowledge or validity about any information about cancers. As far as we have read research papers about cancer risk factors, they do not focus much on the harsh effects of pesticides and insecticides on people's health. This is further verified by our analysis that the share of chemicals as a risk factor in Wikipedia is just 1.83 % (Fig. 20). However, we have observed from the pieces of evidence collected from Twitter (Fig. 19) that chemicals are indeed an essential risk factor (5.51 %) for cancer, whose adverse effect is increasing year by year (Fig. 15).

5.2 Conclusion and Future Works

In this paper, we collected a vast data set of Wikipedia articles and Twitter tweets related to cancer. We show that for every cancer, how the distribution of risk factors changes in two information sources. We have also analyzed cancer and its risk factors temporally (2009–2018) by doing a computational and syntactic analysis on Twitter. We observed some uniform trends in both Wikipedia and Twitter about some cancers, while some are trends specific to a source. We also report the correlation between information diffusion in two parallel sources.

In the future, we would like to extend this work in multiple directions. One immediate task would be to study an additional number of social signals like

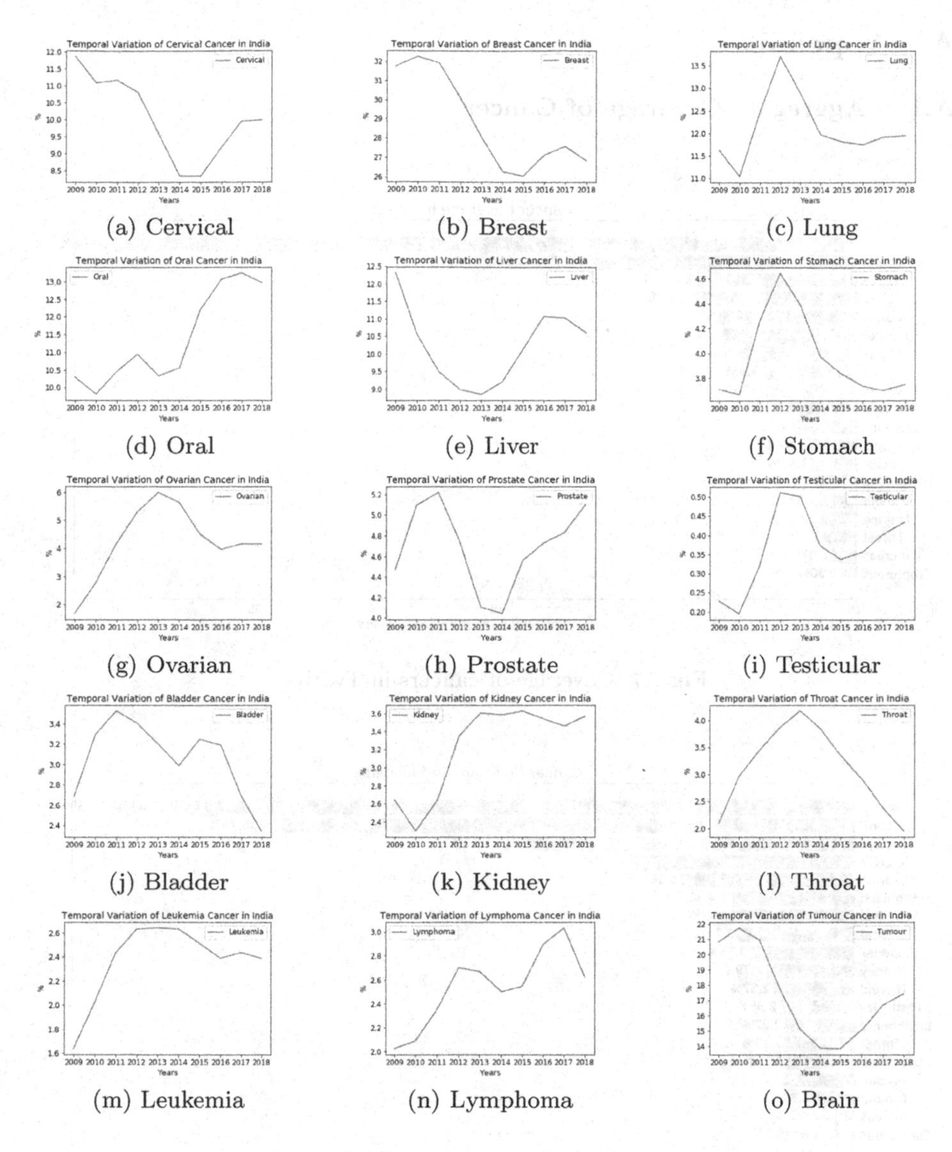

(a) Cervical (b) Breast (c) Lung

(d) Oral (e) Liver (f) Stomach

(g) Ovarian (h) Prostate (i) Testicular

(j) Bladder (k) Kidney (l) Throat

(m) Leukemia (n) Lymphoma (o) Brain

Fig. 16. Temporal variations of cancers

Facebook and take a large number of medical research papers into account to perform LDA. This will give us a wide perspective and a relatively ample amount of buzz about cancer among people. Further, it will also make our analysis more robust and generalized. The second thing would be building a statistical model to predict the probability of cancer due to a particular risk factor. Finally, we would also like to define a metric that will determine the evolution of truth regarding disease and its risk factors.

A Appendix

A.1 Aggregate Coverage of Cancer

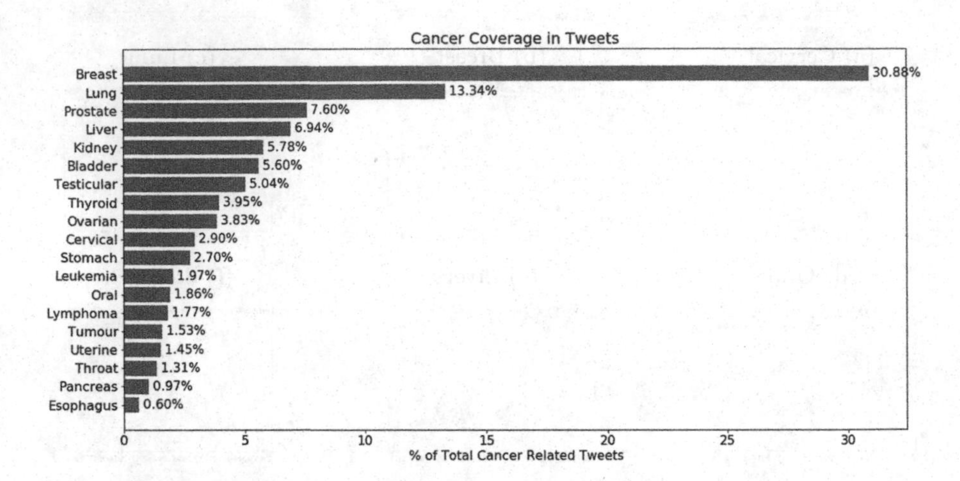

Fig. 17. Coverage of cancers in Twitter

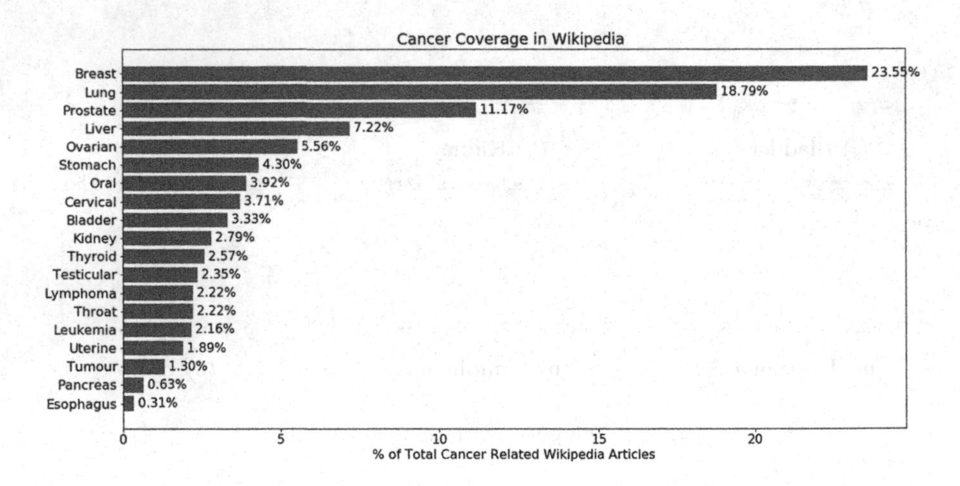

Fig. 18. Coverage of cancers in Wikipedia

A.2 Aggregate Coverage of Risk Factors

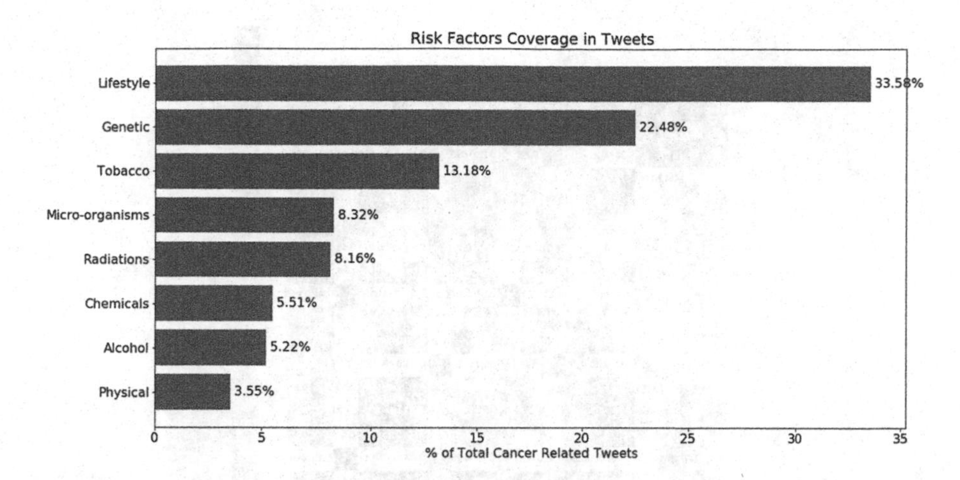

Fig. 19. Coverage of risk factors in Twitter

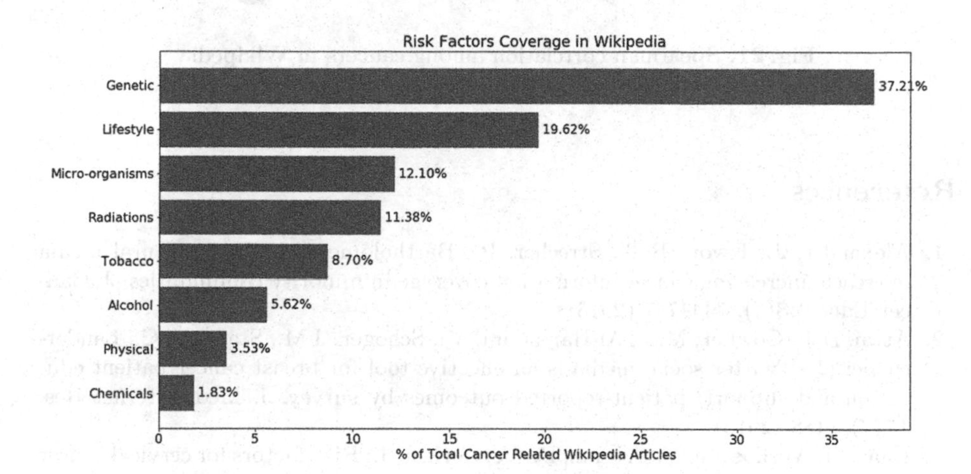

Fig. 20. Coverage of risk factors in Wikipedia

A.3 Correlation Among Cancers

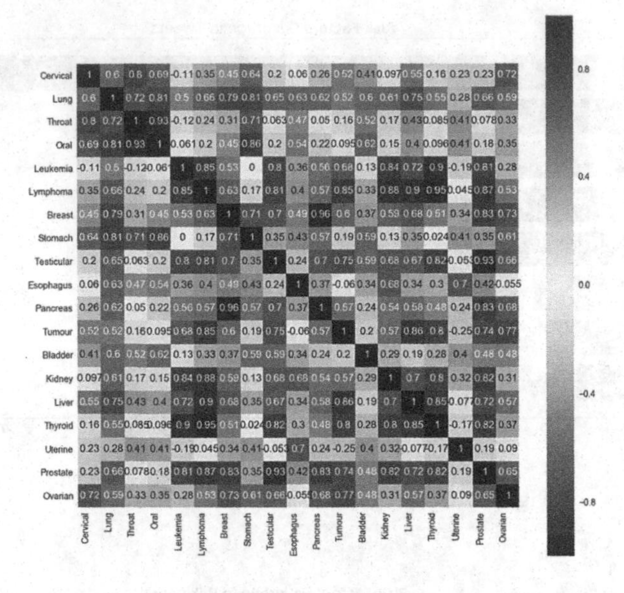

Fig. 21. Spearman correlation among cancers in Wikipedia

References

1. Alexander, J., Kwon, H.T., Strecher, R., Bartholomew, J.: Multicultural media outreach: increasing cancer information coverage in minority communities. J. Cancer Educ. **28**(4), 744–747 (2013)
2. Attai, D.J., Cowher, M.S., Al-Hamadani, M., Schoger, J.M., Staley, A.C., Landercasper, J.: Twitter social media is an effective tool for breast cancer patient education and support: patient-reported outcomes by survey. J. Med. Internet Res. **17**(7), e188 (2015)
3. Baay, M., Verhoeven, V., Avonts, D., Vermorken, J.: Risk factors for cervical cancer development: what do women think? Sexual Health **1**(3), 145–149 (2004)
4. Basch, C.H., Basch, C.E., Ruggles, K.V., Hammond, R.: Coverage of the ebola virus disease epidemic on Youtube. Disaster Med. Public Health Preparedness **9**(5), 531–535 (2015)
5. Blei, D.M., Ng, A.Y., Jordan, M.I.: Latent dirichlet allocation. J. Mach. Learn. Res. **3**(Jan), 993–1022 (2003)
6. Broniatowski, D.A., Paul, M.J., Dredze, M.: National and local influenza surveillance through Twitter: an analysis of the 2012–2013 influenza epidemic. PLoS ONE **8**(12), e83672 (2013)
7. Charles-Smith, L.E., et al.: Using social media for actionable disease surveillance and outbreak management: a systematic literature review. PLoS ONE **10**(10), e0139701 (2015)

8. Corley, C.D., Cook, D.J., Mikler, A.R., Singh, K.P.: Text and structural data mining of influenza mentions in web and social media. Int. J. Environ. Res. Public Health **7**(2), 596–615 (2010)
9. Corley, C.D., Cook, D.J., Mikler, A.R., Singh, K.P.: Using web and social media for influenza surveillance. Advances in Computational Biology. AEMB, vol. 680, pp. 559–564. Springer, New York (2010). https://doi.org/10.1007/978-1-4419-5913-3_61
10. Della Vedova, M.L., Tacchini, E., Moret, S., Ballarin, G., DiPierro, M., de Alfaro, L.: Automatic online fake news detection combining content and social signals. In: 2018 22nd Conference of Open Innovations Association (FRUCT), pp. 272–279. IEEE (2018)
11. El-Serag, H.B., Mason, A.C.: Risk factors for the rising rates of primary liver cancer in the united states. Arch. Intern. Med. **160**(21), 3227–3230 (2000)
12. Ganesh, B., Saoba, S.L., Sarade, M.N., Pinjari, S.V.: Risk factors for prostate cancer: an hospital-based case-control study from Mumbai, India. Indian J. Urol.: IJU: J. Urol. Soc. India **27**(3), 345 (2011)
13. Generous, N., Fairchild, G., Deshpande, A., Del Valle, S.Y., Priedhorsky, R.: Global disease monitoring and forecasting with Wikipedia. PLoS Comput. Biol. **10**(11), e1003892 (2014)
14. Hariharan, K., Padmanabha, V.: Demography and disease characteristics of prostate cancer in India. Indian J. Urol.: IJU: J. Urol. Soc. India **32**(2), 103 (2016)
15. Lee, K., Agrawal, A., Choudhary, A.: Real-time disease surveillance using twitter data: demonstration on flu and cancer. In: Proceedings of the 19th ACM SIGKDD International Conference on Knowledge Discovery and Data Mining, pp. 1474–1477 (2013)
16. Malhotra, J., Malvezzi, M., Negri, E., La Vecchia, C., Boffetta, P.: Risk factors for lung cancer worldwide. Eur. Respir. J. **48**(3), 889–902 (2016)
17. Pentland, A.: Social signal processing [exploratory DSP]. IEEE Signal Process. Mag. **24**(4), 108–111 (2007)
18. Rajagopalan, M.S., et al.: Patient-oriented cancer information on the internet: a comparison of Wikipedia and a professionally maintained database. J. Oncol. Pract. **7**(5), 319–323 (2011)
19. Ram, H., Sarkar, J., Kumar, H., Konwar, R., Bhatt, M., Mohammad, S.: Oral cancer: risk factors and molecular pathogenesis. J. Maxillofac. Oral Surg. **10**(2), 132 (2011)
20. Sinha, R., Anderson, D., McDonald, S., Greenwald, P.: Cancer risk and diet in India. J. Postgrad. Med. **49**(3), 222 (2003)
21. Thackeray, R., Burton, S.H., Giraud-Carrier, C., Rollins, S., Draper, C.R.: Using twitter for breast cancer prevention: an analysis of breast cancer awareness month. BMC Cancer **13**(1), 508 (2013)
22. Vinciarelli, A., Pantic, M., Bourlard, H.: Social signal processing: survey of an emerging domain. Image Vis. Comput. **27**(12), 1743–1759 (2009)
23. Zali, H., Rezaei-Tavirani, M., Azodi, M.: Gastric cancer: prevention, risk factors and treatment. Gastroenterol. Hepatol. Bed Bench **4**(4), 175 (2011)
24. Zou, B., Lampos, V., Gorton, R., Cox, I.J.: On infectious intestinal disease surveillance using social media content. In: Proceedings of the 6th International Conference on Digital Health Conference, DH 2016, pp. 157–161. Association for Computing Machinery, New York (2016). https://doi.org/10.1145/2896338.2896372

Detecting Engagement Bots on Social Influencer Marketing

Seungbae Kim[1] and Jinyoung Han[2(✉)]

[1] Department of Computer Science, University of California, Los Angeles, Los Angeles, USA
sbkim@cs.ucla.edu

[2] Department of Applied Artificial Intelligence, Sungkyunkwan University, Seoul, South Korea
jinyounghan@skku.edu

Abstract. Influencer fraud, which can significantly damage authentic influencers and companies, has become one of the most important problems that can adversely affect the influencer marketing industry. Fraudulent influencers obtain fake engagements on their posts by purchasing engagement bots that automatically generate likes and comments. To identify bots that make fake engagements to influencers, we perform an in-depth analysis on the social network of influencer engagements, which consists of 14,221 influencers, 9,290,895 users, and 65,848,717 engagements. We find that bots tend to have low local clustering coefficients and write short comments which are similar to each other. Based on the analysis results of the unique engagement behavior of bots, we propose a neural network-based model that learns text, behavior, and graph representations of social media users to detect the engagement bots from audiences of influencers. The experimental results show that the proposed model outperforms well-known baseline methods by achieving 80% accuracy.

Keywords: Influencer fraud · Engagement network · Influencer marketing · Bot detection · Fake engagement · Engagement bot

1 Introduction

Influencer marketing has become an effective word-of-mouth method that increases brand awareness among numerous potential customers [4,9]. The effectiveness of the influencer marketing can be suffered from "influencer fraud" that can damage marketing campaigns of companies [11]. That is, influencers may manipulate their influence by obtaining more engagements on their posts using engagement bots that automatically generate likes or comments on social media. Such influencer fraud needs to be seriously considered by marketers since likes or comment counts are used directly as a measure of the success of influencer marketing campaigns [7]. If an influencer hired by a company manipulates his/her

S. Aref et al. (Eds.): SocInfo 2020, LNCS 12467, pp. 124–136, 2020.
https://doi.org/10.1007/978-3-030-60975-7_10

influence using engagement bots, the company obtains an advertising effect that is less than what it paid to hire the influencer. In addition, users may lose their trust in influencers, adversely affecting the influencer marketing industry.

Since bots often adversely affect users, companies, and marketers around social media [6], many researchers have investigated the characteristics of bots on social media, such as conversational patterns and network properties [14], user profile and content sentiment [15], which can shed light on differences between bots and authentic users. Also, there have been attempts to develop bot detection models by learning graph similarity [16] or a sequence of contextualized information [10]. However, little attention has been paid to developing bot detection models for influencer marketing, where such a model is useful to accurately assess the influence of each influencer.

To understand the behavior characteristics of bots in social media marketing, we first conduct an empirical study on social media bots by comparing them with authentic audiences of influencers. To this end, we build an influencer engagement network that is a social network of influencers and their audience connected through social engagements, e.g., via liking posts or commenting. In our study, the constructed influencer engagement network is composed of 14,221 influencers, 9,290,895 users, and 65,848,717 engagements.

By analyzing the influencer engagement network, we identify a group of potential engagement bots who generate tons of engagements to a large number of influencers but have zero followers. We perform an in-depth analysis on the distinct characteristics of the engagement bots and reveal that the engagement bots have lower local clustering coefficients in the influencer engagement network than normal users since the bots tend to engage in the posts of random influencers while normal users usually follow influencers with similar interests. Our analysis further reveals that the identified engagement bots tend to write short comments, which are similar to each other, because they are likely to use a set of pre-populated comments to automatically write comments.

Based on lessons learned, we propose a deep learning model that can detect the engagement bots from influencers' audiences. Our model learns contextualized information from comments, engagement behavior, and structural information of the network by taking multi-modal inputs including text features, behavior features, and graph features, respectively. The experimental results show that the proposed model outperforms well-known baseline methods by achieving about 80% accuracy. The results also reveal that all three input feature sets are useful to represent social media users while the behavior features play the most important role in detecting engagement bots among the three input feature sets.

The remaining of the paper is organized as follows. We first describe previous work on bot detection on social media and social networks of influencer marketing in Sect. 2. In Sect. 3, we define the influencer engagement network and analyze characteristics of the engagement bots. We then introduce the deep learning model for detecting engagement bots, and present experimental results in Sect. 4. Finally, we conclude the paper with future work in Sect. 5.

2 Related Work

Social Bot Detection. Bot detection in social media has been extensively investigated since bots can adversely affect authentic users and companies [6]. Schuchard et al. [14] analyzed the properties of user conversation networks such as centrality, node degrees, and communities to detect bots. Varol et al. [15] presented a large number of high-level user representations including friendship, network, sentiment, temporal, and language features, and evaluated the proposed features using well-known learning algorithms. Wang et al. [16] presented a clustering model to group users with similar network structures. They used a sequence of events to build a user network. Kudugunta et al. [10] proposed a neural network-based bot detection model that learns sequential embeddings of contextualized tweets by using a long short-term memory (LSTM). Our work focuses on detecting bots in social influencer marketing; we propose a bot detection model to address influencer fraud problems.

Social Networks of Influencers. Influencer marketing is known as an effective word-of-mouth marketing method that utilizes popular individuals for advertising products to a large number of audiences [4]. As influencer marketing receives tremendous attention from marketers due to its effectiveness, many researchers have studied various aspects of influencer marketing. Rios et al. [13] presented a method to identify influencers in the social network that is built on user activities. They used semantic analysis to filter out futile links to find influencers. Arenas et al. [1] also proposed to use social networks of consumer reviews to find influencers. They found that identified influencers have a larger scope on various categories with high centrality than normal users. Kim et al. [8] built a social network of influencers based on their followers and followees to understand social relationships among influencers. They found that influencers with similar topics make clusters in the network and have more common followers than other influencers with no common interests. Yang et al. [17] presented the brand mentioning network that connects influencers and brands based on their brand mentioning tags. They proposed a neural network model that can predict sponsorship of social media posts published by influencers by learning graphical features from the brand mentioning network. This paper focuses on analyzing engagement behavior around social influencers and proposing a model to detect fake engagements in social influencer marketing.

3 Influencer Engagement Network

3.1 Definition of Influencer Engagement Network

We define the notion of the influencer engagement network as a directed weighted graph $G = (V, E, W)$. In the given network, a node $v \in V$ represents a user who is an influencer or an audience. Two nodes v_i and v_j in the network are connected if node v_i engages in the activities by node v_j. That is, a directed edge $e_{ij} \in E$ from v_i to v_j exists when the user v_i engages in a post published by the influencer v_j through liking or commenting. Note that an influencer can also engage in another influencer's activities, therefore, edges can exist among influencer nodes. Each edge e_{ij} in the network has a weight value $w_{ij} \in W$ which represents the amount of engagements between two nodes v_i and v_j. We calculate the total number of likes and comments that are generated by a node v_i to a node v_j to derive a weight value w_{ij}.

3.2 Engagement Data

To build the influencer engagement network, we use the Instagram Influencer Dataset [9]. The dataset contains 33,935 influencers and their 10 M Instagram posts published from 2010 to 2018, where 87% of the posts had been published in 2017 and 2018. In this study, we only use the posts that were published after January 1st in 2017 from the original dataset. After filtering out influencers who have less than 100 posts during the given time period, we use the 6,244,555 Instagram posts published by 14,221 influencers who are followed by at least 10,000 users. We then identify a list of users who have engaged in the posts through liking or writing comments. More specifically, we collect 65,848,717 engagements (21,374,920 likes and 44,473,797 comments) generated by 9,290,895 unique users from the Instagram posts. Note that each post in Instagram provides a partial list of associated users, and we exclude all self-engagements in our study. In addition to the engagement data, we further collect the number of followers for each user by crawling his/her Instagram profile page to examine the characteristics of the users.

3.3 Network Analysis

Based on the constructed influencer engagement network, we analyze (i) the number of followers of users who engage in this network, (ii) the amount of engagements (i.e., weights in the network), and (iii) the connectivity of the network, i.e., how a user or an influencer is connected with other users and/or influencers, to understand engaging behavior of users in the influencer engagement network. We further seek to identify a set of suspicious audience in terms of engaging behavior, who can be considered as engagement bots.

Number of Followers of the Engaged Users. We first investigate the number of followers of the influencers' audiences since influencers' posts can be further propagated to the followers of the audience. In other words, influencers can have more influence if their audiences have a large number of followers. Therefore, understanding the size of the social networks of the audience is important in influencer marketing. Figs. 1(a) and 1(b) show the distributions of the number of followers of the engaged users by the cumulative distribution function and probability density function, respectively. As shown in Fig. 1(a), we first find that around 20% of the engaged users do not have followers. We denote

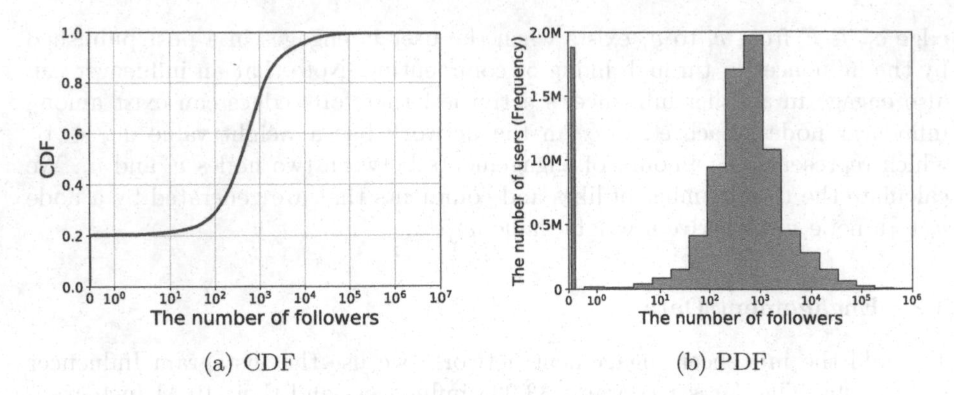

(a) CDF (b) PDF

Fig. 1. Distributions of the number of followers of the engaged users. 20% of the engaged users do not have followers while 4% of the engaged users have more than 10,000 followers. Note that the median and average values of the number of followers of the engaged users is 557 and 3,745, respectively, after disregarding the users who do not have followers.

this user group having no follower as the zero-follower audience. The users in the zero-follower audience are active in engaging in influencers' activities but inactive in having social relationships with other users. We particularly analyze their engaging behavior in Sect. 3.4 since some of the users in the zero-follower audience are suspicious as 'bot accounts' [6]. The remaining audience excluding the zero-follower audience follows a normal distribution whose median value is 557 as shown in Fig. 1(b). Note that 46% of the engaged users have between 100 and 1,000 followers, and only 4% of the engaged users have more than 10,000 followers who can be also considered as influencers.

Amount of Engagements. We next examine the amount of engagements (i.e., weights in the network) for all unique pairs of influencers and users. The weight value refers to the total number of engagements, thus, high weights indicate consistent and loyal engaging relationships between audience and influencers. Figure 2 shows the weights in different colors where red represents higher weight values while purple represents lower weight values. Each dot in Fig. 2 indicates engagements between a user and an influencer. We find that micro-influencers who have less than 100,000 followers, in general, tend to have more active engagements with their audience than macro-influencers who are followed by millions of users. This suggests that audience of micro-influencers show more consistent and loyal engaging behavior than the audience of macro-influencers. We also find that influencers actively interact with other influencers through engagements. On one hand, most of the active interactions between influencers can be found from influencers less than a million followers. On the other hand, influencers with more than a million followers tend to receive engagements from normal users. This can be due to that influencers with common interests tend to have interactions [8] whereas macro-influencers, who are more likely the celebrities, are

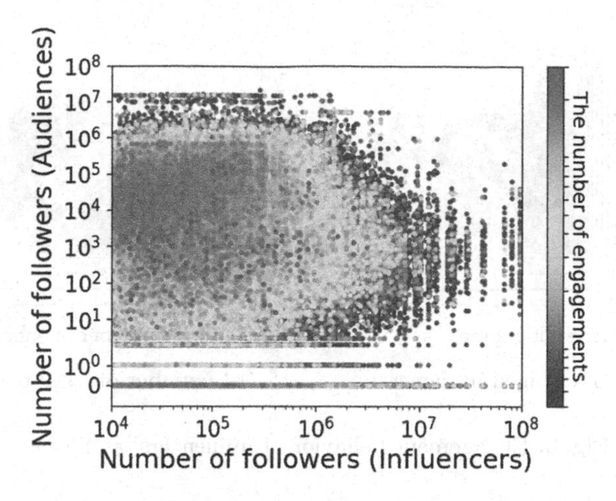

Fig. 2. Amount of engagements (i.e., weights in the given network) for all pairs of influencers and users.

likely to focus on their own activities. The result implies that advertising through micro-influencers can be more effective than hiring a few macro-influencers since advertising posts published by micro-influencers are further propagated to the social networks of engaged influencers. Additionally, we find some users in the zero-follower audience show a substantially large amount of engagements. This suspicious users with abnormal engagement behavior might be the social bots [6].

The Degree of Engagements. To understand how a user engages in multiple influencers, we analyze the node degree of users and their total number of engagements in Figs. 3(a) and 3(b), respectively. Since a user node has directional edges from itself to engaged influencer nodes, its out-degree represents the total number of influencers engaged by the corresponding user. As shown in Fig. 3(a), we find that users tend to have higher out-degrees if they have more followers. In other words, influencers are densely connected with other influencers in the engagement network, whereas normal users usually engage in activities of a relatively smaller number of influencers thereby having lower out-degrees. The same tendency can be found in Fig. 3(b). Despite only 4% of the total audiences have more than 10,000 followers as shown in Fig. 1, they account for 78% of engagement relationships with more than 100 engagements.

Figure 3 shows that there are a set of users with high out-degrees and a large number of engagements, but having no followers, unlike influencers. More specifically, this group of users has different engagement behavior compared to other users as they are not socially connected with other users but contribute a huge number of engagements to numerous influencers. This unique engagement behavior can be generated by social media bots [6]. To conduct an in-depth study on these suspicious user accounts, we set a threshold on the node out-degree as 300, and assume that the zero-follower audience whose node out-degree is higher

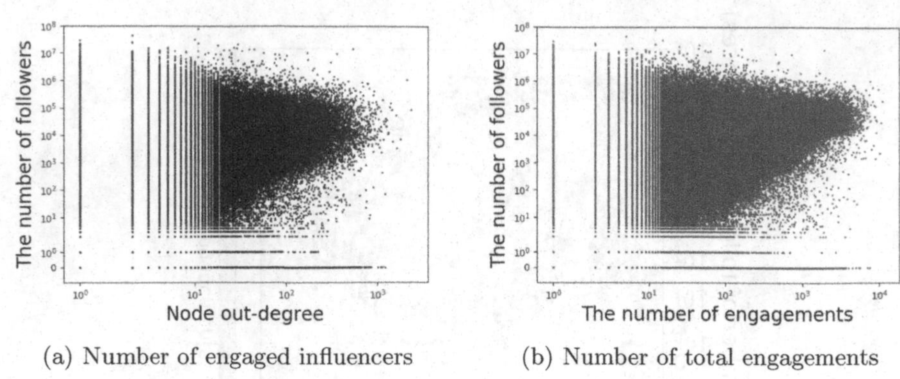

(a) Number of engaged influencers (b) Number of total engagements

Fig. 3. Engagement behavior of influencers' audience.

than 300 can be considered as bots. In our dataset, 206 users in the network are classified as (potential) bots.

3.4 Fake Engagement by Bots

Understanding the characteristics of the potential engagement bots can be useful in identifying influencers who are involved in fake engagements. To this end, we analyze distinct attributes of the potential bots in terms of their (i) local clustering coefficient and (ii) commenting behavior.

Local Clustering Coefficient. It is a widely observed property of social networks that a node's friends are often themselves friends in the network. This property is also observed in the influencer engagement network where a user engages in the influencers who themselves have interactions through engagements. To examine how normal audiences and potential bots show different engagement behavior, we calculate the local clustering coefficient of each node in the engagement network. Note that the local clustering coefficient of a node represents the ratio of friendship between the node's neighbors to all possible pairs of neighbors. Suppose user i engages in a set of influencers $N(i) = \{j|(i,j) \in E\}$ where E is the edge set in the engagement graph G. Then the local clustering coefficient of user i can be calculated as follows:

$$F(i) = \frac{|\{e_{jk} \in E | j,k \in N(i)\}|}{\{|N(i)| \times (|N(i)| - 1)\} \div 2}$$

Figure 4 shows the local clustering coefficients of the influencers, the normal audiences, and the potential bots, respectively. We first find that the potential bots tend to have lower clustering coefficients than normal users. Note that most of the potential bots have the clustering coefficients lower than 0.2. This is because normal users tend to follow or engage in a group of influencers with similar interests, hence having active social interactions among the influencers.

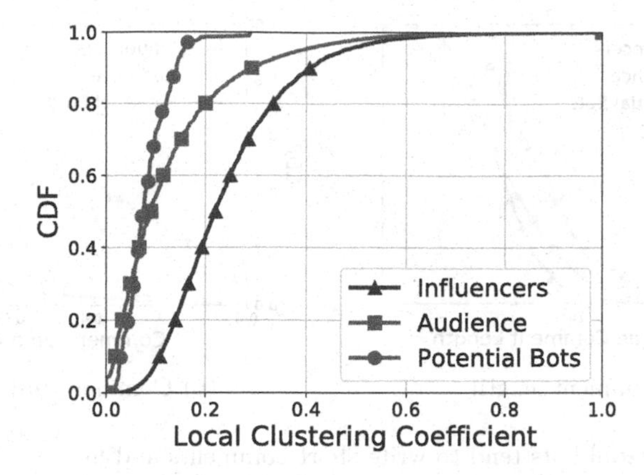

Fig. 4. The potential bots have relatively lower local clustering coefficient values than the normal audience of the influencers while the influencers tend to have higher clustering coefficients than the normal users.

On the other hand, the potential bots are likely to engage in random influencers thereby having little interactions between their neighbor nodes. We also find that the influencers have significantly higher clustering coefficients than other users. This confirms that influencers form clusters in the engagement network by liking or writing comments to posts published by other influencers. As reported in [8], influencers actively tend to interact with other influencers to get attention from followers of the other influencers and ultimately increase their followers. The active engagement behavior of the influencers also helps their posts to be exposed to a set of users who may like or comment on the same posts that the influencers already posted.

Commenting Behavior. To analyze distinct commenting behavior by potential bots, we next investigate all the comments on the influencers' posts. Note that we have 44,473,797 comments in our dataset and we discard influencers' self comments on their own posts in this analysis. We first analyze the length of the comments as shown in Fig. 5(a). We find that potential bots write short comments on the influencers' posts than other normal users. The average length of comments written by the influencers, the normal audience, and the potential bots are 27.0, 29.7, and 17.5, respectively. This implies that the potential bots write comments with simple expressions such as "`Nice picture!`" whereas normal users tend to have more information or their feelings on their comments, for example, "`Like your dress! Where did you buy it?`" thereby having longer comments than the bots. We also find that the influencers have smaller variance on the average comment length than the normal audience, although they have similar average values. This suggests that the influencers are not likely to use simple emojis or a single word on their comments, which are often used by the normal audience, to interact with other influencers.

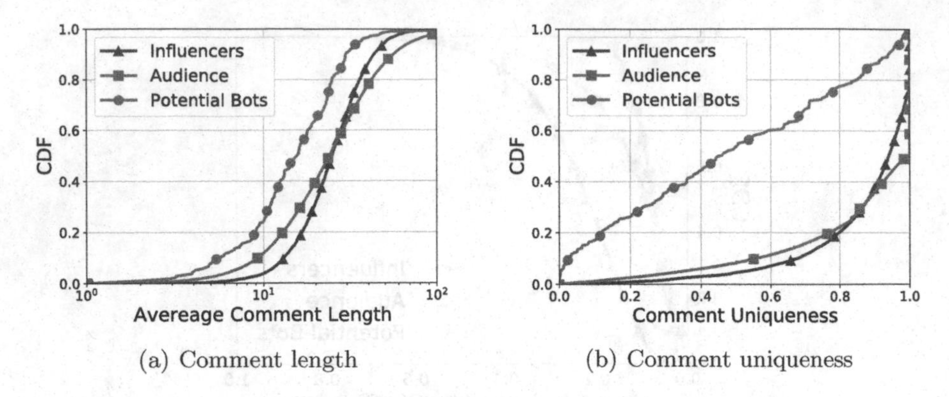

(a) Comment length (b) Comment uniqueness

Fig. 5. The potential bots tend to write short comments and have very low comment uniqueness since they use a set of pre-populated comments to automatically make engagements to other social media users.

We next investigate the uniqueness of the comments written by users and bots. Since social media bots usually use a set of pre-populated comments, the exact same comments by a single bot can be attached to the influencers' posts. The normal users, on the other hand, write comments by themselves, therefore comments are likely to be different from each other. Suppose user i writes a set of comments $C(i) = \{c_n\}_{n=1}^{|C^i|}$, where $|C^i|$ indicates the number of comments written by user i. We then define the comment uniqueness of the user $H(i)$, the rate of the number of unique comments to the total number of comments, as follows:

$$H(i) = 1 - \frac{|\{c_x \in C^i | c_x = c_y, c_y \in C^i\}|}{|C^i|}$$

For example, a user has the comment uniqueness as 1 if all of the comments written by the user are different from each other. If a user uses the same set of comments repeatedly then the user has the comment uniqueness as 0. Figure 5(b) shows the distributions of the comment uniqueness values for the influencer, the normal audience, and the potential bots, respectively. We find that the potential bots have significantly lower comment uniqueness than the normal users. Note that the average comment uniqueness of the influencers, the normal audience, and the potential bots are 0.88, 0.87, and 0.47, respectively. This result confirms that the potential bots engage in influencers by using the same set of comments rather than creating a new comment for each post.

4 Engagement Bot Detection

In this section, we propose a deep learning model that can detect engagement bots from a numerous audience who engages in the posts of influencers.

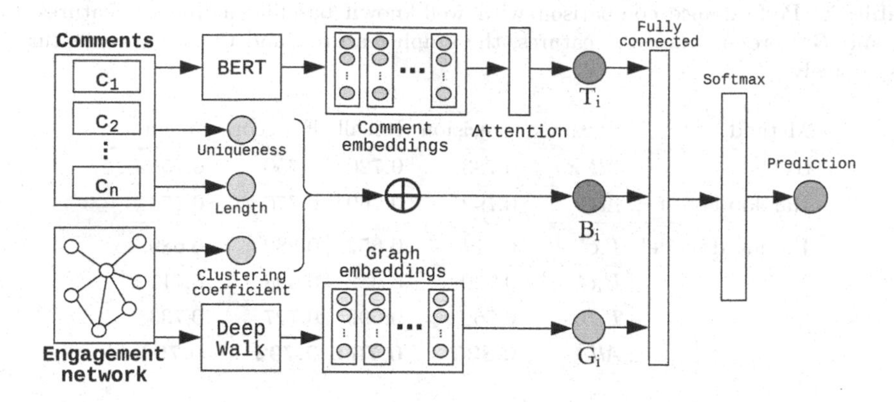

Fig. 6. The overall architecture of the proposed model for detecting engagement bots.

4.1 Model Description

Model Architecture. Figure 6 illustrates the overall architecture of the proposed model for detecting potential engagement bots. The aim of the model is to detect engagement bots from audience data by learning their social relationships and engagement behaviors. The proposed model utilizes three sets of input features including text features, behavior features, and graph features.

To model contextualized information from comments written by the audience, we apply the pre-trained neural language model, BERT [5]. The BERT takes all comments of a user i, $C(i) = \{c_1, c_2, \cdots, c_n\}$, and generates the comment embedding whose dimension is 768. Note that we set the sequence length of an input comment as 100 since most comments have their lengths shorter than 100. We then apply the attention mechanism [2] to give more weights on more important comments to represent each user. The attention layer projects the comment embeddings into a hidden space and estimates the importance score of each comment. We finally obtain the text features, T_i, by using a weighted combination of comment embeddings and corresponding scores. We set the dimension of the text features as 128.

The proposed model also exploits the behavior features, B_i, to capture distinct engagement behavior of bots from the normal audience. The behavior features contain the three engagement properties, the comment uniqueness, the average comment length, and the local clustering coefficient, analyzed in Sect. 3.4. We compute the property values of each user by taking user comments and the structural information of the engagement network, then concatenate the values to generate the behavior features.

In addition to the text and behavior features, we also encode the influencer engagement network to learn social relationships based on the user engagements. To generate a graph representation, G_i, we use DeepWalk [12] that employs short random walks from nodes in the network. Note that we set the number of walks,

Table 1. Performance comparison with well-known baseline methods. Features T, G, and B represent the text features, the graph features, and the behavior features, respectively.

Method	Features	Precision	Recall	F-1 score	Accuracy
SVC	*All*	0.783	0.720	0.750	0.755
RandomForest	*All*	0.792	0.760	0.776	0.776
Proposed model	T,G	0.714	0.652	0.682	0.689
	B,G	0.750	0.652	0.698	0.711
	T,B	0.762	0.696	0.727	0.733
	All	**0.826**	**0.760**	**0.792**	**0.796**

the length of the random walk, and the dimension of the graph embedding as 10, 20, and 64, respectively.

We concatenate the three sets (text, behavior, and graph) of each user representation, and add a fully connected layer with the Rectified Linear Unit (ReLU) as the activation function. The dimension of the hidden layer is set to 128. Finally, we utilize the cross-entropy for the loss function to detect the engagement bots.

Training Procedure. Since the number of users who are considered as potential bots is highly imbalanced to the total number of audience in our data, we use both oversampling and undersampling techniques to have balanced training data. More specifically, we first apply SMOTE [3] to the 206 bot accounts to make 10,000 samples. We then undersample normal users to have the same number of samples with the bot accounts. After making the balanced dataset, we split our data into training and testing sets with 8:2 ratio. We set the number of epochs and the learning rate as 100 and 0.0001, respectively. Note that we do not use follower counts of the users in the engagement network, which are directly implicated for identifying bot accounts, to prevent information leak in the training procedure.

4.2 Experiment Results

Table 1 shows the performances of the proposed model and two well-known baseline methods including the Support Vector Classifier (SVC) and the Random forest. To evaluate the performance, we use the precision, recall, F-1 score, and accuracy as evaluation metrics. As shown in Table 1, we find that our model outperforms the baseline models that use the same sets of features with the proposed model. This suggests that the proposed neural network is more effective in capturing useful features for detecting possible engagement bots than others.

In addition to the performance comparison with the baseline methods, we also conduct experiments with different sets of input features to analyze the importance of the different input features. We measure the performance of the model by excluding one set of features from the text, behavior, and graph features, respectively. The results show that the performances of the proposed model without one set of features are lower than the baseline methods. This implies that all three sets of features efficiently represent the distinct characteristics of the engagement bots from normal users. We find that the behavior features are more effective for detecting the engagement bots than the text and graph features. Note that the performance losses are observed in the F-1 score of the behavior, text, and graph features as 13.9%, 11.9%, and 8.1%, respectively. This reveals that the comment uniqueness, the average comments length, and the clustering coefficient values in the behavior features are important predictors in capturing the engagement behavior of bots. Moreover, the text features, which are the attentive comment embeddings, also can exhibit distinctive commenting patterns of the engagement bots. The graph features, on the other hand, show relatively less performance loss than the other features. That is probably because a few normal users tend to have high out-degrees thereby having similar graph representations with the engagement bots.

5 Conclusion

In this paper, we conducted the empirical study on the engagement behavior of influencers' audiences. We proposed the influencer engagement network based on liking and commenting behavior on the posts of the influencers in social media. We summarize the findings of our analysis on the constructed influencer engagement network as follows: (i) 80% of audiences have less than 1,000 followers and 20% of audiences have zero followers, and (ii) only 4% of audiences have more than 10,000 followers but they constantly make interactions with other influencers thereby having more engagements than normal audiences. We then performed an in-depth analysis on the engagement behavior of bot accounts that generate fake likes and comments on the posts of influencers. We found that the engagement bots tend to have (i) lower local clustering coefficients, (ii) shorter average comment lengths, and (iii) lower comment uniqueness than normal users. Based on lessons learned, we proposed the neural network model that can detect engagement bots by learning the text, behavior, and graph representations of users. The proposed model achieved about 80% accuracy, which outperforms other well-known baseline methods. As a future work, we plan to apply the graph convolutional networks (GCNs) to represent the influencer engagement network. Applying the GCNs can further enhance the model performance by learning the structural information and the node features together in detecting engagement bots. To further improve the model performance in detecting bots, we plan to use temporal features such as posting time of influencers and engagement time of users in the model.

Acknowledgements. This research was supported by Basic Science Research Program through the National Research Foundation of Korea (NRF) funded by the Ministry of Education (NRF-2018R1D1A1A02085647).

References

1. Arenas-Marquez, F.J., Martínez-Torres, M.R., Toral, S.: Electronic word-of-mouth communities from the perspective of social network analysis. Technol. Anal. Strateg. Manag. **26**(8), 927–942 (2014)
2. Bahdanau, D., Cho, K., Bengio, Y.: Neural machine translation by jointly learning to align and translate. arXiv preprint arXiv:1409.0473 (2014)
3. Chawla, N.V., Bowyer, K.W., Hall, L.O., Kegelmeyer, W.P.: Smote: synthetic minority over-sampling technique. J. Artif. Intell. Res. **16**, 321–357 (2002)
4. De Veirman, M., Cauberghe, V., Hudders, L.: Marketing through instagram influencers: the impact of number of followers and product divergence on brand attitude. Int. J. Advert. **36**(5), 798–828 (2017)
5. Devlin, J., Chang, M.W., Lee, K., Toutanova, K.: Bert: pre-training of deep bidirectional transformers for language understanding. arXiv preprint arXiv:1810.04805 (2018)
6. Ferrara, E., Varol, O., Davis, C., Menczer, F., Flammini, A.: The rise of social bots. Commun. ACM **59**(7), 96–104 (2016)
7. Hoffman, D.L., Fodor, M.: Can you measure the ROI of your social media marketing? MIT Sloan Manag. Rev. **52**(1), 41 (2010)
8. Kim, S., Han, J., Yoo, S., Gerla, M.: How are social influencers connected in instagram? In: Ciampaglia, G.L., Mashhadi, A., Yasseri, T. (eds.) SocInfo 2017. LNCS, vol. 10540, pp. 257–264. Springer, Cham (2017). https://doi.org/10.1007/978-3-319-67256-4_20
9. Kim, S., Jiang, J.Y., Nakada, M., Han, J., Wang, W.: Multimodal post attentive profiling for influencer marketing. In: Proceedings of The Web Conference 2020, pp. 2878–2884 (2020)
10. Kudugunta, S., Ferrara, E.: Deep neural networks for bot detection. Inf. Sci. **467**, 312–322 (2018)
11. Marciano, J.: The real economic losses from influencer fraud. Marketing Technology Insights (2019)
12. Perozzi, B., Al-Rfou, R., Skiena, S.: Deepwalk: online learning of social representations. In: Proceedings of the 20th ACM SIGKDD International Conference on Knowledge Discovery and Data Mining, pp. 701–710 (2014)
13. Rios, S.A., Aguilera, F., Nuñez-Gonzalez, J.D., Graña, M.: Semantically enhanced network analysis for influencer identification in online social networks. Neurocomputing **326**, 71–81 (2019)
14. Schuchard, R., Crooks, A., Stefanidis, A., Croitoru, A.: Bots in nets: empirical comparative analysis of bot evidence in social networks. In: Aiello, L.M., Cherifi, C., Cherifi, H., Lambiotte, R., Lió, P., Rocha, L.M. (eds.) COMPLEX NETWORKS 2018. SCI, vol. 813, pp. 424–436. Springer, Cham (2019). https://doi.org/10.1007/978-3-030-05414-4_34
15. Varol, O., Ferrara, E., Davis, C.A., Menczer, F., Flammini, A.: Online human-bot interactions: detection, estimation, and characterization. In: Eleventh International AAAI Conference on Web and Social Media (2017)
16. Wang, G., Konolige, T., Wilson, C., Wang, X., Zheng, H., Zhao, B.Y.: You are how you click: clickstream analysis for sybil detection. In: Presented as part of the 22nd {USENIX} Security Symposium ({USENIX} Security 2013), pp. 241–256 (2013)
17. Yang, X., Kim, S., Sun, Y.: How do influencers mention brands in social media? Sponsorship prediction of instagram posts. In: Proceedings of the 2019 IEEE/ACM International Conference on Advances in Social Networks Analysis and Mining, pp. 101–104 (2019)

Social Capital as Engagement and Belief Revision

Gaurav Koley, Jayati Deshmukh, and Srinath Srinivasa(✉)

International Institute of Information Technology, Bangalore, India
sri@iiitb.ac.in

Abstract. Social Capital or "goodwill" is an essential ingredient of any collective activity – be it commercial, cultural or administrative activity. In online environments, several models have been pursued for recording and utilizing social capital based on signals including likes or upvotes. Such explicitly stated signals are susceptible to impulsive behavior and hyperinflation. In this paper, we develop an implicit model for social capital based on the extent of *engagement* generated by any participant's activities, and the way this engagement leads to a *belief revision* about the participant from other members of the community. Two kinds of social capital measures are proposed: an *authority* score that indicates engagement, and a *citizenship* score that calibrates value-addition made by a user as a result of engaging with others' content. The proposed model is implemented in two online communities showing different kinds of content authorities, supported by a strong community of engaged citizens.

Keywords: Social capital · Engagement · Belief revision · Social networks

1 Introduction

Social capital or "goodwill" is an important element of social interactions. Conventionally, social transactions were analyzed from the lens of rational utility functions. However, rational utility is itself known to be affected by cognitive factors like trust, reputation, goodwill, aversion to risk, etc. [2,14,15].

In order to account for such factors, online social spaces typically have some form of a reputation or rating model based on explicitly elicited signals like star ratings, upvotes/downvotes, karma points, user endorsements, etc. The underlying paradigm for interpreting these signals, is based on the assumption of *cross-rating* where, stakeholders with conflicting objectives, balance out biases in the ratings on an aggregate level [12]. Sometimes, the base signals are further used as input for some form of spreading activation algorithm to diffuse and aggregate the scores across the population.

However, there are some recurring challenges with explicitly stated signals. Explicit signal models would typically result in a dearth of representative data,

S. Aref et al. (Eds.): SocInfo 2020, LNCS 12467, pp. 137–151, 2020.
https://doi.org/10.1007/978-3-030-60975-7_11

because many users may simply ignore providing a rating. In addition, users tend to speak up with bad ratings and other forms of impulsive responses to bad experiences; and tend to take good experiences for granted, resulting in an overall bias in the rating. Other forms of racial, gender and ethnic biases may also influence explicit rating mechanisms. There is also a danger of "mob rating" by groups of disgruntled participants, or even by a posse of fans, that can greatly skew actual signals from that of the population.

There is hence, a need for computational modeling of goodwill, that is based on implicit evidence and better theoretical underpinnings of what constitutes a signal for social capital.

It is generally acknowledged that social capital can be seen as a function of positive engagement [5,22,26]. People having goodwill towards one another, tend to be "strongly engaged"– they easily disclose information to one another, entrust one another with something of value, and even develop a shared sense of identity.

Engagement can be of a variety of forms [18,23,27]. It represents people investing something of value to them, in the work of somebody else. Online social networks typically have their own custom signals to indicate engagement. For instance, engagement on Twitter is often measured using metrics like number of mentions, retweets and likes.

However, to build our theory of social capital, we first develop an application agnostic model of engagement based on the extent of *sustained attention* generated by any user's activities. Sustained attention indicates a recurrent and immersive form of interest. This becomes a precursor for other forms of engagement. Sustained attention may be contrasted with *cursory attention*, which may only elicit impulsive responses.

Based on this model of engagement, we use Degroot Belief revision to propose two new network metrics: Authority Rank and Citizen Rank. *Authority Rank* shows the relative command of an actor over its network for a topic. This can be used to find influencers in a network. *Citizen Rank* ranks actors in a network based on their engagement or interaction characteristics. This can be used to find highly engaged users in a network.

2 Related Work

Several scholars have proposed the use of social network analysis to study the social capital of groups and individuals; to emphasise the value of relationships and networks to maintain social capital [7,13].

The individualist approach explores the idea that an actor's network or their position within the network affects their social capital. Individuals benefit from the following structural features of their networks: size of an actor's ego network, where small networks with strong ties provide material and emotional support and large networks with weak ties provide access to information and resources [16,25]; and actor's position within the network, where having a broker position leads to ability to access a greater diversity of resources than others [6,7].

The groupist perspective focuses on a network's structural features and how the rise and maintenance of reciprocity and trust is enabled by that network structure. Coleman [8] posits that trust and the feeling of mutual obligation is higher among members of a complete network where all the actors are connected to each other.

Empirical research has shown social capital to correlate with increased perceived credibility of the flow of information within a network and people trusting each other on a long-term basis [7]. In addition, Granovetter [9] notes that this closed network structure persuades friends to behave honestly amongst themselves.

In recent times, researchers have looked at Sen's Capability Approach [20,21] to explain certain properties exhibited by social capital. Sen calls social capital as an endowment – or, "a set of means to achieve a life people reason to value".

The potential positive effects of social capital can be seen through the influence of social connections. Friends and family helps individuals in lots of ways – economically, socially and emotionally. Several social scientists including Granovetter [9] emphasise the role of close friends and family, as well as casual acquaintances, in finding jobs. Burt [6] focuses on the lack of close ties as a motivator for knowledge sharing and mobility of individuals. Mwangi and Ouma [17] show that social capital increases the network reach of individuals and thereby enhances financial inclusion through increased access to informal loans in Kenya. Social capital encourages social trust and membership, reduces health risks amongst children and adolescents [28] and discourages individuals from engaging in harmful activities like smoking and binge drinking [4]. Social capital is also linked with greater well-being according to self-reported survey measures [10] and reduced crime [1]. Social capital also helps businesses – for example in Bowling Alone [19], Putnam observes that the formal and informal cooperation between startup companies in the Silicon Valley has led to their collective success.

However highly entrenched networks can hinder people as well, despite having high extents of social capital. Close knit communities usually have strong social bonds, with the individuals relying heavily on relatives and others of the same community for support. The lack of social bridges that can connect them to the wider society, can also turn them into outsiders and hinder their social development and economic upliftment [10]. Social capital can also be put to harmful use as well. The trust and reciprocity that allows close-knit networks like mafia, criminal gangs and cults to operate is also a form of social capital [10].

3 Approach and Model

In this work, our focus is on building a computational model for measuring social capital. As noted earlier, a common aspect in the disparate theories of social capital surveyed, is the element of engagement generated by a person's activities. Engagement in turn, leads to other elements like reciprocity, trust, credibility, etc. We define engagement itself, in terms of consistent receipt of an abstract concept called "stroke"– that refers to positive affinity received from others.

3.1 Engagement and Strokes

Engagement manifests with a *creator* who creates resources, and one or more *consumers* who consume the resources. A specific instance of consumption by consumer u engaging with a resource created by creator v is called a *stroke* from u to v. The term "stroke" is borrowed from Transaction Analysis in psychology [3], which in our case, represents an episode of psychological engagement from the recipient to the creator of a resource.

Each element of a stroke is formally represented as follows:

$$s = \langle\, u,\ v,\ t,\ \epsilon_{u \to v}\,\rangle$$

where u is the stroke initiator, v is the stroke receiver, t is the time of stroke and $\epsilon_{u \to v}$ is the amount of engagement in the stroke. In different scenarios, the act of engagement or stroke refers to different kind of social actions, e.g., reading a research paper in scholarly networks and (optionally) subsequently citing it, or responding to a tweet on Twitter. The action represents a directional influence and we consider that there is an attention flow from the consumer to the creator.

The model is heavily dependent on what is meant by engagement $\epsilon_{u \to v}$ and how it is measured. Any measurable resource can be used in our model that has following properties:

Limited: An agent should have only a finite quantity of resources that characterize the given mode of engagement.

Conserved: The resource should also be *conserved* i.e. once the measurable resource is allocated to someone as engagement, the same cannot be assigned to anyone else.

Renewable: The available resource should be renewed after some fixed duration for all users.

There are several examples of resources which have the above properties and can be used as a measure of engagement. One of the simplest is *time spent.* The amount of time an actor spends engaging with another actor through their tweets, posts, etc. can be tracked and used as a measure of engagement since the amount of time an agent can spend is limited in a day and is also renewed everyday. Another similar measure on Twitter can be the number of retweets as a user can only post a maximum of 2400 tweets including retweets and this limit is reset everyday.

The amount of engagement $\epsilon_{u \to v}$ is typically a function of *contiguous* expending of scarce resources. In the case of time spent, engagement is a function of contiguous units of time that was spent, leading to *sustained attention* by the attention-giver. A user spending n units of time contiguously on a resource is said to be more engaged than a user spending n units of time on a resource in several smaller chunks.

The above nature of engagement is also corroborated by studies from prospect theory, specifically the cognitive models of System 1 and System 2 [24]. Engagement is a characteristic feature of System 2 cognition, that indulges in deliberative and conscious reasoning. Merely glancing through the content and/or reacting impulsively to a post, cannot be considered as engagement. Engagement requires deeper cognitive involvement of the attention-giver. Hence, engagement also results in saturation, fatigue, or diminishing returns, when sustained for too long a time.

Given these, we characterize the engagement function as an S-shaped sigmoid curve, on the extent of contiguous time spent. This is elaborated in the subsections on computing Authority and Citizenship ranks.

A Social Network can now be represented as a directed graph depicting engagement pathways between users. A social graph can be formulated as a weighted graph (V, E, w), where V is the set of users and E is the pair-wise engagement between users. The weight on an edge (u, v) is the total engagement obtained from u to v across all strokes from u to v.

3.2 Belief Revision

Strokes and engagement between pairs of actors, changes the nature of their relationship. A strong engagement from one actor to another would result in greater familiarity, influence, affinity, trust, and other possibilities. This change in the nature of relationship between any two actors also diffuses through the network, affecting to different extents, others' beliefs about these actors. We model this dynamic as a social learning process using the DeGroot Belief Revision model [11]. Let us take a social network of n agents where everybody has an opinion or belief about other actors at any given time t. In the simplest sense, suppose we are considering a specific topic or interest p, this belief is represented by a probability representing the belief holder's confidence about the other actor, regarding interest p. A belief $b^p_{u \mapsto v}$ represents the goodwill or credibility that u has towards v, regarding topic p. Unless u is directly stroking v, u does not receive any new information to change its opinions. But, actors can interact and communicate with their neighbors due to which, beliefs diffuse through the network. Each acquaintance link between actors is represented as a belief matrix T, where $T_{i,j}$ represents the belief actor i has about actor j. The belief matrix is also modeled as a row-stochastic matrix, where the belief values are all non-negative, and the belief values by one actor about all other actors in the network, add up to 1.

As engagements change over time due to strokes between interacting actors, belief values changes across the population. Let b^t_i represent column i indicating incoming belief values for actor i from all other actors, at time t. Belief revision for actor i happens as follows:

$$b_i^{t+1} = T \cdot b_i^t \tag{1}$$

Unravelling the recursion, the tth period opinions can be computed by

$$b_i^t = T^t \times b_i^0 \quad (2)$$

Since we assume that every actor has a belief about every other actor for any given topic, the network represented by T is fully connected, aperiodic and irreducible. This makes the stationary distribution of T independent of the initial distributions of belief vectors and as t becomes very large, all columns converge to the same belief vector **b**.

$$\mathbf{b} = \lim_{t \to \infty} T^t \times b_i^0 \quad (3)$$

We use the above DeGroot Belief Revision model to compute two kinds of scores: Authority Rank and Citizen Rank for any actor.

3.3 Authority and Citizen Rank

Authority Rank is a measure of the social impact of an actor in the social network. This can be interpreted as the network's belief about an actor's credibility about a given topic. For example, AR(5) = 0.09 can be understood as the network's belief that the social capital of actor 5 extends up to 9% of the network.

We say that an actor v's influence on actor u depends on the relative slice of attention paid by u to v. For example, let us consider actor A who spends 4 hours engaging equally with actor C and 3 other actors on a social networking site and actor B who engages with only actor C for 1 hour. Although, the time (engagement measure in this scenario) spent by A and B on C is the same, A will be less influenced by C as compared to B.

We construct the belief matrix T from the engagement strokes as defined earlier. Relative engagement is computed as per Eq. 4 where the values are normalized such that $T_{i,j} \mapsto [0, 1]$.

$$T_{i,j} = \frac{w(i,j)}{\sum_{\forall k, i \to k} w(i,k)} \quad (4)$$

We want to reduce the influence of low engagement and dampen the effect of extremely high engagement as well. This is due to the law of diminishing returns, as well as entrenchment effects arising due to very high engagement. For someone who is already highly engaged, any higher engagement doesn't lead to a linear increase in influence. Therefore, we use a softmax function to dampen lower engagement values and skew the engagement to give more weight to higher engagement values using Eq. 5, that gives us the authority matrix.

$$A_{i,j} = \begin{cases} \frac{\exp(\lambda \cdot T_{i,j})}{\sum_{k \neq i} \exp(\lambda \cdot T_{i,k})} & \text{if } i \neq j \\ 0 & \text{if } i = j \end{cases} \quad (5)$$

Here $A_{i,j} \mapsto [0, 1]$ and λ is a scaling parameter that is representative of the level of engagement required to judge another actor's worth as an authority. λ therefore depends on the measure of engagement being used as well as the network structure. Higher the median value of engagement for a kind of engagement measure, higher should be the λ. For example, if time is being used a measure of engagement, for a social network of videos, average time of engagement will be in 10s of minutes while for a twitter like network it will be lesser. Therefore, λ will be smaller for a twitter like network as compared to a video network.

Thus A represents a strongly connected, aperiodic and irreducible graph, therefore, a convergence point for the network's belief exists.

$$A^* = \lim_{i \to \infty} A^i \tag{6}$$

and the left diagonal of A^* gives us the Authority Rank (AR) for the social network as:

$$AR(i) = A^* \times \mathbf{1} \tag{7}$$

Here, A is a non-Ergodic system, thus the initial state of A determines the stable point of the system at the end of Eq. 6. Therefore, by varying λ, different results could be found for AR.

To find the ideal λ, we search the space $[1, \infty)$ and settle for a value of λ for which subsequent increase in value doesn't change the AR values.

We pick the λ for which Authority Ranks stabilize. This stable value of λ changes across different social networks and different underlying notions of engagement. For a network, the value at which λ stabilizes also indicates the depth at which people prefer engaging with others on the network. A small value of stable λ represents that actors prefer a number of connections with cursory engagement, over deep engagement. A large value of stable λ means that actors prefer engaging deeply than merely interacting with lots of other actors. The stable value for λ hence represents a characteristic feature of the network dynamics.

Citizen Rank is a measure of the perceived value accrued due to an actor's participation in the social network. It can be interpreted as the network's belief about the actor's contributions to preserving and improving the network. For example, CR(5) = 0.09 can be understood as the network's belief that actor 5 is a stakeholder of the network with their stake being worth 9%.

The Citizen Rank is computed in an analogous fashion to that of Authority Rank, but with some differences. We start with the T matrix that represents belief values. We define participation matrix as follows:

$$p_{i,j} = \frac{w(j,i)}{\sum\limits_{\forall k, i \to k} w(k,i)} \tag{8}$$

Here $p_{i,j}$ is the participation or engagement i has received from j. We run it through a similar softmax scaling treatment to give higher weightage to high engagement as follows:

$$C_{i,j} = \begin{cases} \frac{\exp(\beta \cdot p_{i,j})}{\sum_{k \neq i} \exp(\beta \cdot p_{i,k})} & \text{if } i \neq j \\ 0 & \text{if } i = j \end{cases} \tag{9}$$

Although both A and C matrices look similar, there are some fundamental differences between the interpretation of the scores. The value $A_{i,j}$ represents the total proportion of engagement given by i that was directed at j. In contrast, the value $C_{i,j}$ represents the engagement given by i to j, as a proportion of the total engagement received by j from all others. Hence, i may have given all its strokes to j, but if j were to be receiving a lot of strokes from several others, its citizenship contribution would still be low, while its contribution to the authority of i would be high.

As with the authority equation, the term β is the scaling parameter that represents point of valuable participation through engagement. It controls the point at which the creator believes that the reader is engaged. β therefore depends on the measure of engagement being used, the kind of content on the network and the network structure. Similar to λ, a higher median value of engagement should reflect a higher β. Thus β would be smaller for a Twitter-like network than for a video network.

Similar to A, C is also strongly connected, aperiodic and irreducible, therefore, a convergence point for the network's belief exists at:

$$C^* = \lim_{i \to \infty} C^i \tag{10}$$

and the left diagonal of C^* gives us the Citizen Rank (CR) for the social network as:

$$CR(i) = C^* \times \mathbf{1} \tag{11}$$

Here, C, like A, describes a non-Ergodic system. Therefore, we search the space $[1, \infty)$ and settle for a value of β for which subsequent increase in value doesn't change the CR values. This stable value of β changes across different social networks and different measures of engagement. For a network, the stable point of β value indicates whether for the community the level of engagement is valuable compared to the number of interactions. A small β represents that more interactions are better for network maintenance over deep engagement. A large stable value of β means that rich meaningful, highly engaging interactions are preferred.

The complete algorithm for computing Authority Rank and Citizen Rank from engagement values along with the ideal λ and β values is described in Algorithm 1 in the Appendix section.

4 Experiments

Our measure of engagement is based on the amount of time a consumer spends on a creator either reading, viewing or responding to their works. We design several experiments to capture user engagement in different online social networks and use this data to show the value derived from using the AR and CR metrics. We also compare our engagement model across different networks as well as with other importance metrics like PageRank.

4.1 Gratia

The proposed model were tested on two online social environments, to which we had access. The first was called Gratia, which is an academic pre-print management system[1], where users can share pre-prints of their papers as well as read pre-prints of others' technical reports and papers, in an online reader. We capture the amount of time spent by the users viewing each paper, which gives us an estimate of the sustained attention and hence, the engagement received by the said paper. We have designed mechanisms to attribute the attention received by papers to its creators.

The portal revolves around resources in the form of PDF documents. Each resource has one or more "creators" who are the owners of that resource and any engagement with that resource will be attributed to them.

Each resource is also indexed with one or more topical tags representing its contents. When users view resources, the portal captures the amount of time users spend on engaging with a resource. This amount of time is then attributed to the creators of that resource by organising them as strokes as defined earlier. In the experimental models, the total amount of engagement received by a resource, was divided equally among all its creators.

The time spent is tracked on the server side, with restrictions in place to ensure that a particular user can have only one resource open at a time. It is also ensured that the amount of time being counted towards engagement is only when the user is actively interacting with the open resource. This prevents the system from being gamed by false engagement attacks.

For the experiments, the data was collected from the activities of 68 users. A total of 1534 min of time were spent by users, collectively, engaging with 15 users across 23 resources leading to a graph made up of 253 strokes.

4.2 CircuitVerse

The second online portal where the proposed model was tested, is called CircuitVerse[2]. CircuitVerse is an easy to use digital logic circuit simulator which provides a platform to create, share and learn digital circuits.

For our data collection procedure, the same backend and data model as for Gratia was used. For the experiments, data was collected over a sample of 711 users. This led to 1540 strokes across 554 resources. A total of 343 users spent 6928 min engaging with 456 users.

4.3 Twitter

The proposed model was also tested on Twitter. Here, tweets were scraped for a particular hashtag and then for each twitter handle, a retweet or mention was taken as 1 stroke. For the experiments, data was collected over a sample of 2251 users. This dataset consisted of 6278 strokes over 6873 sampled tweets.

---

[1] https://goodwill.zense.co.in.
[2] https://circuitverse.org/.

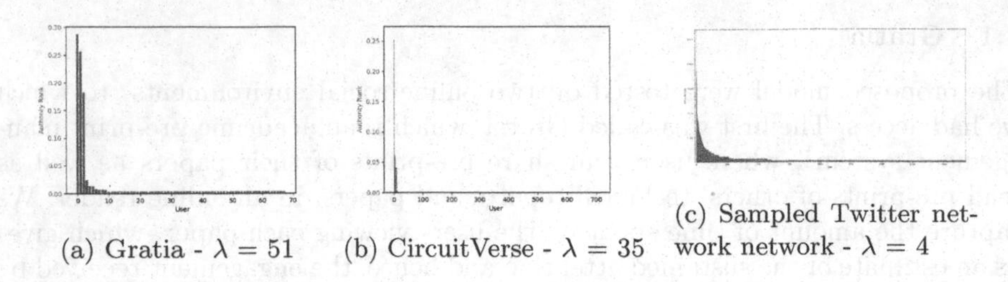

(a) Gratia - $\lambda = 51$ (b) CircuitVerse - $\lambda = 35$ (c) Sampled Twitter network network - $\lambda = 4$

Fig. 1. Authority Ranks for networks, sorted in decreasing order along with its stable λ value

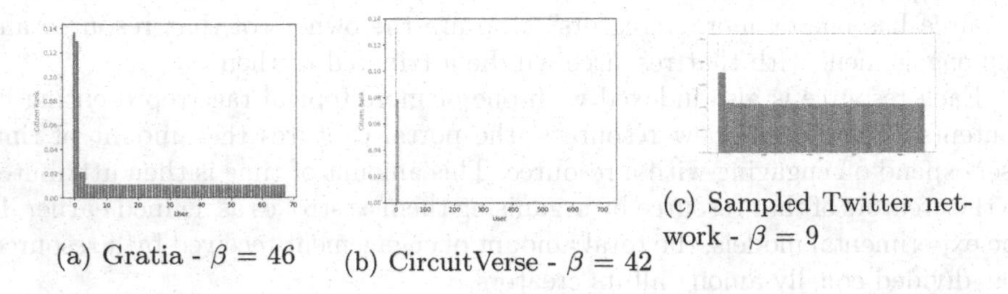

(a) Gratia - $\beta = 46$ (b) CircuitVerse - $\beta = 42$ (c) Sampled Twitter network - $\beta = 9$

Fig. 2. Citizen Ranks for networks, sorted in decreasing order along with its stable β value

5 Results

Figure 1 shows the plots of Authority Rank across the different social networks. We observe that CircuitVerse has a much more skewed distribution of AR values than Gratia, while Gratia took much longer for its λ value to stabilize. This indicates that the Gratia community tended to spend more time on its resources, than CircuitVerse. The skewed nature of CircuitVerse indicates the formation of few authorities or participants who generated a high impact across the community.

Figure 2 shows the plots of Citizen Rank for the networks. Even here, CircuitVerse has a much more skewed distribution of CR values than Gratia.

Table 1. Comparing λ across networks

Network	Avg time per stroke	Stable λ
Gratia	6.06	51
CircuitVerse	4.49	35
Twitter	–	4

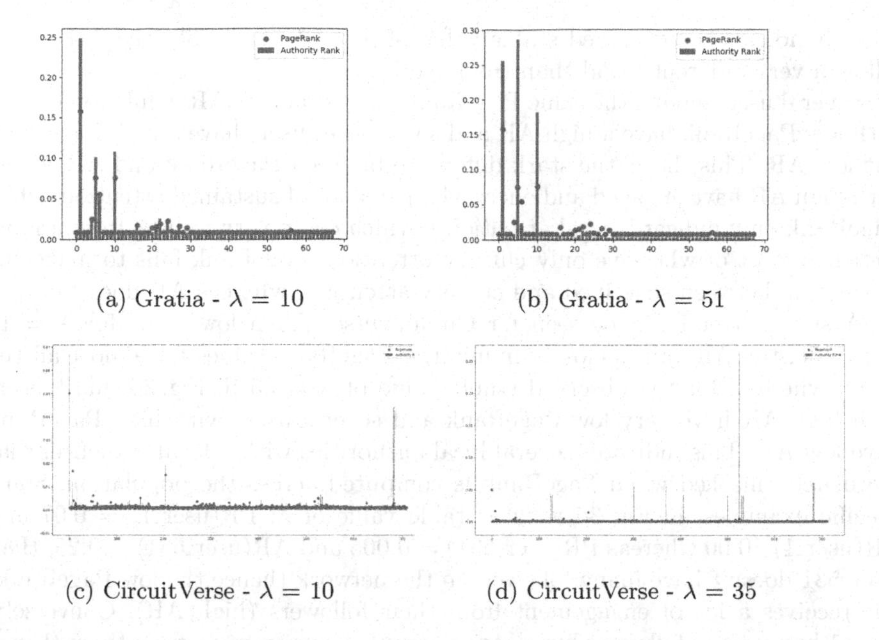

Fig. 3. PageRank and Authority Ranks comparison with low and stable λ values

5.1 Comparing Across Networks

Hypothesis H1: *Average value of engagement is correlated to stable λ value across different networks.*

Table 1 shows the correlation seen between average time per stroke and the number of iterations it took for λ to stabilize. Gratia has a higher average time per stroke/interaction and also has the highest λ value. Comparatively, CircuitVerse has a lower average time and also a much lower stable λ value. This suggests that Gratia's users are more engaged with the other user's they interact with than CircuitVerse's users.

Inference: *The stabilized values of λ can be seen as an indicator of the levels of engagement present in the population.*

5.2 Comparing with PageRank

Hypothesis H2: *Authority Ranks closely mimic PageRank for λ values which are much smaller than the stable λ, but diverge from PageRank as the number of iterations increase.*

Figures 3a and 3b compare the PageRank for a user to their Authority Rank for the network of users of Gratia. When we supply a low value for $\lambda = 10$ in Fig. 3a, the AR values closely mimic the PageRank values for all the users. The exact values might be different but the relative distribution is similar. On the other hand, for the observed stable value of $\lambda = 51$ in Fig. 3b, the AR values follow a very different trend than the PageRank.

Several users having the same PageRank, have different AR; while some users with low PageRank have a high AR and some other users have a high PageRank but low AR. This shows the stark difference between PageRank and AR. Users with high AR have engaged audiences who pay a lot of sustained attention, while PageRank only indicates total attention – which could very well be from a large audience most of who give only cursory attention. PageRank fails to make the distinction between sustained and cursory attention, whereas AR does.

A similar trend can be seen for CircuitVerse. For a low value for $\lambda = 10$ in Fig. 3c, the AR values closely mimic the PageRank values for almost all the users, whereas, for the observed stable value of $\lambda = 35$ in Fig. 3d, most users with high AR have very low PageRank and several users with high PageRank have low AR. This indicates several local authorities with a loyal audience, that are usually masked when PageRank is computed across the population. Some specific examples, in Fig. 3d with a stable value of λ: PR(user_1) = 0.07 and AR(user_1) 0.00 whereas PR(user_531) = 0.005 and AR(user_531) = 0.25. Here User_531 doesn't have many followers in the network (hence the low PageRank) but receives a lot of engagement from their followers (high AR). Conversely, User_1 has a many followers but doesn't get much engagement from them (hence the low AR but high PageRank value).

Inference: *Hence, while AR starts off similar to PageRank, it diverges from it to reveal significantly different semantics, by the time its λ values stabilize. AR computations can be used not only to get a sense of the level of engagement in a given social network, but also local authorities with a loyal engaged audience, within the population.*

6 Conclusions

To the best of our knowledge, the concept of social capital has lacked computational underpinnings, particularly in the field of social network analysis. This work is an effort to address this gap and model social capital in terms of engagement and belief revision.

Tracking the dynamics of engagement can help us understand not just how communities are formed, but also how influence and tractable social changes happen in populations. The proposed model provides quantifiable metrics like λ and β, as well as the AR and CR scores, using which, we can reason about levels of engagements and their underlying dynamics.

Acknowledgments. We thank Prof. Sridhar Mandyam K (IIIT Bangalore) for his help and work on Belief revision model; Dr. Prasad Ram (Gooru) whose work inspired AR and CR metrics; Anshumaan Agrawal (IIIT Bangalore) for countless fruitful discussions and Satvik Ramaprasad (CircuitVerse) for providing data from CircuitVerse.

A Appendix: Algorithm for Computing Authority and Citizen Ranks

Algorithm 1: Computing AR and CR using Engagement

1 let T be a $|V| \times |V|$ matrix of engagement.
2 T[u][v] = $\sum \epsilon_{u \to v}$ $\forall$ strokes between u and v; $u, v \in V$
3 $A = T$, $C = T^\top$
4 **for** $i \leftarrow 1$ **to** $|V|$ **do**
5   **for** $j \leftarrow 1$ **to** $|V|$ **do**
6         $A[i][j] = \frac{A[i][j]}{\sum_j A[i][j]}$
7         $C[i][j] = \frac{C[i][j]}{\sum_j C[i][j]}$
8   **end**
9 **end**
10 $AR\_old = []$, $CR\_old = []$
11 **for** $\lambda \leftarrow 1$ **to** ∞ **do**
12   **for** $i \leftarrow 1$ **to** $|V|$ **do**
13   **for** $j \leftarrow 1$ **to** $|V|$ **do**
14   **if** $i == j$ **then**
15                 $A[i][j] = 0$
16   **else**
17                 $A[i][j] = \frac{\exp(\lambda \cdot A[i][j])}{\sum_{k \neq i} \exp(\lambda \cdot A[i][k])}$
18   **end**
19   **end**
20   **end**
21     $A^* = \lim_{i \to \infty} A^i$
22     $AR = diagonal(A^*)$
23   **if** $AR = AR\_old$ **then**
24   **break**
25   **else**
26         $AR\_old = AR$
27   **end**
28 **end**
29 **for** $\beta \leftarrow 1$ **to** ∞ **do**
30   **for** $i \leftarrow 1$ **to** $|V|$ **do**
31   **for** $j \leftarrow 1$ **to** $|V|$ **do**
32   **if** $i == j$ **then**
33                 $C[i][j] = 0$
34   **else**
35                 $C[i][j] = \frac{\exp(\beta \cdot C[i][j])}{\sum_{k \neq i} \exp(C[i][k])}$
36   **end**
37   **end**
38   **end**
39     $C^* = \lim_{i \to \infty} C^i$
40     $CR = diagonal(C^*)$
41   **if** $CR = CR\_old$ **then**
42   **break**
43   **else**
44         $CR\_old = CR$
45   **end**
46 **end**

References

1. Akcomak, T.S., ter Weel, B.: The impact of social capital on crime: evidence from the Netherlands. Reg. Sci. Urban Econ. **42**(1–2), 323–340 (2012). https://doi.org/10.1016/j.regsciurbeco.2011.09.008. https://www.sciencedirect.com/science/article/pii/S0166046211001165
2. Ariely, D.: Predictably irrational. Harper Audio (2008)
3. Berne, E.: Games People Play: The Basic Handbook of Transactional Analysis. Ballantine Books, New York (1964)
4. Bolin, K., Lindgren, B., Lindström, M., Nystedt, P.: Investments in social capital - implications of social interactions for the production of health. Soc. Sci. Med. **56**(12), 2379–2390 (2003). https://doi.org/10.1016/S0277-9536(02)00242-3. http://www.ncbi.nlm.nih.gov/pubmed/12742602
5. Brehm, J., Rahn, W.: Individual-level evidence for the causes and consequences of social capital. Am. J. Polit. Sci. **41**(3), 999–1023 (1997)
6. Burt, R.S.: Structural holes vs. network closure as social capital. In: Lin, N., Cook, K.S., Burt, R.S. (eds.) Social Capital, pp. 31–56. Aldine de Gruyter, New York (2001). https://doi.org/10.1002/jps.2600650914
7. Burt, R.S.: Brokerage and Closure, vol. 1. Oxford University Press, Oxford (2014). https://doi.org/10.1007/s13398-014-0173-7.2
8. Gibbs, J.P., Coleman, J.S.: Foundations of Social Theory. Soc. Forces **69**(2), 625 (2006). https://doi.org/10.2307/2579680. http://www.jstor.org/stable/2579680?origin=crossref
9. Grusky, D., Granovetter, M.S.: The strength of weak ties. In: Inequal. 21st Century, pp. 249–252. Elsevier (2018). https://doi.org/10.4324/9780429499821-43
10. Healy, T., Côté, S.: The well-being of nations: the role of human and social capital. In: Education and Skills, vol. 47. ERIC (2001). https://doi.org/10.1111/1467-954X.00194. http://www.oecd.org/site/worldforum/33703702.pdf
11. Jadbabaie, A., Molavi, P., Sandroni, A., Tahbaz-Salehi, A.: Non-bayesian social learning. Games Econ. Behav. **76**(1), 210–225 (2012)
12. Jøsang, A., Ismail, R., Boyd, C.: A survey of trust and reputation systems for online service provision. Decis. Support Syst. **43**(2), 618–644 (2007)
13. Kadushin, C.: Too much investment in social capital? Soc. Netw. **26**(1), 75–90 (2004). https://doi.org/10.1016/j.socnet.2004.01.009
14. Kahneman, D.: A perspective on judgment and choice: mapping bounded rationality. Am. Psychol. **58**(9), 697 (2003)
15. Kahneman, D., Tversky, A.: Prospect theory: an analysis of decision under risk. In: Handbook of the Fundamentals of Financial Decision Making: Part I, pp. 99–127. World Scientific (2013)
16. Lin, N., Cook, K., Burt, R.S., Wellman, B., Frank, K.A.: Network capital in a multilevel world: getting support from personal communities. In: Lin, N., Cook, K.S., Burt, R.S. (eds.) Social Capital, pp. 233–273. Aldine de Gruyter, New York (2019). https://doi.org/10.4324/9781315129457-10
17. Mwangi, I., Ouma, S.: Social capital and access to credit in Kenya. Am. J. Soc. Manag. Sci. **3**(1), 8–16 (2012). https://doi.org/10.5251/ajsms.2012.3.1.8.16. http://www.scihub.org/AJSMS/PDF/2012/1/AJSMS-3-1-8-16.pdf
18. Peng, S., Zhou, Y., Cao, L., Yu, S., Niu, J., Jia, W.: Influence analysis in social networks: a survey. J. Netw. Comput. Appl. **106**, 17–32 (2018)
19. Putnam, R.D.: Bowling alone: the collapse and revival of American community (2000). https://doi.org/10.1145/358916.361990

20. Sen, A.: Rational fools: a critique of the behavioral foundations of economic theory. Philos. Public Aff. **6**(4), 317–344 (1977). http://www.jstor.org/stable/2264946
21. Sen, A.K.: Development as Freedom. Oxford University Press, Oxford (2017). https://doi.org/10.4324/9781912281275
22. Shah, D.V.: Civic engagement, interpersonal trust, and television use: an individual-level assessment of social capital. Polit. Psychol. **19**(3), 469–496 (1998)
23. Tussyadiah, S.P., Kausar, D.R., Soesilo, P.K.: The effect of engagement in online social network on susceptibility to influence. J. Hospit. Tour. Res. **42**(2), 201–223 (2018)
24. Tversky, A., Kahneman, D.: Advances in prospect theory: cumulative representation of uncertainty. J. Risk Uncertainty **5**(4), 297–323 (1992)
25. Van Der Gaag, M., Snijders, T.A.: The resource generator: social capital quantification with concrete items. Soc. Netw. **27**(1), 1–29 (2005). https://doi.org/10.1016/j.socnet.2004.10.001
26. Veenstra, G.: Social capital, ses and health: an individual-level analysis. Soc. Sci. Med. **50**(5), 619–629 (2000)
27. Voorveld, H.A., van Noort, G., Muntinga, D.G., Bronner, F.: Engagement with social media and social media advertising: the differentiating role of platform type. J. Advert. **47**(1), 38–54 (2018)
28. Vyncke, V., et al.: Does neighbourhood social capital aid in levelling the social gradient in the health and well-being of children and adolescents? A literature review. BMC Public Health **13**(1), 65 (2013). https://doi.org/10.1186/1471-2458-13-65. http://www.ncbi.nlm.nih.gov/pubmed/23339776. http://www.pubmedcentral.nih.gov/articlerender.fcgi?artid=PMC3574053

Employee Satisfaction in Online Reviews

Philipp Koncar(✉) and Denis Helic

Graz University of Technology, Graz, Austria
{philipp.koncar,dhelic}@tugraz.at

Abstract. Employee satisfaction impacts the efficiency of businesses as well as the lives of employees spending substantial amounts of their time at work. As such, employee satisfaction attracts a lot of attention from researchers. In particular, a lot of effort has been previously devoted to the question of how to positively influence employee satisfaction, for example, through granting benefits. In this paper, we start by empirically exploring a novel dataset comprising two million online employer reviews. Notably, we focus on the analysis of the influencing factors for employee satisfaction. In addition, we leverage our empirical insights to predict employee satisfaction and to assess the predictive strengths of individual factors. We train multiple prediction models and achieve accurate prediction performance (ROC AUC of best model = 0.89). We find that the number of benefits received and employment status of reviewers are most predictive, while employee position has less predictive strengths for employee satisfaction. Our work complements existing studies and sheds light on the influencing factors for employee satisfaction expressed in online employer reviews. Employers may use these insights, for example, to correct for biases when assessing their reviews.

Keywords: Employee satisfaction · Employer reviews · Kununu

1 Introduction

Employee satisfaction contributes to better employee engagement [56,60] and is strongly connected to the overall performance and productivity of businesses [32]. Further, employee satisfaction impacts the lives of many employees as they spend substantial amounts of their time at work [20,22]. Hence, employee satisfaction has attracted much attention in existing research [1,2,4,10,35,40,54]. While some studies considered the advantage of high employee satisfaction for businesses (e.g., increased performance, reduced employee turnover or absences) [49, 50,55], other studies focused on how to increase and foster employee satisfaction (e.g., through changing positions or granting benefits) [12,51]. Here, we set our focus on the latter and extend existing research by conducting a large-scale analysis of online employer reviews contained in an unexplored dataset. The potential of such reviews to complement traditional management measurements has been depicted in previous research [11,21,45], allowing to overcome problems with traditional assessment methods (e.g., annual employee surveys), such

S. Aref et al. (Eds.): SocInfo 2020, LNCS 12467, pp. 152–167, 2020.
https://doi.org/10.1007/978-3-030-60975-7_12

as employees that are reticent about feedback out of fear for consequences [46] or managers not open for criticism [29].

Research Question. Based on findings in previous research, in this paper we ask how employee benefits [1,15,48,57], employee positions [10,13,14,18] as well as employment status [5,23,25] interact with employee satisfaction expressed in online employer reviews.

Approach. To answer our research question, we empirically analyze online employer reviews found on kununu, a so far unexplored reviewing platform where employees can anonymously rate their employers. In particular, we adhere to existing research [1,5,10,13–15,18,23,25,48,57] and set our focus on the influence of employee benefits, position, and employment status on employee satisfaction. Our dataset comprises more than 2 200 000 reviews of more than 380 000 employers operating in 43 different industries. On kununu, reviews comprise an overall rating ranging between one ("very bad") and five ("very good") stars as well as additional details, such as the position of reviewers and the benefits they had received. We interpret these overall ratings as an expression of employee satisfaction. Finally, we conduct a logistic regression to predict employee satisfaction, allowing us to assess the predictive strength of individual influencing factors.

Findings and Contributions. Overall, our results empirically confirm previous findings in existing research. For example, we observe that higher numbers of employee benefits positively influence employee satisfaction expressed in online employer reviews [1,15,48,57]. Further, we find that former employees review more negatively as compared to current employees, which reflects previous results suggesting that dissatisfaction causes employees to quit [5,23,25]. However, we also find that employees of higher positions (e.g., managers) review more positively, suggesting that employee position may have an influence on employee satisfaction, contradicting previous findings stating the opposite [10,18]. Lastly, with our prediction experiment, we find that the number of benefits granted to and the employment status of employees have the highest predictive strengths for employee satisfaction. When we combine all individual features, we achieve a mean ROC AUC of 0.89.

Our work contributes to the understanding of the influencing factors for employee satisfaction expressed in online employer reviews. Thus, we add fruitful input to the debate of employee satisfaction in social and management sciences.

2 Related Work

Employee Satisfaction. Employee satisfaction has been interpreted in a number of ways in previous research. For example, Blood [4] stated that employee satisfaction is depending on the values one brings to the job. On the contrary, Schneider and Schmitt [54] defined it as entirely depending on organizational conditions and not on predispositions of employees. Locke [35] thought of it more as an interaction between work conditions and individual employees and stressed that satisfaction is relevant to the engagement of the latter. More precisely, he defined employee satisfaction as a positive or pleasant emotional state resulting from people's appreciations of their own job.

Previous studies focused on the positive influence of high employee satisfaction on commitment and engagement of employees. For example, researchers investigated the positive relationship of employee satisfaction and engagement of hotel managers [24] or teachers [40]. Other works studied the positive influence of employee satisfaction on turnover [2] or on organizational commitment [36].

On the contrary to these studies, we analyze employee satisfaction as a dependent variable (i.e., aspects such as pay or rewards that benefit employee satisfaction). Existing studies demonstrated, for example, how to effectively reward employees for increased employee satisfaction [1,12,57] or how employee position influences employee satisfaction [10,18]. We base our work on these previous studies to analyze a novel dataset of more than two million online employer reviews, allowing us to find new insights regarding the influencing factors of employee satisfaction expressed in online reviews.

Online Reviews. Numerous works have studied online employer reviews and most works focused on the reviewing platform *glassdoor*[1]. For example, Marinescu et al. [41] described a selection bias in online reviews, meaning that people with extreme opinions are more motivated to share their experiences as compared to people with moderate opinions. Chandra [6] depicted differences in work-life balance between eastern and western countries using reviews written on glassdoor. Luo [37] inferred important characteristics impacting employee satisfaction and named innovation and quality as one of the most important. In a more recent work, Green et al. [21] analyzed employer reviews on glassdoor and their influence on stock returns. Their results indicate that companies with improvements in reviews (i.e., reviews becoming more positive over time) significantly outperform companies with declines (i.e., reviews becoming more negative over time). Dabirian et al. [11] extracted 38,000 reviews of highest and lowest ranked employers on glassdoor in order to identify what employees care about. Contrary to these works, we investigate online employer reviews found on kununu and also consider a larger quantity of reviews.

Online reviews have also been studied extensively in other contexts. As early as in 2001, Chatterjee [7] studied how negative online reviews influence consumers. Chen et al. [9] investigated online reviews of automobiles and depicted that the intention to write reviews is depending on the price and quality of products. Li and Hitt [34] as well as Chen et al. [8] studied online book reviews, where the former found that the average rating declines over time and the latter found that recommendations in reviews can have a positive impact on sales. Other works focused on the impact of online reviews on hotel business performance [59,61]. More recent and established works investigated the helpfulness of online reviews mostly written on Amazon [17,31,38,39], but also on

[1] https://glassdoor.com.

the video games platform Steam [3,19] or the review aggregation website Metacritic [30,53]. Our work separates from these studies, which analyze reviews of material goods, as we focus on reviews of work experiences.

3 Dataset and Empirical Results

3.1 Dataset

We conduct our analysis on a novel dataset comprising reviews found on kununu, a platform that offers employees the possibility to anonymously review their employer. The platform is provided in English as well as in German for employers located in Austria, Germany, Switzerland (all since 2007) and the USA (since 2013). Reviews written on kununu consist of 13 individual aspect ratings, such as *company culture* or *teamwork*, each ranging between 1 ("very bad") and 5 ("very good"). These aspect ratings are aggregated into an *overall rating* for each review. Additionally, reviews contain a headline (120 characters at a maximum) and information about the job status of a reviewer (i.e, whether they are a current or former employee). The position of employees (i.e., either "employee", "management", "temporary", "freelancer/self-employed", "co-op", "apprentice" or "other"), the benefits and perks (e.g., life insurance or free parking) as well as free-form text are optionally disclosed by reviewers.

We automatically crawled[2] all reviews present on kununu up to the end of September 2019. Since kununu is bilingual, we had to normalize German names

Table 1. Dataset statistics. The table lists descriptive statistics of our preprocessed dataset comprising employer reviews written on kununu up to the end of September 2019.

# Industries	43
# Employers	385 736
# Reviews	2 240 276
... # thereof for Austria (since 2007)	139 760
... # thereof for Germany (since 2007)	1 255 641
... # thereof for Switzerland (since 2007)	114 514
... # thereof for USA (since 2013)	730 361
... # thereof including positions	2 239 189
... # thereof including perks or benefits	1 855 491
Overall rating median	3.85
Review length (in words) median	24
# Benefits median	5

[2] We used multiprocessing and multiple web proxies to shorten the execution time of the crawling and implemented other measures (e.g., testing for missing or duplicate data) as well as manually checked a selection of reviews to assure data integrity.

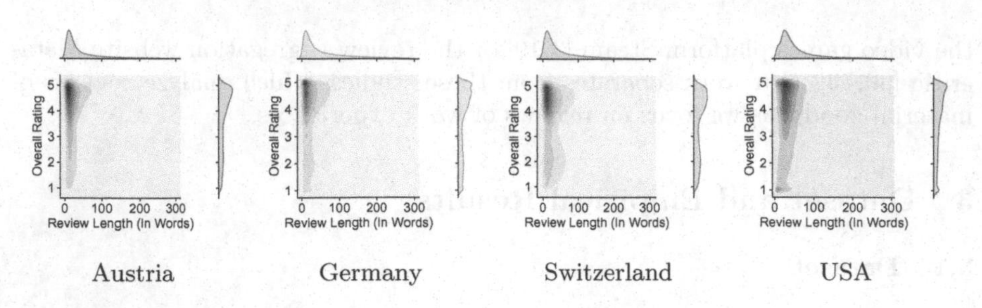

Fig. 1. Review length and overall ratings. The figure depicts bivariate KDEs of review length in words (distributions truncated at 300 words which is still above the 95th percentile for all languages) and overall ratings of reviews. We observe that most of the reviews contain none or only a few words as well as weak negative correlations between the two, suggesting that dissatisfied employees rather devote time and effort to write optional text.

(e.g., of positions) to English ones in order to compare reviews across languages. We list descriptive statistics of our preprocessed dataset, comprising 2 240 276 reviews of 385 736 employers operating in 43 different industries, in Table 1.

Preliminary Descriptive Analysis. We observe a larger number of reviews with positive overall ratings as compared to reviews with negative overall ratings (overall rating median = 3.85). Most notably, reviews of employers located in the USA seem to be more controversial with slightly higher probabilities for one star ratings as compared to European countries. We report a slight increase in the mean overall rating over time for the three European countries, whereas this mean for the USA decreases during the first years only to catch up with other countries in 2017. Regarding the length (in words) of optional review texts, we report long-tailed distributions for each of the four countries, indicating that the majority of reviews contains none or only a few words, whereas only a small number of reviews contain longer texts (review length median = 24). To see if reviews with optional texts are rather positive or negative, we compute Spearman's rank correlation coefficients between the number of characters in reviews with optional text and their respective overall ratings for each of the four countries. We depict bivariate kernel density estimations (KDE) between these two in Fig. 1. Here, we find a weak negative correlation (all p-values < 0.0005) for each European country with ρ ranging between -0.20 and -0.17. In case of reviews of employers located in the USA, the negative correlation is much weaker with $\rho = -0.08$ (p-value < 0.0005). This suggests that dissatisfied employees may rather invest time to address issues as compared to satisfied employees.

Overall, we report cultural, temporal and textual differences across reviews and an in-depth analysis of these differences might be promising for future work.

3.2 Empirical Results

Next we study the relationship between employee satisfaction and employee benefits, employee position as well as employment status.

Employee Benefits. To investigate the influence of benefits on employee satisfaction, we compute Spearman's rank correlation coefficients between the number of benefits received and the overall rating of reviews, respectively for each of the four countries in our dataset.

We depict bivariate KDEs of the number of benefits and overall ratings of reviews in Fig. 2. Overall, we observe a positive correlation between the number of benefits received and overall ratings for each of the four countries. This suggests that the more benefits employees receive, the higher is their overall rating in their reviews. However, strength of correlation varies across countries. While the mean Spearman's rank correlation coefficient for European countries is 0.46 (all p-values < 0.0005), this correlation is smaller for reviews of employers located in the USA with $\rho = 0.29$ (p-value < 0.0005), despite the fact that USA reviews have, on average, the second largest number of benefits per review (Austria: 6.32, Germany: 5.59, Switzerland: 5.38, USA: 6.30). Hence, we observe cultural differences, indicating that benefits for employees working for companies in the USA are not as influential as for employees of European companies.

Our findings on the positive influence of employee benefits for employee satisfaction are similar to those in existing research [57]. Further, studies reported that benefits become more important as part of the compensation of employees [15] and have grown in relevance to employees as well as in their variety [48]. Also, not all benefits positively correlate with satisfaction to the same extent and some of them even negatively correlate with each other [1]. Thus, we analyze what benefits were granted most to satisfied employees (reviews with overall rating ≥ 4) and compare that to benefits granted to dissatisfied employees (reviews with overall rating ≤ 2) in order to infer if benefits are equally relevant.

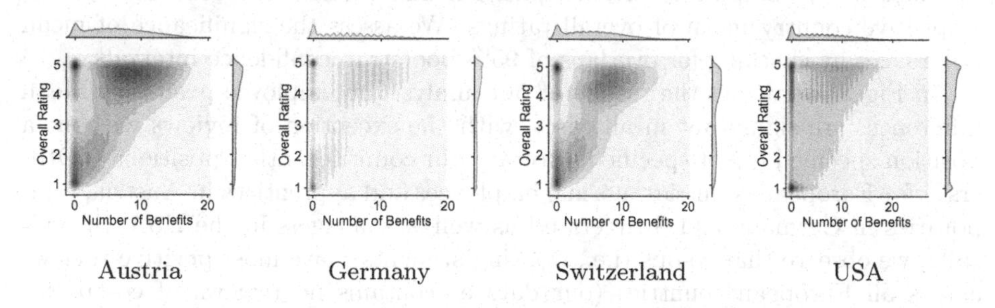

Fig. 2. Influence of benefits on employee satisfaction. We plot bivariate KDEs of the number of benefits and overall ratings of reviews, respectively for each country. Overall, we observe positive correlations between the two for all countries, with the exception of the USA for which we find weaker correlations. This observation supports previous findings [1,15,48,57] and indicates the positive influence of employee benefits on employee satisfaction.

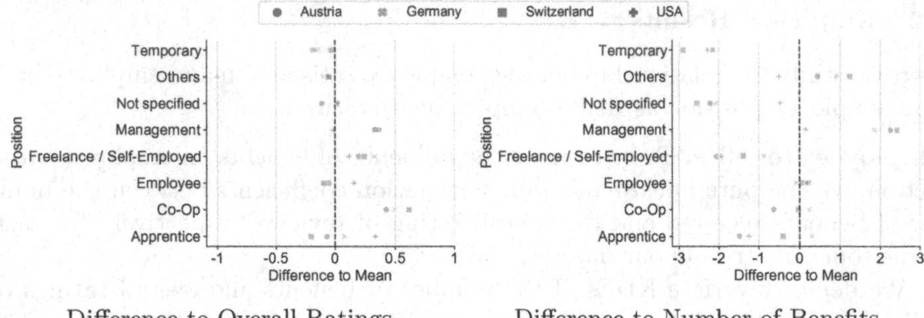

Fig. 3. Influence of position on employee satisfaction. In this figure we illustrate how employee position influences employee satisfaction in online employer reviews. In Figure a, we depict differences in overall ratings compared to country means, respectively for each position (note that we have no co-ops in the USA based version of kununu). Overall, we observe that, contrary to our expectations, employee positions have influence on overall ratings. For example, managers give a higher overall rating compared to the mean overall rating in all European countries. In Figure b, we depict differences in the number of benefits received compared to country means, respectively for each position. Again, we observe that benefits depend on position. For example, managers get, on average, two additional benefits compared to the country mean.

Here, we observe that satisfied employees receive different benefits as compared to dissatisfied employees. In particular, satisfied employees receive benefits, such as *flexible working hours*, *401k* or *fitness programs*, that have a high impact on work and life quality. On the contrary, dissatisfied employees receive benefits, such as *parking* or *discounts*. Hence, online employer reviews reflect similar behavior as reported in previous research [1].

Employee Position. We study the influence of employee positions on employee satisfaction by computing for each position and country the difference to the respective country mean of overall ratings. We assess the significance of mean differences by checking for overlaps of 95% bootstrap confidence intervals.

In Fig. 3, we depict the results of our analysis on employee positions. Mean differences are significant in all cases, with the exception of reviews without a position specified ("Not specified") for all four countries, other positions ("Others") for European countries, normal employees and apprentices in Austria, temporaries in Germany and Switzerland as well as managers in the USA. Specifically, we observe that co-ops (i.e., working students) have most positive reviews across all European countries (our dataset contains no reviews of co-ops for the USA), rating significantly more positive compared to the respective country mean (about half a star; 0.50 in numbers). Managers rate second most highest in Austria, Germany and Switzerland (all three significant), whereas managers in the USA rate slightly more negative compared to the country mean

Fig. 4. Influence of employment satisfaction on employment status. The figure illustrates how employee satisfaction potentially influences employment status. We plot KDEs of overall ratings for current (a) and former (b) employees, indicating that current employees have higher probabilities for positive overall ratings as compared to former employees who have higher probabilities to review more negatively. Further, we depict the time (in years) that lies between writing a review and termination of former employees (with 95% bootstrap confidence intervals), suggesting that overall ratings increase with time passed.

(non-significant). Apprentices rate significantly higher in the USA compared to European Countries, suggesting that interns and trainees are more satisfied with their education and instruction in the USA.

Our findings contradict existing research which suggests that employee position does not significantly influence employee satisfaction [10,18]. Additionally, existing studies showed that higher positions (e.g., managers) often receive more compensation [13,14]. To investigate whether this behavior is reflected in online employer reviews, we compare the number of benefits granted to employee positions to the respective country means. Again, we assess significance of mean differences by checking for overlaps of 95% bootstrap confidence intervals.

We depict results for the influence of position on the number of benefits received in Fig. 3b. Differences are significant in all cases, except for normal employees in all four countries, co-ops in Germany and Switzerland as well as managers in the USA. We observe that managers in German speaking countries, on average, receive two benefits more compared to the respective country means. This behavior is different to managers in the USA, who, on average and according to our dataset, receive the same number of benefits as the country mean and perhaps suggesting higher levels of equality among different employee positions in the USA. Unsurprisingly, temporary and self-employed personnel receive fewer benefits in all four countries as they are not permanently employed.

Overall, our findings contradict existing results [10,18] that suggest no influence of position on employee satisfaction. However, we confirm existing studies suggesting that higher positions receive more compensation [13,14].

Employment Status. We analyze the potential impact of employee satisfaction on employment status by computing two-sample Kolmogorov–Smirnov tests between overall rating distributions of current and former employees, respectively for each country. Further, we investigate the impact of time lying between creations of reviews and terminations of former employees (up to 8 years as data becomes sparse with longer time periods).

We illustrate the results for this analysis in Fig. 4. In general, current employees rate rather positively, having a mean overall rating of 3.86 across all countries, whereas possible frustration of former employees is clearly reflected in their mean overall rating of 2.81. We depict the distribution of overall ratings for current and for former employees in Fig. 4a and in Fig. 4b respectively. In any case, distributions are significantly different (all p-values < 0.0005) according to the two-sample Kolmogorov–Smirnov tests. For European countries, probabilities are similar for either negative or positive reviews, while probabilities for one star reviews are much higher for reviews of employers in the USA, suggesting higher frustration levels for their employees after termination as compared to employees of European companies.

Our findings reflect reports in existing research which suggest that employee satisfaction is the main reason for staying with or leaving an employer [25]. Further, employee satisfaction is strongly, positively correlated with organization commitment [5] and employee dissatisfaction is antecedent to forming an intention to quit [23], providing a potential explanation for our findings.

In Fig. 4c, we depict mean overall ratings according to the time that lies between the terminations of former employees and the creations of their reviews, respectively for each country. Our results suggest that the longer a termination lies in the past, the more positive are the reviews. While the mean overall rating of reviews written one month after termination is 2.59, this value is 3.91 after ten months, and mean values increase even more with longer time deltas.

Overall, our findings support our initial expectation that former employees review more negatively as compared to current employees. This effect, however, seems to weaken with more time lying between the termination of former employees and creation of their reviews.

4 Predicting Employee Satisfaction

We conduct a logistic regression to predict employee satisfaction based on our empirical results. We distinguish between reviews from satisfied and dissatisfied employees based on the overall rating contained in reviews. More precisely, we consider reviews with overall ratings less than or equal to the first quartile (overall rating ≤ 2.42) as expressions of dissatisfaction and reviews with overall ratings equal to or greater than the third quartile (overall rating ≥ 4.54) as expressions of satisfaction. We create classes this way to counteract the general bias towards more positive reviews (cf. Table 1 for the overall rating median).

We implement our logistic regression with default parameters[3] and train multiple models respectively with the following feature spaces[4]:

(i) *Country:* The country the reviewed employer is located in.
(ii) *Year:* The year the review was written in.
(iii) *Review Length:* The length of reviews defined by the number of words.
(iv) *Benefits:* The number of benefits the reviewing employee received.
(v) *Position:* The position of the reviewing employee.
(vi) *Employment Status:* The employment status of the reviewing employee.
(vii) *All:* The combination of all the above feature spaces.

Note that we use one-hot encoding to transform categorical features (country, year, position and employment status) and that we standardize numerical features (review length and benefits). Further, we remove all reviews that are missing any of the described features. This leaves us with 430 998 positive and 419 309 negative reviews. To evaluate and compare performance of our models, we use ten-fold cross validation and report mean ROC AUC values over the folds.

Results. In Fig. 5 we illustrate the mean ROC AUC respectively for each feature space. When we consider individual feature spaces, the number of benefits is most

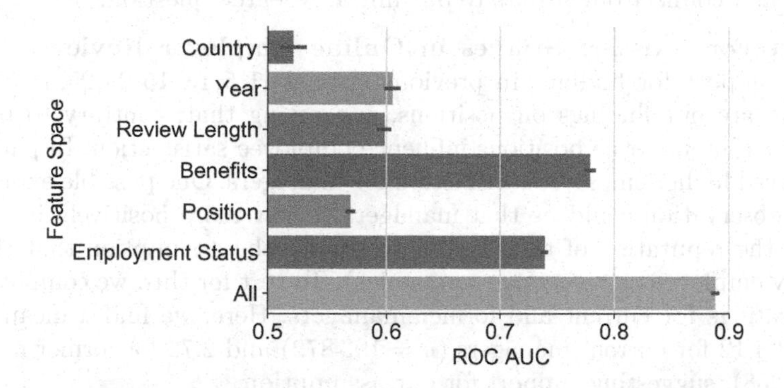

Fig. 5. Prediction results. In this figure we depict the results of our prediction experiment with which we assess the predictive strength of individual feature spaces and the combination of those for the prediction of employee satisfaction. The error bars indicate 95% bootstrap confidence intervals. We observe that the number of benefits received and the employment status of employees have the highest predictive strengths, while other feature spaces perform only minimally better than the random baseline (ROC AUC of 0.5). The combination of all features further improves the performance to a mean ROC AUC of 0.89, demonstrating that we can accurately predict employee satisfaction and that the consideration of all features is reasonable.

[3] As implemented in scikit-learn 0.22.2 (https://scikit-learn.org/0.22/).

[4] We do not consider the time delta between termination of employees and creation of their reviews as this feature is only available for 14% of reviews in our dataset.

predictive for employee satisfaction with a mean ROC AUC of 0.78, followed by the employment status with a value of 0.74. Our remaining feature spaces perform notably worse. The model we train with the year the review was written results in a mean ROC AUC of 0.61, similar to the model using the length of reviews which results in a value of 0.60. We report that the position of employees has very low predictive power with a mean ROC AUC of 0.57, only followed by the country the reviewed employer is located in, which performed the worst with 0.52 and only minimally better than a random baseline (0.5). However, when we combine all the individual feature spaces, we can further increase performance by 0.11 (compared to the model trained on the number of benefits) to 0.89, indicating that the features provide complementary information on employee satisfaction.

5 Discussion

With our empirical analysis of online employer reviews we provide new insights into the influencing factors for employee satisfaction expressed in such reviews. Our prediction experiment demonstrates the predictive strengths of individual features for predicting employee satisfaction. In the following, we discuss our findings and connect our results to our initial research question.

Support For Existing Studies in Online Employer Reviews. Overall, we find support for findings in previous studies [1,5,13–15,23,25,48,57]. One exception are our findings on positions, suggesting that, contrary to previous works [10,18], employee positions influence employee satisfaction. In particular, we observed higher employee satisfaction for managers. One possible explanation for this observation could be that managers review more positively in order to increase the reputation of their company (under the assumption that they are currently employed at respective companies). To test for this, we compute mean overall ratings for current and former managers. Here, we find a mean overall rating of 4.12 for current managers ($n = 193\,872$) and 2.72 for former managers ($n = 72\,258$), suggesting support for our assumption.

We observed that benefits have less impact on overall ratings for reviews of employers located in the USA as compared to European countries. After further investigating this result, we notice that USA based reviews include fewer benefits related to work-life balance (e.g., *flexible workhours* or *home office allowed*) as compared to German speaking countries. Thus, one possible explanation for the weaker correlation could be the lack of work-life benefits, which according to existing research are important for high employee satisfaction [47,58].

In the case of the influence of employee positions on overall ratings, we found that managers of European companies are more satisfied as compared to managers of companies located in the USA. This observation can be a reflection of different leadership styles between the USA and European countries, where for the former pressure on managers might be higher because decision making is much more egalitarian in the USA compared to European countries [44]. However, this could also be due to the fact that managers in German speaking countries enjoy more advantages as compared to those working for companies located in the USA (cf Fig. 3b). Also notable is that freelancers rate more positive in all four countries compared to respective country means, supporting exiting work indicating that freelancers are more satisfied with their job because of higher levels of freedom [16,42,52].

We found that former employees rate more positive the more time lies between the creation of reviews and their termination. This suggests that frustration of reviewers dissipates over time ("time heals wounds"), which follows the intuition that forgiving becomes easier over time [43]. However, we suggest a more detailed analysis of this observation for future work.

Prediction. Finally, our prediction experiment revealed that the number of benefits and the position of employees are most predictive for employee satisfaction. However, only when considering the combination of all features, we achieved the best prediction performance. Note that more sophisticated approaches, such as deep learning, might further improve the prediction performance. Based on our observations, we suggest that employers should assign more responsibilities to their employees as well as grant them more freedom, especially related to work-life balance. Overall, these observations are strongly related to long existing theories of management sciences, such as Herzberg's Two-Factor Theory [26–28] which is based on similar suggestions.

Limitations. We based our work on employer reviews found on kununu, only one platform among many others providing similar reviewing possibilities. Despite the large amount and the variety of data, the quality of our analysis may be improved by considering additional platforms, such as *glassdoor.com*. However, note that our analysis requires adjustments to other platforms as they use different rating mechanics and consider other employer characteristics. Thus, our analysis is biased towards the particularities of kununu. Further, we acknowledge a potential bias introduced by reviewers, such as different interpretations of rating scales or herding behavior, as suggested by existing research [33]. We leave an in-depth analysis of this phenomenon on kununu for future work. For our prediction experiment, we defined employee satisfaction based on the quartiles of overall rating distributions. While small adjustments to this definition did not noticeably alter our results, other definition may result in different findings.

6 Conclusion

In this paper, we investigated online employer reviews comprised in an unexplored dataset to shed light on the influencing factors for employee satisfaction. We obtain comparable results from online employer reviews to results from existing research based on e.g., survey data. The only exception to this are our results regarding the influence of employee position on employee satisfaction where we find that they are more important in online reviews as compared to previous findings. Further, we observe cultural differences across employers, for example, benefits have less impact on employee satisfaction in the USA as compared to

European countries. With our prediction experiment we depicted the predictive strengths of our individual findings, suggesting that the number of benefits and the employment status convey the most information for predicting employee satisfaction. When we combined all different features, we achieved a mean ROC AUC of 0.89, demonstrating the we can accurately predict satisfied and dissatisfied employees based on only a handful of features. Employers may use our findings to correct for biases when assessing their reviews or adapt management measures, such as shifting parts of compensation towards more benefits, as we demonstrated that they are most influential for employee satisfaction in reviews.

For future work, we plan to extend our analysis to learn more about the reviewing behavior of employees, for example, by considering the individual review aspects or by investigating industrial differences. Further, we want to adapt our analysis methods to other datasets comprising online employer reviews.

Acknowledgments. Parts of this work were funded by the go!digital Next Generation programme of the Austrian Academy of Sciences.

References

1. Artz, B.: Fringe benefits and job satisfaction. Int. J. Manpower (2010)
2. Aydogdu, S., Asikgil, B.: An empirical study of the relationship among job satisfaction, organizational commitment and turnover intention. Int. Rev. Manage. Mark. **1**(3), 43 (2011)
3. Barbosa, J.L., Moura, R.S., Santos, R.L.d.S.: Predicting Portuguese steam review helpfulness using artificial neural networks. In: Proceedings of the 22nd Brazilian Symposium on Multimedia and the Web, pp. 287–293. ACM (2016)
4. Blood, M.R.: Work values and job satisfaction. J. Appl. Psychol. **53**(6), 456 (1969)
5. Brown, S.P., Peterson, R.A.: Antecedents and consequences of salesperson job satisfaction: meta-analysis and assessment of causal effects. J. Mark. Res. **30**(1), 63–77 (1993)
6. Chandra, V.: Work-life balance: eastern and western perspectives. Int. J. Hum. Resour. Manage. **23**(5), 1040–1056 (2012)
7. Chatterjee, P.: Online reviews: do consumers use them? (2001)
8. Chen, P.Y., Wu, S.y., Yoon, J.: The impact of online recommendations and consumer feedback on sales. In: ICIS 2004 Proceedings, p. 58 (2004)
9. Chen, Y., Fay, S., Wang, Q.: Marketing implications of online consumer product reviews. Bus. Week **7150**, 1–36 (2003)
10. Cornelißen, T.: The interaction of job satisfaction, job search, and job changes. An empirical investigation with German panel data. J. Happiness Stud. **10**(3), 367–384 (2009)
11. Dabirian, A., Kietzmann, J., Diba, H.: A great place to work!? Understanding crowdsourced employer branding. Bus. Horiz. **60**(2), 197–205 (2017)
12. Darling, K., Arn, J., Gatlin, R.: How to effectively reward employees. Ind. Managem.-Chicago Atlanta- **39**, 1–4 (1997)
13. De Cremer, D.: How self-conception may lead to inequality: effect of hierarchical roles on the equality rule in organizational resource-sharing tasks. Group Organ. Manage. **28**(2), 282–302 (2003)

14. De Cremer, D., van Dijk, E., Folmer, C.R.: Why leaders feel entitled to take more. Psychol. Perspect. Ethical Behav. Decis. Making 107–119 (2009)
15. DeCenzo, D.A., Robbins, S.P., Verhulst, S.L.: Fundamentals of Human Resource Management. Wiley, Hoboken (2016)
16. Deprez, A., Raeymaeckers, K.: A longitudinal study of job satisfaction among flemish professional journalists. J. Mass Commun. **2**(1), 1–15 (2012)
17. Diaz, G.O., Ng, V.: Modeling and prediction of online product review helpfulness: a survey. In: Proceedings of the 56th Annual Meeting of the Association for Computational Linguistics (Volume 1: Long Papers), pp. 698–708 (2018)
18. Dienhart, J.R., Gregoire, M.B.: Job satisfaction, job involvement, job security, and customer focus of quick-service restaurant employees. Hospit. Res. J. **16**(2), 29–43 (1993)
19. Eberhard, L., Kasper, P., Koncar, P., Gütl, C.: Investigating helpfulness of video game reviews on the steam platform. In: 2018 Fifth International Conference on Social Networks Analysis, Management and Security (SNAMS), pp. 43–50. IEEE (2018)
20. Ernst Kossek, E., Ozeki, C.: Work-family conflict, policies, and the job-life satisfaction relationship: a review and directions for organizational behavior-human resources research. J. Appl. Psychol. **83**(2), 139 (1998)
21. Green, T.C., Huang, R., Wen, Q., Zhou, D.: Crowdsourced employer reviews and stock returns. J. Financ. Econ. **134**, 236–251 (2019)
22. Greenhaus, J.H., Beutell, N.J.: Sources of conflict between work and family roles. Acad. Manag. Rev. **10**(1), 76–88 (1985)
23. Griffeth, R.W., Hom, P.W., Gaertner, S.: A meta-analysis of antecedents and correlates of employee turnover: update, moderator tests, and research implications for the next millennium. J. Manag. **26**(3), 463–488 (2000)
24. Gunlu, E., Aksarayli, M., Perçin, N.Ş.: Job satisfaction and organizational commitment of hotel managers in Turkey. Int. J. Contemp. Hospit. Manage. (2010)
25. Hausknecht, J.P., Rodda, J., Howard, M.J.: Targeted employee retention: performance-based and job-related differences in reported reasons for staying. Hum. Resour. Manage.: Publ. Cooper. School Bus. Adm. Univ. Michigan Alliance Soc. Hum. Resour. Manage. **48**(2), 269–288 (2009)
26. Herzberg, F.: Motivation to Work. Routledge, Abingdon (2017)
27. Herzberg, F., Mausner, B., Snyderman, B.: The motivation to work (1959)
28. Herzberg, F.I.: Work and the nature of man (1966)
29. Holland, P., Cooper, B.K., Hecker, R.: Use of social media at work: a new form of employee voice? Int. J. Hum. Resour. Manage. **27**(21), 2621–2634 (2016)
30. Kasper, P., Koncar, P., Santos, T., Gütl, C.: On the role of score, genre and text in helpfulness of video game reviews on metacritic. In: 2019 Sixth International Conference on Social Networks Analysis, Management and Security (SNAMS), p. todo. IEEE (2019)
31. Kim, S.M., Pantel, P., Chklovski, T., Pennacchiotti, M.: Automatically assessing review helpfulness. In: Proceedings of the 2006 Conference on Empirical Methods in Natural Language Processing, pp. 423–430. Association for Computational Linguistics (2006)
32. Kumar, V., Pansari, A.: Measuring the benefits of employee engagement. MIT Sloan Manage. Rev. **56**(4), 67 (2015)
33. Lauw, H.W., Lim, E.P., Wang, K.: Quality and leniency in online collaborative rating systems. ACM Trans. Web (TWEB) **6**(1), 1–27 (2012)
34. Li, X., Hitt, L.M.: Self-selection and information role of online product reviews. Inf. Syst. Res. **19**(4), 456–474 (2008)

35. Locke, E.A.: The nature and causes of job satisfaction. Handb. Ind. Organ. Psychol. (1976)
36. Lumley, E., Coetzee, M., Tladinyane, R., Ferreira, N.: Exploring the job satisfaction and organisational commitment of employees in the information technology environment. Southern Afr. Bus. Rev. **15**(1) (2011)
37. Luo, N., Zhou, Y., Shon, J.: Employee satisfaction and corporate performance: mining employee reviews on glassdoor.com (2016)
38. Malik, M., Iqbal, K.: Review helpfulness as a function of linguistic indicators. Int. J. Comput. Sci. Netw. Secur. **18**, 234–240 (2018)
39. Malik, M., Hussain, A.: An analysis of review content and reviewer variables that contribute to review helpfulness. Inf. Process. Manage. **54**(1), 88–104 (2018)
40. Malik, M.E., Nawab, S., Naeem, B., Danish, R.Q.: Job satisfaction and organizational commitment of university teachers in public sector of Pakistan. Int. J. Bus. Manage. **5**(6), 17 (2010)
41. Marinescu, I., Klein, N., Chamberlain, A., Smart, M.: Incentives can reduce bias in online reviews. Technical report, National Bureau of Economic Research (2018)
42. Massey, B.L., Elmore, C.J.: Happier working for themselves? Job satisfaction and women freelance journalists. J. Pract. **5**(6), 672–686 (2011)
43. McCullough, M.E., Fincham, F.D., Tsang, J.A.: Forgiveness, forbearance, and time: the temporal unfolding of transgression-related interpersonal motivations. J. Pers. Soc. Psychol. **84**(3), 540 (2003)
44. Meyer, E.: Being the boss in Brussels, Boston, and Beijing. Harv. Bus. Rev. **95**, 70–77 (2018)
45. Miles, S.J., Mangold, W.G.: Employee voice: untapped resource or social media time bomb? Bus. Horiz. **57**(3), 401–411 (2014)
46. Milliken, F.J., Morrison, E.W., Hewlin, P.F.: An exploratory study of employee silence: Issues that employees don't communicate upward and why. J. Manage. Stud. **40**(6), 1453–1476 (2003)
47. Muse, L., Harris, S.G., Giles, W.F., Feild, H.S.: Work-life benefits and positive organizational behavior: is there a connection? J. Organ. Behav.: Int. J. Ind. Occup. Organ. Psychol. Behav. **29**(2), 171–192 (2008)
48. Newman, J.M., Gerhart, B., Milkovich, G.T.: Compensation. McGraw-Hill Higher Education, New York (2016)
49. Podsakoff, P.M., Williams, L.J.: The relationship between job performance and job satisfaction. Generalizing Lab. Field Sett. **207**, 253 (1986)
50. Ramayah, T., Nasurdin, A.: Integrating importance into the relationship between job satisfaction and commitment: a conceptual model. ICFAI J. Organ. Behav. **5**(2), 20–27 (2006)
51. Rathi, N., Rastogi, R.: Job satisfaction and psychological well-being. Icfai Univ. J. Organ. Behav. **7**(4), 47–57 (2008)
52. Ryan, K.M.: The performative journalist: job satisfaction, temporary workers and american television news. Journalism **10**(5), 647–664 (2009)
53. Santos, T., Lemmerich, F., Strohmaier, M., Helic, D.: What's in a review: discrepancies between expert and amateur reviews of video games on metacritic. In: Proceedings of the ACM on Human-Computer Interaction, vol. 3, no. CSCW, p. 140 (2019)
54. Schneider, B., Schmitt, N.: Staffing Organizations. Goodyear Publishing Company (1976)
55. Smith, P.C., et al.: The measurement of satisfaction in work and retirement: a strategy for the study of attitudes (1969)

56. Steinhaus, C.S., Perry, J.L.: Organizational commitment: does sector matter? Public Product. Manage. Rev. 278–288 (1996)
57. Tessema, M.T., Ready, K.J., Embaye, A.B.: The effects of employee recognition, pay, and benefits on job satisfaction: cross country evidence. J. Bus. Econ. **4**(1), 1–12 (2013)
58. Thompson, C.A., Beauvais, L.L., Lyness, K.S.: When work-family benefits are not enough: the influence of work-family culture on benefit utilization, organizational attachment, and work-family conflict. J. Vocat. Behav. **54**(3), 392–415 (1999)
59. Vermeulen, I.E., Seegers, D.: Tried and tested: the impact of online hotel reviews on consumer consideration. Tour. Manag. **30**(1), 123–127 (2009)
60. Weiss, H.M.: Deconstructing job satisfaction: separating evaluations, beliefs and affective experiences. Hum. Resour. Manage. Rev. **12**(2), 173–194 (2002)
61. Ye, Q., Law, R., Gu, B.: The impact of online user reviews on hotel room sales. Int. J. Hospit. Manage. **28**(1), 180–182 (2009)

Stable Community Structures and Social Exclusion

Boxuan Li[1], Martin Carrington[2], and Peter Marbach[2](✉)

[1] The University of Hong Kong, Pok Fu Lam, Hong Kong
liboxuan@connect.hku.hk
[2] University of Toronto, Toronto, Canada
{carrington,marbach}@cs.toronto.edu

Abstract. In this paper we study social exclusion in social (information) networks using a game-theoretic approach, and study the stability of a certain class community structures that are a Nash equilibrium. The main result of our analysis shows that all stable community structures (Nash equilibria) in this class are community structures under which some agents are socially excluded, and do not belong to any of the communities. This result is quite striking as it suggests that social exclusion might be the "norm" (an expected outcome) in social networks, rather than an anomaly.

Keywords: Social networks · Social exclusion · Stable Nash equilibria

1 Introduction

Social exclusion has been recognized as an important aspect in understanding social networks [9]. Roughly, social exclusion can be characterized by a distribution of goods and services that excludes certain (groups of) individuals from having any access at all to these goods and services. Or the goods and services that they have access to do not match their needs and as such do not provide any "benefit". Two key questions for understanding social exclusion are a) "what is the process that leads to social exclusion?", and b) "what (social) policies can be put in place in order to re-integrate excluded (marginalized) individuals back into a community (society)?". In this paper we focus on the first question and study the process that leads to social exclusion. While concentrating on this question, we believe that the models and insights into the process of how social exclusion occurs can potentially also be used to analyze and design (social) policies to counteract social exclusion.

The question of how social exclusion occurs has been studied in the literature where the focus has been on studying "who is doing the excluding?" [1]. In this paper we take a different approach and rather than focusing on the question "who is doing the exclusion", we study whether social exclusion might be in fact a structural property of communities in social networks. To do this, we use a game-theoretic approach to model the interaction among agents in a social

S. Aref et al. (Eds.): SocInfo 2020, LNCS 12467, pp. 168–180, 2020.
https://doi.org/10.1007/978-3-030-60975-7_13

network that has been studied by Carrington and Marbach [5]. Using this game-theoretic framework, we study a) which community structures (Nash equilibria) that emerge in a social network are stable, and b) whether there exist stable community structures under which some agents are excluded (marginalized) and do not belong to any community.

The key result that we obtain through our analysis is that the only community structures (Nash equilibria) that are stable are community structures with social exclusion. That is, the only community structures that are stable are structures under which some agents are marginalized and do not belong to any community. This is a striking result that suggests that social exclusion is indeed a "reality" (inevitable) in social networks. In this sense, the results obtained by our analysis suggest a new, and fundamentally different, understanding of social exclusion. That is rather than focusing on the question "who is doing the exclusion", one should focus on how do we deal (as a society) with social exclusion as a systematic property of social networks? We discuss this in more details in Sect. 6.

The rest of the paper is organized as follows. In Sect. 3 we describe the model we adopt to study community networks. In Sect. 4 we introduce the perturbation model that we use to study the stability of Nash equilibria. In Sect. 5 we present our main results. The proofs of the results can be found in [8].

2 Related Work

Studying social exclusion has received considerable attention in economic, social and health sciences; we refer to [9] for an overview of this vast literature. However studying social exclusion using a formal, model-based approach, has obtained much less attention. The only existing literature that uses a formal model-based approach to study social exclusion that we are aware of is the paper by Carrington and Marbach [5].

Using a game-theoretic approach, the research in [5] formally studies whether in social (information) networks there exist Nash equilibria under which some agents are marginalized, and not included in any of the communities. To do that, the work in [5] considers a particular type of social networks where agents (individuals) share/exchange information. Each agent chooses to join communities that maximizes its own utility obtained from content obtained, and shared, within the community. The analysis in [5] shows that there exists a class of Nash equilibria for the resulting game under which some agents are excluded (marginalized) from the community structure, i.e. they do not belong to (join) any of the communities. The reason for this is that these agents would have a negative utility in all of the communities that exist in the Nash equilibrium. These agents then have the choice to either join a community where the utility they receive is negative, or not join any community at all (and obtain a utility of zero). In this situation agents are better off not joining any community, and they become marginalized.

While the analysis in [5] shows that Nash equilibria with marginalized agents do exist, it does not address the question whether these Nash equilibria are indeed likely to occur and persist in a social network. Note that communities in social network constantly change (are perturbed) as some agents leave, and other agents join, the community. For this situation, we are interested in studying whether a Nash equilibrium is stable in the sense that it is robust to small changes (perturbations). Stable Nash equilibria are likely to persist and hence to be observed in social networks. On the other hand, Nash equilibria that are not stable will eventually vanish, and as a result are not likely to occur. In the following we are interested in whether Nash equilibria with marginalized agents are robust to perturbations, and hence likely to occur and persist in social networks.

In our analysis we use the concept of a stable Nash equilibrium that we formally define in Sect. 4. Roughly, we define a stable Nash equilibrium as a Nash equilibrium that is robust to (local) perturbations. There is an extensive literature in game-theory that studies the properties of Nash equilibria under perturbations. These perturbations could either be perturbations to the agents' value function, or to the agents' (players') strategies [2,10]. The approach that we consider is most closely related to stochastic fictitious play [4,6,7] where each player's payoffs are perturbed in each period by random shocks. The convergence results of stochastic fictitious play has been analyzed for games with an interior evolutionary stable strategy (ESS), zero sum games, potential games, near potential games and supermodular games. These results are obtained using techniques from stochastic approximation theory that show that one can characterize the perturbed best response dynamic of stochastic fictitious games by a differential equation defined by the expected motion of the stochastic process. We use the same approach in this paper where we model the perturbed best response dynamics of the agents by a differential equation as given in Sect. 4.

3 Background

In this section we describe the model and results of [5] that we use for our analysis. The model considers the situation where agents produce (generate) and consume (obtain) content in a social (information) network. Agents can form communities in order to share/exchange content more efficiently, and obtain a certain utility from joining a given community. Using a game-theoretic framework, the community structure that emerges is characterized by a Nash equilibrium.

More precisely, the model in [5] is given as follows. Assume that each content item that is being produced in the social (information) network can be associated with a particular topic, or content type. Furthermore assume that there exists a structure that relates different content topics with each other. In particular, assume there exists a measure of "closeness" between content topics that characterizes how strongly related two content topics are. To model this situation, the topic of a content item is given by a point x in a metric space, and the closeness between two content topics $x, x' \in \mathcal{M}$ is then given by the distance measure $d(x, x')$, $x, x' \in \mathcal{M}$, for the metric space $\mathcal{M}$.

The set of agents in the network, and agents' interests as well as agents' ability to produce content are then given as follows. Assume that there is a set $\mathcal{A}_d$ of agents that consume content, and a set $\mathcal{A}_s$ of agents that produce content, where the subscripts stand for "demand" and "supply". Furthermore, associate with each agent that consumes content a center of interest $y \in \mathcal{M}$, i.e. the center of interest y of an agent is the topic that the agent is most interested in. The interest in content topic x of an agent with center of interest y is given by

$$p(x|y) = f(d(x,y)), \qquad x, y \in \mathcal{M}, \tag{1}$$

where $d(x,y)$ is the distance between the center of interest y and topic x, and $f : [0,\infty) \mapsto [0,1]$ is a non-increasing function. The interpretation of the function $p(x|y)$ is as follows: when an agent with center of interest y consumes (reads) a content item on topic x, then it finds it interesting with probability $p(x|y)$ as given by Eq. (1). As the function f is non-increasing, this model captures the intuition that the agent is more interested in topics that are close to its center of interest y.

Similarly, given an agent that produces content, the center of interest y of the agent is the topic for which the agent is most adept at producing content. The ability of the agent to produce content on topic $x \in \mathcal{M}$ is given by

$$q(x|y) = g(d(x,y)), \tag{2}$$

where $g : [0,\infty) \mapsto [0,1]$ is a non-increasing function.

In the following we identify an agent by its center of interest $y \in \mathcal{M}$, i.e. agent y is the agent with center of interest y. As a result we have that $\mathcal{A}_d \subseteq \mathcal{M}$ and $\mathcal{A}_s \subseteq \mathcal{M}$.

3.1 Community $C = (C_d, C_s)$

A community $C = (C_d, C_s)$ consists of a set of agents that consume content $C_d \subseteq \mathcal{A}_d$ and a set of agents that produce content $C_s \subseteq \mathcal{A}_s$. Let $\beta_C(x|y)$ be the rate at which agent $y \in C_s$ generates content items on topic x in community C. Let $\alpha_C(y)$ be the fraction of content produced in community C that agent $y \in C_d$ consumes. To define the utility for content consumption and production, assume that when an agent consumes a single content item, it receives a reward equal to 1 if the content item is of interest and relevant, and pays a cost of $c > 0$ for consuming the item. The cost c captures the cost in time (energy) to read/consume a content item. Using this reward and cost structure, the utility rate ("reward minus cost") for content consumption of agent $y \in C_d$ is given by

$$U_C^{(d)}(y) = \alpha_C(y) \int_{x \in \mathcal{M}} \Big[Q_C(x) p(x|y) - \beta_C(x) c \Big] dx,$$

where

$$Q_C(x) = \int_{y \in C_s} \beta_C(x|y) q(x|y) dy, \text{ and } \beta_C(x) = \int_{y \in C_s} \beta_C(x|y) dy.$$

Similarly, the utility rate for content production of agent $y \in C_s$ is given by

$$U_C^{(s)}(y) = \int_{x \in \mathcal{M}} \beta_C(x|y) \Big[q(x|y) P_C(x) - \alpha_C c \Big] dx,$$

where

$$P_C(x) = \int_{y \in C_d} \alpha_C(y) p(x|y) dy, \text{ and } \alpha_C = \int_{z \in C_d} \alpha_C(z) dz.$$

The utility rate for content production captures how "valuable" the content produced by agent y is for the set of content consuming agents C_d in the community C [5].

3.2 Community Structure and Nash Equilibrium

A community structure defines how agents organize themselves into communities, where in each community agents produce and consume content as described in the previous subsection.

A community structure is given by a triplet $(\mathcal{C}, \{\alpha_{\mathcal{C}}(y)\}_{y \in \mathcal{A}_d}, \{\beta_{\mathcal{C}}(\cdot|y)\}_{y \in \mathcal{A}_s})$, where $\mathcal{C}$ is the set of communities C that exist in the structure, and

$$\alpha_{\mathcal{C}}(y) = \{\alpha_C(y)\}_{C \in \mathcal{C}}, y \in \mathcal{A}_d, \text{ and } \beta_{\mathcal{C}}(\cdot|y) = \{\beta_C(\cdot|y)\}_{C \in \mathcal{C}}, y \in \mathcal{A}_s,$$

are the consumption fractions and production rates, respectively, that agents allocate to the different communities $C \in \mathcal{C}$.

Assume that the total consumption fractions and production rates of each agent are bounded by $0 < E_p \leq 1$, and $E_q > 0$, respectively, i.e. we have that

$$\|\alpha_{\mathcal{C}}(y)\| = \sum_{C \in \mathcal{C}} \alpha_C(y) \leq E_p, \ y \in \mathcal{A}_d, \text{ and } \|\beta_{\mathcal{C}}(y)\| = \sum_{C \in \mathcal{C}} \|\beta_C(\cdot|y)\| \leq E_q, \ y \in \mathcal{A}_s,$$

where $\|\beta_C(\cdot|y)\| = \int_{x \in \mathcal{M}} \beta_C(x|y) dx$.

To analyze the interaction among agents, assume that agents join communities in order to maximize their own utility rates. That is, agents join communities, and choose allocations $\alpha_{\mathcal{C}}(y)$ and $\beta_{\mathcal{C}}(\cdot|y)$, in order to maximize their own consumption, and production utility rates, respectively.

A Nash equilibrium is then given by a community structure $(\mathcal{C}^*, \{\alpha^*_{\mathcal{C}}(y)\}_{y \in \mathcal{A}_d}, \{\beta^*_{\mathcal{C}}(\cdot|y)\}_{y \in \mathcal{A}_s})$ such that for all agents $y \in \mathcal{A}_d$ we have that

$$\alpha^*_{\mathcal{C}}(y) = \underset{\alpha_{\mathcal{C}}(y): \|\alpha_{\mathcal{C}}(y)\| \leq E_p}{\arg\max} \sum_{C \in \mathcal{C}} U_C^{(d)}(y),$$

and for all agents $y \in \mathcal{A}_s$, we have that

$$\beta^*_{\mathcal{C}}(\cdot|y) = \underset{\beta_{\mathcal{C}}(\cdot|y): \|\beta_{\mathcal{C}}(y)\| \leq E_q}{\arg\max} \sum_{C \in \mathcal{C}} U_C^{(s)}(y).$$

3.3 Community Structure $\mathcal{C}(L_C, l_d)$

The above model was analyzed in [5] for the case of a specific metric space, and a specific family of information communities. More precisely, the analysis in [5] considers the one-dimensional metric space given by an interval $\mathcal{R} = [-L, L) \subset \mathbb{R}$, $L > 0$, with the torus metric, i.e. the distance between two points $x, y \in \mathcal{R}$ is given by

$$d(x, y) = ||x - y|| = \min\{|x - y|, 2L - |x - y|\},$$

where $|x|$ is the absolute value of $x \in \mathbb{R}$. The metric space $\mathcal{R}$ with the torus metric is the simplest (non-trivial) one-dimensional metric space for the analysis of community structures in information networks. The reason for this is that the torus metric is "symmetric" and does not have any "border effects", which simplifies the analysis.

Furthermore, the analysis in [5] assumes that $\mathcal{A}_d = \mathcal{A}_s = \mathcal{R}$, i.e. for each content topic $x \in \mathcal{R}$ there exists an agent in $\mathcal{A}_d$ who is most interested in content of type x, and there exists an agent in $\mathcal{A}_s$ who is most adept at producing content of type x.

In addition, the analysis in [5] considers a particular family of community structures $\mathcal{C}(L_C, l_d)$, $L_C > 0$ and $0 < l_d \leq L_C$, given as follows.

Let $N \geq 2$ be a given integer. Furthermore, let $L_C = \frac{L}{N}$, where L is the half-length of the metric space $\mathcal{R} = [-L, L)$, and let l_d be such that $0 < l_d \leq L_C$. Finally, let $\{m_k\}_{k=1}^N$ be a set of N evenly spaced points on the metric space $\mathcal{R} = [-L, L)$ given by $m_{k+1} = m_1 + 2L_C k$, $k = 1, ..., N-1$.

Given L_C, l_d, and m_k, $k = 1, ..., N$, as defined above, the set of communities $\mathcal{C} = \{C^k = (C_d^k, C_s^k)\}_{k=1}^N$ of the structure $\mathcal{C}(L_C, l_d)$ is then given by the intervals

$$C_d^k = [m_k - l_d, m_k + l_d) \text{ and } C_s^k = [m_k - L_C, m_k + L_C).$$

Furthermore, the allocations $\{\alpha_\mathcal{C}(y)\}_{y \in \mathcal{R}}$ and $\{\beta_\mathcal{C}(\cdot|y)\}_{y \in \mathcal{R}}$ of the community structure $\mathcal{C}(L_C, l_d)$ are given by

$$\alpha_{C^k}(y) = \begin{cases} E_p & y \in C_d^k \\ 0 & \text{otherwise} \end{cases}, \qquad k = 1, ..., N, \tag{3}$$

and

$$\beta_{C^k}(\cdot|y) = \begin{cases} E_q \delta(x - x^*(y)) & y \in C_s^k \\ 0 & \text{otherwise} \end{cases}, \qquad k = 1, ..., N, \tag{4}$$

where

$$x^*(y) = \arg\max_{x \in \mathcal{R}} q(x|y) P_{C^k}(x).$$

Note that for $l_d = L_C$, the community structure $\mathcal{C}(L_C, l_d) = \mathcal{C}(L_C, L_C)$, and all agents belong to at least one community in $\mathcal{C}(L_C, L_C)$. On the other hand if we have that $l_d < L_C$ then community structure has "gaps", and there are agents that do not belong to any community. In particular, the content consuming agents in the sets

$$D^k = [m_k + l_d, m_{k+1} - l_d), \qquad k = 1, ..., N-1,$$

and $D^N = [m_N + l_d, m_1 - l_d)$, do not belong to any community in $\mathcal{C}(L_C, l_d)$. This means that these agents are marginalized, and excluded from the community structure.

3.4 Nash Equilibria $\mathcal{C}(L_C^*, l_d^*)$

To analyze the existence of Nash equilibria within the family $\mathcal{C}(L_C, l_d)$ of community structures as defined in the previous section, [5] made the following assumptions for the functions f and g that are used in Eq. (1) and Eq. (2).

Assumption 1. *The function* $f : [0, \infty) \mapsto [0, 1]$ *is given by* $f(x) = \max\{0, f_0 - ax\}$, *where* $f_0 \in (0, 1]$ *and* $a > 0$. *The function* $g : [0, \infty) \mapsto [0, 1]$ *is given by* $g(x) = g_0$, *where* $g_0 \in (0, 1]$. *Furthermore, we have that* $f_0 g_0 > c$.

The condition in Assumption 1 that $f_0 g_0 > c$ is a necessary condition for a Nash equilibrium to exist, i.e. if this condition is not true, then there does not exist a Nash equilibrium [5]. Under the above assumptions, the following two results regarding the existence, and properties of, Nash equilibria within the family $\mathcal{C}(L_C, l_d)$ of community structures were obtained in [5].

The first result states that there always exists a Nash equilibrium with $l_d = L_C$ within the family $\mathcal{C}(L_C, l_d)$ of community structures as defined above.

Proposition 3.1. *Let the functions f and g be as given in Assumption 1. Then the community structure* $\mathcal{C}(L_C^*, L_C^*)$ *is a Nash equilibrium if, and only if,*

$$L_C^* \leq \frac{f_0}{a} - \frac{c}{a g_0}.$$

Proposition 3.1 states that there always exists a Nash equilibrium under which no agents are marginalized.

The next result provides characterization of a Nash equilibrium with marginalized agents.

Proposition 3.2. *Let the functions f and g be as given in Assumption 1. Then the community structure* $\mathcal{C}(L_C^*, l_d^*)$ *with*

$$0 < l_d^* < L_C^*, \quad \text{and} \quad L_C^* = \frac{L}{N}$$

where $N \geq 2$ *is an integer, is a Nash equilibrium if, and only if,*

$$l_d^* = \frac{f_0}{a} - \frac{c}{a g_0}.$$

Note that Proposition 3.2 that there always exists a community structure $\mathcal{C}(L_C^*, l_d^*)$ that is a Nash equilibrium with marginalized agents if we have that

$$L > 2\left(\frac{f_0}{a} - \frac{c}{a g_0}\right),$$

i.e. if the content space $(-L, L]$ is large enough.

4 Stable and Neutral-Stable Nash Equilibria $\mathcal{C}(L_C^*, l_d^*)$

The results from [5] presented in the previous section show that there always exists a Nash equilibrium $\mathcal{C}(L_C^*, l_d^* = L_C^*)$ under which no agents are marginalized. Furthermore, it shows that if the content space $(-L, L]$ is large enough then there always exists a Nash equilibrium $\mathcal{C}(L_C^*, l_d^*)$, $0 < l_d^* < L_C^*$ under which some agents are marginalized.

The goal of this paper is to characterize which of the Nash equilibria $\mathcal{C}(L_C^*, l_d^*)$ obtained in the previous section are stable, i.e. robust to small perturbations to the community structure. This question is motivated by the following observation. As over time agents may join, or leave, communities in a social network, community structure is not static but changes over time. For this situation, we are interested in studying whether a community structure is stable in the sense that it is robust to small changes (perturbations) to the structure.

In order to study this question, we use the following approach. For a given Nash equilibrium $\mathcal{C}(L_C^*, l_d^*)$, we consider two adjacent communities $C^1 = (C_d^1, C_s^1)$ and $C^2 = (C_d^2, C_s^2)$. Without loss of generality we assume that the two communities are given by

$$C_d^1 = [-L_C - l_d^*, -L_C + l_d^*) \tag{5}$$

and

$$C_d^2 = [L_C - l_d^*, L_C + l_d^*), \tag{6}$$

as well as

$$C_s^1 = [-2L_C, 0) \tag{7}$$

and

$$C_s^2 = [0, 2L_C). \tag{8}$$

Note that by the definition of the community structure $\mathcal{C}(L_C, l_d^*)$ given in Sect. 4, there always exist at least two communities in a Nash equilibrium $\mathcal{C}(L_C^*, l_d^*)$.

We then perturb the "boundary" between the two communities by a small amount, and study the dynamics of the boundaries between the two communities after this perturbation. In particular we study whether over time the perturbation vanishes and the original community structure is again obtained. If the perturbation vanishes over time, then this suggests that the Nash equilibrium is stable in the sense that the Nash equilibrium is resistant to (local) perturbations. On the other hand if the communities are no restored after the perturbation, then this suggests that the Nash equilibrium is not stable.

More precisely, we consider the following perturbation model to study the stability of a Nash equilibrium. For $t = 0$ we set

$$C_d^1(t = 0) = [-L_C - l_d^*, -L_C + l_d^* + \delta_{dl}(0))$$

and

$$C_d^2(t = 0) = [L_C - l_d^* + \delta_{dr}(0), L_C + l_d^*),$$

as well as

$$C_s^1(t=0) = [-2L_C, \delta_{sl}(0)) \text{ and } C_s^2(t=0) = [\delta_{sr}(0), 2L_C),$$

where

$$\delta_{dl}(0) = \delta_{dr}(0), \quad \text{if } l_d^* = L_C$$

and

$$\delta_{dl}(0) < 2L_C - 2l_d^* + \delta_{dr}(0), \quad \text{if } l_d^* < L_C,$$

as well as $\delta_{sl}(0) = \delta_{sr}(0)$.

We analyze the dynamics of the boundaries between the two communities after the initial perturbation, i.e. we characterize the trajectory of the perturbation $\delta_{dl}(t)$, $\delta_{dr}(t)$, $\delta_{sl}(t)$, and $\delta_{sr}(t)$, over time $t \geq 0$. The intuition behind the dynamics of the boundaries is as follows. Note that the community boundaries represent the agents at the border between the two communities. We refer to these agents as the border agents. We then assume that a border agent will join the community which provides the highest utility, given that the highest utility is non-negative. If the highest utility is negative, then the border agent will leave the community and not join any community (and obtain a utility equal to 0).

Using this intuition, we then assume that the rate at which the boundaries move is given by the difference between the utilities in the two communities. That is, the higher the difference of the two utilities, the higher the rate (the faster) with which border agents move from one community to the other. The corresponding differential equations for the dynamics of the boundaries for $t \geq 0$ are then given by

$$\frac{d\delta_{dl}(t)}{dt} = U^{(d)}_{C_1(t)}(-L_C + l_d^* + \delta_{dl}(t)) - \max\left\{0, U^{(d)}_{C_2(t)}(-L_C + l_d^* + \delta_{dl}(t))\right\} \quad (9)$$

and

$$\frac{d\delta_{dr}(t)}{dt} = \max\left\{0, U^{(d)}_{C_1(t)}(L_C - l_d^* + \delta_{dr}(t))\right\} - U^{(d)}_{C_2(t)}(L_C - l_d^* + \delta_{dr}(t)) \quad (10)$$

Similarly, we have that

$$\frac{d\delta_{sl}(t)}{dt} = U^{(s)}_{C_1(t)}(-L_C + l_d^* + \delta_{sl}(t)) - \max\left\{0, U^{(s)}_{C_2(t)}(-L_C + l_d^* + \delta_{sl}(t))\right\} \quad (11)$$

and

$$\frac{d\delta_{sr}(t)}{dt} = \max\left\{0, U^{(s)}_{C_1(t)}(L_C - l_d^* + \delta_{sr}(t))\right\} - U^{(s)}_{C_2(t)}(L_C - l_d^* + \delta_{sr}(t)). \quad (12)$$

Note that in the above differential equation the border agent will never move to a community that provides a negative utility.

Using this model, we say that the Nash equilibrium is stable if the following is true.

Definition 4.1. *Let $\mathcal{C}(L_C^*, l_d^*)$ be a given Nash equilibrium, and let $C_1 = (C_d^1, C_s^1)$ and $C_2 = (C_d^2, C_s^2)$ be two communities in $\mathcal{C}(L_C^*, l_d^*)$ as given by Eq. (5)–(8).*

We say that the Nash equilibrium $\mathcal{C}(L_C^*, l_d^*)$ is stable if there exists a $\delta > 0$ such that for

$$0 < \delta_{dl}(t=0), \delta_{dr}(t=0), \delta_{sl}(t=0), \delta_{sr}(t=0) \leq \delta$$

we have for the differential equations given by Eq. (9)–(12) that

$$0 \leq \delta_{dl}(t), \delta_{dr}(t), \delta_{sl}(t), \delta_{sr}(t) \leq \delta, \qquad t \geq 0,$$

and

$$\lim_{t\to\infty} \delta_{dl}(t) = \lim_{t\to\infty} \delta_{dr}(t) = \lim_{t\to\infty} \delta_{sl}(t) = \lim_{t\to\infty} \delta_{sr}(t) = 0.$$

This definition captures the intuition that under a stable Nash equilibrium we have that small perturbations to the community boundaries will vanish (become equal to 0) over time.

In addition, we use a weaker notion of stability to which we refer to as a neutral-stable Nash equilibrium. To define a neutral-stable Nash equilibrium, we use the following notation.

Let $C_1(t) = (C_d^1(t), C_s^1(t))$ and $C_2(t) = (C_d^2(t), C_s^2(t))$, be the structure of two communities under the above perturbation model at time t. For an agent $y \in C_s^1(t) \cup C_s^2(t)$, let

$$x_1^*(y,t) = \arg\max_{x\in\mathcal{R}} q(x|y) \int_{z\in C_d^1(t)} p(x|z)dz$$

be the optimal content for agent y to produce in community $C_1(t)$, and let

$$x_2^*(y,t) = \arg\max_{x\in\mathcal{R}} q(x|y) \int_{z\in C_d^2(t)} p(x|z)dz$$

be the optimal content for agent y to produce in community $C_2(t)$. Using this definition, let the utilities of agents $y \in C_d^1(t) \cup C_d^2(t)$ be given as follows,

$$U_{C_1(t)}^{(d)}(y) = E_p E_q \int_{z\in C_s^1(t)} \Big[q(x_1^*(z,t)|z)p(x_1^*(z,t)|y) - c\Big] dz$$

and

$$U_{C_2(t)}^{(d)}(y) = E_p E_q \int_{z\in C_s^2(t)} \Big[q(x_2^*(z,t)|z)p(x_2^*(z,t)|y) - c\Big] dz.$$

Similarly, let the utilities of agents $y \in C_s^1(t) \cup C_s^2(t)$ be given as follows,

$$U_{C_1(t)}^{(s)}(y) = E_p E_q \int_{z\in C_d^1(t)} \Big[q(x_1^*(y,t)|y)p(x_1^*(y,t)|z) - c\Big] dz$$

and

$$U_{C_2(t)}^{(s)}(y) = E_p E_q \int_{z\in C_d^2(t)} \Big[q(x_2^*(y,t)|y)p(x_2^*(y,t)|z) - c\Big] dz.$$

Definition 4.2. *Let $\mathcal{C}(L_C^*, l_d^*)$ be a given Nash equilibrium, and let $C_1 = (C_d^1, C_s^1)$ and $C_2 = (C_d^2, C_s^2)$ be two communities in $\mathcal{C}(L_C^*, l_d^*)$ as given by Eq.* (5)–(8). *We say that the Nash equilibrium $\mathcal{C}(L_C^*, l_d^*)$ is neutral-stable if one of the following is true for the differential equations given by Eq.* (9)–(12).

1. *There exists a $\delta > 0$ such that for*

$$0 < \delta_{dl}(t=0), \delta_{dr}(t=0), \delta_{sl}(t=0), \delta_{sr}(t=0) \leq \delta$$

we have

$$0 \leq \delta_{dl}(t), \delta_{dr}(t) \leq \delta, \qquad t \geq 0, \ \ and \ \lim_{t\to\infty} \delta_{dl}(t) = \lim_{t\to\infty} \delta_{dr}(t) = 0,$$

as well as

$$\lim_{t\to\infty} U^{(s)}_{C_1(t)}(y) = \lim_{t\to\infty} U^{(s)}_{C_2(t)}(y) > 0, \qquad y \in C_s^1 \cup C_s^2.$$

2. *There exists a $\delta > 0$ such that for*

$$0 < \delta_{dl}(t=0), \delta_{dr}(t=0), \delta_{sl}(t=0), \delta_{sr}(t=0) \leq \delta$$

we have

$$0 \leq \delta_{sl}(t), \delta_{sr}(t) \leq \delta, \qquad t \geq 0, \ \ and \ \lim_{t\to\infty} \delta_{sl}(t) = \lim_{t\to\infty} \delta_{sr}(t) = 0,$$

as well as

$$\lim_{t\to\infty} U^{(d)}_{C_1(t)}(y) = \lim_{t\to\infty} U^{(d)}_{C_2(t)}(y) > 0, \qquad y \in C_d^1 \cup C_d^2.$$

This definition captures the case where either the content producers, or content consumers, are neutral (indifferent) regarding which community to join as they obtain the same utility rate in both communities.

5 Results

We obtain the following results for the perturbation model of Sect. 4. The first result states that (almost) all Nash equilibrium $\mathcal{C}(L_C^*, l_d^* = L_C^*)$ as given by Proposition 3.1 under which no agents are marginalized, are not stable.

Proposition 5.1. *All Nash equilibria $\mathcal{C}(L_C^*, l_d^* = L_C^*)$ with $L_C^* < \left(\frac{f_0}{a} - \frac{c}{ag_0}\right)$ are neither stable nor neutral-stable.*

The next result states that all Nash equilibria $\mathcal{C}(L_C^*, l_d^*)$ as given by Proposition 3.2 under which some agents are marginalized, are neutral-stable.

Proposition 5.2. *All Nash equilibria $\mathcal{C}(L_C^*, l_d^*)$ with $l_d^* = \left(\frac{f_0}{a} - \frac{c}{ag_0}\right) < L_C^*$ as given by Proposition 3.2 are neutral-stable.*

The above results state that Nash equilibria with marginalized agents are (neutral) stable, but (almost) all Nash equilibria with no marginalized agents are not stable.

6 Conclusions

We have studied social exclusion in social networks using a game-theoretic framework. In particular, we asked and analyzed the question whether social exclusion might be a structural property of communities in social networks. The results obtained under the model considered in this paper show that all (neutral) stable Nash equilibria are Nash equilibria with social exclusion. This is quite a striking result as it suggests that having marginalized individuals (i.e. social exclusion) is the "norm" in social networks, rather than an anomaly or exception. In this sense, the obtained results provide a new understanding of social exclusion, where rather than asking "who is doing the exclusion" the correct question to ask is how do we deal with social exclusion as a systematic property of social networks (society)?

We note that the results in this paper were obtained under a particular game-theoretic model. In that sense, rather than providing a definite "proof" that having marginalized individuals (i.e. social exclusion) is the "norm" in social networks, rather than an anomaly or exception, the results that we obtain is a first evidence that suggests that this might be the case. Given the potential impact and implications of this result in our understanding of social exclusion, important and interesting follow-up research is to verify whether these results a) are true under more general models than the one considered in this paper, and b) can be verified/observed in real-life case studies? We discuss possible directions to study these two research questions in more details below.

The results obtained in this paper were for a particular case where the functions f and g are as given by Assumption 1. A natural question to ask is whether these results extend to more general functions f and g. An extension of the formal analysis to more general functions f and g seems challenging and it is not clear whether it can be done. As a result, one might need to resort to numerical case studies to investigate this question. We carried out such numerical case studies for more general functions, and the initial results that we obtained suggest that the results indeed carry over to more general settings.

An interesting and important aspect of the results obtained in this paper is whether they indeed provide the correct insight into the process of how social exclusion occurs in real-life social networks and communities. One potential avenue for exploring this question could be using the effect of globalization on social exclusion. This topic has been extensively studied and documented [9]. In particular, the work by Beall [3] provides a concrete study of the influence of globalization on local communities in Pakistan and South Africa. An interesting question is whether the empirical results in [3] could be explained by, and match, the model-based results obtained in this paper.

Finally, an interesting direction for future research is to study whether the models used for the analysis in this paper could be used to design (social) policies to re-integrate marginalized individuals into a community/society.

References

1. Aktinson, A.B.: Social exclusion, poverty and unemployment. In: Aktinson, A.B., Hills, J. (eds.) Exclusion, Employment and Opportunity. CASE paper 4. London School of Economics (1998)
2. Balcan, B., Braverman, M.: Nash equilibria in perturbation-stable games. Theory Comput. **13**(13), 1–31 (2017)
3. Beall, J.: Globalization and social exclusion in cities: framing the debate with lessons from Africa and Asia. Environ. Urban. **14**(1), 41–51 (2002)
4. Candogan, O., Ozdaglar, A., Parrilo, P.A.: Near-potential games: geometry and dynamics. ACM Trans. Econ. Comput. **1**(2), 1–32 (2013)
5. Carrington, M., Marbach, P.: Community structures in information networks. In: Avrachenkov, K., Huang, L., Marden, J.R., Coupechoux, M., Giovanidis, A. (eds.) GameNets 2019. LNICST, vol. 277, pp. 119–127. Springer, Cham (2019). https://doi.org/10.1007/978-3-030-16989-3_9
6. Fudenberg, D., Kreps, D.M.: Learning mixed equilibria. Games Econ. Behav. **5**, 320–367 (1993)
7. Hofbauer, J., Sandholm, W.: On the global convergence of stochastic fictitious play. Econometrica **70**(6), 2265–2294 (2002)
8. Li, B., Carrington, M., Marbach, P.: Stable community structures and social exclusion. arxiv:2007.14515 (2020)
9. Mathieson, J., et al.: Social exclusion: meaning, measurement and experience and links to health inequalities - a review of literature. WHO Soc. Exclus. Knowl. Netw. **1**, 91 (2008)
10. Jackson, M.O., Rodriguez-Barraque, T.: Epsilon-equilibria of perturbed games. Games Econ. Behav. **75**(1), 198–216 (2012)

Inside the X-Rated World of "Premium" Social Media Accounts

Nikolaos Lykousas[1(✉)], Fran Casino[1], and Constantinos Patsakis[1,2]

[1] University of Piraeus, Piraeus, Greece
{nlykousas,fran.casino,kpatsak}@unipi.gr
[2] Athena Research Center, Marousi, Greece

Abstract. During the last few years, there has been an upsurge of social media influencers who are part of the adult entertainment industry, referred to as *Performers*. To monetize their online presence, Performers often engage in practices which violate community guidelines of social media, such as selling subscriptions for accessing their private "premium" social media accounts, where they distribute adult content. In this paper, we collect and analyze data from FanCentro, an online marketplace where Performers can sell adult content and subscriptions to private accounts in platforms like Snapchat and Instagram. Our work aims to shed light on the semi-illicit adult content market layered on the top of popular social media platforms and its offerings, as well as to profile the demographics, activity and content produced by Performers.

Keywords: Influencers · Marketplace · Performers · Adult content · Premium accounts · Community guidelines

1 Introduction

In the world of social media, content creators play a central role in shaping a global online culture. The content creators who raise in popularity can attain the status of online micro-celebrities, and they are commonly characterized as *influencers* [9]. The main objective of influencers is to produce digital content which attracts users' attention and rapidly gains popularity, often becoming 'viral', in platforms such as Instagram and YouTube [7,13]. In this regard, influencers leverage focused visual content and targeted communication techniques to capture and sustain the attention of social media users, thus building large follower bases and attaining organic social reach. Social media content creators can thus monetize their reach in various ways, such as using word-of-mouth marketing techniques and promoting brands and campaigns [11,16].

One of the most prevalent strategies employed by influencers to entice followers towards heightened forms of emotional engagement is sexualized labour [5]. Posting sexualized images in social media is a popular form of self-presentation for young adults [1,3,14,15], and it is outlined as the core tactic to attract followers for a particular type of influencers, which are categorized as *"Performers"* in [5].

This category of influencers includes adult performers/entertainers, sex workers and models. In all cases, after building an audience in mainstream social media, Performers redirect their followers to external outlets for purchasing exclusive content, often

S. Aref et al. (Eds.): SocInfo 2020, LNCS 12467, pp. 181–191, 2020.
https://doi.org/10.1007/978-3-030-60975-7_14

pornographic in nature. Notable examples of such outlets are platforms like OnlyFans[1] (effectively an 'adult' version of Instagram), and "premium" Snapchat accounts, offering a lucrative income stream for Performers looking to monetize their online presence [2]. For social media platforms like Snapchat, the community guidelines[2] explicitly *prohibit accounts that promote or distribute pornographic content.* Nonetheless, it has been shown that community guidelines cannot be effectively enforced to ban adult content in social media [12]. As such, Performers who systematically violate community guidelines by posting overtly sexual content, have to use external means for managing transactions with their client base, as well as maintaining their digital presence in multiple social outlets, in case their accounts get suspended.

In this paper, we analyze data collected from *FanCentro*[3], a platform where Performers can monetize their fan base via selling subscriptions to their private social media accounts. Additionally, FanCentro enables Performers to directly sell private content through a media feed, as well as chatting functionality between Performers and their subscribers. As a requirement for opening an account in FanCentro, Performers have to provide a digital copy of government-issued ID for age verification purposes. After this verification step, FanCentro, for a fraction of the paid subscriptions, handles all of the necessary transactions and administrative activities.

There are two main reasons we chose FanCentro over other similar platforms such as OnlyFans, which have gained wide mainstream media attention [4]. First, its primary focus is selling access to "premium" accounts in social platforms which, strictly, are not content marketplaces (i.e. Snapchat and Instagram). Second, FanCentro website provides a complete listing[4] of Performer profiles, enabling us to collect data without having to employ sampling techniques which could potentially bias our findings. Our work aims to shed light on the mechanics of the semi-illicit industry of premium social media subscriptions and services offered by Performers, in the context of adult content marketplaces such as FanCentro.

2 Data Collection

We constructed a *complete* dataset with the profiles of Performers registered in FanCentro as of April 5th, 2020. In Fig. 1, we provide an illustrative example of a Performer's profile page. We note that only Performers have public profiles and can post content, while regular users/subscribers can only interact with Performers (i.e. follow, message, like/comment to their posts) and not other users. In total, we collected the profile attributes, published content metadata, and offered products for 16, 488 users. For this, we created a crawler which consumes the API used by FanCentro's website, enabling us to collect the relevant data. Despite the "public" nature of collected information, we follow Zimmer's approach [17]. In this regard, the data remains anonymized during all the steps of our analysis, and we report only aggregate findings.

1 https://onlyfans.com.
2 https://www.snap.com/en-US/community-guidelines.
3 https://fancentro.com.
4 In contrast, OnlyFans platform does not have such functionality.

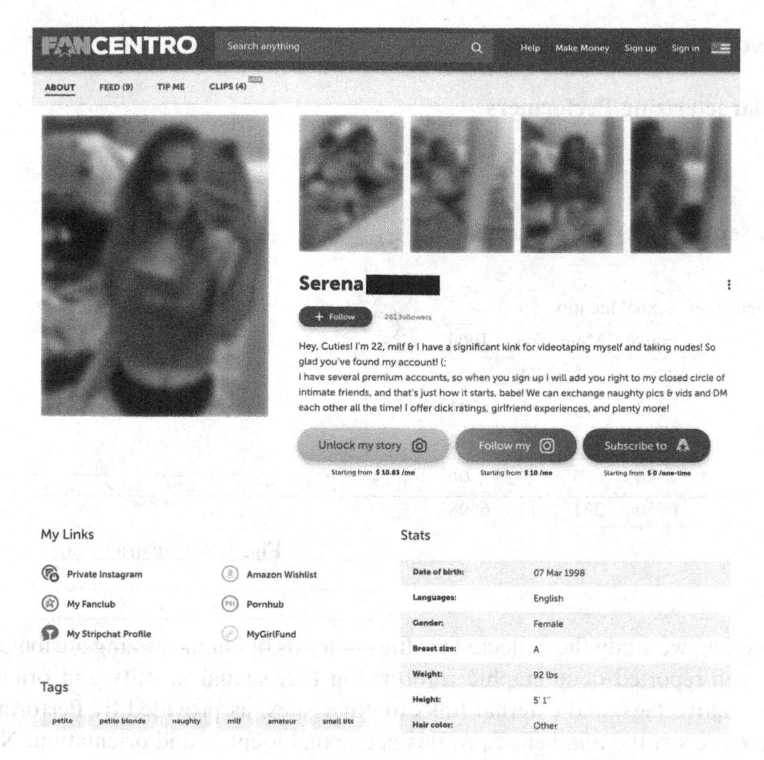

Fig. 1. An example Performer's profile page in FanCentro.

In order to measure the activity in FanCentro in terms of new registrations of Performers, in Fig. 2, we plot the number of accounts created each week since the launch of the platform. From January of 2017 (FanCentro launch), the weekly registrations show an increasing trend until a peak was reached in November of 2018. Since then the registration rate has been generally sustained, until we observe a spike in registrations the last week of March 2020, followed by the first week of April 2020, with 196 and 161 new users, respectively. This sharp increase in new users towards the end of March 2020 is also reflected in other similar sites, and it can be linked to the coronavirus pandemic, the consequent lockdowns, and its implications for sex work [6,10].

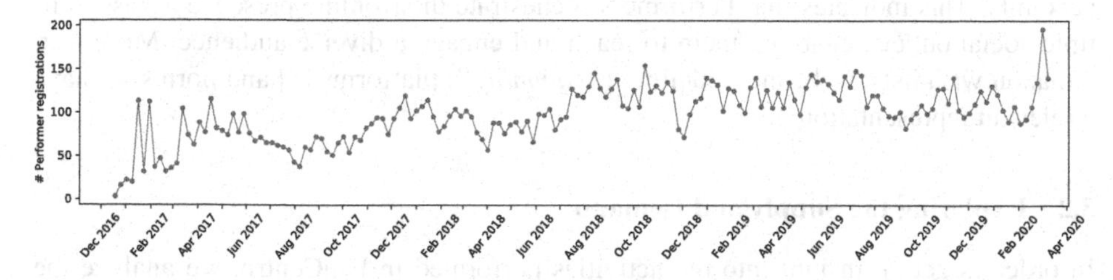

Fig. 2. Weekly registrations

3 Results and Discussion

3.1 Characterizing Performers

Table 1. Sexual identity and orientation

Sexual orientation	Sexual identity			
	Female	Male	Trans	Total
Bisexual	2486	52	41	2579
Gay	202	34	5	241
Straight	3544	141	27	3712
Trans	18	4	44	66
Total	6250	231	117	6598

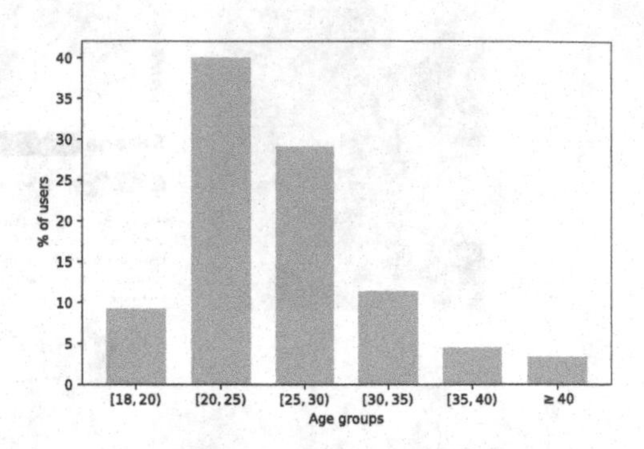

Fig. 3. Age distribution

In this section, we study the collected profiles in terms of characterizing attributes. This includes self-reported demographic information (i.e. sexual identity and orientation, age), descriptive tags, and external links to other sites, as provided by Performers. In Table 1, we report the number of profiles per sexual identity and orientation. Notably, 9, 879 profiles did not include this information. Nevertheless, after analyzing the rest of the profiles, we can conclude that the majority of Performers identify as straight females. In Fig. 3, we depicted the age distribution for the profiles containing the birth-date attribute (4, 526 profiles). We observe that the most common age group is 20–25 years (1, 857 profiles), followed by 25–30 (1, 347 profiles). The latter means that the 70% of Performers who reported their birthday are within the age bracket of 20 to 30 years. The next step of our analysis focused on the tags used by the Performers. In this regard, Fig. 4a shows a WordCloud representation of the most frequent tags used by Performers (found in 4, 558 profiles). We observe that they mostly include pornographic terms, with "sexy" and "ass" being the most popular (1, 472 and 928 occurrences, respectively). The outcomes of the analysis of the external links are depicted in Fig. 4b. We can observe that the most common external links from the profiles collected in our dataset are Instagram and Twitter, closely followed by public Snapchat accounts. This indicates that Performers orchestrate their online presence across multiple social outlets, enabling them to reach and engage a diverse audience. Moreover, Amazon wish lists, webcam modelling (*"camming"*) platforms [8] and porn sites have a relevant representation.

3.2 Exploring the Supply and Demand

In order to get an insight into the activities performed in FanCentro, we analyze the metadata information related to the collected profiles, including the amount of funds

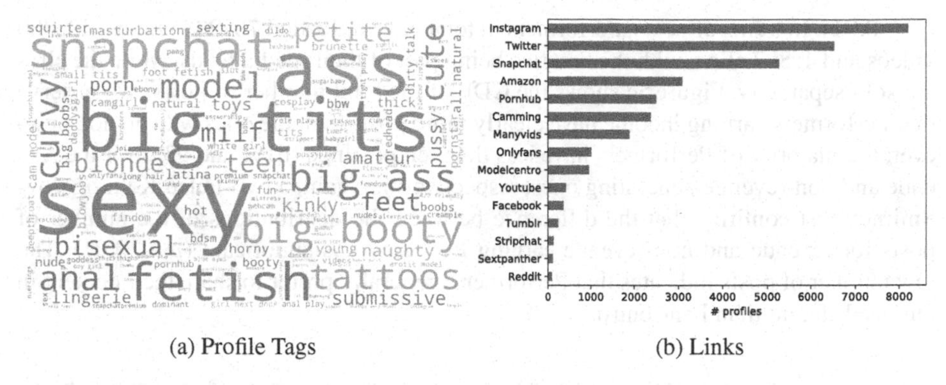

(a) Profile Tags

(b) Links

Fig. 4. Descriptive characteristics of Performers profiles: tags and links to external sites.

payable to the Performer in the next payout (revenue), the followers and the content posted by Performers.[5]

The *revenue* reflects the monetary sum of recurring sales (i.e. subscriptions) at crawl time, plus any income from one-off payments (including gratuity/tips, video clip sales and 'lifetime access' services) that are on hold by FanCetro until the next payout to the content creator.[6] Provided the dynamic nature of subscriptions and content produced by Performers, revenue is a quantity that fluctuates due to a variety of reasons, including cancellation of subscriptions, chargebacks, external factors governing Performers' popularity, etc. To assess the extent to which the revenue fluctuates over time, we use a snapshot of FanCentro profiles that we collected on March 2nd, 2020. To this end, a two-tailed Kolmogorov-Smirnov test was used, revealing no significant differences in Performers' revenues between two consecutive months ($p = 0.44$). We found that the revenue distribution is extremely skewed, with the overwhelming majority of the Performers (96.4%) generating zero revenue within the aforementioned period. In Fig. 5a, we plot the revenue cumulative distribution function (CDF) for the 602 revenue-earning Performers (3.6% of profiles). We observe that 80% are below the minimum payout threshold of 100 USD[7], meaning that only a negligible fraction of the performers in our dataset (0.8% approx.) would be certain to receive income by FanCentro during the next payout. Nonetheless, the revenues for the period between 23 March - 5 April 2020 period reach up to $12,615$ USD. In total, the gross earnings of Performers amount to $73,607$ USD for the payout period captured in our dataset.

Next, in Fig. 5b, we show the CDF of the number of followers. Contrary to the revenue, 78.5% of the profiles have followers. However, the revenue-generating Performers have up to two orders of magnitude more followers than the rest. The statistical significance of this difference was also confirmed by a two-tailed Kolmogorov-Smirnov test

[5] Revenue is personal in nature and is normally visible only via the dashboard of each Performer. We have contacted FanCentro regarding this matter, and it has been removed from the data delivered via the public API.

[6] FanCentro pays Influencers once a week after two weeks of the revenue generation date according to the license agreement.

[7] https://centroprofits.com/faq.

($p < 0.01$). In terms of posts, Performers in total have uploaded 73,233 photos, 43,860 videos and 4,867 clips, with the first two being part of their media feeds, while the clips are sold separately. Figure 5c shows the CDF of the total number of posts. We observe that Performers earning income have clearly more posts than the ones who do not, however, the majority of Performers have less than ten postings (61% and 93% for the revenue and non-revenue generating ones, respectively). Again, a two-tailed Kolmogorov-Smirnov test confirms that the difference between the distributions of the number of posts for revenue and non-revenue earning Performers is significant ($p < 0.01$). The low number of posts indicates that Performers, generally prefer to share their content in outlets different than FanCentro.

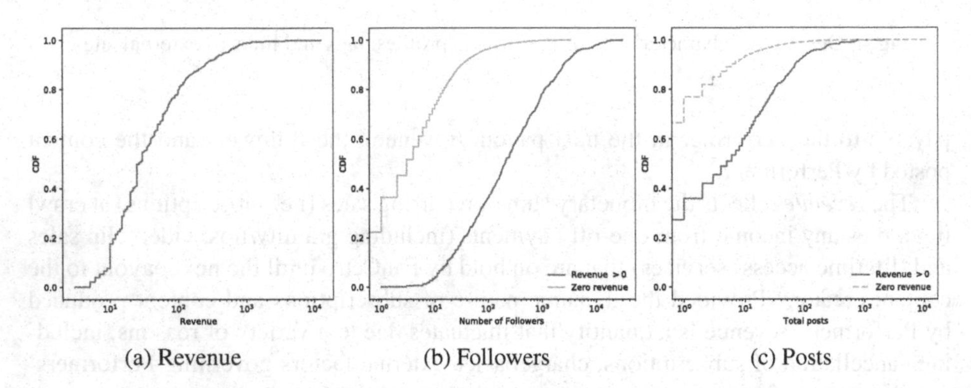

(a) Revenue (b) Followers (c) Posts

Fig. 5. Cumulative distribution functions (CDFs) of (non-zero) revenue, number of followers and posts.

3.3 FanCentro Content

To get a better understanding of the content Performers upload in FanCentro, we analyze their media feeds which, in terms of access, can contain two kinds of posts: *private* (only accessible by paying subscribers to their media feed) and *public* (freely accessible). In our dataset, the majority (89%) of posts are private (104,737 posts), while the rest are public (12,356 posts).

In Fig. 6, we depict the number of posts per month. We observe a consistently increasing trend in the number of posts, with a spike of 21,300 posts in December 2018, followed by March 2020 (7,325 posts), which is the second most active month in terms of posting activity. Next, we examine the characteristics of Performers' posts in terms of text content (titles) and user reactions, which results are depicted in Fig. 7. In our dataset, user reactions to Performers' posted content are relatively scarce, with 79% and 92% of the posts receiving zero likes and comments, respectively. This behaviour can be observed in Fig. 7a, which shows the CDFs of the reactions per post.

Notably, the majority of these posts received just one reaction, while the most popular post in our dataset has 316 likes and 55 comments. The low number of reactions

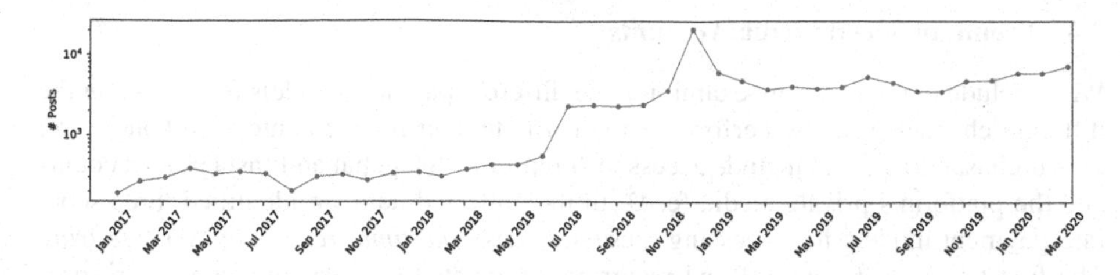

Fig. 6. Monthly posting activity

in posts comes in contrast with the relatively large numbers of followers that Performers attract, as showcased previously. In fact, there exists only a moderate correlation between the number of reactions per post and the total number of a Performer's followers (Spearman's $\rho = 0.48$). In Fig. 7b, we created a WordCloud of the post titles. It is apparent that, apart from terms of endearment and sexual terms, the phrase "subscriber benefits" is prevalent, which could provide an explanation for our previous observation: to a significant extend, Performers might use FanCentro media feed posts as an additional means to promote their premium content in other channels.

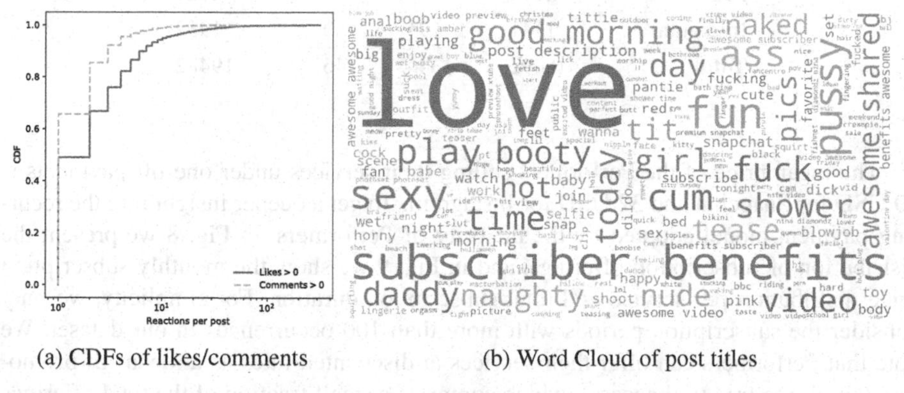

(a) CDFs of likes/comments (b) Word Cloud of post titles

Fig. 7. Characteristics of posts in terms of text content and reactions (likes, comments).

Finally, we study the characteristics of the video clips uploaded by Performers, which are sold separately. The 4,867 clips in our dataset were produced by 920 Performers, 285 (31%) of which had non-zero revenue. A subset of 1,078 clips is categorized as "free for followers", meaning that the Performers' followers can view these clips for free. This could explain the high numbers of followers that some of the Performers attract since this is a characterizing behaviour of the consumers of adult content [12]. Clips in our dataset have a mean duration of approximately 8 min and an average price of 11 USD per clip, while clip duration and price are moderately correlated (Spearman's $\rho = 0.45$).

3.4 Premium Social Media Accounts

We conclude our analysis by examining the different payment models for accessing the different channels used by Performers to distribute their private content. In FanCentro, the purchasable services include access to "premium" Snapchat and Instagram accounts and the platform's private media feed[8]. In the collected data we identified three separate payment models for accessing these services: *one-time*, *recurring* and *free trial*. The first two refer to one-off and recurring payments to access new content, respectively, while the "free trial" model allows customers to have a month of free access to the specific service, before reverting to recurring subscription payment. In Table 2, we present the distribution of the different payment models for the offered services. Private Snapchat is by far the most popular premium service, and the majority of Performers prefer offering their services as subscriptions.

Table 2. Premium services

Premium service	Payment model			
	Reccuring	One-time	Free Trial	Total
Snapchat	11635	1153	41	12829
FanCentro	4716	0	5	4721
Instagram	1741	191	0	1932
Total	18092	1344	46	19482

The mean price of the Performers selling their services under one-off payments is 30 USD for Snapchat and 32 USD for Instagram. To get a deeper insight into the recurring payment model adopted by the majority of Performers, in Fig. 8 we present the distribution of subscription offerings, and in Fig. 9 we show the monthly subscription price distribution per service and total subscription duration. For simplicity, we only consider the subscription periods with more than 100 occurrences in our dataset. We note that Performers can offer their services at discounted rates as a means of promotion (similar to free trial access), which comprise a small fraction of the total offerings (2, 004 in total).

In Fig. 8, we observe that the most popular service is the yearly Snapchat subscription, offered by 5, 892 performers, followed by monthly Snapchat subscription (3, 828 offerings) and yearly access to FanCentro feed (2, 921 offerings). While three-month and half-year subscriptions exist, they are not common, accounting only for 25% of total offerings. The subscription fee is calculated on the total subscription period. As such, the monthly price generally decreases as the subscription duration increases. The monthly subscription to Performers' premium accounts, which is the pricier option in

[8] Recently FanCentro has introduced a purchasable direct messaging service enabling direct communication between users and Performers. Nonetheless, we excluded it from our analysis due to the low number of observations in our dataset.

all cases, on average costs 21.7 USD and 58 USD for Snapchat (Fig. 9a) and Instagram (Fig. 9c) accounts, respectively. Notably, in the first case, the price can go up to 5,000 USD, and in the second case up to 8,000 USD. In this regard, the lowest priced service is access to FanCentro media feed (μ = 17.3 USD), which can cost up to 500 USD monthly (Fig. 9b). Nevertheless, the most common subscription duration is one year, priced on average 10 USD/month for Snapchat and FanCentro feed, and 14 USD/month for Instagram. Additionally, discounted rates show an average decrease of 6 USD/month for Snapchat, 14 USD/month for Instagram and 4 USD/month for FanCentro feed, when compared to the normal prices of each service, respectively.

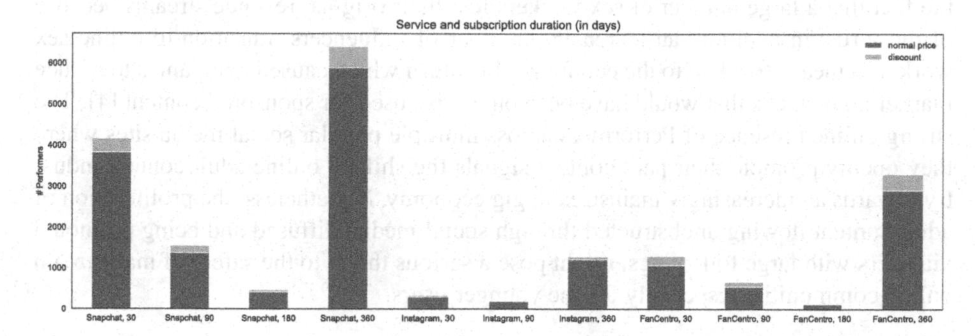

Fig. 8. Number of offerings per service and subscription duration.

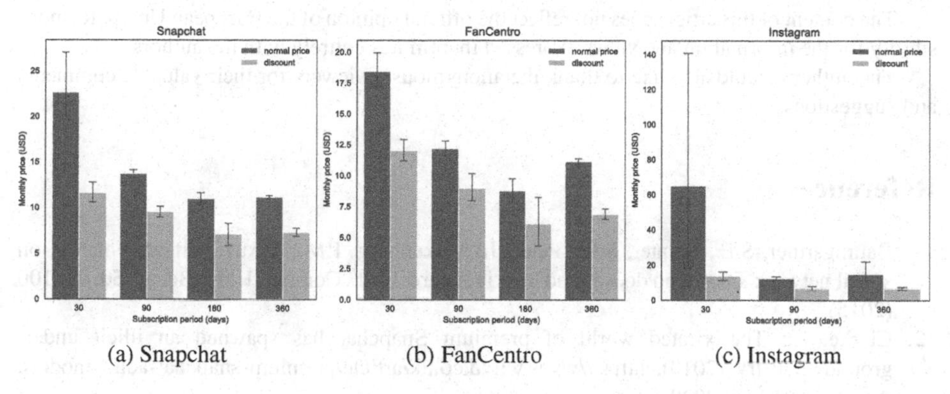

(a) Snapchat (b) FanCentro (c) Instagram

Fig. 9. Bar plots of monthly subscription price (normal and discounted) per service and subscription duration.

4 Conclusions

In this work, we performed the first quantitative analysis of the semi-illicit adult content market layered on the top of popular social media platforms like Snapchat and Instagram. To this end, we studied the demographics and activity of the selling users in FanCentro, as well as some descriptive characteristics of the content they upload. The existence of sites like FanCentro where Performers can openly sell and promote premium social media accounts indicate that the industry built on the inefficacy of social media platforms to enforce community guidelines for effectively banning adult content is here to stay. This inefficacy is exploited and monetized in large scale, exacerbated by the fact the explicit content is staying "hidden" in private accounts, access to which is sold through the different models studied.

Moreover, our findings indicate that the coronavirus-induced lockdowns have accelerated the growth of this marketplace. This phenomenon is also reflected by the rise of other influencer-centric adult content markets, such as OnlyFans, which observed a major increase in traffic during the coronavirus pandemic [6, 10]. In part, this is due to the fact that a large number of sex workers lost their original revenue streams because of the virus; in addition, an increasing number of influencers transition to online sex work as a means to adapt to the economic downturn which caused companies to reduce marketing budgets, that would have been otherwise used for sponsored content [4]. The strong online presence of Performers across multiple popular social media sites where they openly promote their paid content signals the shift of online adult content industry towards an increasingly mainstream, gig economy. Nonetheless, the proliferation of adult content flowing unobstructed through social media, diffused and being promoted via users with large followings, might pose a serious threat to the safety of mainstream online communities, especially for the younger users.

Acknowledgements. This work was supported by the European Commission under the Horizon 2020 Programme (H2020), as part of the project *LOCARD* (https://locard.eu) (Grant Agreement no. 832735).

The content of this article does not reflect the official opinion of the European Union. Responsibility for the information and views expressed therein lies entirely with the authors.

The authors would also like to thank the anonymous reviewers for their valuable comments and suggestions.

References

1. Baumgartner, S.E., Sumter, S.R., Peter, J., Valkenburg, P.M.: Sexual self-presentation on social network sites: who does it and how is it perceived? Comput. Hum. Behav. **50**, 91–100 (2015)
2. Clarke, L.: The x-rated world of premium Snapchat has spawned an illicit underground industry (2019). https://www.wired.co.uk/article/premium-snapchat-adult-models. Accessed 04 May 2020
3. Daniels, E.A.: Sexiness on social media: the social costs of using a sexy profile photo. Sex. Media, Soc. **2**(4), 2374623816683522 (2016)
4. Downs, C.: OnlyFans, Influencers, and the Politics of Selling Nudes During a Pandemic (2020). https://www.elle.com/culture/a32459935/onlyfans-sex-work-influencers/. Accessed 14 May 2020
5. Drenten, J., Gurrieri, L., Tyler, M.: Sexualized labour in digital culture: Instagram influencers, porn chic and the monetization of attention. Gender Work Org. **27**(1), 41–66 (2020)
6. Drolet, G.: Sex Work Comes Home (2020). https://www.nytimes.com/2020/04/10/style/camsoda-onlyfans-streaming-sex-coronavirus.html. Accessed 04 May 2020

7. Gómez, A.R.: Digital fame and fortune in the age of social media: A classification of social media influencers. aDResearch: Revista Internacional de Investigación en Comunicación (19), 8–29 (2019)
8. Henry, M.V., Farvid, P.: "Always hot, always live": computer-mediated sex work in the era of camming. Women's Stud. J. **31**(2), 113–18 (2017)
9. Khamis, S., Ang, L., Welling, R.: Self-branding, "micro-celebrity" and the rise of social media influencers. Celebr. Stud. **8**(2), 191–208 (2017)
10. Lee, A.: Coronavirus is bad news for Big Porn but great news for OnlyFans (2020). https://www.wired.co.uk/article/coronavirus-porn-industry-onlyfans. Accessed 04 May 2020
11. Lou, C., Yuan, S.: Influencer marketing: how message value and credibility affect consumer trust of branded content on social media. J. Interact. Advert. **19**(1), 58–73 (2019)
12. Lykousas, N., Gómez, V., Patsakis, C.: Adult content in social live streaming services: characterizing deviant users and relationships. In: 2018 IEEE/ACM International Conference on Advances in Social Networks Analysis and Mining (ASONAM), pp. 375–382. IEEE (2018)
13. Nandagiri, V., Philip, L.: Impact of influencers from Instagram and Youtube on their followers. Int. J. Multidisc. Res. Modern Educ. **4**(1), 61–65 (2018)
14. van Oosten, J.M., Peter, J., Boot, I.: Exploring associations between exposure to sexy online self-presentations and adolescents' sexual attitudes and behavior. J. Youth Adolesc. **44**(5), 1078–1091 (2015)
15. van Oosten, J.M., Vandenbosch, L.: Sexy online self-presentation on social network sites and the willingness to engage in sexting: a comparison of gender and age. J. Adolesc. **54**, 42–50 (2017)
16. Terranova, T.: Attention, economy and the brain. Cult. Mach. **13** (2012)
17. Zimmer, M.: "But the data is already public": on the ethics of research in Facebook. Ethics Inf. Technol. **12**(4), 313–325 (2010)

Impact of Online Health Awareness Campaign: Case of National Eating Disorders Association

Yelena Mejova[1(✉)] and Víctor Suarez-Lledó[2]

[1] ISI Foundation,Turin, Italy
yelenamejova@acm.org
[2] University of Cadiz,Cadiz, Spain
victor.sanz@uca.es

Abstract. National Eating Disorders Association conducts a NEDAwareness week every year, during which it publishes content on social media and news aimed to raise awareness of eating disorders. Measuring the impact of these actions is vital for maximizing the effectiveness of such interventions. This paper is an effort to model the change in behavior of users who engage with NEDAwareness content. We find that, despite popular influencers being involved in the campaign, it is governmental and nonprofit accounts that attract the most retweets. Furthermore, examining the tweeting language of users engaged with this content, we find linguistic categories concerning women, family, and anxiety to be mentioned more within the 15 days after the intervention, and categories concerning affiliation, references to others, and positive emotion mentioned less. We conclude with actionable implications for future campaigns and discussion of the method's limitations.

Keywords: Health informatics · Health interventions · Twitter · Social media · Mental health · Eating disorders

1 Introduction

The National Eating Disorders Association (NEDA) conducts a "NEDAwareness" week every year at the end of February[1]. NEDA is a nonprofit organization dedicated to supporting individuals and families affected by eating disorders, focusing on prevention, cures and access to quality care. Millions of people in US are at some point affected by eating disorders, which have a second highest mortality rate for mental disorders [17]. During NEDAwareness week, NEDA publishes content on social media and news, promoting awareness and linking to resources.

[1] https://www.nationaleatingdisorders.org/blog/announcing-national-eating-disorders-awareness-week-2020.

S. Aref et al. (Eds.): SocInfo 2020, LNCS 12467, pp. 192–205, 2020.
https://doi.org/10.1007/978-3-030-60975-7_15

As health intervention campaigns on social media are becoming more prevalent [4], the evaluation of their impact is becoming imperative in conducting effective interventions. This project is an effort to understand the impact NEDAwareness content has on social media users, in particular those using Twitter. We collect a dataset of NEDA-related tweets during the 2019 NEDAwareness week, capturing 20,197 tweets from 11,470 users. We extend the dataset with historical tweets of 7,870 users in order to capture their behavior before and after the event, and complement this collection with a baseline control group of 1,668 users. We begin by analyzing the dissemination and scope of this campaign, finding that, despite having influential accounts involved, it is the government agencies and nonprofits that achieve the most retweets. Secondly, we attempt to capture to what extent NEDA content achieves significant changes in the conversation topics of its audience. Compared to baseline users who did not engage with NEDA content, we find the users who did engage significantly change their language in the two weeks after the intervention, focusing more on womanhood, family, and anxiety, and sharing fewer social experiences and positive emotions.

2 Related Work

Social media has been widely employed to study mental health, ranging from fitness enthusiasts [13] and pro-eating disorder communities [3] to depression [9] and suicidal ideation [10]. Language of social media posts has been successfully used to predict their authors' internal states, achieving, for instance, precision of 0.74 on CES-D depression scale questionnaire [9,26]. Images have also been used to characterize those with depression and anxiety [11].

Meanwhile, health authorities began employing social media as a platform for behavior change and health awareness campaigns, especially in the domain of cancer awareness. Yearly drives #WorldCancerDay and National Breast Cancer Awareness Month (NBCAM) have been shown to attract engagement especially around women's cancers, and especially on Twitter [31]. Cancer discussions have been analyzed in the form of a follow network, showing distinct communities of breast and prostate cancer conversations [12]. Communities around pro- and anti- eating disorder have also been examined on Flickr [33], Tumblr [8] and Instagram [3]. Most of these works, however, measure the interaction of the audience with the campaign material, failing to follow up on the potential changes in behavior after the intervention. A notable exception is the measurement of whether those posting to pro-anorexia Flickr communities continue to do so after being exposed to anti-anorexia content [33]. They find that, unlike the intended effect, these users would post longer to pro-anorexia communities. Thus, it is important to verify the effects of online interventions. In this study, we attempt to quantify the change in posting activity of those engaged in NEDAwareness week in the time following the intervention.

Beyond campaign engagement, recent causality methods have encouraged the measurement of post-intervention behavior. A general framework for event outcome analytics through social media has been proposed by [15,23], and has been employed to, for instance, gauge the impact of alcohol use in students [14] and psychopathological effects of psychiatric medication [27]. Additionally, West et al. [32] used search queries to track the behavior of users after signaling weight loss intention. In this work, we employ temporal analysis of daily online interactions, compared to a non-treatment control, to uncover outcomes of a communication campaign.

Fig. 1. Example post during NEDAwareness week.

3 NEDAwareness Week Data Collection

Primary data collection happened during the intervention week February 25 - March 3 using Twitter Streaming API with the following hashtags: `NEDAstaff`, `NEDAwareness`, `NEDA`, `ComeAsYouAre`, `SOSChat` (compiled with the assistance of NEDA staff). The resulting collection comprises of 20,197 tweets from 11,470 users. We then performed another collection of the historical tweets on April 18–21, 2019, resulting in up to 3,200 tweets for 7,870 of the users (some accounts were closed or private). Thus, we obtained approximately a total of 12 million tweets from all users.

An example tweet is shown in Fig. 1. As the theme of 2019 was "Come as you are", the content often deals with mental health issues of people from a plurality of ages, races, genders and gender identities. As can be seen in Fig. 2, the campaign spans several weeks, with much of the content being produced not by the official NEDA account (red line), but by collaborators and audience retweets (blue line). The largest number of tweets come from `@NEDAstaff`, and two private accounts, with each tweeting (or retweeting) around 200–300 tweets. Beside the official `@NEDAstaff` account, the others do not have an official NEDA affiliation. In general terms, they are activists who relate to the content of other sources besides NEDA. They often interact with other institutional accounts such as "Eating Disorder Hope", "ADAA" (Anxiety and Depression Association of America), or "Mental Health America".

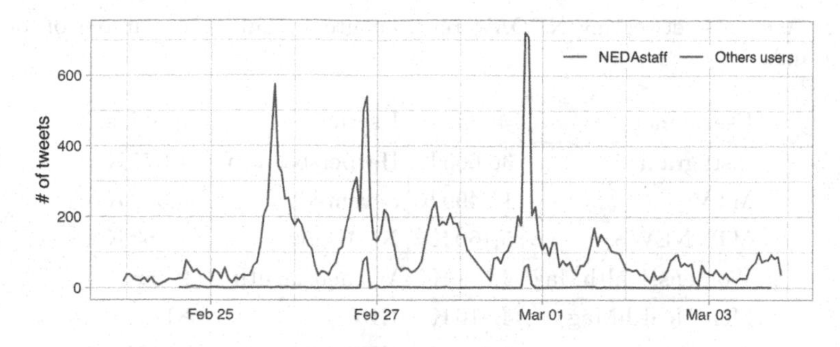

Fig. 2. Tweets per hour during the NEDAwareness week.

Finally, in order to compare the users who have engaged with NEDA content, we consider a "baseline" set of users who do not necessarily engage with NEDA. We collect a sample of users who have tweeted on February 25 - March 3 on any of diet and health related words. Similarly to the NEDA dataset, we collect the historical tweets of the captured users, resulting in 539,844 tweets of 1,668 different users.

4 Results

4.1 Reach

Although `@NEDAstaff` account had 37,567 followers during the NEDAWareness month, their content was retweeted by several popular accounts, dramatically expanding its potential reach. Table 1 shows the 20 accounts with the largest number of followers who have retweeted NEDA content. Note that many are accounts of media companies (`@instagram`, `@MTV`, `@MTVNEWS`, `@Pinterest`), others are magazines (`@WomensHealthMag`, `@MensHealthMag`, `@TeenVogue`), with additional engagement from governmental institutions including The National Institute of Mental Health (`@NIMHgov`) and U.S. Department of Health & Human Services (`@HHSGov`), as well as the Human Rights Campaign (`@HRC`).

Besides potential views, we examine the retweet and like statistics of the content in Fig. 3. Both are heavy tailed distributions, having a median of 2 likes and 4 retweets, with most posts getting little interaction. Furthermore, the Giant Connected Component of the retweet network, in which nodes are users and edges the retweet relationship, can be seen in Fig. 4, colored with five communities identified via the Walktrap algorithm [7,24]. NEDA account is at

Table 1. Accounts retweeting NEDAwareness content, ranked by number of followers (in thousands, K).

Username	# fol.'s	Username	# fol.'s
Instagram	36,665 K	Harpersbazaarus	1,677 K
MTV	15,499 K	Seventeen	1,359 K
MTVNEWS	5,160 K	NIMHgov	1,153 K
WomensHealthMag	4,581 K	Womenshealth	936 K
MensHealthMag	4,516 K	HRC	811 K
TeenVogue	3,340 K	HHSGov	754 K
Inquirerdotnet	2,792 K	Dosomething	750 K
Ginger_Zee	2,340 K	ABC7NY	653 K
Pinterest	2,337 K	Teddyboylocsin	646 K
Jimparedes	1,751 K	Allure_magazine	576 K

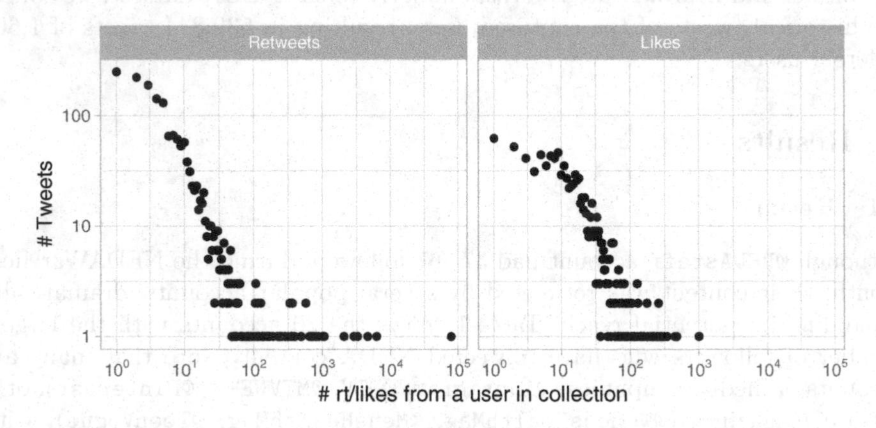

Fig. 3. Distribution of tweets having certain number of retweets (left) and likes (right), log scale.

the center of the largest community, followed by @NIMHgov and @MentalHealthAm (Mental Health America, a nonprofit organization). Note that the best reach in terms of retweets was achieved via governmental and nonprofit accounts, despite the more popular media accounts being involved in the conversation, putting in question whether such influencers result in wider reach.

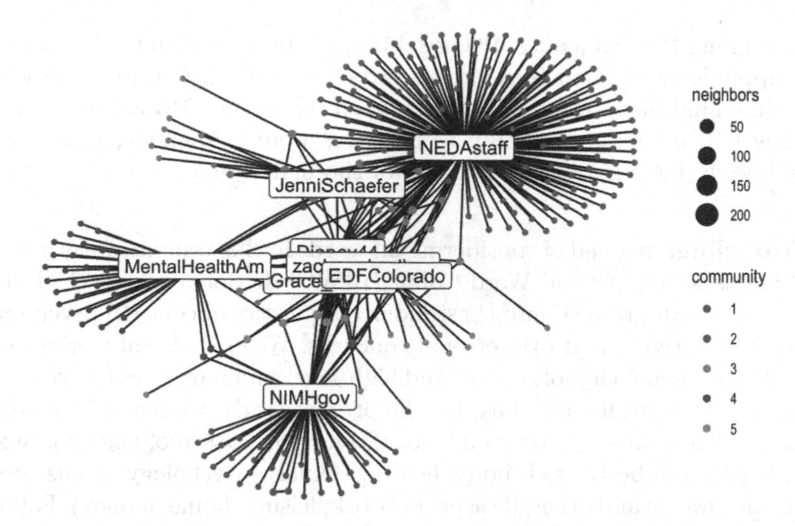

Fig. 4. Retweet network GCC with communities.

4.2 Impact

Timeline Partitioning. The continuous nature of Twitter allows us not only to see the unfolding of the NEDAwareness campaign, but also to capture the posting behavior of users involved, both before and after. Here, we ask, to what extent has the language of Twitter users changes after interacting with NEDAwareness content? We take an exploratory approach wherein we examine the posting behavior 15 days before and 15 days after such interaction, as defined below.

Following previous work on health behavior change [32], we define "day 0" as the first time a user engages with NEDA content by retweeting it or posting a related tweet. Note that we cannot track users who merely saw the NEDAwareness content, as this information is not available through Twitter API. Then we select all users who have tweets published in the 15 days before and after interaction with the NEDA content. We call the 1,746 users who have at least 3 tweets in the 15 days before and 3 tweets in the 15 days after "day 0" as "target users". For the 2,991 "baseline users", we define the 0 day as the first day of NEDAwareness campaign, loosely coupling the time span with that of target users.

Gender. To distinguish impact by gender of the users, we compile a (human) name dictionary with associated genders by combining names extracted from a large collection of Google+ accounts [19] with baby names published by National Records of Scotland[2] and United States National Security[3], resulting in a dictio-

[2] https://www.nrscotland.gov.uk/statistics-and-data/statistics/statistics-by-theme/vital-events/names/babies-first-names/full-lists-of-babies-first-names-2010-to-2014.

[3] https://www.ssa.gov/oact/babynames/limits.html.

nary containing 106,683 names. We use this name list to match to user names, as well as apply heuristics (such as having "Mrs." or "Mr."). Out of those selected for timeline analysis, 762 users were detected as female, 313 as male, and the remaining 671 as unknown. The baseline users were 855 female, 748 male and 1388 unknown, having a better coverage of the male gender.

Text Modeling. Instead of considering all words in this content, we group them using Linguistic Inquiry and Word Count (LIWC) dictionary [25], which has 72 categories of words grouped into (1) standard linguistic process, (2) psychological process, (3) relativity, and (4) personal concerns. We exclude categories dealing with basic grammar (parts of speech) and high-level summary ones for which more focused ones were available. Thus, for the present study, we select 51 categories including self-references (I, we, you, shehe), emotion (posemo, negemo, anxiety, anger), health and body (feel, body, health, sexual), psychology (focus present, focus future, swear) and other life aspects (work, leisure, home, money). Following this categorization, each tweet is represented as a 51-dimensional vector.

Effect Estimation. Using this vector, we measure whether a user's language changes from before to after interacting with NEDAwareness content. We use the Causal Impact Package[4] which estimates the causal effect of some intervention on a behavior over time. This method compares the changes between a response time series (our target users) and a set of control time series (baseline). Given these two series, the package constructs a Bayesian structural time-series model that builds a prediction of the time series if the intervention had never occurred, and compares it to the actual outcome [2]. For instance, the first panel Fig. 5 shows the actual tweet rate for the Female LIWC category as a solid line and the baseline tweet rate as a dashed line. The second shows the difference between observed data and baseline, and the third shows the cumulative effect of the intervention. In this example, we can observe that the rate of tweets having Female LIWC category is higher than that for the baseline.

Table 2 shows the categories which have the magnitude of relative effect of 1% or more and *p*-value <0.05 in the overall dataset. For these categories, users who have interacted with the NEDA content have changed the way they tweeted after the intervention beyond the changes in the general trend. We also show the effects for each gender separately – we observe that the effect is not evenly distributed between the genders.

Changes in Language. As can be seen from the table, the category showing most change after the intervention is *Female*, containing words such as *women*, *she*, *her*, etc. For example, the following tweet talks about the trans woman identity and emphases the word *women*: "*rt (USER): trans women are women. trans women are women. trans women are women. trans women are women. trans women are women. trans women are...*". In particular, the time series for

[4] https://google.github.io/CausalImpact/CausalImpact.html.

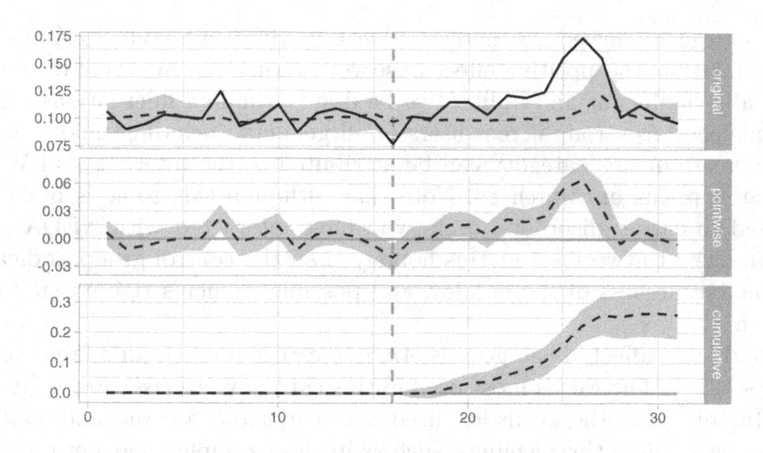

Fig. 5. Time series in causal impact analysis for Female category, top: observed tweet rate (solid) and baseline (dashed), middle: difference between the two, bottom: cumulative effect after intervention.

Table 2. Relative effect of interaction with NEDA content upon users' use of LIWC categories (with example words in parentheses). Significance: . $p < 0.05$, * $p < 0.01$.

Word category	All	Female	Male	Unkn.
Female (women, her, she)	17.4*	13.1*	24.3*	24.0*
Anxiety (risk, stress, upset)	7.6*	13.0*	−2.7	10.9*
Family (family, daughter, families)	6.5*	7.0*	4.2	15.2*
Money (donate, donation)	6.0*	4.9*	14.0*	4.1
Religion (church, goddess)	5.2	0.7	8.3	12.5*
Achievement (team, queen, celebrat*)	3.8	5.2	−2.9	4.0
They (they)	3.4	5.8*	1.6	0.7
Negate (don't)	2.9*	4.4*	−6.6	6.5
Health (maternity)	2.5*	4.2	9.0*	−2.5
Power (help, threat, terror)	2.5	1.0	1.3	4.5*
Negative emotions (risk, stress, upset)	2.1	3.9*	−3.4	0.6
Informal (retweet, twitter, fb)	1.1	0.0	3.2*	0.7
See (look, bright, show)	−2.0	2.0	0.2	−7.9*
Ipron (I)	−1.5	−0.2	−3.0	−1.2
Discrepancy (inadequa*)	−2.0	−0.6	−9.5*	1.2
You (You)	−2.2	−3.1	1.0	−2.9
Different (different, didn't)	−2.7*	−0.5	−5.6	−0.8
Positive emotion (share, sharing, help)	−3.3*	−3.1*	−7.5*	−1.4
Tentative (unsure, confusing, confused)	−3.3*	−0.9	−3.1	−6.0*
She/he (his, he, her, she)	−7.0	−7.9	−1.8	−4.5
Affiliation (we, our, us)	−7.2*	−5.9*	−8.9*	−6.9*

Female category showed an increase of +17% (95% interval [+11%, +23%]). This means that the positive effect observed during the intervention period is statistically significant and unlikely to be due to random fluctuations. Interestingly, this effect is strong across users of all genders, including unknown. Some of the increase in this category can be attributed to the International Women's Day that happens on March 8[5]. Note that although this topic is not directly connected to eating disorders, users who have interacted with NEDA content are more likely to tweet about this holiday than the control group, indicating a heightened awareness of the holiday, and possible women's rights issues associated with it.

Second most affected category is *Anxiety* category, with an increase of +8% ([+3%, +12%]). The words most used in this category are *risk*, *stress*, *upset* and *worry*. Interestingly, the words less used were *confusing*, *horrible* and *doubts*. For instance, users share their feelings, such as in this example: "*currently I am restless, scared, mistrustful, rattled, insecure, frightened, impatient, anxious.*". The category is significantly different for female users (as well as unknown gender), and not male. Figure 6(c,d) shows the top 10 words associated with anxiety category words for female and male users. We find female users mentioning words "struggling" and "struggle", as well as "mental", "depression" and "eating". On the other hand, many keywords on male side are associated with finances, such as "@financialbuzz" and "$safe" (by convention, "$" precedes tickers or financial information), as well as "cse" (National Child Sexual Exploitation Awareness).

Third most affected category is *Family*, with an increase of +6% ([+2%, +11%]). This category includes words related to the family. Interestingly, the words within that category which are most used are again closely related to women, which are *ma*, *daughter*, and *family*. Note that *wife*, *bro* and *daddy* are the least used within this category. For example, the following tweet emphasizes the female members of families and their needs: "*rt (user): they are our sisters, nieces, cousins, daughters, aunts, granddaughters, wives, mothers, grandmothers, friends... they need...*". Figure 6(a,b) shows the top 10 words most associated with the family category (those found in the context of tweets also containing words from family category) for the male and female gender. We find that, whereas women mention "kids" and "pregnant", for men the emphasis is more on "friends", "life", and "time". We also notice a keyword related to "FathersRightsHQ", which posts a mixture of political news and mental issues associated with family.

On the bottom of Table 2, we find categories which are used less by the users after interacting with NEDA content. It is more difficult to provide concrete examples of content *not* posted, but we draw the reader's attention to the fact that these categories are *Affiliation* (ex: *boyfriend*, *our*, *together*), *She/he* (ex: *he*, *she*, *herself*), and *Positive emotion* (ex: *amazing*, *favorite*, *sharing*).

Finally, we model the text produced by all of these users in the 15 days after NEDAwareness event via LDA at k = 15 topics (selected manually for greatest cohesion). Figure 7 shows the prevalence of each topic in each detected gender,

---

[5] https://en.wikipedia.org/wiki/International_Women's_Day.

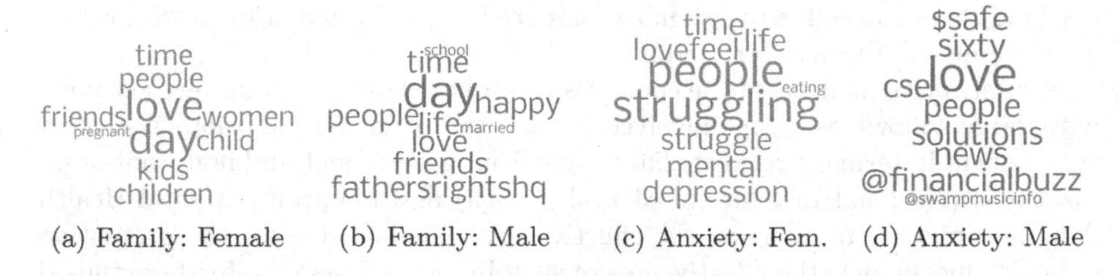

(a) Family: Female (b) Family: Male (c) Anxiety: Fem. (d) Anxiety: Male

Fig. 6. Top 10 words associated with family and anxiety categories, by gender.

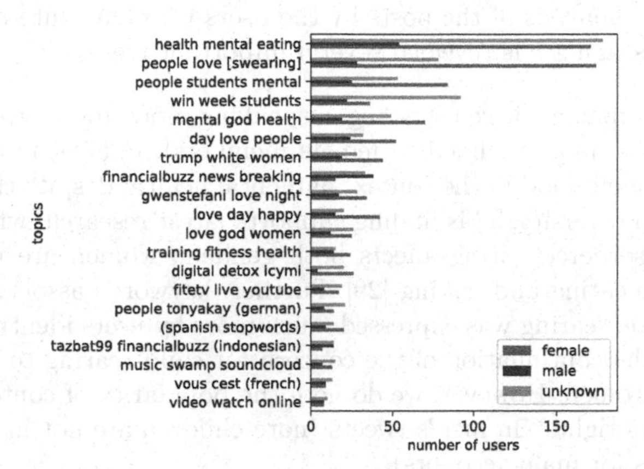

Fig. 7. Number of users assigned to LDA topics, per gender.

with each topic signified by the top descriptive terms. Topics are assigned by the greatest likelihood to the aggregated tweets for each user. As there are more female users, we find most topics to be dominated by them, except the topics around "financialbuzz". Other topics having a larger proportion of males are around "students", "training" and "fitness", as well as around music ("gwenstefani"). As we can see, the main topics of family, anxiety, and womanhood do not show up in these topics, illustrating the need for finer-grained analysis above.

5 Discussion and Conclusions

Modeling user behavior through online self-expression is an important complement to the traditional survey-based methods of behavior change evaluation, extending the reach of analysis at a low cost. Social informatics community has recently focused on health recommender systems and gamification, for instance, estimating the factors determining a mobile user's perception of the recommendation [30], testing gamified persuasive messaging for behavior change [6], and developing game design guidelines for improving subjective wellbeing [5]. This work extends the purview of the intervention to social media – a platform increasingly used for health messaging – and proposes an unobtrusive methodology for tracking change in self-expression, as compared to pre-intervention levels, as well as in comparison to a control group.

Quantitative analysis of the content's reach has shown that, despite accounts with large followings being involved in the campaign's promotion, the most engagement in terms of retweets have come from government and nonprofit organizations, (National Institute of Mental Health) and nonprofit (Mental Health America) putting in question the effectiveness of influencers for the promotion of health messaging (the effectiveness of such influencers may be further studied through the lens of social-psychological theory [21]).

Further, our analysis of the posts by the users who have interacted with the campaign's messaging has revealed several important trends:

- After the campaign, these users began speaking more about *women* and *family* (latter also more inclined to female members), indicating a general concern over womanhood in the context of mental health, despite the focus of the campaign on diversity. This finding supports latest research, which finds that although disordered eating affects both genders, women are more likely to report binge eating and fasting [29]. Further, keywords associated with children and childbearing was expressed much more by users identified as female, suggesting the continuation of the concern for child-rearing to be largely the purview of women. However, we do note the popularity of content associated with "fathers rights" in men's tweets (note children are not in the list of top family terms for male accounts).
- Secondly, despite positive body acceptance messaging of the campaign, we find a marked increase in anxiety-related and decrease in positive emotion words, suggesting the audience of the intervention has the need to share negative experiences. Indeed, mental health self-disclosure has been observed on several social media platforms, and been compared to a virtual "support group" [22]. How large organizations like NEDA fit into such a community is an interesting research question. In particular, the rise in anxiety-related words was more pronounced for users identified as female, instead of male ones (although the smaller number of male users may have played a role in statistical significance calculations). The outright mentions of "depression" and "mental" keywords by female users may point to *psychological openness* that has been recorded in women to seek help from mental health professionals [18].
- This leads to our third observation: the decrease in the use of words such as *boyfriend*, *our*, and *together* and others in the *affiliation* category, indicating a comparative lack of social engagement, as expressed by the users. Unlike in other word categories discussed above, this change is more prevalent for male users. As social isolation has been shown to affect vulnerable youths [28], it may be an important component of well-being to track.

A number of notable limitations must be mentioned. First, all purely observational studies are limited to public behaviors people choose to share with others. To complement the public view of an individual with the private, we plan on extending this study with surveying, diaries, and other traditional techniques (such as in [1] which evaluated the impact of a breast cancer education drive on Twitter). Second, the time span of the analysis should be lengthened beyond 15 days studied here to measure the long-term behavior of the subjects, as well as tracking their "information diet" [16] that may reinforce or undermine the desired behavior. Third, the demographics of the impacted population are unclear, with limited location and gender information available on Twitter, and further studies in technology usage will allow for a more precise estimates of message exposure [20]. Finally, this work does not utilize the images posted by the users, and in future work we plan to extract further features from multimedia, following recent work in [11].

Importantly, mental health research deals with potentially vulnerable populations, and whereas in this work only the largest accounts were revealed and example tweets rephrased as much as possible for de-identification, privacy is an ongoing concern. To limit the exposure of the individuals involved, the data will be made available to other researchers in anonymized fashion, and in accordance with EU General Data Protection Regulation (GDPR).

References

1. Attai, D.J., Cowher, M.S., Al-Hamadani, M., Schoger, J.M., Staley, A.C., Landercasper, J.: Twitter social media is an effective tool for breast cancer patient education and support: patient-reported outcomes by survey. J. Med. Internet Res. **17**(7), e188 (2015)
2. Brodersen, K.H., Gallusser, F., Koehler, J., Remy, N., Scott, S.L.: Inferring causal impact using Bayesian structural time-series models. Ann. Appl. Stat. **9**, 247–274 (2014). http://research.google.com/pubs/pub41854.html
3. Chancellor, S., Lin, Z.J., Goodman, E.L., Zerwas, S., De Choudhury, M.: Quantifying and predicting mental illness severity in online pro-eating disorder communities. In: Conference on Computer-Supported Cooperative Work and Social Computing (CSCW) (2016)
4. Chou, W., Oh, A., WP, K.: Addressing health-related misinformation on social media. JAMA **320**, 2417–2418 (2018). http://dx.doi.org/10.1001/jama.2018.16865
5. Ciocarlan, A., Masthoff, J., Oren, N.: Qualitative study into adapting persuasive games for mental wellbeing to personality, stressors and attitudes. In: Adjunct Publication of the 25th Conference on User Modeling, Adaptation and Personalization. UMAP 2017, pp. 402–407. Association for Computing Machinery, New York (2017). https://doi.org/10.1145/3099023.3099111
6. Ciocarlan, A., Masthoff, J., Oren, N.: Kindness is contagious: study into exploring engagement and adapting persuasive games for wellbeing. In: Proceedings of the 26th Conference on User Modeling, Adaptation and Personalization. UMAP 2018, pp. 311–319. Association for Computing Machinery, New York (2018). https://doi.org/10.1145/3209219.3209233
7. Csardi, G., Nepusz, T.: The igraph software package for complex network research. InterJ. Complex Syst. **1695**, 1–9 (2006). http://igraph.org
8. De Choudhury, M.: Anorexia on tumblr: a characterization study. In: Proceedings of the 5th International Conference on Digital Health 2015, pp. 43–50. ACM (2015)

9. De Choudhury, M., Gamon, M., Counts, S., Horvitz, E.: Predicting depression via social media. ICWSM **13**, 1–10 (2013)
10. De Choudhury, M., Kiciman, E., Dredze, M., Coppersmith, G., Kumar, M.: Discovering shifts to suicidal ideation from mental health content in social media. In: Proceedings of the 2016 CHI Conference on Human Factors in Computing Systems, pp. 2098–2110. ACM (2016)
11. Guntuku, S.C., Preotiuc-Pietro, D., Eichstaedt, J.C., Ungar, L.H.: What Twitter profile and posted images reveal about depression and anxiety. In: Proceedings of the International AAAI Conference on Web and Social Media, vol. 13, pp. 236–246 (2019)
12. Himelboim, I., Han, J.Y.: Cancer talk on Twitter: community structure and information sources in breast and prostate cancer social networks. J. Health Commun. (2014). https://doi.org/10.1080/10810730.2013.811321
13. Holland, G., Tiggemann, M.: "Strong beats skinny every time": disordered eating and compulsive exercise in women who post fitspiration on Instagram. Int. J. Eat. Disord. (2017). https://doi.org/10.1002/eat.22559
14. Kiciman, E., Counts, S., Gasser, M.: Using longitudinal social media analysis to understand the effects of early college alcohol use. In: Twelfth International AAAI Conference on Web and Social Media (2018)
15. Kıcıman, E., Thelin, J.: Answering what if, should i and other expectation exploration queries using causal inference over longitudinal data. In: Conference on Design of Experimental Search and Information Retrieval Systems (DESIRES) (2018)
16. Kulshrestha, J., Zafar, M.B., Noboa, L.E., Gummadi, K.P., Ghosh, S.: Characterizing information diets of social media users. In: Ninth International AAAI Conference on Web and Social Media (2015)
17. Levine, A.S.: Eating disorders and obesity-a comprehensive handbook. Am. J. Clin. Nutr. (2003). https://doi.org/10.1093/ajcn/77.5.1343
18. Mackenzie, C.S., Gekoski, W., Knox, V.: Age, gender, and the underutilization of mental health services: the influence of help-seeking attitudes. Aging Ment. Health **10**(6), 574–582 (2006)
19. Magno, G., Weber, I.: International gender differences and gaps in online social networks. In: Aiello, L.M., McFarland, D. (eds.) SocInfo 2014. LNCS, vol. 8851, pp. 121–138. Springer, Cham (2014). https://doi.org/10.1007/978-3-319-13734-6_9
20. McCloud, R.F., Kohler, R.E., Viswanath, K.: Cancer risk-promoting information: the communication environment of young adults. Am. J. Prev. Med. **53**(3), S63–S72 (2017)
21. McNeill, A.R., Briggs, P.: Understanding Twitter influence in the health domain: a social-psychological contribution. In: Proceedings of the 23rd International Conference on World Wide Web, pp. 673–678 (2014)
22. Naslund, J., Aschbrenner, K., Marsch, L., Bartels, S.: The future of mental health care: peer-to-peer support and social media. Epidemiol. Psychiatr. Sci. **25**(2), 113–122 (2016)
23. Olteanu, A., Varol, O., K\ic\iman, E.: Distilling the outcomes of personal experiences: a propensity-scored analysis of social media. In: Proceedings of The 20th ACM Conference on Computer-Supported Cooperative Work and Social Computing (2017)
24. Pedersen, T.L.: ggraph: an implementation of grammar of graphics for graphs and networks (2019). https://CRAN.R-project.org/package=ggraph, r package version 2.0.0

25. Pennebaker, J.W., Francis, M.E., Booth, R.J.: Linguistic inquiry and word count: Liwc 2001Lawrence Erlbaum Associates, Mahway, vol. 71, no. 2001, p. 2001 (2001)
26. Radloff, L.S.: The CES-D scale: a self-report depression scale for research in the general population. Appl. Psychol. Meas. **1**(3), 385–401 (1977)
27. Saha, K., Sugar, B., Torous, J., Abrahao, B., Kıcıman, E., De Choudhury, M.: A social media study on the effects of psychiatric medication use. In: Proceedings of the International AAAI Conference on Web and Social Media, vol. 13, pp. 440–451 (2019)
28. Storch, E.A., Milsom, V.A., DeBraganza, N., Lewin, A.B., Geffken, G.R., Silverstein, J.H.: Peer victimization, psychosocial adjustment, and physical activity in overweight and at-risk-for-overweight youth. J. Pediatr. Psychol. **32**(1), 80–89 (2007)
29. Striegel-Moore, R.H., et al.: Gender difference in the prevalence of eating disorder symptoms. Int. J. Eat. Disord. **42**(5), 471–474 (2009)
30. Torkamaan, H., Ziegler, J.: Rating-based preference elicitation for recommendation of stress intervention. In: Proceedings of the 27th ACM Conference on User Modeling, Adaptation and Personalization. UMAP 2019, pp. 46–50. Association for Computing Machinery, New York (2019). https://doi.org/10.1145/3320435.3324990
31. Vraga, V., et al.: Cancer and social media: a comparison of traffic about breast cancer, prostate cancer, and other reproductive cancers on Twitter and Instagram. J. Health Commun. (2018). https://doi.org/10.1080/10810730.2017.1421730
32. West, R., White, R.W., Horvitz, E.: From cookies to cooks: insights on dietary patterns via analysis of web usage logs. In: Proceedings of the 22nd international conference on World Wide Web, pp. 1399–1410 (2013)
33. Yom-Tov, E., Fernandez-Luque, L., Weber, I., Crain, S.P.: Pro-anorexia and pro-recovery photo sharing: a tale of two warring tribes. J. Med. Internet Res. **14**(6), e151 (2012)

Moral Framing and Ideological Bias of News

Negar Mokhberian[1(✉)], Andrés Abeliuk[1], Patrick Cummings[2], and Kristina Lerman[1]

[1] Information Sciences Institute, University of Southern California, Marina del Rey, CA, USA
{nmokhber,aabeliuk,lerman}@isi.edu
[2] Aptima, Woburn, MA, USA
pcummings@aptima.com

Abstract. News outlets are a primary source for many people to learn what is going on in the world. However, outlets with different political slants, when talking about the same news story, usually emphasize various aspects and choose their language framing differently. This framing implicitly shows their biases and also affects the reader's opinion and understanding. Therefore, understanding the framing in the news stories is fundamental for realizing what kind of view the writer is conveying with each news story. In this paper, we describe methods for characterizing moral frames in the news. We capture the frames based on the Moral Foundation Theory. This theory is a psychological concept which explains how every kind of morality and opinion can be summarized and presented with five main dimensions. We propose an unsupervised method that extracts the framing Bias and the framing Intensity without any external framing annotations provided. We validate the performance on an annotated twitter dataset and then use it to quantify the framing bias and partisanship of news.

Keywords: Framing · Bias · Moral foundation theory · News

1 Introduction

A growing share of Americans receives vital information and news from online sources [8]. In recent years, however, the online information environment has grown increasingly polarized along political and ideological lines [21]. Anecdotal evidence suggests that the shift to online news consumption has accelerated during the uncertainty and the 'fog of war' created by the Covid19 pandemic,[1] accompanied by growing ideological polarization, which colors how information about the pandemic is produced and consumed [24]. These converging developments suggest a need for accurate methods to quantify the ideological bias of

[1] https://www.wsj.com/articles/trump-mobilizes-the-white-house-to-tackle-coronavirus-but-adds-to-the-fog-of-war-11585934823.

S. Aref et al. (Eds.): SocInfo 2020, LNCS 12467, pp. 206–219, 2020.
https://doi.org/10.1007/978-3-030-60975-7_16

news sources. The resulting metrics could help the public become more responsible consumers of news by cutting through polarization.

Measuring ideological biases from text is a problem that has only recently attracted attention from the Natural Language Processing (NLP) community. Beyond identifying topics in the text, this research attempts to uncover how implicit biases manifest themselves through language and how to infer them from the text. Research has shown that language captures implicit associations that shape how people perceive the world and relationships within it, including gender and career representations that lead people to associate women with artistic careers and men scientific careers [3]. In cognitive linguistics and communication theory, the different perspectives and meanings that texts carry are captured by "semantic frames". When applied to news media, framing captures how news stories express meaning and cultural values and how they evoke an emotional reaction from readers. Media framing has been studied as a tool to influence the opinion of newsreaders [18].

Political scientists have characterized the framing of text using Moral Foundations Theory (MFT) [11]. MFT defines five moral foundations which categorize the intuitive ethics and automatic emotional reactions to various social situations. These foundations concern dislike for the suffering of others (care/harm), dislike of cheating (fairness/cheating), group loyalty (loyalty/betrayal), respect of authority and tradition (authority/subversion), and concerns with purity and contamination (sanctity/degradation). Across cultures, assessments of moral foundations reveal correlations between political ideology and the five foundations [10], as well as interesting relationships between gender, culture, religion, and personality. For example, consistently and across studies, liberals and conservatives draw upon these foundations to different degrees, measured via questionnaires [9]: liberals assign more weight to the Harm and Fairness foundations, whereas conservatives are more sensitive to the Ingroup, Authority, and Purity.

We use MFT as a generalizable framework to quantify the framing Bias in the news. The five moral foundations (care/harm, fairness/cheating, loyalty/betrayal, authority/subversion, sanctity/degradation) give a high-level understanding of the values promoted by news sources, which we use to quantify the biases inherent in how the news is framed by different sources.

In this paper, we propose a framework to quantify the moral framing of texts. The framework leverages a large corpus of tweets annotated with their moral foundations [12]. We use sequence embeddings as features to train a classifier to predict the scores of text corresponding to the moral dimensions. Additionally, instead of using the embedding directly, we evaluate a low-dimensional feature representation of the text based on the Frame Axis approach [16]. The Frame Axis method computes the Bias and Intensity of each moral frame based on alignments of the text relative to specific target words. We show that these simple features capture the moral frames implicit in the text, at least as well as much more complex (high-dimensional) representations. Finally, we show that moral frames help with the prediction of the partisanship of news based on the headlines. Our work demonstrates the feasibility of automatically classifying the moral framing and political partisanship of news.

2 Background

Moral Foundations Theory. Moral Foundation Theory (MFT) has been first introduced by Haidt and Joseph to explain moral differences across cultures [11]. The theory introduces five basic moral foundations which are the basis of many intuitive and cultural human psychological values. These five dimensions consist of Care/Harm, Fairness/Cheating, Loyalty/Betrayal, Authority/Subversion, and Purity/Degradation. Later, Graham and Haidt showed that Liberals and Conservatives have substantial variation in their moral concerns across these five Moral Foundations [9]. This triggered future works analyzing the political rhetoric based on MFT and characterizing the political ideologies using the five Moral Foundations.

Other than the works in the psychology domain, there are other works considering MFT in computational linguistic approaches. They are mostly using the lexical resource Moral Foundation Dictionary (MFD) [9]. This dictionary consists of words regarding virtues and vices of the five Moral Foundation dimensions and a sixth dimension regarding general morality terms. Studies rely on the usage of words with respect to the Moral Foundation dictionary terms. For example, to analyze temporal changes in the frequency of MFT terms in English books for the years 1900 to 2007 [25], or to demonstrate changes in the Moral Foundation language regarding the word 'gay' in the US political Senate speeches from 1988 to 2012 [7]. The study showed that republicans were significantly using more Purity words than Democrats.

Recently, novel methods have been developed to recognize if a text is relevant to any of the MFT dimensions. This problem has been formulated as a classification problem on tweets in [7,14] and on news in [6].

NLP. Some of the previous work on analyzing MFT in text corpora have relied on the word counts [6]. Other studies [7,15] have used features based on word embeddings [20] or sequence embeddings [5] to obtain more robust models and better performance. Recent work has proposed methods for analyzing rhetorical frames in text. In SemAxis [1], the authors introduce semantic axis which are word-level domain semantics structured on word antonym pairs. The similarity of a word with respect to different predefined antonymous axes can capture the semantics of the word in various contexts. (e.g., the word 'soft' can be a negative word in the context of sports and positive in the context of toys). The word and the semantic axis are represented in the same representation vector space trained on a corpora. The authors of SemAxis later introduced the concept "Frame Axis" which is a method of characterizing the framing of a text by identifying the most relevant semantic axis [16]. Similarly, studies have explored using groups of words with opposite meanings to define semantic dimensions to improve the interpretability of text representations [19].

3 Methods

In this section, we describe our computational framework to measure the framing bias of news sources based on the text of their headlines. The framework is composed of two tasks: (1) An unsupervised method to learn a low-dimensional representation characterizing the text with respect to a set of target words; (2) A supervised classification task using Twitter data with human-annotated MF framing, with contextualized language representation models.

In the first task, we represent the text with scores according to the moral dimensions, computing framing Bias and Intensity scores for each MF using the Frame Axis [16] method. This approach projects the words on micro-frame dimensions characterized by two sets of opposing words. These MF framing scores capture the ideological slant of the news source and can be used as features for predicting partisanship. Details are given in Sect. 3.2.

In the second task, we leverage the annotated Twitter dataset [12] to develop a supervised model to classify the MFT related micro-frames in the news headlines (see Sect. 3.3). In Sect. 4.1, we validate the accuracy of different latent representations for the text on the annotated Twitter data set.

Finally, in Sect. 4.3 we present a case study of our framework analyzing the moral framing differences between Liberal and Conservative media in USA news sources.

3.1 Data

News Articles. We use "All the News" dataset from Kaggle[2]. The data primarily falls between the years 2016 and July 2017. The news sources include the New York Times, Breitbart, CNN, Business Insider, the Atlantic, Fox News, Buzzfeed News, National Review, New York Post, the Guardian, NPR, Reuters, Vox, and the Washington Post. For each news story, we have worked with its headline and publication (the news outlet). Also, we have looked up the political leaning of each source from allsides[3] website and have added a column indicating the political side of each news. We have eliminated the news stories from central sources and only kept the liberal and conservative leaning sources. We have narrowed down the headlines to the ones related to immigration and elections topics. We did this by checking the headlines to include some hand-picked words regarding each topic. Among several topics, we chose these two because the frequency of articles falling in these two topics was larger. After all the above steps, the data consisted of 3242 news articles about immigration and 29345 regarding elections.

Annotated MF Data. We use the annotated twitter dataset [12] to train and test classifiers for each MF micro-frame. Several trained human annotators determined which of the MF virtues or vices are most relevant for each tweet,

[2] https://www.kaggle.com/snapcrack/all-the-news.
[3] https://www.allsides.com/media-bias/media-bias-ratings.

Table 1. Some of the positive and negative words (virtues and vices) associated with the five dimensions of the Moral Foundations Theory and general morality.

Care	Fairness	Ingroup	Authority	Purity	General morality
Virtues					
Care	Fair	Ally	Abide	Austerity	Blameless
Benefit	Balance	Cadre	Allegiance	Celibate	Canon
Amity	Constant	Clique	Authority	Chaste	Character
Caring	Egalitarian	Cohort	Class	Church	Commendable
Compassion	Equable	Collective	Command	Clean	Correct
Empath	Equal	Communal	Compliant	Decent	Decent
Guard	Equity	Community	Control	Holy	Doctrine
Peace	Fairminded	Comrade	Defer	Immaculate	Ethics
Protect	Honest	Devote	Father	Innocent	Exemplary
Safe	Fair	Familial	Hierarchy	Modest	Good
Secure	Fairly	Families	Duty	Pious	Goodness
Shelter	Impartial	Family	Honor	Pristine	Honest
Shield	Justice	Fellow	Law	Pure	Legal
Sympathy	Tolerant	Group	Leader	Sacred	Integrity
Vices					
Abuse	Bias	Deceive	Agitate	Adultery	Bad
Annihilate	Bigotry	Enemy	Alienate	Blemish	Evil
Attack	Discrimination	Foreign	Defector	Contagious	Immoral
Brutal	Dishonest	Immigrant	Defiant	Debase	Indecent
Cruelty	Exclusion	Imposter	Defy	Debauchery	Offend
Crush	Favoritism	Individual	Denounce	Defile	Offensive
Damage	Inequitable	Jilt	Disobey	Desecrate	Transgress
Destroy	Injustice	Miscreant	Disrespect	Dirt	Wicked
Detriment	Preference	Renegade	Dissent	Disease	Wretc.hed
Endanger	Prejudice	Sequester	Dissident	Disgust	Wrong
Fight	Segregation	Spy	Illegal	Exploitation	
Harm	Unequal	Terrorist	Insubordinate	Filth	
Hurt	Unfair		Insurgent	Gross	
Kill	Unjust		Obstruct	Impiety	

or if there are no moral concepts related to that tweet. In total there are 11 dimensions for each tweet (virtues and vices of the five MFT dimensions and also non-moral dimension). Each annotator can assign more than one MF dimension to a single tweet. To aggregate the votes for a single tweet, we have assigned 1 to the dimensions having at least two votes and 0 to the dimensions that have less than two votes. There were 35k tweet ids provided in this dataset and at least three annotators per tweet.

3.2 Quantifying Moral Frames with Frame Axis

For the unsupervised method, we quantify the strength of the moral framing of text along the dimensions of MFT using the Frame Axis approach [16]. Frame

Axis proposes two measures—*Intensity* and *Bias*—to capture the document-level framing based on the word contributions [16]. Intensity and Bias for a text are calculated as the weighted average of mapping of its words towards the desired semantic axis.

Each semantic axis (also called micro-frame) builds on a set of antonyms, i.e., words with opposite meaning [1]. In our case, we choose the vices and virtues of the Moral Foundations as opposites of a word axis, e.g.. Care terms vs Harm terms from the MFT. Some of these words are shown in Table 1. For each moral foundation (MF) dimension, the axis is calculated by subtracting the average vector of the embeddings of positive words (virtues) and the average vector of the embeddings of negative words (vices) of that MF dimension. Formally, let m be one of the MF dimensions (e.g.. Care) and V_m^+ denote the set of embedding vectors of virtue words (e.g.. vectors for Care words) and V_m^- denote to set of vectors of vice words (e.g. vector for Harm words), then the semantic axis corresponding to this MF dimension is:

$$A_m = mean(V_m^+) - mean(V_m^-) \quad (1)$$

For the computations of this part, the embeddings of words are obtained from the pretrained GloVe model [20] called "Common Crawl" which includes 2.2M vocab[4]. Following [16], we define the *framing Bias* B_m^D of a document D along a semantic axis m as:

$$B_m^D = \frac{\sum_{d \in D} f_d \; s(A_m, d)}{\sum_{d \in D} f_d}, \quad (2)$$

where a document $D = \{d_1, \ldots, d_n\}$ is defined as a set of embeddings of its words; $s(A_m, d)$ is the cosine similarity between the semantic axis A_m and word d; f_d is the frequency of word d in the document. In other words, the Bias of a text with regard to a moral foundation axis m is the weighted average of the cosine similarity of its words with that axis. Notice that, if the embeddings represent sentences then there are no repetitions, i.e., $f_d = 1$. The absolute value of the Bias captures the relevance of the document to the moral dimension, while the sign of the similarity captures a bias toward one of the poles in the moral dimension. The positive sign of Bias will shows the document is aligned with the positive pole of the Axis and negative sign shows the opposite.

Second, we use *framing Intensity* on a moral dimension to capture how heavily that moral dimension appears in the document with respect to the background distribution:

$$I_m^D = \frac{\sum_{d \in D} f_d \left(s(A_m, d) - B_m^T\right)^2}{\sum_{d \in D} f_d}, \quad (3)$$

where B_m^T is the baseline framing Bias of the entire text corpus T on a moral dimension m. Intensity doesn't reveal information about the polarization. However, in situations that both poles of an axis actively appear in a text, the positive and negative terms will cancel out each other, and the document wouldn't

[4] https://nlp.stanford.edu/projects/glove/.

Table 2. Evaluation of moral foundation classifiers on annotated tweets.

Moral foundation	Precision	Recall	F1	Accuracy	Baseline F1	Baseline Acc.
BERT embedding features						
AVG	0.771	0.822	0.775	0.822	0.705	0.705
Authority	0.808	0.875	0.817	0.875	0.778	0.776
Fairness	0.662	0.774	0.681	0.774	0.655	0.655
Harm	0.746	0.768	0.734	0.768	0.613	0.613
Ingroup	0.802	0.873	0.816	0.873	0.779	0.781
Purity	0.910	0.935	0.908	0.935	0.527	0.527
Morality	0.698	0.705	0.694	0.705	0.879	0.879
Frame axis features						
AVG	0.787	0.818	0.773	0.818	0.705	0.705
Authority	0.883	0.888	0.851	0.888	0.778	0.776
Fairness	0.770	0.795	0.743	0.795	0.655	0.655
Harm	0.694	0.740	0.666	0.740	0.613	0.613
Ingroup	0.799	0.873	0.816	0.873	0.779	0.781
Purity	0.898	0.933	0.907	0.933	0.879	0.879
Morality	0.676	0.683	0.655	0.683	0.527	0.527

have a significant Bias toward any pole of that axis, but Intensity will show the relevance to that axis.

As a result, each document can be represented by 12 dimensions, each representing the Bias and Intensity scores for each of the 6 MF dimensions in Table 1.

3.3 Classifying Moral Frames from Text

In the supervised approach, we leverage the corpus of tweets, manually annotated with their moral foundations, as train data to learn a classifier model on MF frames from text.

We trained a binary classifier on the twitter dataset to learn the relevance of the Moral Foundations. Specifically, each MF is considered as a label that can get values 0 or 1, and for each MF we train a binary classifier. A Logistic Regression classifier is used for this part.

For creating text features, we use a contextualized sequence encoding method known as Bidirectional Encoder Representations from Transformers (BERT) [5] to obtain the embeddings for each tweet. We encode each tweet in a 768-dimensional encoding using a pre-trained BERT model.

In the inference phase, we convert the text in the test dataset to the same feature space, and each classifier gives a likelihood showing how much the given text is related to the corresponding moral foundation.

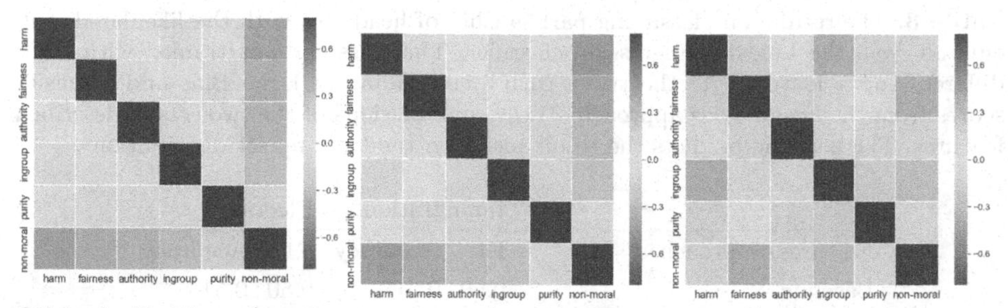

Original Twitter Annotations | Twitter learned frame scores | News learned frame scores

Fig. 1. The correlation (a) among the count of annotators selected a MF for each tweet (b) among the MF frame likelihoods learned from the supervised method on the tweets and (c) on the news headlines.

4 Results

4.1 Measuring Moral Framing from Text

First, we evaluate the ability of the two latent representations to measure MF frames from the text on the annotated twitter dataset. We compare the classification performance of the tweet BERT embeddings to the Frame Axis features on the annotated tweets. We run the classification task repeatedly on random 0.75/0.25 train/test splits. Classification results in Table 2 show that both approaches dramatically outperform the baseline. The baseline predicts each moral foundation according to its frequency distribution in the training set. They also outperform the method from [12], which reported $F1 < 0.5$ on a subset of the twitter dataset. Remarkably, Frame Axis achieves comparable performance to the embedding-based approach using only two features (Bias and Intensity) for each moral dimension.

4.2 Relationships Between Moral Framings

In this section, we explore the empirical correlations among the different MF frames learned from the data. We take inspiration from previous work studying correlations using surveys based on the Moral Foundations Questionnaire [9]. For this purpose, we compute the correlation matrix using two approaches: 1) based on the original hand-annotated MF frames on the Twitter dataset; and 2) based on the inferred frames that we obtained from our supervised method. With the latter method, we can infer the MF frames correlation matrix for the Twitter data and the news articles corpus. To evaluate the method, we compare the inferred and hand-annotated correlation of MF frames. Finally, we explore the correlations learned from the news headlines.

The annotated Twitter dataset contains the number of human annotators who have selected each tweet as relevant to a MF frame. We can compare the

Table 3. The results on classifying partisanship of headlines with the likelihoods calculated from the Logistic Regression classifier. The classifier was trained with three different input features: 1) the tweet BERT embeddings; 2) the Bias and Intensity scores from the Frame Axis approach; 3) the combinations of the two sets of described features. The baseline predicts the result based on the training set distributions.

	Immigration		Election	
Features/approach	F1	Accuracy	F1	Accuracy
Baseline	0.50	0.51	0.50	0.51
MF Likelihoods	0.61	0.63	0.60	0.61
Frame axis	0.68	0.69	0.66	0.66
MF Likelihoods + Frame Axis	0.69	0.70	0.67	0.67

correlation matrix of these expert annotations to the likelihoods calculated by the classification task for each MF frame. Figure 1 shows that the correlations between moral foundation dimensions on the raw annotations (Fig. 1a) are very similar to the correlations between MFs likelihoods predicted by the supervised classifier on the same Twitter data (Fig. 1b). Even though we are using distinct classifiers for each MF and our classifier does not see different labels at the same time, still the correlations in (Fig. 1b) are comparable to (Fig. 1a), showing that the automatically detected frames are consistent with human judgment. For example, we see that similar to the original Twitter dataset, the correlation of non-moral and other moral foundations stay negative, which makes sense because if a text is not showing any MF frame, then the highest score for it must be the non-moral label and all the other frame scores must be low. However, when we use the supervised classifier to compute MF frame scores on the news headlines, we see different correlations between the moral dimensions (Fig. 1c). Here, Purity is more correlated with the Care/Harm dimension, and Ingroup is more correlated with Fairness and Authority than in Twitter data.

4.3 Moral Framing and Partisanship of News

Predicting Partisanship of News. Using moral frames as features, we classify the partisanship of news headlines. We have chosen Immigration and Elections as two categories of news for experimenting. Our dependent variable is the partisanship of the news source (Liberal or Conservative), and we use Bias and Intensity scores for each moral foundation and the likelihoods obtained from the classifier trained on the tweets as features to test for systematic differences in the moral framing of news. We test whether moral frame scores can distinguish between ideologies of news sources from different political sides. The Bias and Intensity are unsupervised measures since for calculating those, no annotations are needed. Another feature set for representing MF framing can be obtained providing the news headlines as test data to the classifier previously trained on the annotated twitter data. This supervised classifier gives likelihoods corresponding to each of the MF dimensions which we use to classify the partisanship.

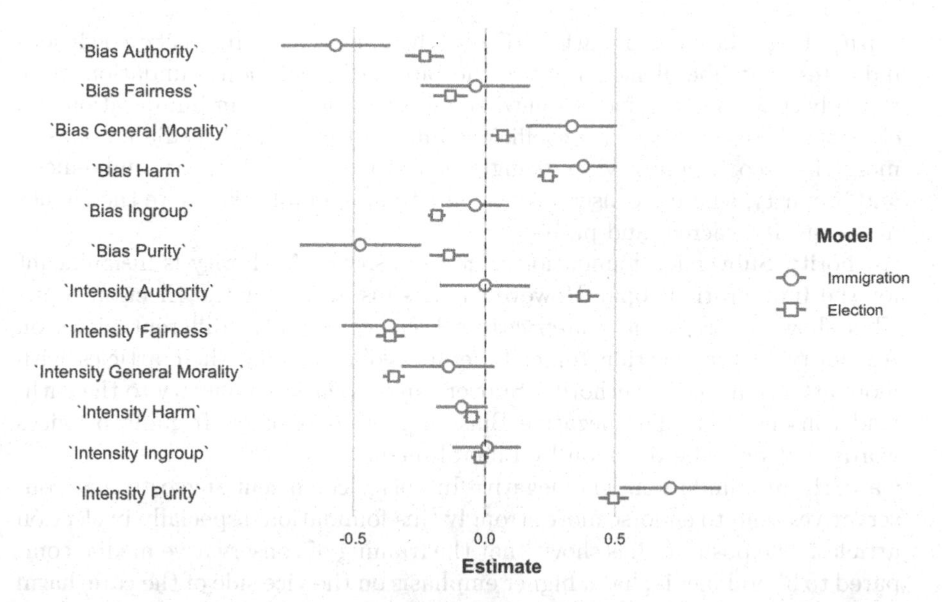

Fig. 2. Coefficients of Logistic Regression classification task using moral frames to predict liberal partisanship. The coefficients whose 0.95 confidence intervals exclude 0 are significant. The binary classifier target label is the partisanship with Liberal as 1 and Conservative as 0.

Table 3 shows F1 and accuracy of classifying the partisanship based on supervised MF likelihoods and unsupervised Frame Axis scores used as features. The row 'combine' denotes concatenating these two feature sets. The baseline is a simple classifier that learns the distribution of partisanship from the training data and uses it to make predictions for the test data. Features based on moral frames outperform the baseline by a wide margin.

Moral Framing of News. Next, to quantify systematic differences in moral framing between liberal and conservative news articles, we inspect the model coefficients learned by the partisanship classifier. One key aspect of using the Frame Axis approach is that the model coefficients are straightforward to interpret (see Sect. 3.2) and have a competitive performance. Figure 2 reports the Frame Axis coefficients and their 0.95 confidence intervals for Intensity and Bias features for each moral foundation. Since in our setting, we coded the label for Conservative partisanship as 0 and Liberal partisanship as 1, the interpretation of a positive coefficient is that all else being equal, Liberal news articles are more likely than Conservative articles to exhibit the corresponding attribute. We highlight the following findings:

- The sign of the coefficients are consistent across the two topics analyzed and show significant differences across partisanship for most moral foundations.

- Purity/Degradation foundation. The high and positive Intensity coefficient indicates that liberal media stress the purity/degradation foundation more strongly than conservative media when reporting news about immigration and elections. The negative bias coefficient implies that liberal media emphasize more the recognition of vices being violated regarding dirtiness, unholiness, and impurity, whereas conservative media tend to emphasize more the virtues like austerity, sacred, and pure.
- Authority/Subversion foundation. The Intensity of Authority is insignificant for the Immigration topic. However, it is substantial for the Election topic. This shows liberals and conservatives hold significantly different views on Authority in the Election topic. Liberal media, framing their articles with more attention to the Authority/Subversion foundation, contrary to the withheld consensus [9]. The negative Bias suggests a stronger framing on vices words that describe rebellion by Liberal media.
- Care/Harm foundation. The negative Intensity coefficient suggests that conservatives tend to endorse more strongly this foundation, especially in election articles. The positive Bias shows that the framing of conservative media, compared to liberal media, has a higher emphasis on the vice side of the care/harm moral dimension, which condemns malice, abuse, and inflicting suffering.
- The large positive coefficient of Bias for General Morality suggests that liberal media tends to frame their news about immigration with a more positive moralistic view, indicative of normative judgments (e.g., good, moral, noble). While Conservative media use a more negative moral judgment based on words like bad, incorrect, or offensive.

5 Conclusion

Recent research puts in evidence a change in partisanship among the general electorate, where the number of issues with partisan conflict has increased [2]. Studies suggest a link between partisan media exposure and polarization [22], driven by motivated reasoning to explain why partisan media polarizes viewers [17]. An alternative approach emphasizes the impact on how the news media frame political discourse. In their seminal study, Kahneman and Tversky, demonstrate the impact of framing in human decisions [23]. Since then, there has been vast experimental evidence on how moral framing can influence attitudes towards polarizing issues like climate change [13].

In this paper, we focused on developing computational methods to detect moral framing Biases in news media using NLP techniques. Inspired by the work done on Frame Axis [16], identifying meaningful micro-frames from antonym word pairs, coupled with the principled moral foundation theory, we choose the vices and virtues of the five Moral Foundations as polar opposites of a micro-Frame Axis. We proceed to first, validate our approaches on an annotated Twitter dataset, and second, study the moral framing Bias on partisan news articles related to immigration and elections topics. Our findings reveal systematic differences across liberal and conservative media. It supports the correlations between political ideology and the five foundations which have been shown in the empirical evidence based on surveys [9]. In particular, our results suggest that rhetoric on Purity is different between liberal media compared to conservative media, where liberals tend to use more rhetoric towards the violations of the Purity moral foundation. This observation, supporting the growing evidence that Purity is among the most differentiating moral dimension between conservatives and liberals [4].

Our work has some limitations that should be noted. First, our methods rely on the dataset, and changing the dataset might change the results. We have demonstrated our methods on two different topics immigration and elections. Second, in the calculation of Bias, as defined in Eq. (2), the negations in the sentence which can change the polarization are not considered. e.g.., "This is not fair" would have a positive Bias in Fairness micro-frame because the presence of the word "not" is not considered in the definition of Bias. However, this problem does not appear in Intensity because that measures the frequency of usage of words from both poles of each axis. Lastly, even though there is a twitter dataset annotated with moral foundations, there is no annotated dataset of news articles. In the supervised method described in Sect. 3.3, we have trained the classifier on the labeled tweets and then predicted the moral foundations of news headlines using that model.

As future work, we plan to study different topics and compare framing across different news sources. Most of the studies have paid attention to liberal and conservative news sources. An interesting question is how central news sources MF framing would be different from the polarized ones? Another path to continue this work will be leveraging the BERT text encodings by fine tuning it according to our MF frame classification task.

Acknowledgements. This work was conducted in connection with Contracts #W56KGU-19-C-0004 with the U.S. Army Combat Capabilities Development Command C5ISR Center. The views, opinions, and findings contained in this document are those of the authors and should not be construed as an official position of the United States Army.

References

1. An, J., Kwak, H., Ahn, Y.Y.: SemAxis: a lightweight framework to characterize domain-specific word semantics beyond sentiment. In: Proceedings of the 56th Annual Meeting of the Association for Computational Linguistics (Volume 1: Long Papers), pp. 2450–2461, July 2018
2. Brewer, M.D.: The rise of partisanship and the expansion of partisan conflict within the American electorate. Polit. Res. Q. **58**(2), 219–229 (2005)
3. Caliskan, A., Bryson, J.J., Narayanan, A.: Semantics derived automatically from language corpora contain human-like biases. Science **356**(6334), 183–186 (2017)
4. Dehghani, M., et al.: Purity homophily in social networks. J. Exp. Psychol. Gen. **145**(3), 366 (2016)

5. Devlin, J., Chang, M.W., Lee, K., Toutanova, K.: BERT: pre-training of deep bidirectional transformers for language understanding. In: Proceedings of the 2019 Conference of the North American Chapter of the Association for Computational Linguistics: Human Language Technologies, Volume 1 (Long and Short Papers), pp. 4171–4186, June 2019
6. Fulgoni, D., Carpenter, J., Ungar, L., Preoţiuc-Pietro, D.: An empirical exploration of moral foundations theory in partisan news sources. In: Proceedings of the Tenth International Conference on Language Resources and Evaluation (LREC 2016), pp. 3730–3736, May 2016
7. Garten, J., Boghrati, R., Hoover, J., Johnson, K.M., Dehghani, M.: Morality between the lines: detecting moral sentiment in text. In: Proceedings of IJCAI 2016 Workshop on Computational Modeling of Attitudes (2016)
8. Gottfried, J., Shearer, E.: News Use Across Social Medial Platforms 2016. Pew Research Center (2016)
9. Graham, J., Haidt, J., Nosek, B.A.: Liberals and conservatives rely on different sets of moral foundations. J. Pers. Soc. Psychol. **96**(5), 1029 (2009)
10. Graham, J., Nosek, B.A., Haidt, J., Iyer, R., Koleva, S., Ditto, P.H.: Mapping the moral domain. J. Pers. Soc. Psychol. **101**(2), 366 (2011)
11. Haidt, J., Joseph, C.: Intuitive ethics: how innately prepared intuitions generate culturally variable virtues. Daedalus **133**(4), 55–66 (2004)
12. Hoover, J., et al.: Moral foundations twitter corpus: a collection of 35k tweets annotated for moral sentiment. Soc. Psychol. Pers. Sci. (2020)
13. Hurst, K., Stern, M.J.: Messaging for environmental action: the role of moral framing and message source. J. Environ. Psychol. **68**, 101394 (2020)
14. Johnson, K., Goldwasser, D.: Classification of moral foundations in microblog political discourse. In: Proceedings of the 56th Annual Meeting of the Association for Computational Linguistics (Volume 1: Long Papers), pp. 720–730, July 2018
15. Kennedy, B., Atari, M., Davani, A.M., Hoover, J., Omrani, A., Graham, J., Dehghani, M.: Moral concerns are differentially observable in language (2020). https://doi.org/10.31234/osf.io/uqmty, https://doi.org/10.31234/osf.io/uqmty
16. Kwak, H., An, J., Ahn, Y.Y.: FrameAxis: characterizing framing bias and intensity with word embedding (2020)
17. Levendusky, M.S.: Why do partisan media polarize viewers? Am. J. Polit. Sci. **57**(3), 611–623 (2013)
18. Liu, S., Guo, L., Mays, K., Betke, M., Wijaya, D.T.: Detecting frames in news headlines and its application to analyzing news framing trends surrounding U.S. gun violence. In: Proceedings of the 23rd Conference on Computational Natural Language (CoNLL), pp. 504–514 (2019)
19. Mathew, B., Sikdar, S., Lemmerich, F., Strohmaier, M.: The polar framework: Polar opposites enable interpretability of pre-trained word embeddings. In: Proceedings of The Web Conference 2020, WWW 2020, p. 1548–1558 (2020)
20. Pennington, J., Socher, R., Manning, C.D.: Glove: global vectors for word representation. In: Empirical Methods in Natural Language Processing (EMNLP), pp. 1532–1543 (2014)
21. Prior, M.: Media and political polarization. Annu. Rev. Polit. Sci. **16**, 101–127 (2013)
22. Sunstein, C.R.: Going to Extremes: How Like Minds Unite and Divide. Oxford University Press, Oxford (2009)

23. Tversky, A., Kahneman, D.: Rational choice and the framing of decisions. In: Karpak, B., Zionts, S. (eds.) Multiple Criteria Decision Making and Risk Analysis Using Microcomputers. NATO ASI Series (Series F: Computer and Systems Sciences), vol. 56. Springer, Heidelberg (1989). https://doi.org/10.1007/978-3-642-74919-3_4
24. Van Bavel, J.J., et al.: Using social and behavioural science to support COVID-19 pandemic response. Nat. Hum. Behav. **4**, 464–471 (2020)
25. Wheeler, M.A., McGrath, M.J., Haslam, N.: Twentieth century morality: the rise and fall of moral concepts from 1900 to 2007. PLoS one **14**(2) (2019)

Malicious Bot Detection in Online Social Networks: Arming Handcrafted Features with Deep Learning

Guanyi Mou(✉) and Kyumin Lee(✉)

Worcester Polytechnic Institute, Worcester, MA 01609, USA
{gmou,kmlee}@wpi.edu

Abstract. Online social networks (OSNs) have long been suffering from various types of malicious bots (e.g., spammers, fake followers, social bots, and content polluters). Recent studies show that they have also been actively involved in delivering hate speeches and disseminating misinformation. Over several years, researchers have proposed multiple approaches to identify some types of them to lower their impact on the OSNs. However, their strategies mostly focused on handcrafted features to capture characteristics of malicious users, or their deep learning approaches may only work under certain situations (e.g., under the dense retweets/sharing behavior). To overcome the limitation of the prior work, in this paper, we propose a novel framework that incorporates handcrafted features and automatically learned features by deep learning methods from various perspectives. It automatically makes the balance between them to make the final prediction toward detecting malicious bots. In particular, we (i) combine publicly available 15 Twitter user datasets and categorize these accounts into two groups (i.e., legitimate accounts and malicious bot accounts); and (ii) propose a deep learning framework that jointly learns various features and detects malicious accounts. Our experimental results show that our proposed model outperforms 7 state-of-the-art methods, achieving 0.901 accuracy. Our ablation study shows that all types of our features positively contribute to enhancing the model performance.

Keywords: Malicious bot detection · Deep learning

1 Introduction

Malicious bots have misused the power of Online social networks (OSNs) such as Twitter, Facebook, and Weibo, continuously caused significant disturbance to the overall online social environment, and shaped unhealthy trends, bias, and misbelief in societies (e.g., COVID-19 related misinformation [27]). Their accounts[1] have made severe impact and damage to the OSNs by causing inconveniences, intensifying contradictions, and aggravating prejudices [1].

[1] We use terms user and account, interchangeably.

S. Aref et al. (Eds.): SocInfo 2020, LNCS 12467, pp. 220–236, 2020.
https://doi.org/10.1007/978-3-030-60975-7_17

Despite the long history of causing ongoing negative impact, malicious bots did not quit on being the Grand Villain on the OSNs. They have been emerging, evolving, and participating in new types of destructive activities. Reports[2] and analysis of malicious bots involved in hate speech dissemination [2,43] and fake news propagation [35,36] show their seemingly ever-lasting significant impact. Efficiently and accurately detecting them is still a crucial problem.

In recent years, OSN service providers established policies for warning, blocking, and suspending malicious accounts[3]. According to our study described in Appendix A.1, some of these malicious accounts are still alive for years without any suspension or proper treatment.

Researchers have proposed approaches to detect specific types of malicious bots [6,10,14,18,29]. Even though these approaches identified some malicious bots, we are still facing new challenges with new malicious bots such as hashtag promoters and social spambots, especially, political bots and even extremists such as ISIS recruiters [3,4]. Most of the existing frameworks identify new groups of useful handcrafted features and then apply them to traditional machine learning classifiers for satisfying results. They are thus placing the performance and robustness on an intuitively vulnerable position, as manipulators can play with those handcrafted features and deploying direct adversarial attacks against them.

Deep learning techniques, however, were less addressed in this domain. To the best of our knowledge, existing deep learning frameworks for malicious bot detection are, to some extent, limited in analyzing and using some particular perspectives of OSN accounts, usually only focusing on capturing temporal patterns [11,32] or simple single tweet patterns [28].

To fill this gap, we propose a unified deep learning framework, which analyzes both temporal patterns and posting contents, and also incorporates handcrafted features. There are a few challenges. First, how to collect information for various types of malicious bots? Second, how can we extract features which distinguish between malicious users and legitimate users? Third, how can we create a unified framework that is capable of effectively detect malicious bots?

By keeping these challenges in mind, in this paper, we combine publicly available Twitter datasets, which contain accounts of content polluters, fake followers, traditional spambots, social spambots, and legitimate users. Then, we extract handcrafted features, and automatically learned features by deep learning methods. Finally, we combine both features and make a balance between them toward building a malicious bot detection model.

In this paper, we make the following contributions:

- We propose a novel joint learning framework that is capable of detecting various malicious bots altogether and distinguishing them against legitimate accounts. It combines both handcrafted features (i.e., profile and activity features and LIWC-based personality features) and automatically learned features (i.e., temporal behavior related features and text related features).

[2] https://bit.ly/39mGlnm and https://bit.ly/3hlpt38.
[3] https://help.twitter.com/en/rules-and-policies/twitter-rules.

- Our model outperforms 7 state-of-the-art methods, and we analyze intuitively and logically for the good performance.
- We conduct an ablation study, which shows that all components/feature types positively contribute to our proposed model's performance.

2 Related Work

Specific types of malicious bots were studied in the past. Some researchers focused on analyzing and detecting content polluters and spammers in OSNs [7,21,29,33]. Their classification methods focused on different perspectives such as temporal patterns of behaviors [10], social networks [6] and others [21,34]. DARPA held a twitter bot challenge [37] for better understanding and detecting bots. Davis *et al.* [18] proposed a framework *BotOrNot* (later on evolved and renamed as *Botometer*) which was trained on a dataset of malicious users. Adewole *et al.* [1] made a thorough review of 65 bot detection papers. Alfifi *et al.* [3,4] studied the behavior of long-lived and eventually suspended Arabic Twitter accounts in social media (especially ISIS accounts), and also discussed the level of automatic/botness of these accounts. Their results showed that the percentage of automated posting behaviors was relatively high. There were also researches revealing that malicious bots were involved in trending topics by disseminating misinformation and hate speech [2,35,36,43].

In malicious bot detection architectures, researchers have developed many traditional machine learning models [12,30]. Cresci *et al.* [14,17] proposed a DNA inspired model that produced a relatively good result without much detailed information from users. Yang *et al.* [42] proposed methods from another perspective, where they enhanced the performance with data selection.

To the best of our knowledge, although models relying on handcrafted features provide somehow compelling and convincing results in their previous experiments, they may face two vital challenges: 1) *Handcrafted features are not entirely scalable:* as the number of proposed features is increasing; it is getting harder to discover novel and helpful features. Thus the performance improvement/gain is reduced as researchers add additional handcrafted features into their models. For example, the *Botometer* [18] used more than one thousand handcrafted features. It is intuitively way too challenging for human intelligence to come up with new ways/ideas of inventing additionally useful features. Many researchers thus turn to focus on more complicated features such as the features extracted from network information based on trust propagation theories. The trend here also went down deeper and thus more computation consuming: for example, to achieve better performance, Beskow *et al.* [6] categorized information into four tiers, and eventually used all of them. The last tier involved using friends' timeline information. This network information is large in size and collection time, so not being feasible in many practical cases. 2) *Handcrafted features provided a clear target for adversarial attacks:* Manipulators of malicious bots can play with their profiles, posting contents and patterns to reduce distinguishing power of the handcrafted features and avoid the detection. Thus, the arms race might get harder for classifiers and their robustness gets decreased as malicious bots change their behavior [40].

Some deep learning frameworks are applied in learning from temporal behavior information [11,32] based on the assumption that malicious bots' behaviors are hard to hide as automated mechanisms follow the designed patterns that are relatively regular and not as random as human. Other deep learning frameworks are applied in learning from text information [28]. But, the deep learning frameworks do not focus on enough scope of the whole picture of each user. Some frameworks require the datasets to satisfy specific properties, to enable them to come up with effect: for example, *RTBust* [32] focuses on the retweet-tweet patterns of user timelines, so if the user's retweet counts are way too sparse, the framework will intuitively not work well. We show the performance of this model as a baseline in the experiment section. As certain special-purpose bots, such as fake followers, may not have enough posting timestamps to reveal non-human-like patterns, deep learning models (e.g., Chavoshi *et al.* [11]) only encoding temporal patterns would not work well. Overall, the performance of existing frameworks, which rely on learned features, often cannot reach as high accuracy as handcrafted features. We conjecture one of the main reasons is that those frameworks did not include wide scope of user information.

Our framework differs from the prior researches in the following ways: (1) We combine the advantages of both handcrafted features and automatically learned features via deep learning, thus being scalable and performance promising; (2) Under the limited information in the publicly available datasets, we use only limited user profile information, posting content information, and posting timestamps, while not expanding to the expensive network information; and (3) We design our framework in learning both temporal posting features and language model features, and the mechanism in handling the balance of these features to achieve desirable performance.

Table 1. Dataset status.

Source	Lee	Cresci	Cresci	Gilani	Varol	Cresci	Midterm	Botometer	Botwiki	Celebrity	Cresci	Political	Pronbots	Vendor	Verified	Sum
Year	'11	'15	'17	'17	'17	'18	'18	'19	'19	'19	'19	'19	'19	'19	'19	
Legit	18537	-	1077	1140	1343	5269	7027	61	-	1293	296	-	-	-	1898	37941
Bot	17241	4685	5465	970	665	5875	25	18	98	-	269	13	1568	605	-	37497
Sum	35778	4685	6542	2110	2008	11144	7052	79	98	1293	565	13	1568	605	1898	75438

3 Dataset and Account Types

We chose Twitter as the primary social networking site in this study because Twitter has a generous data sharing policy to the third party researchers, and some public datasets with rich information are available. But, our proposed framework is generally applicable to other social networking sites with minor modification. To avoid potential bias and subjective labeling caused by collecting and labeling data by ourselves and include various types of malicious bots, we used 15 Twitter benchmark datasets from Botometer's repository[4]. They come from various sources [13–16,23,29,32,38,41,42] and contain many types of users (content polluters, fake accounts, traditional & social spambots, stock related bots, political bots, fake followers, verified accounts, celebrity accounts, legitimate users, etc.). Each dataset's name as well as their original user types are described in Appendix A.2.

In the datasets, *Lee'11*, *Cresci'15*, and *Cresci'17* included the original profile information, posted tweets, and timestamps, while the other datasets only included the user IDs on Twitter. Thus we kept the above mentioned three datasets' original data and collected the other datasets' user information via Twitter API in April 2020. We grouped legitimate users, verified accounts and celebrity accounts as legitimate, while other types of accounts as malicious bots.

We then filtered out accounts with inconsistent labels, no posting information, or non-English posting. Table 1 shows the final detailed number of accounts that we used for the experiment. Some datasets originally only contain one type of user accounts (e.g., Vendor'19 and Botwiki'19). Overall, our final dataset consists of 37,941 legitimate accounts and 37,497 malicious bots. The dataset is thus almost naturally and perfectly balanced. The alive malicious bots that we successfully fetc.hed from Twitter API are intuitively harder to detect as they managed to evade the platform's detection mechanisms. In other words, detecting these long-surviving malicious bots is a hard and important problem.

4 Our Framework

In this section, we introduce our framework, which combines handcrafted features and learned features through deep neural networks. We show a general view of our framework in Fig. 1. The framework is composed of three parts:

- ***Feature Extraction***: handles with handcrafted features.
- ***Feature Learning***: learns useful embeddings from multiple perspectives.
- ***Decision Making***: combines all the features and embeddings together and makes the final prediction.

We show some of the hyper-parameters in the figure. The detailed settings are described in Sect. 5. Table 2 presents a list of our extracted/learned features.

4.1 Feature Extraction

We extract handcrafted features in two ways:

We first extracted 20 widely used handcrafted features (called traditional features) from each account. They have been proven to be useful, and positively contributed to the performance of models in the literature [29,38]. Those features mainly come from user profile information, and some counting based frequency/ratio from user posting contents. As user profile information is naturally

[4] https://botometer.iuni.iu.edu/bot-repository/datasets.html.

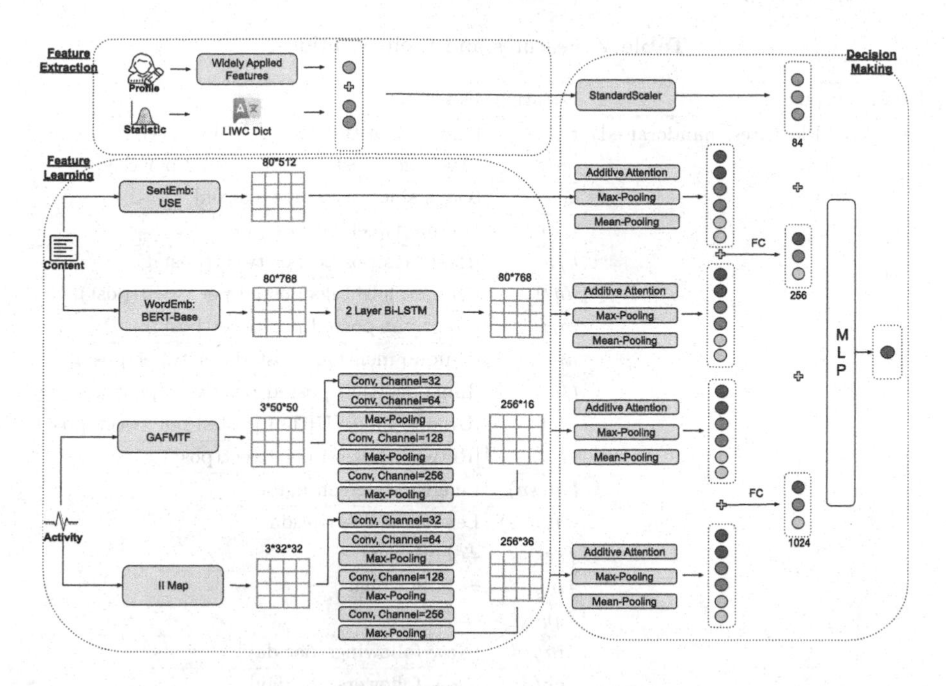

Fig. 1. Our overall framework. *FC* stands for fully connected layer.

categorized, it is intuitively straightforward to use handcrafted features for better measurement. For unique posts, we first translate all links to the same word "URL", anonymize mentions, hashtags, and special tags, and then count the number of unique posts based on these transformed data. For the compression ratio of user tweets, we used Python's zip package with its default zip setting. For extreme cases where a user has no follower, we adopt the #Followings as the ff ratio. We transform the existing/living seconds of accounts into the unit of days in floating number, and as described above, we guarantee every account in our dataset has at least one posting record. Thus the ratios of the features will not face NaN or Inf problems.

We also extracted Linguistic Inquiry and Word Count (LIWC)[5], a dictionary for text analysis and personality analysis. It categorizes words into each meaningful types/groups. For each account in our dataset, we concatenate their tweets, count the number of meaningful words, and then calculate the occurrence of words belonging to each category. We naturally treat the proportion of these occurrences as features. Thus, the number of features we extract from LIWC equals to the number of categories of LIWC. We extracted 64 features by using the LIWC 2007 dictionary. From a high level of view, LIWC features capture the general statistics of each user's profile, activities, and their preferences in terms of word usage. Malicious bots serve for different purposes against legitimate

[5] http://liwc.wpengine.com/.

Table 2. Features and their notations.

Feature type	Notation	Description
Traditional features (handcrafted)	t	|Number of tweets posted by user|
	t/d	|Average tweets(posts) posted per day|
	dd	|Days since account creation|
	ut/t	Unique tweet(post) ratio
	h/t	|Hashtags posted per tweet(post)|
	uh/t	|Unique hashtags posted per tweet(post)|
	m/t	|Mentions posted per tweet(post)|
	um/t	|Unique mentions posted per tweet(post)|
	l/t	|Links (URLs) posted per tweet(post)|
	ul/t	|Unique links (URLs) posted per tweet(post)|
	rt/t	|Retweet posted per tweet(post)|
	$len(sn)$	Length of screen name
	$len(des)$	Length of description
	fer	# Followers
	$fing$	# Followings
	fav	# Favorites
	$fing/d$	|New followings per day|
	fer/d	|New followers per day|
	ff	Following follower ratio
	cr	Content compression ratio
LIWC features (handcrafted)	-	64 LIWC features
Text features	-	Sentence level embedding
	-	Word level embedding + Bi-LSTM
Temporal behavior	-	Inter-posting-time difference pair features
	-	GASF, GADF, MTF features

accounts. Since malicious bots disseminate specially designed messages, these features may reveal the difference.

The aforementioned ones sum up to 84 handcrafted features.

4.2 Feature Learning

We automatically learned useful embeddings/features from posting contents and temporal behaviors by using neural networks. Thus, the learning part is divided into two components: *Text Embeddings* and *Temporal Behavior Embeddings*.

Text Embeddings. Given strings $S = [s_1, s_2, \ldots, s_m]$, we encode them into 2D matrices $TextEmb(S) = [Encoder(s_i)]$, where $i = 1, 2, \ldots, m$. Each string in this context is a tweet's content. The Encoder differs in the following two ways:

1. **Sentence level embeddings (*SentEmb*)**: For all the tweets posted by the same user, each tweet is encoded by a fixed-length vector, which

represents its general information. A sentence embedding of each tweet would be $SentEmb(s_i) = v_i$, which is a 1D vector. By stacking all tweet vectors together, we derive a 2D matrix that contains all high-level information of user postings. These postings do not necessarily have any sequential relatedness, as each one is a unique sentence embedding.

2. **Word-level embeddings (*WordEmb*)**: For all tokens in all tweets, each word can be encoded by another fixed-length vector, which represents its semantic meaning. Word embeddings of each tweet would be $WordEmb(s_i) = W_i$, which is a 2D matrix. By concatenating all 2D matrices together, we also derive a 2D matrix. This matrix does contain sequential information as words' semantics are connected. Together they form unique synthetics.

SentEmb contains high-level information and thus can be used for possible sentence-level similarity comparison or repeated pattern learning. In contrast, *WordEmb* contains more detailed, more abundant information, which may help learn the difference between malicious or legitimate accounts in terms of frequent word usage difference, synthetic structure difference, sentiment difference, etc. By combining both embeddings, we derive better text representation. We used Universal Sentence Encoder (USE) [9] for SentEmb and BERT-Base [19] + Bi-LSTM for WordEmb representation.

Temporal Behavior Embeddings. Given the sequences of each user's posting timestamp $T = [t_0, t_1, \ldots, t_n]$, we applied two methods for mapping these sequences into 3D images: **GAFMTF** [39] and **II Map** [11] for informative pattern recognition. Then, we design two convolutional neural networks for learning features out of them.

1. **GAFMTF**: *Gramian Angular Field (GAF)* methods were proposed for a better encoding of timeseries in polar coordinate system than Cartesian coordinate representations. The authors [39] claimed their mapping method is invertible and also preserves absolute temporal relations. We first encode time difference sequences $TD = [t'_1, t'_2, \ldots, t'_n]$ into polar coordinates, where $t'_i = t_i - t_{i-1}$. Then, we map it into a 2D plane. The GAF method provides summation-graph (GASF) and difference-graph (GADF),

$$GAF(TD) = \begin{bmatrix} <\widetilde{t'_1}, \widetilde{t'_1}> <\widetilde{t'_1}, \widetilde{t'_2}> \cdots <\widetilde{t'_1}, \widetilde{t'_n}> \\ <\widetilde{t'_2}, \widetilde{t'_1}> <\widetilde{t'_2}, \widetilde{t'_2}> \cdots <\widetilde{t'_2}, \widetilde{t'_n}> \\ \vdots \qquad \vdots \qquad \ddots \qquad \vdots \\ <\widetilde{t'_n}, \widetilde{t'_1}> <\widetilde{t'_n}, \widetilde{t'_2}> \cdots <\widetilde{t'_n}, \widetilde{t'_n}> \end{bmatrix}$$

$$<\widetilde{t'_i}, \widetilde{t'_j}> = \widetilde{t'_i} \cdot \widetilde{t'_j} - \sqrt{1 - \widetilde{t'_i}^2} \cdot \sqrt{1 - \widetilde{t'_j}^2}, \text{ for GASF}$$

$$<\widetilde{t'_i}, \widetilde{t'_j}> = \sqrt{1 - \widetilde{t'_i}^2} \cdot \widetilde{t'_j} - \widetilde{t'_i} \cdot \sqrt{1 - \widetilde{t'_j}^2}, \text{ for GADF}$$

and $\widetilde{t'_i}$ is t'_i in polar coordinates.

Markov Transition Field (MTF) method analyzes the timeseries change from another perspective: the transition probability of value changes across the time series. The authors also reported its effectiveness, especially stacked with GAF outcome matrices.

We first split TD into Q quantile bins (here we set $Q = 2$), and then assign t'_i into corresponding quantile bins $q_i, i \in Q$. Next, we measure the transition probability of sequence value change in terms of different bins.

$$MTF(TD) = \begin{bmatrix} w_{ij|t'_1 \in q_i, t'_1 \in q_j} & \cdots & w_{ij|t'_1 \in q_i, t'_n \in q_j} \\ w_{ij|t'_2 \in q_i, t'_1 \in q_j} & \cdots & w_{ij|t'_2 \in q_i, t'_n \in q_j} \\ \vdots & \ddots & \vdots \\ w_{ij|t'_n \in q_i, t'_1 \in q_j} & \cdots & w_{ij|t'_n \in q_i, t'_n \in q_j} \end{bmatrix}$$

where w_{ij} is a transition probability of t'_{k+1} in quantile bin q_j, given t'_k in quantile bin q_i, for all $k \in N$.

We stacked 3 2D matrices (i.e., GASF, GADF & MTF) to produce a 3D matrix.

2. **II Map**: Unlike GAFMTF as a general sequence encoding method, this method was explicitly proposed for bot detection. This method tends to focus more on the pairwise sequence neighborhood pattern.

$IPT(T, lag) = [(t_0, t_{0+lag}), (t_1, t_{1+lag}), \cdots, (t_{n-lag}, t_n)]$,

where pairs of tweet posting timestamp differences are mapped into 2D planes. Upon the control of different "lag" values, we also come up with a 3D matrix. In this paper, we set $lag = 1, 2, 3$. We can interpret this method as inter-posting-time (IPT) mapping or IPT embedding.

Given the generated two 3D matrices (by *GAFMTF* and *II Map*) as inputs, we design two similar, but independent convolutional neural networks for learning useful features. Each convolutional neural network consists of four convolutional layers and three max-pooling layers as shown in Fig. 1. In all convolutional layers, the filter size is $3 * 3$ with $stride = (1, 1)$ and $padding = (1, 1)$. In all max-pooling layers, the window size is $2 * 2$ and $stride = (2, 2)$. We used batch normalization after each convolution layer, and LeakyReLU [31] as the default activation function. To the best of our knowledge, we are the first applying GAFMTF in bot detection domain.

4.3 Decision Making

Given extracted features and learned embeddings/features from the different components, we have to unify them and make full use of them for the final prediction. The design of decision making is non-trivial in three reasons:

- Learned embeddings are so far matrices, while handcrafted features are 1D vectors. A good mechanism for flattening those matrices is needed. A simple direct flattening may create too much redundant information. How to balance the relative size (number of features) among different parts of features? Auto learned features are scalable but maybe way larger than handcrafted features

in terms of the number of features. To avoid handcrafted features being overwhelmed, we have to design mechanisms to balance their contribution and significance well.
- How to handle the possible redundant information/noise inside each part of learned features? Especially, learned embeddings/features are large in size, so there may be possible correlations among them. A good design should have mechanisms to reduce the size of each part as well as to reduce correlations of each feature toward improving the decision making performance.
- Handcrafted features may have greatly varying value scale, which might not fit well with the other features. We had to handle such a problem to enable our framework to have a smoother learning curve.

To overcome these difficulties, we design decision making part as follows:

- **To rescale a value scope of the handcrafted features**, we normalize them in the training set and apply the normalizer in the validation set and testing set, to ensure not having information leak across experiment sets. The details of data split are described in Sect. 5.
- **For flattening 2D matrices into 1D vectors**, those representation matrices were then fed into two independent additive attention [5] + max-pooling + mean-pooling mechanisms. This was partly inspired by ULMFit [24], where the authors of this language model reported such design helped increasing performance. The two vectors of each part were then concatenated together as the final text embedding representation.
- **For balancing the size of different components**, after doing matrix flattening and concatenation, text embeddings and temporal behavior embeddings go through two different fully connected layers for resizing the feature dimensions, where the output size of those two layers are tunable as hyperparameters.
- Eventually, we concatenate all three parts of features together and feed them into multiple fully connected layers (MLP) for the final prediction. To reduce feature co-adapting and model overfitting, we applied dropout at all layers except the last fully connected layer as implicit Bayesian approximation [22].

5 Experiment

5.1 Experimental Setting

Dataset. Given 37,497 malicious bots and 37,941 legitimate accounts. We randomly split the dataset with 60% for training, 20% for validation, and 20% for testing. Through the dataset splitting procedure, we manually guarantee that all source datasets have at least one account shown up in each split, so to ensure that the experiment results are fair to all sources.

Table 3. A general view of 7 baselines and our model (*JntL*).

Models	Using info.			Feature		Algo.		Domain specific		
Notation	Profile	Text	Timeseries	Handcraft	Auto	NN.	Trad. ML	Classify	Twitter	Bot detect
Lee'11	✓	✓		✓			✓	✓	✓	✓
Kim'14		✓			✓		✓	✓		
Tweet2Vec'16		✓			✓	✓		✓	✓	
Chavoshi'18			✓		✓	✓		✓	✓	✓
Kudugunta'18		✓		✓	✓	✓		✓	✓	✓
RTBust'19			✓		✓	✓	✓	✓	✓	✓
Yang'19	✓	✓	✓	✓			✓	✓	✓	✓
JntL	✓	✓	✓	✓	✓	✓	✓	✓	✓	✓

Data Coverage and Goal. As we mentioned in Sects. 2 and 3, baselines and our approach only access limited user profile information, posting content information, and posting timestamps without other expensive information such as social network information because of the limited information commonly available across the 15 public datasets. Our goal is to maximize the malicious bot detection performance under the available information. Additional information like the social network could potentially further improve model performance.

Baselines and Our Model. We implemented the 7 state-of-the-art baselines based on various perspectives of user account information. The baselines consist of *Lee'11* [29], *Kim'14* [25], *Tweet2Vec'16* [20], *Chavoshi'18* [11], *Kudugunta'19* [28], *RTBust'19* [32], *Yang'19* [42]. Our proposed framework called *JntL*, which is a joint learning model shown in Fig. 1. Table 3 shows a general view of the baselines and our model in terms of used user information, feature types and algorithm types. The detailed description of the baselines is presented in Appendix A.3. Our source code is available at https://github.com/GMouYes/MaliciousBotDetection.

Parameter Tuning and Evaluation Metrics. We used the reported best hyper-parameters of the baselines. If the authors of them did not report the best hyper-parameters, we conducted grid search to obtain the optimal baselines.

To help the reproducibility of our model, we report our settings and hyper-parameters other than those already shown in Fig. 1. Cross Entropy was chosen as the default loss function, and we used Adam as our optimizer. LeakyReLU with default parameters was chosen as the activation function for decision making layers. The Batch size is 128. The two-layer Bi-LSTM has hidden dimension of 256, to produce WordEmb based features, so the output of two directions is 512. The dropout between them is 0.2. The resizing fully connected layer for the text embedding layer is 256, while the resizing fully connected layer for the temporal behavior embedding layer is 1024. The decision making component (i.e., MLP) has two hidden layers, which have 512 nodes and 128 nodes. Its output layer produces a probability of being a malicious bot. The dropout rate of each layer in the decision making component is both 0.05. We evaluate all models based on Precision(Pre), Recall(Rec), F1, and Accuracy(Acc).

Table 4. Experimental results.

Class	Overall		Legitimate users			Malicious bots		
Measure	ACC.	F1	Pre.	Rec.	F1	Pre.	Rec.	F1
Lee'11	.874	.874	.875	.875	.875	.874	.874	.874
Kim'14	.829	.829	.838	.818	.828	.821	.841	.830
Tweet2Vec'16	.660	.660	.652	.696	.673	.670	.624	.646
Chavoshi'18	815	.815	.809	.828	.818	.822	.803	.812
Kudugunta'18	.837	.837	.854	.816	.834	.822	.859	.840
RTBust'19	.497	.342	.500	**.979**	.662	.322	.010	.019
Yang'19	.872	.872	.834	.922	.878	.912	.822	.865
JntL	**.901**	**.901**	**.886**	.922	**.903**	**.918**	**.880**	**.898**

5.2 Experimental Results

Our Model vs. Baselines. We compared our model against 7 baselines. Table 4 presents the experimental results. The best result of each column is marked in bold. Our model outperformed the baselines achieving 0.901 accuracy, improving 3% compared with the best baselines (*Lee'11* and *Yang'19*). This result indicates that our proposed framework is better than the baselines in terms of identifying both types of accounts.

Table 5. Ablation study result. "-" represents removing a corresponding feature type.

Class	Overall		Legitimate users			Malicious bots		
Model	Acc.	F1	Pre.	Rec.	F1	Pre.	Rec.	F1
JntL	.901	.901	.886	.922	.903	.918	.880	.898
-SentEmb	.889	.889	.901	.876	.888	.878	.902	.890
-WordEmb	.892	.892	.878	.912	.895	.907	.872	.890
-IPTEmb	.899	.899	.899	.899	.899	.898	.898	.898
-GAFMTF	.886	.886	.905	.864	.884	.868	.908	.888
-TraditionalFeatures	.887	.887	.882	.895	.888	.892	.880	.886

From the results, we conclude that jointly and thoroughly learned features provide better results than partially observed user's information. While analyzing user's postings provide rich information of malicious intent, text features like *Kim'14* take only the order of postings but does not pay attention to the exact posting time. Thus, incorporating temporal behavior would be complementary and helpful. However, given less active malicious users, posted messages and posting behavior may not be sufficient, so incorporating profile information would be helpful and necessary. Statistical information provides another general view of user behavior without caring about specific activities. Scalable auto learned features help boost the detection performance. *RTBust'19* does not perform well enough, as it actually requires the retweet behavior to reach a certain threshold. However this requirement is not generally met across all source datasets. In other words, the approach may require collecting each user's data with a longer time span. Baselines such as *Lee'11* and *Yang'19* using handcrafted features were relatively performed well, confirming usefulness of handcrafted features despite two drawbacks (i.e., not entirely scalable and a clear target for adversarial attacks) mentioned in Sect. 2. The other deep learning based baselines reached comparable results but not sufficient to beat the handcrafted features based baselines. We conjecture that this is mainly due to the fact that their work only focused on a part of user behaviors like retweets. On the contrary, our model's higher performance is because we considered a wider scope of user information, incorporated both handcrafted features and auto learned features, and made balance between them by using our decision making component.

5.3 Ablation Study

We conducted an ablation study to understand the contribution of each type of the features. Table 5 presents the experimental results when we remove one of the five feature types from our proposed model (*JntL*). For example, *-SentEmb* means excluding sentence level embeddings from *JntL*. We notice that (1) all types of features/embeddings positively contributed to the final prediction performance. (2) Even if we exclude traditional features (handcrafted features), ablation results still outperform all the baselines. The results reflect the success and effectiveness of our joint learning approach, where multi-perspective information based auto-learning provides a unique scalable advantage. GAFMTF features contribute the most, while WordEmb features contribute the least. This result is mainly because Sentence level embedding already captures part of the content information, while GAFMTF and IPTEmb encode temporal behaviors a lot differently. Automatically learned features easily scale better and thus provide helpful support to handcrafted features. Future work can be to explore other ways to learn scalable features through deep learning frameworks automatically.

6 Conclusion

In this paper, we aimed to detect various types of malicious bots altogether and distinguish them against legitimate users. In particular, we combined 15 publicly available Twitter datasets. We grouped accounts into two classes: (1) malicious bots; and (2) legitimate accounts. Then, we proposed a novel joint learning framework based on handcrafted features and auto learned features toward detecting malicious bots. Experimental results showed that our framework outperformed all baselines, achieving 0.901 accuracy, improving 3% against the best baseline. The ablation study provided supporting information indicating all parts of our framework non-trivially contributed to performance improvement.

Acknowledgements. This work was supported in part by NSF grant CNS-1755536, AWS Cloud Credits for Research, and Google Cloud.

A Appendix

A.1 Account Status

As we keep the original information of *Lee'11*, *Cresci'15* and *Cresci'17*, we checked the current status of those malicious bots as shown in Table 6. Overall 68.3% malicious bots are still alive on Twitter, some of which lived more than ten years. This fact indicates that there is a great room to improve the current Twitter's bot detection system.

A.2 Source Dataset Details

We list the original user types that each dataset contains as follows:
Lee'11 [29]: content polluters, and legitimate users
Cresci'15 [13]: various kinds of fake accounts
Cresci'17 [14]: traditional & social spambots, and legitimate users
Cresci'18 [15,16]: stock related bots, and legitimate users
RTBust'19 [32]: retweet bots, and legitimate users
Gilani'17 [23]: bots, and legitimate users
Varol'17 [38]: bots, and legitimate users
Midterm'18 [42]: political bots, and legitimate users
Botwiki'19 [42]: social and bots
Political'19 [41]: political bots
Pronbots'19 [41]: bots advertising scam sites
Vendor'19 [41]: fake followers
Verified'19 [41]: verified legitimate users
Celebrity'19 [41]: celebrity accounts (legitimate)
Botometer'19 [41]: bots and legitimate users

We grouped legitimate users, verified accounts and celebrity accounts as legitimate, while other types of accounts as malicious bots.

Table 6. Recent status of malicious accounts.

Source	Deleted	Suspended	Alive	Sum
Lee'11	1,417 (8.2%)	3,868 (22.4%)	11,956 (69.3%)	17,241
Cresci'15	282 (6.0%)	2,336 (49.9%)	2,067 (44.1%)	4,685
Cresci'17	344 (6.3%)	443 (8.1%)	4,678 (85.6%)	5465
Overall	2,043 (7.5%)	6,647 (24.3%)	18,701 (68.3%)	27,391

A.3 Detailed Baseline Descriptions

Lee'11 [29]. Authors proposed handcrafted features extracted from user profiles, posting contents and the change of following/follower list over time. We built their best Random Forest model without the network features.
Kim'14 [25]. This is a convolutional text classification architecture that achieved comparable performance against state-of-the-art models. Its hyper-parameters are stable across different domains. We applied his work in using the tweets posted by each user for classifying the accounts.
Tweet2Vec'16 [20]. Tweet2Vec was proposed as a general-purpose tweet embedding framework, trained with neural networks for the hashtag prediction subtask. This work generates domain-specific feature representations of tweets. We constructed a bot detection model, following the proposed architecture, where the embedding layer is followed with fully connected layers.
Chavoshi'18 [11]. Authors proposed a method for mapping the posting timestamp pairs into 2D images to make better use of the temporal posting behavior information of each account. Convolutional neural networks can be applied for the downstream bot detection task.
Kudugunta'19 [28]. This is a framework using LSTM for learning content features and then combine them with several handcrafted features.
RTBust'19 [32]. RTBust is a framework using temporal retweet/tweet patterns for bot detection. Such a framework captures the information in tweet/retweet sequences and extracted features using the variational autoencoder (VAE) [26]. Then the feature embedding generated by the encoders is fed into HDBSCAN [8], an unsupervised clustering method. Outliers are treated as malicious bots.
Yang'19 [42]. Random Forest built on various authors' proposed features.

References

1. Adewole, K.S., Anuar, N.B., Kamsin, A., Varathan, K.D., Razak, S.A.: Malicious accounts: dark of the social networks. J. Netw. Comput. Appl. **79**, 41–67 (2017)
2. Albadi, N., Kurdi, M., Mishra, S.: Hateful people or hateful bots? detection and characterization of bots spreading religious hatred in Arabic social media. CSCW (2019)
3. Alfifi, M., Caverlee, J.: Badly evolved? Exploring long-surviving suspicious users on twitter. In: Ciampaglia, G.L., Mashhadi, A., Yasseri, T. (eds.) SocInfo 2017. LNCS, vol. 10539, pp. 218–233. Springer, Cham (2017). https://doi.org/10.1007/978-3-319-67217-5_14
4. Alfifi, M., Kaghazgaran, P., Caverlee, J., Morstatter, F.: Measuring the impact of ISIS social media strategy. In: MIS2 (2018)
5. Bahdanau, D., Cho, K., Bengio, Y.: Neural machine translation by jointly learning to align and translate. In: ICLR (2015)
6. Beskow, D.M., Carley, K.M.: Bot conversations are different: leveraging network metrics for bot detection in twitter. In: ASONAM (2018)
7. Bhat, S.Y., Abulaish, M.: Community-based features for identifying spammers in online social networks. In: ASONAM (2013)

8. Campello, R.J.G.B., Moulavi, D., Sander, J.: Density-based clustering based on hierarchical density estimates. In: Pei, J., Tseng, V.S., Cao, L., Motoda, H., Xu, G. (eds.) PAKDD 2013. LNCS (LNAI), vol. 7819, pp. 160–172. Springer, Heidelberg (2013). https://doi.org/10.1007/978-3-642-37456-2_14
9. Cer, D., et al.: Universal sentence encoder for English. In: EMNLP (2018)
10. Chavoshi, N., Hamooni, H., Mueen, A.: Temporal patterns in bot activities. In: WWW (2017)
11. Chavoshi, N., Mueen, A.: Model bots, not humans on social media. In: ASONAM (2018)
12. Conroy, N.J., Rubin, V.L., Chen, Y.: Automatic deception detection: methods for finding fake news. In: Proceedings of the 78th ASIS&T Annual Meeting (2015)
13. Cresci, S., Di Pietro, R., Petrocchi, M., Spognardi, A., Tesconi, M.: Fame for sale: efficient detection of fake twitter followers. Decis. Support Syst. **80**, 56–71 (2015)
14. Cresci, S., Di Pietro, R., Petrocchi, M., Spognardi, A., Tesconi, M.: The paradigm-shift of social spambots: evidence, theories, and tools for the arms race. In: WWW (2017)
15. Cresci, S., Lillo, F., Regoli, D., Tardelli, S., Tesconi, M.: $ fake: Evidence of spam and bot activity in stock microblogs on twitter. In: ICWSM (2018)
16. Cresci, S., Lillo, F., Regoli, D., Tardelli, S., Tesconi, M.: Cashtag piggybacking: uncovering spam and bot activity in stock microblogs on twitter. ACM Trans. Web (TWEB) **13**(2), 1–27 (2019)
17. Cresci, S., Petrocchi, M., Spognardi, A., Tognazzi, S.: Better safe than sorry: an adversarial approach to improve social bot detection. In: WebSci (2019)
18. Davis, C.A., Varol, O., Ferrara, E., Flammini, A., Menczer, F.: Botornot: a system to evaluate social bots. In: WWW (2016)
19. Devlin, J., Chang, M.W., Lee, K., Toutanova, K.: BERT: pre-training of deep bidirectional transformers for language understanding. In: NAACL (2019)
20. Dhingra, B., Zhou, Z., Fitzpatrick, D., Muehl, M., Cohen, W.: Tweet2vec: character-based distributed representations for social media. In: ACL (2016)
21. Ferrara, E.: Measuring social spam and the effect of bots on information diffusion in social media. In: Lehmann, S., Ahn, Y.-Y. (eds.) Complex Spreading Phenomena in Social Systems. CSS, pp. 229–255. Springer, Cham (2018). https://doi.org/10.1007/978-3-319-77332-2_13
22. Gal, Y., Ghahramani, Z.: Dropout as a Bayesian approximation: representing model uncertainty in deep learning. In: ICML (2016)
23. Gilani, Z., Farahbakhsh, R., Tyson, G., Wang, L., Crowcroft, J.: Of bots and humans (on twitter). In: ASONAM (2017)
24. Howard, J., Ruder, S.: Universal language model fine-tuning for text classification. In: ACL, July 2018
25. Kim, Y.: Convolutional neural networks for sentence classification. In: EMNLP (2014)
26. Kingma, D.P., Welling, M.: Auto-encoding variational Bayes. In: ICLR (2014)
27. Ko, R.: Social media is full of bots spreading Covid-19 anxiety. Don't fall for it (2020). https://www.sciencealert.com/bots-are-causing-anxiety-by-spreading-coronavirus-misinformation
28. Kudugunta, S., Ferrara, E.: Deep neural networks for bot detection. Inf. Sci. **467**, 312–322 (2018)
29. Lee, K., Eoff, B.D., Caverlee, J.: Seven months with the devils: a long-term study of content polluters on twitter. In: ICWSM (2011)
30. Ma, J., Gao, W., Wei, Z., Lu, Y., Wong, K.F.: Detect rumors using time series of social context information on microblogging websites. In: CIKM (2015)

31. Maas, A.L., Hannun, A.Y., Ng, A.Y.: Rectifier nonlinearities improve neural network acoustic models. In: ICML (2013)
32. Mazza, M., Cresci, S., Avvenuti, M., Quattrociocchi, W., Tesconi, M.: RTbust: exploiting temporal patterns for botnet detection on twitter. In: WEBSCI (2019)
33. Miller, Z., Dickinson, B., Deitrick, W., Hu, W., Wang, A.H.: Twitter spammer detection using data stream clustering. Inf. Sci. **260**, 64–73 (2014)
34. Morstatter, F., Wu, L., Nazer, T.H., Carley, K.M., Liu, H.: A new approach to bot detection: striking the balance between precision and recall. In: ASONAM (2016)
35. Ruths, D.: The misinformation machine. Science **363**(6425), 348 (2019)
36. Shao, C., Ciampaglia, G.L., Varol, O., Yang, K.C., Flammini, A., Menczer, F.: The spread of low-credibility content by social bots. Nat. Commun. **9**(1), 4787 (2018)
37. Subrahmanian, V., et al.: The darpa twitter bot challenge. Computer **49**(6), 38–46 (2016)
38. Varol, O., Ferrara, E., Davis, C.A., Menczer, F., Flammini, A.: Online human-bot interactions: detection, estimation, and characterization. In: ICWSM (2017)
39. Wang, Z., Oates, T.: Encoding time series as images for visual inspection and classification using tiled convolutional neural networks. In: AAAI-W (2015)
40. Yang, C., Harkreader, R., Gu, G.: Empirical evaluation and new design for fighting evolving twitter spammers. IEEE Trans. Inf. Forens. Secur. **8**(8), 1280–1293 (2013)
41. Yang, K.C., Varol, O., Davis, C.A., Ferrara, E., Flammini, A., Menczer, F.: Arming the public with artificial intelligence to counter social bots. Hum. Behav. Emerg. Technol. **1**(1), 48–61 (2019)
42. Yang, K.C., Varol, O., Hui, P.M., Menczer, F.: Scalable and generalizable social bot detection through data selection. In: AAAI (2020)
43. Young, L.Y.: The effect of moderator bots on abusive language use. In: ICPRAI (2018)

Spam Detection on Arabic Twitter

Hamdy Mubarak, Ahmed Abdelali, Sabit Hassan(✉), and Kareem Darwish

Qatar Computing Research Institute, Doha, Qatar
{hmubarak,aabdelali,sahassan2,kdarwish}@hbku.edu.qa

Abstract. Twitter has become a popular social media platform in the Arab region. Some users exploit this popularity by posting unwanted advertisements for their own interest. In this paper, we present a large manually annotated dataset of advertisement (Spam) tweets in Arabic. We analyze the characteristics of these tweets that distinguish them from other tweets and identify their targets and topics. In addition, we analyze the characteristics of Spam accounts. We utilize Support Vector Machines (SVMs) and contextual embedding based models to identify these Spam tweets with macro averaged F1 score above 98%.

Keywords: Spam filtering · Advertisement detection · Social media analysis · Arabic social media

1 Introduction

Different social media platforms employ different policies for promoting content legitimately. Twitter, for example, charges users for promoting tweets/accounts[1]. Such promotion of content is necessary for the survival of social media platforms. However, some advertising outfits seek a different path and attempt to exploit the popularity of certain accounts or hashtags on Twitter to promote products and services, often through multiple accounts that they control. Piggybacking on popular hashtags is typically referred to as *hashtag hijacking* and pollutes users' Twitter feeds with Spam. Another spamming approach involves mentioning popular accounts or replying to them, which potentially annoys the followers of such accounts. Figure 1 gives examples of both types in Arabic tweets. As the examples show, the ads often contain potentially illegal products or scams. Most of the Spam that we saw in the Arabic Twitter involves advertisements, and this kind of Spam will be our focus in this paper.

Although there is significant work on spam for English [13,16,17,20,22,24], it is relatively less explored for other languages such as Arabic [7]. Though English models may be used for Arabic, the divergent properties of Arabic, including different orthography and syntax and its complex morphology may lead to poor results [2]. The limited work on Arabic spam detection on Twitter is mostly focused on Saudi Arabian tweets [4,5,12]. However, Arabic is spoken across the Middle East and North Africa (MENA) region, and Arabic

[1] ads.twitter.com.

S. Aref et al. (Eds.): SocInfo 2020, LNCS 12467, pp. 237–251, 2020.
https://doi.org/10.1007/978-3-030-60975-7_18

(a) (b)

Fig. 1. Examples of **(a) hashtag hijacking**, where #COVID19 hashtag is used to advertise bootlegged movies, and **(b) popular account spamming**, where a reply to an Aljazeera.net tweet (about unrest in Baghdad) is an ad a drug to enhance erections.

dialects across the region can differ significantly [1]. To avoid such geographical limitations, we present a Spam dataset composed of tweets that are collected from the MENA region, without targeting any specific country. Our dataset is comprised of 12K manually annotated advertisement tweets (Spam) along with 90K non-advertisement tweets (Ham). We analyze distinguishing characteristics of Spam in contrast to Ham to identify salient features such as words, tweet lengths, emojis, hashtags, and URLs. We further extract topics and targets of the Spam tweets. Further, we employ state-of-the-art classification techniques, including the use of fine-tuned contextualized embeddings based on transformer models and Support Vector Machines (SVM) classification trained on word and character n-grams and static embeddings. Our best performing model surpasses 98% macro-averaged F1-score. The contributions of this paper are:

- We built a large and geographically distributed Arabic Spam/Ham dataset, which we make freely available for research[2].
- We thoroughly analyzed Spam tweets and Spam accounts to identify their most distinguishing properties.
- We employed the latest classification techniques to effectively identify Spam in the Arabic Twitter sphere.

2 Related Work

Spam detection is well explored for the English Twitter sphere (e.g., [13,16,17, 20,22,24]). Prior work employed different models, ranging from traditional bag-of-word models to sophisticated deep learning models. Bag-of-word models yield

[2] Dataset can be downloaded from http://alt.qcri.org/resources/SpamArabic-Twitter.tgz.

suboptimal results due to the short length of tweets (originally 140 characters and later increased to 280 characters[3]) [22]. The availability of many Spam detection datasets and advances in machine learning and deep learning approaches have resulted in highly effective spam detectors. For example, McCord et al. [17] achieved 95.7% F1-score in spam detection on Twitter using a random forest classifier. Other related works focused on detecting bots on Twitter since bots may generate spams. Botometer uses features extracted from 200 latest tweets and recent mentions from other users for each account to decide if an account is a bot or not [23,25,26].

On the other hand, the literature on Arabic spam detection is rather sparse. Nonetheless, there has been work on detecting undesirable Arabic content such as offensive or vulgar language [6,14,15,18,19] and adult-content [2,8]. These works are related to spam detection because Spam tweets often contain vulgar language, profanity, and sexually suggestive content. Saeed et al. [21] used an ensemble method composed of both rule-based and machine-learning classifiers to detect Spam in Arabic reviews. However, reviews can be significantly different from Spam on social media platforms such as Twitter. Alharbi and Aljaedi [7] conducted a survey on Saudi locals to identify risks associated with spams and used a random forest classifier for spam detection. Al-Khalifa [4] analyzed the content and behavior of Twitter spam accounts from Saudi Arabia, and identified different types of spam accounts and observed that they are generally socially disconnected. Other works on automatic detection of Arabic spam include the use of different classifiers such as Naïve Bayes, SVM, and Rule-based classifiers, and mostly focused on Twitter usage in Saudi Arabia [5,12].

3 Data Description

In this section, we describe our methodology for constructing our Spam/Ham dataset, the resultant dataset, and an analysis of the properties of Arabic spam.

3.1 Data Collection

As mentioned earlier, many advertising outfits attempt to exploit the popularity of websites, YouTube channels, and social media accounts to spread their Spam to large audiences. Typically, such spamming is prohibited. For example, rules on commenting on the BBC news website do not allow posing comments for users' financial gain (advertising, sponsorship etc.)[4]. Though such rules also exist on Twitter, Spam accounts do not find difficulties in posting many product or service advertisements while mentioning or replying to popular Twitter accounts. Further, the people behind them often use bots to increase the volume of their posts. As shown in Fig. 2, a user comments that replies to Aljazeera tweets are infested with Spam, many of which are inappropriate, and another

[3] https://developer.twitter.com/en/docs/basics/counting-characters.
[4] https://www.bbc.co.uk/usingthebbc/terms/what-are-the-rules-for-commenting/.

Fig. 2. Partial translation: (a) Most comments on Aljazeera topics are ads about sexual ability, ejaculation, and diets ... **(b)** ... Alarabiya has become only for Ads about impotence. **(c)** I expected to find comments. Nothing except ads. Find a solution!

user replies that such is the case of Alarabiya news channel. Yet a third user complains about the annoying Spam. Such is typical of popular Arabic Twitter accounts and may repulse their audiences.

In our study, we targeted replies to the **Aljazeera** news channel account (@AJArabic), which is one of the most popular Arabic Twitter accounts with more than 15M followers. We anticipated that the Spam accounts would post on this account as well as other popular accounts. As we show later in Table 3, Spam accounts that target @AJArabic do indeed target other popular accounts. We collected replies to @AJArabic for one week during December 2019 ($\approx$ 300 parent tweets). We sorted accounts by the percentage of @AJArabic tweets to which the accounts reply. Then, we manually inspected the accounts from the top of the list, until we found 100 unique Spam accounts, which collectively commented on nearly 70% of all @AJArabic tweets regardless of topic. Typically, Spam accounts advertise products or services and their tweets are irrelevant to parent tweets. These instructions were provided to a native speaker for annotation. Figure 1 (b) shows such a reply.

Using Twitter APIs, we obtained all the timeline tweets of these accounts. In all, we collected 46k timeline tweets. Though the most recent 3.2k timeline tweets are obtainable via the Twitter APIs, many of these accounts had fewer tweets in their timelines. Since Spam accounts may repeat tweets, we selected a random sample of 200 unique tweets at most from each account, leading to a set of 12,541 Spam tweets.

For Ham tweets, we randomly selected 121k unique tweets (almost 10 times larger than our Spam tweets) from a set of 50 million tweets that we collected between March and September 2018 using the query "lang:ar" (language is Arabic). Given that the vast majority of tweets are not Spam, we assumed that all the tweets in our sample were Ham. To verify this assumption, we manually annotated a random sample of 1,000 tweets, and we found that 53 tweets out of them can be considered as Spam. Since this assumption is overwhelmingly correct, we opted to avoid manually labeling Ham tweets.

To construct the final dataset, we split tweets into 70% training, 10% development, and 20% testing. We split the Ham dataset randomly. However since individual accounts in the Spam dataset may have distinctive features (e.g. specific emojis, hashtags, words), we split the Spam tweets for training, development, and testing by users. Even though our classification would be at tweet level, splitting by users would help us avoid having tweets from a single user in both training and testing, which would affect the generalization of the classifier. Table 1 lists the breakdown of the details of the final dataset.

Table 1. Spam dataset statistics

Split	Spam accounts	SpamTweets	Total Tweets	Tokens	Spam %
Train	70	8,680	94,680	1.5M	9.2%
Dev	10	1,159	11,159	182K	10.4%
Test	20	2,702	28,383	466K	9.5%
Total	100	12,541	134,222	2.2M	9.3%

3.2 Properties of Spam Tweets

In this subsection, we analyze the distinguishing properties of Spam tweets.

Topics: We randomly sampled 1,000 tweets and manually annotated them for their topics. Table 2 lists these topics, their percentages and examples. Advertisements about diets and sexual products account for 85% of all Spam. The remaining Spam mostly advertised medicines for different ailments. For example, honey is widely consumed in the Arab World for its medicinal properties.

Targets: Next, given the timeline tweets of Spam accounts, we examined the accounts they target with replies or mentions. In all, they targeted 2200+ different accounts. We annotated the most frequent 200 targets, which accounted for 75% of all replies and mentions. The top 10 targets are shown in Table 3. Also, the country distribution of all targets in our sample is shown in Fig. 3. We used the ISO 3166-1 alpha-2 for country codes [5]. We observed from the geographical distribution of target accounts that half of the targets are from Saudi Arabia, and the majority of the remaining targets are from other Gulf countries, such as Kuwait, United Arab Emirates, and Qatar.

Distinctive Words: We contrasted our Spam and Ham tweets in our dataset to identify the most distinctive words and emojis. To do so, we computed the so-called valence score for each token in each dataset [10]. The score helps determine the distinctiveness of a given token in a specific set S_0 in reference to another set S_1. Given $N(t, S_0)$, which is the frequency of the term t in set S_0, valence is computed as follows:

[5] https://en.wikipedia.org/wiki/List_of_ISO_3166_country_codes.

Table 2. Topics of Spam tweets

Topic	%	Example and Translation
Diet	48	اخسر أكثر من ١٢ كيلو خلال ٣٠ يوم الحل السريع لمشاكل السمنة والترهلات. Quick solution to obesity and flabbiness. Lose more than 12 kilos in 30 days
Sex	37	أقوى منتجات القدرة الجنسية: انتصاب قوي وتأخير قذف Sexual ability products: strong erection and delayed ejaculation
Multiple	9	تخسيس، تسمين، قدرة جنسية، تساقط شعر، تفتيح بشرة Weight loss, fattening, sexual ability, hair loss, skin lightening
Bones	3	تخلص من ألم المفاصل وخشونه الركبة Get rid of joint pain and knee roughness
Digest	1	وداعا للغازات، تخلص من آلام القولون Say goodbye to gases, get rid of colon pains
Skin	1	لعلاج تساقط الشعر وحب الشباب For the treatment of hair loss and acne
Food	1	عسل ملكات أصلي. جميع الأنواع من أفخر المناحل Pure royal honey. All kinds from the finest apiaries
Edu.	1	متخصصون في أبحاث التخرج لكافة التخصصات بأسعار مرضية Specialized in graduation research for all disciplines at satisfactory prices

Table 3. Top Spam targets (popular with millions or thousands of followers).

Account	Description	%	Country	Followers
@AlArabiya	Alarabiya news channel	2.4	SA	14.1M
@sabqorg	Sabq electronic newspaper	2.2	SA	13.1M
@alqabas	Alqabas newspaper	2.0	KW	476.8K
@ajlnews	Ajel electronic newspaper	2.0	SA	2.5M
@naharkw	Alnahar newspaper	1.9	KW	54.5K
@emaratalyoum	Emirates Today newspaper	1.7	AE	2.2M
@AlraiMediaGroup	Alrai TV and newspaper	1.7	KW	899.7K
@AJArabic	Aljazeera news channel	1.6	QA	15M
@aawsat	Alsharq Alawsat news agency	1.6	UK	34.5K
@riyadiyatv	Saudi Sports Channels	1.5	SA	995.9K

$$V(t, S_0) = 2\frac{\frac{N(t,S_0)}{N(S_0)}}{\frac{N(t,S_0)}{N(S_0)} + \frac{N(t,S_1)}{N(S_1)}} - 1 \tag{1}$$

where $N(S_i)$ is the total number of occurrences of all tokens in the set S_i. Figure 4 shows a tag cloud of the most distinctive words in the Spam tweets, where valence $\varepsilon > 0.9$ and the size of the words corresponds to their frequency. The most

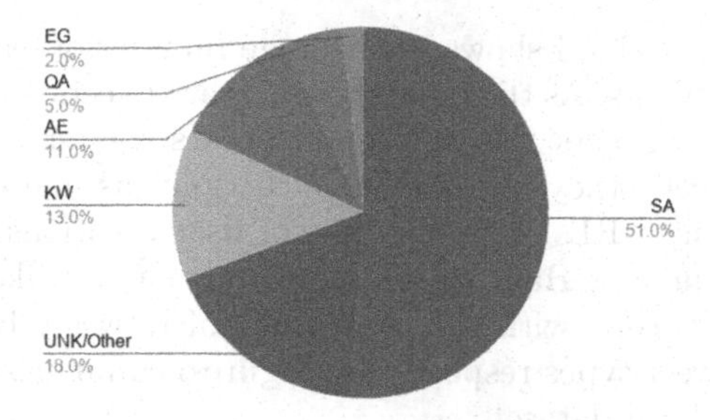

Fig. 3. Distribution of the countries of target accounts.

Fig. 4. Tag cloud of most distinctive words for Spamwith valence score $\vartheta >= 0.9$.

Fig. 5. Top emojis for **(a) Spam, (b) Ham** datasets with valence score $\vartheta >= 0.9$.

prominent words are about urging customers to order (e.g. للطلب– "to order"), communication methods (e.g. واتساب– "WhatsApp"), or related to weight loss (e.g. وزنك– "your weight") and sexual ability (e.g. الانتصاب– "erection"). As for emojis, Fig. 5 shows the 20 most distinctive emojis for both Spam and Ham tweets. When comparing them, we can ascertain that the top emojis for the Spam tweets contain: a) no sad emojis; b) multiple emojis expressing strength (e.g. flexed arms, fire, 100%, check mark, etc.); and c) positive expressions of love and beauty (e.g. heart with cupid arrow and hearts). The Ham tweet emojis express larger varieties of emotions, including sad ones.

Surface Features: Further, we analyzed the surface differences between Spam and Ham in terms of length of tweets, number of hashtags, emojis, user mentions, and URLs per tweet. Table 4 shows the average number of features per tweet in the Spam and Ham tweets. As the table shows, Spam tweets are on average nearly twice as long as Ham tweets. Though Spam tweets use far fewer hashtags (one fifth) than Ham tweets, they use more than 6 times as many emojis and more than 4 times as many URLs. The numbers of user mentions are sightly higher for Spam tweets. Further, Ham tweets were much more likely to be retweets compared to Spam tweets, with the percentage of retweets being 51% and less than 1% for both tweet types respectively. Figure 6 summarizes the distribution of these features in both datasets.

Table 4. Average number of features per tweets in Spam and Ham datasets.

	Average number of					
	Words	Hashtags	Emojis	@mentions	URLs	Retweets
Spam	25.44	0.08	6.05	1.23	1.14	0.01
Ham	14.33	0.40	0.91	0.94	0.27	0.51

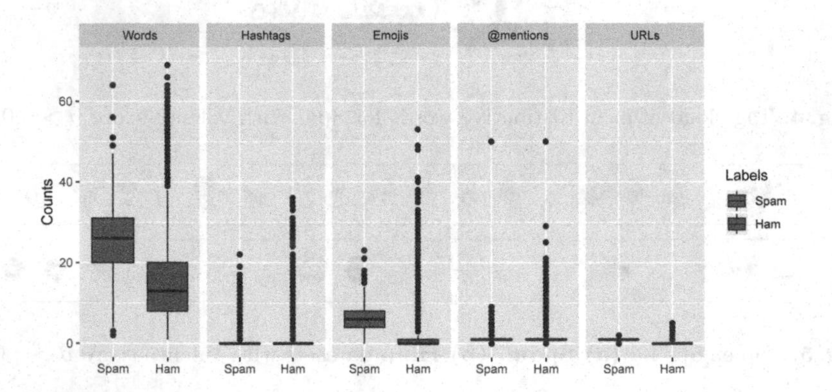

Fig. 6. Properties of Spam and Ham tweets.

Engagement: We observed that the vast majority of Spam tweets (∼97%) didn't receive interaction from other users, such as likes, retweets, or comments. For comparison, 45.8% of the Ham tweets were liked and 14.1% were retweeted.

Dialect (Language Varieties): We examined the dialect used in the sampled tweets and found that 87% of tweets are written completely in Modern Standard Arabic (MSA), the rest have words written in dialects, namely: Levantine dialect (6%), Egyptian dialect (5%) and Gulf dialect (2%). As Spam accounts want to reach a large number of audiences, they use either the standard language or simple dialects that are widely understood in most Arab countries.

3.3 Properties of SpamAccounts

In this subsection, we report on some of the characteristics Spam accounts, namely profile information (verification, description, and location), followers and followees, creation date, and possible suspension.

User Profiles: Based on the profile information of Spam accounts, we found that none of them was verified, 40 accounts provided user descriptions, and locations and URLs were available for only 21 and 14 accounts respectively. For the accounts that authored the Ham tweets (52,569 accounts), 0.8% of them were verified, 79.5% provided a description, and 55.9% provided a location.

Followers and Followees: Figure 7 shows the details of follower and followee counts for Spam and Ham accounts. We can see that almost half of the Spam accounts have very few followers (11 accounts have 0 followers, 8 accounts have 1 follower, etc.), and a quarter of them have very few followees. For Ham accounts, only 6% and 4% of accounts have less than 10 followers and followees respectively. Further, around 29% of Ham accounts have more than 1,000 followers as opposed to only 3% for Spam accounts. For the accounts in our Ham dataset, the average number of followers and followees were 10,307 and 1,343 respectively.

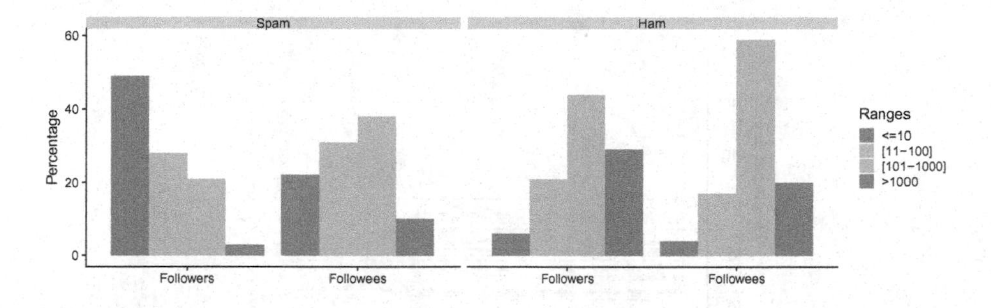

Fig. 7. Followers and followees counts for Spam and Ham accounts

Suspension of Spam accounts by Twitter: We wanted to examine the lifespan of Spam accounts and how quickly Twitter suspends them. To do so, we checked the availability of these accounts after six months of our collection date (i.e. in May 2020), and we found that Twitter has suspended 79% of them as they violated Twitter usage rules[6], and the remaining 21% were still active. It is noteworthy that the bulk of the suspended accounts produced higher tweet volumes compared to the unsuspended ones, which tweeted roughly once a day on average. Twitter can seemingly detect spammers when they abuse or flood the platform with high volumes of redundant messages. Figure 8 shows the average tweets per day for the different accounts and their respective lifespans. While some

[6] https://help.twitter.com/en/rules-and-policies/twitter-rules.

spammers get detected and suspended, a significant number of Spam accounts continue to operate undetected as shown in the figure. This motivates our work to find and tag these types of accounts. For comparison, after more than six months, the percentage of Ham accounts in our dataset that were suspended, closed, or made private was 5.8%, significantly lower than that of Spam accounts.

Bots and Spam accounts. In July 2020, we used Botomoeter[7] [26] to assess Spam accounts. Botometer checks the activity of a Twitter account and gives it a score based on how likely the account is to be a bot. This score ranges between 0 and 5, with zero being most human-like and five being the most bot-like. We considered scores less than 2 as human, 3 and above as bot, and the values in the middle as uncertain. Out of the 100 Spam accounts, only 21 accounts were still active in the time of testing. Botometer identified 12 accounts (57%) as bots with high confidence, 2 accounts (10%) mistakenly as humans (ex: uDeoFccZPOuxT7b), and in third of the cases, it was uncertain. We took the same number from Ham accounts, and we found that Botometer identified 2 accounts mistakenly as bots (ex: ZXi2i) where the rest were recognized as humans. We can see that there is a high correlation between Bots and Spam accounts, and tools for identifying Bots can help in Spam detection. We observed also that Botometer has difficulties in recognizing bots when they post tweets that toggle between Arabic and English (ex: QatarPrayer).

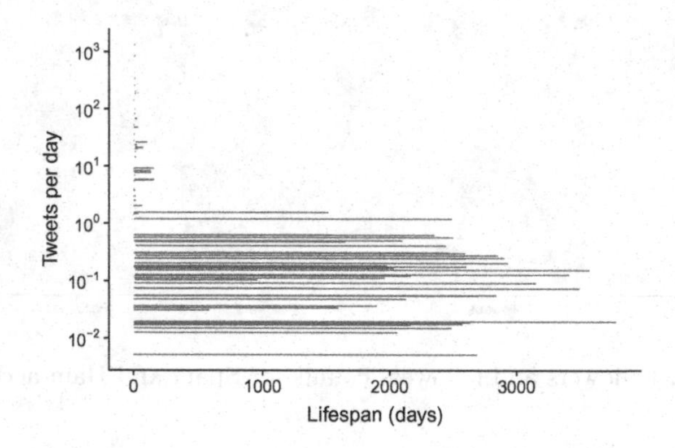

Fig. 8. Lifespan of Spam accounts and their average tweets per day

4 Experimental Setup and Results

We conducted a variety of classification experiments to identify Spam tweets using our training, development, and test splits. We report classification results on the test set, using accuracy (A), precision (P), recall (R), and F1-measure (F1) for the Spam class and also, macro-averaged F1 (mF1), which averages F1-measure for the Spam and Ham classes to account for the class imbalance.

[7] https://botometer.iuni.iu.edu/.

4.1 Classification Models

Baseline. For the baseline, we always predicted the majority class (Ham). Due to the class imbalance, using this baseline yields 90.5% accuracy.

Table 5. Classifier performance for Spam detection

Model	Features	A	P	R	F1	mF1
Majority class	-	90.5	0.00	0.00	0.00	47.5
SVM	Char[2-4]	99.4	**99.6**	94.1	96.8	98.2
SVM	Word[1-5]	99.4	98.6	95.5	97.0	98.3
SVM	Char[2-4]+Word[1-5]	99.4	99.4	94.0	96.7	98.2
SVM	Mazajak Embeddings	98.8	93.8	94.0	93.9	96.6
AraBERT	-	**99.7**	99.2	**97.1**	**98.1**	**98.9**

SVMs. We used an SVM classifier with a linear kernel[8] that we trained using a variety of features, namely:

- character n-grams, including bigrams, trigrams, and 4-grams. We used a bag-of-words model with tf-idf (term frequency-inverse document frequency) term weighting. We vectorized character n-grams using scikit-learn.
- word n-grams, where we varied n between 1 and 5. Similar to character n-grams, we used a bag-of-words model with tf-idf weighting.
- concatenation of word and character n-gram vectors.
- static embeddings representations, where we mapped words into an embeddings space using the pre-trained Mazajak word embeddings [3], which use a skip-grams that is trained on 250M tweets.

Contextual Embeddings: AraBERT. Pre-trained Bidirectional Encoder Representations from Transformers (BERT) [11] models have been shown to perform really well when fine-tuned for downstream natural language processing tasks. AraBERT [9] is a BERT-based model, pre-trained on Arabic news articles and Arabic Wikipedia. We fine-tuned AraBERT for 2 epochs for Spam detection.

4.2 Results

Table 5 reports on the results of all our experiments. As the results show, most of our classification models were capable of identifying Spam reliably. Among the different features for SVM, word n-grams performed the best, followed by a combination of character and word n-grams. Using the static Mazajak embeddings

[8] We used the libSVM implementation in scikit-learn https://scikit-learn.org/.

performed the worst. This may indicate that surface features may be more important than the actual contents of the tweets. Hence, detecting surface features such as emojis, hashtags, and retweets aids classification. Fine-tuned AraBERT outperformed all the other classifiers with mF1 score of 98.9%. This could be attributed to the fact that deep contextualized BERT-based models can capture context-level information that SVMs fail to capture.

4.3 Error Analysis

For error analysis, we manually annotated all errors made by our best system, namely the fine-tuned AraBERT model, on the test set (a total of 100 errors).

False Negatives. Our classifier misclassified 78 Spam tweets as Ham, which accounts for the majority of errors. The most prominent errors (encountered for 41% of errors) are listed in Table 6. For these errors, we can potentially increase the accuracy if we apply more normalization (e.g. replacing repeated letters) and increase the number of Spam accounts and their tweets to have more coverage for styles and topics. We could not identify specific reasons for rest of the errors.

Table 6. Error types of Spam detected as Ham

Error Type	%	Example and Translation
Non-standard spelling	15%	تنحيييييف بوووقت قياااسي بددددوون حرماااان Slimminnng in recorrd tiiime withouttt deprivatiiion (Slimming in record time without deprivation)
Short text with less clues	15%	استمتعي بجمالك مع كلين https://t.co/Gmg... Enjoy your beauty with Clean https://t.co/Gmg...
Specific products or diseases (OOV)	11%	وداعا للتبول اللاارادي عند الأطفال Say goodbye to bedwetting for children

False Positives. Our classifier misclassified 22 Ham tweets as Spam. In 16 of the cases, we found that such errors are due to errors in the training data where we assumed that all random tweets have no Spam. An example of such errors is the following tweet: "نحن نستقدم عمالة : كوافيرات، موظفات صالون. تواصل معنا" ("We bring workers: hairdressers, salon employee. Contact us!"). The remaining errors are due to short tweets, such as: "الخاص @fgArg..." ("Private @fgArg..").

Generalization. In order to ascertain if our best model, namely AraBERT, is able to generalize beyond our dataset, we experimented with a new test set. In May 2020, we queried Twitter for up to 500 tweets containing mentions of the top 5 accounts in Table 3. Of those, 1,980 were replies, which we manually annotated as Spam or Ham. 104 were annotated as Spam. AraBERT was able to correctly identify 100 Spam tweets, meaning that we achieved 100% precision and 96.2% recall. This suggests that our model is able to generalize from our dataset to properly classify new tweets. However, we acknowledge that our work targets spammers who reply to popular accounts. Thus, we may not be capturing the full extent of spams on Twitter. We plan to explore this in the future.

5 Conclusion

We presented a manually annotated dataset for detecting Arabic Spam tweets, and we identified different distinguishing properties of Spam and Ham tweets such as tweet lengths, use of emojis, user mentions, hashtags and URLs. We also identified topics and targets of these Spam tweets and some characteristics of Spam accounts. Our experiments showed that Support Vector Machines (SVMs) and AraBERT can detect these Spam tweets very reliably (macro-F1 > 98%).

References

1. Abdelali, A., Mubarak, H., Samih, Y., Hassan, S., Darwish, K.: Arabic dialect identification in the wild. arXiv abs/2005.06557 (2020)
2. Abozinadah, E., Mbaziira, A., Jones, J.H.: Detection of abusive accounts with Arabic tweets. Int. J. Knowl. Eng. **1**(2), September 2015
3. Abu Farha, I., Magdy, W.: Mazajak: an online Arabic sentiment analyser. In: Proceedings of the Fourth Arabic Natural Language Processing Workshop, pp. 192–198. Association for Computational Linguistics, Florence, August 2019. https://doi.org/10.18653/v1/W19-4621, https://www.aclweb.org/anthology/W19-4621
4. Al-Khalifa, H.S.: On the analysis of twitter spam accounts in Saudi Arabia. Int. J. Technol. Diffus. (IJTD) **6**(1), 46–60 (2015)
5. Al Twairesh, N., Al Tuwaijri, M., Al Moammar, A., Al Humoud, S.: Arabic spam detection in twitter. In: The 2nd Workshop on Arabic Corpora and Processing Tools 2016 Theme: Social Media, p. 38 (2016)
6. Albadi, N., Kurdi, M., Mishra, S.: Are they our brothers? analysis and detection of religious hate speech in the Arabic twittersphere. In: 2018 IEEE/ACM International Conference on Advances in Social Networks Analysis and Mining (ASONAM), pp. 69–76. IEEE (2018)
7. Alharbi, A., Aljaedi, A.: Predicting rogue content and Arabic spammers on twitter. Fut. Internet **11**, 229 (2019). https://doi.org/10.3390/fi11110229
8. Alshehri, A., Nagoudi, E.M.B., Alhuzali, H., Abdul-Mageed, M.: Think Before Your Click: Data and Models for Adult Content in Arabic Twitter (2018)
9. Antoun, W., Baly, F., Hajj, H.M.: Arabert: transformer-based model for Arabic language understanding. arXiv abs/2003.00104 (2020)
10. Conover, M., Ratkiewicz, J., Francisco, M.R., Gonçalves, B., Menczer, F., Flammini, A.: Political polarization on twitter. ICWSM **133**, 89–96 (2011)

11. Devlin, J., Chang, M.W., Lee, K., Toutanova, K.: BERT: pre-training of deep bidirectional transformers for language understanding. In: Proceedings of the 2019 Conference of the North American Chapter of the Association for Computational Linguistics: Human Language Technologies, Volume 1 (Long and Short Papers). pp. 4171–4186. Association for Computational Linguistics, Minneapolis, June 2019. https://doi.org/10.18653/v1/N19-1423, https://www.aclweb.org/anthology/N19-1423
12. El-Mawass, N., Alaboodi, S.: Detecting Arabic spammers and content polluters on twitter. In: 2016 Sixth International Conference on Digital Information Processing and Communications (ICDIPC), pp. 53–58. IEEE (2016)
13. Grier, C., Thomas, K., Paxson, V., Zhang, M.: @spam: the underground on 140 characters or less. In: Proceedings of the 17th ACM Conference on Computer and Communications Security, CCS 2010, pp. 27–37. Association for Computing Machinery, New York (2010). https://doi.org/10.1145/1866307.1866311, https://doi.org/10.1145/1866307.1866311
14. Hassan, S., Samih, Y., Mubarak, H., Abdelali, A.: ALT at SemEval-2020 task 12: Arabic and English offensive language identification in social media. In: Proceedings of the International Workshop on Semantic Evaluation (SemEval) (2020)
15. Hassan, S., Samih, Y., Mubarak, H., Abdelali, A., Rashed, A., Chowdhury, S.A.: ALT submission for OSACT shared task on offensive language detection. In: Proceedings of the 4th Workshop on Open-Source Arabic Corpora and Processing Tools, with a Shared Task on Offensive Language Detection, pp. 61–65. European Language Resource Association, Marseille, May 2020. https://www.aclweb.org/anthology/2020.osact-1.9
16. Herzallah, W., Faris, H., Adwan, O.: Feature engineering for detecting spammers on twitter: modelling and analysis. J. Inf. Sci. **44**(2), 230–247 (2018). https://doi.org/10.1177/0165551516684296
17. McCord, M., Chuah, M.: Spam detection on twitter using traditional classifiers. In: Calero, J.M.A., Yang, L.T., Mármol, F.G., García Villalba, L.J., Li, A.X., Wang, Y. (eds.) ATC 2011. LNCS, vol. 6906, pp. 175–186. Springer, Heidelberg (2011). https://doi.org/10.1007/978-3-642-23496-5_13
18. Mubarak, H., Darwish, K.: Arabic offensive language classification on twitter. In: Weber, I., et al. (eds.) SocInfo 2019. LNCS, vol. 11864, pp. 269–276. Springer, Cham (2019). https://doi.org/10.1007/978-3-030-34971-4_18
19. Mubarak, H., Darwish, K., Magdy, W.: Abusive language detection on Arabic social media. In: Proceedings of the First Workshop on Abusive Language Online, pp. 52–56. Association for Computational Linguistics, Vancouver, August 2017. https://doi.org/10.18653/v1/W17-3008, https://www.aclweb.org/anthology/W17-3008
20. Lin, P.-C., Huang, P.-M.: A study of effective features for detecting long-surviving twitter spam accounts. In: 2013 15th International Conference on Advanced Communications Technology (ICACT), pp. 841–846 (2013)
21. Saeed, R.M., Rady, S., Gharib, T.F.: An ensemble approach for spam detection in Arabic opinion texts. J. King Saud Univ. Comput. Inf. Sci. (2019)
22. Sriram, B., Fuhry, D., Demir, E., Ferhatosmanoglu, H., Demirbas, M.: Short text classification in twitter to improve information filtering. In: Proceedings of the 33rd International ACM SIGIR Conference on Research and Development in Information Retrieval, SIGIR 2010, pp. 841–842. Association for Computing Machinery, New York (2010). https://doi.org/10.1145/1835449.1835643
23. Varol, O., Ferrara, E., Davis, C.A., Menczer, F., Flammini, A.: Online Human-bot Interactions: Detection, Estimation, and Characterization (2017)

24. Yang, C., Harkreader, R., Gu, G.: Empirical evaluation and new design for fighting evolving twitter spammers. IEEE Trans. Inf. Forens. Secur. 8(8), 1280–1293 (2013). https://doi.org/10.1109/tifs.2013.2267732
25. Yang, K.C., Hui, P.M., Menczer, F.: Bot electioneering volume: visualizing social bot activity during elections. In: Companion Proceedings of The 2019 World Wide Web Conference, May 2019. https://doi.org/10.1145/3308560.3316499, http://dx.doi.org/10.1145/3308560.3316499
26. Yang, K.C., Varol, O., Hui, P.M., Menczer, F.: Scalable and generalizable social bot detection through data selection. In: Proceedings of the AAAI Conference on Artificial Intelligence, vol. 34, no. 01, pp. 1096–1103, April 2020. https://doi.org/10.1609/aaai.v34i01.5460, http://dx.doi.org/10.1609/aaai.v34i01.5460

Beyond Groups: Uncovering Dynamic Communities on the WhatsApp Network of Information Dissemination

Gabriel Peres Nobre[1](✉), Carlos Henrique Gomes Ferreira[1,2], and Jussara Marques Almeida[1]

[1] Universidade Federal de Minas Gerais, Belo Horizonte, Brazil
gabriel.nobre@dcc.ufmg.br

[2] Universidade Federal de Ouro Preto, Ouro Preto, Brazil

Abstract. In this paper, we investigate the network of information dissemination that emerges from group communication in the increasingly popular WhatsApp platform. We aim to reveal properties of the underlying structure that facilitates content spread in the system, despite limitations the application imposes in group membership. Our analyses reveal a number of strongly connected user communities that cross the boundaries of groups, suggesting that such boundaries offer little constraint to information spread. We also show that, despite frequent changes in community membership, there are consistent co-sharing activities among some users which, even while holding broad content diversity, lead to high coverage of the network in terms of groups and individual users.

Keywords: WhatsApp · Community detection · Information spread

1 Introduction

Social media applications are known as tools to connect people, enhance human interactions, ultimately contributing to information diffusion on the Internet, a widely explored subject of study [22,25,33]. WhatsApp is one such application with great popularity in many countries such as India, Brazil and Germany [41]. Indeed, the application, which has recently surpassed the mark of 2 billion monthly users worldwide [40], has been shown to play an important role as a vehicle for information dissemination and social mobilization [34].

WhatsApp allows for one-to-one and group conversations, both end-to-end encrypted. WhatsApp groups are structured as private chat rooms, generally under a certain topic and limited to only 256 users who can participate in multiple discussions at the same time. However, a group manager may choose to share an invitation link to join the group on websites and social networks, which effectively makes the group publicly accessible, as anyone with access to the link can join in and become a member. We note that groups are dynamic spaces of conversations, with users joining and leaving over time, at their will. Also, there is no limit in the number of groups a user can participate in at any time.

S. Aref et al. (Eds.): SocInfo 2020, LNCS 12467, pp. 252–266, 2020.
https://doi.org/10.1007/978-3-030-60975-7_19

A number of recent studies have analyzed group conversations in WhatsApp [8,13,28,33,34], often discussing content and temporal properties of information dissemination within groups. A few have hinted at the potential for information virality by taking a bird's eye view of the network that emerges from information exchange in different groups [8,34]. However, an investigation of the properties of this network, which may reveal an underlying structure that facilitates information dissemination at large, is still lacking.

In this paper, we delve deeper into the network of information dissemination in WhatsApp. Specifically, we use a dataset consisting of messages posted in over 150 publicly accessible WhatsApp groups in Brazil during a 6 week period, encompasing the 2018 general elections [34]. We build a sequence of media-centric networks so as to capture the relationships established among users who shared the same piece of content in one or more groups during pre-defined time windows. We then analyze properties of these networks and how they evolve over time. In particular, we are interested in investigating to which extent users in different groups, intentionally or not, build "communities" of content spread, which consistently help speeding up information dissemination within the system.

Specifically, we tackle the following two research questions (RQs):

- **RQ1:** To which extent cross-group user communities emerge from analyzing the WhatsApp media-centric networks?
- **RQ2:** What are key properties of these communities and how they evolve over time?

One challenge when analyzing the media-centric networks is that, by definition, they model interactions established among multiple users, that is, many-to-many interactions, as opposed to traditionally studied binary relationships. As recently argued [4,10], such many-to-many interactions may lead to the formation of networks with multiple spurious and random edges. These spurious edges may ultimately hide the real underlying structure (often called the network backbone) representing the phenomenon under study, in our case, information spread across the monitored groups. Thus, as a first step to our study, we applied a technique [37] to extract the backbone of each media-centric network. We then used the Louvain algorithm [6] to extract communities in the backbone of each network, and analyzed their properties and how they evolve over time.

Our study revealed a number of strongly connected user communities that extrapolate the boundaries of groups, suggesting that such boundaries offer little constraint (if any) to information spread. Indeed, it is often the case that the same community covers a large number of different groups, and users with higher centrality in these communities are the ones with most impact on the community's group coverage and on the content uniqueness spread through the network. We also observed that around 30% of the users persist in the network backbone over time, whereas those with highest activity tend to remain even in the same community. These results suggest that, though WhatsApp groups are limited to small sizes, the platform effectively has an underlying network that greatly facilitates information spread, revealing consistent co-sharing activities among the users which, even while holding broad content diversity, lead to high user and group coverage. While corroborating arguments in [34], our study offers a much deeper analysis, revealing properties that help understand the information dissemination phenomenon in a very popular communication platform.

2 Related Work

There is a rich literature on online information spread and its related phenomena involving the modeling of online user interactions [2,16], the extraction of user communities [6,17], the identification of important users [15] and the analysis of the information diffusion process [8,25,34]. In the context of community extraction, there is a plethora of techniques [12,32] exploiting from statistical inference [1,19] to modularity optimization [44] and dynamic clustering [12,35]. Others have offered analyses of communities' structural properties [17,24] and temporal behavior [2]. In contrast, the authors of [16] described a system that allows the detection of influential users in dynamic social networks. Other authors have examined multi-model networks by proposing a framework for detecting community evolution over time [36,39].

WhatsApp has emerged as a global tool for communication [40] and has driven many recent researches. In [8], the authors analyzed information spread within groups from the perspective of user attention on different topics, while in [7] the authors focused on partisan activities, aiming to distinguish left-wing and right-wing groups. The work in [33] describes a system for collecting shared messages in publicly accessible WhatsApp groups. The collected content was later used in analyses of content properties and temporal dynamics of the dissemination of images [34], textual messages [33] and audio content [28].

Despite WhatsApp group communication being restricted to small groups (up to 256 users), there has been evidence of its use for information dissemination at large [18,27]. The only prior work that hinted at possible reasons for that was [34], which presented the network structure of the monitored groups, briefly analyzing some of its basic structural properties. Also, the authors of [13] showed that recent limitations on messaging forwarding [42] are not effective to block viral content spread across the network (despite contributing to delay it).

Our goal here is to delve deeper into information spread in WhatsApp groups, unveiling the underlying media sharing network that emerges from communication in those groups and investigating the extent to which the formation of user communities that cross group boundaries occur, favoring information spread. We thus complement prior analyses of WhatsApp, focusing on the network structure and how its properties relate to information dissemination in the platform.

3 Methodology

As stated, this work aims to investigate the formation of user communities that may favor information dissemination in WhatsApp. To that end, we adopted the three-step methodology depicted in Fig. 1. First, we collected a dataset containing messages shared in a number of WhatsApp groups during a six-week period. The dataset was then broken into six non-overlapping snapshots, one for each week. The data was then processed to filter less relevant data and to identify (near-)duplicated content (Step 1). Next, we built a *media-centric network* for each snapshot. For each network, we proceeded to retrieve its *backbone* which, as argued later, is a necessary step to remove random and spurious edges (Step 2). Finally, we ran a community detection algorithm [6] to extract communities from each backbone and characterized their structural properties and temporal dynamics (Step 3). These steps are further discussed in the following sections.

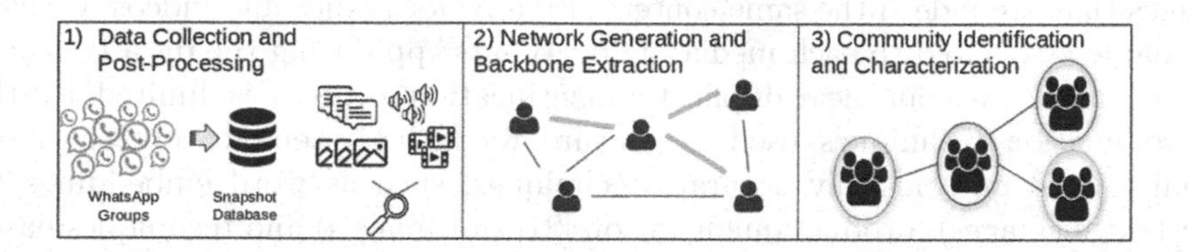

Fig. 1. Overview of our methodology

3.1 Data Collection and Post-processing

We used a dataset gathered by the WhatsApp Monitor [33], consisting of messages shared in political-oriented publicly accessible WhatsApp groups in Brazil. As described in [33], these groups were detected by searching for invitation links on public websites and online social networks. Our dataset covers six weeks around the 2018 general elections in Brazil (1th and 2nd rounds in October 7th and 28th, respectively), ranging from September 17th until October 28th in 2018. The dataset was broken into six non-overlapping one-week snapshots, and we restricted our analyses to 155 groups for which data was available in all snapshots.

Table 1 provides an overview of our dataset, showing the numbers of users, groups and messages (text, images, audios and videos) shared per week in the selected groups. It also shows the average number of users active per group, average number of messages shared per group, and average number of messages shared by a user in a group, for each week. Weeks including election dates (3 and 6) are highlighted in bold. As seen in the table, the numbers of users and messages vary over the snapshots. However, there tends to be an increase in activity around the dates of the two rounds of the election.

The collected data was then processed to extract and store the following information associated with each message: timestamp, anonymized user identifiers[1], group identifier and the media(s) (text, image, audio or video) shared through the message. Next, we filtered out text messages shorter than 180 characters, as suggested in [33], so as to retain only those that most probably carry

[1] Indeed, our data contains only cellular phone numbers. Thus, we are not able to identify the same user with multiple phone numbers.

relevant information. Then, as a final processing, we ran a number of heuristics to identify (near)-duplicate content, a necessary step to build the media-centric networks (next section). The specific heuristic depends on the type of media. For textual content, we used the Jaccard Similarity Coefficient [20] to perform paired comparisons, using a threshold of 70% of similarity to detect near duplicates (as performed in [33]). For images, we followed [34]: we generated the Perceptual Hash (pHash) [43] of each file and grouped together those with same hash. The pHash algorithm works by detecting color variations on the image resulting on a hash value. By comparing hash values we are able to detect resized and modified images that are indeed the same content. Finally, for audios and videos, we used the name associated to each media file by WhatsApp during the data transfer.

We note that our near-duplicate identification process is limited by the approximation techniques used. As future work, we intend to explore more sophisticated and possibly accurate techniques, such as word embeddings[29] (for text messages), product quantization [21] (for images) and techniques based on audio and video content analysis [23,38], which may enhance the generated network.

Table 1. Overview of our dataset (155 WhatsApp groups, 09/17 – 10/28/2018)

Weeks	1	2	3	4	5	6
# Unique users	4,994	4,774	5,115	4,815	4,439	4,914
Average # users/group	34.68	33.21	35.60	33.27	31.14	34.09
Average # messages/group	575	598	590	536	490	599
Avg # msgs/user (in a group)	16.59	18.01	16.58	16.13	15.75	17.59
# Text messages	89,136	92,650	91,438	83,118	75,982	92,840
# Image messages	13,018	13,208	13,274	13,471	11,922	17,113
# Audio messages	1,614	1,644	2,000	1,842	1,621	2,059
# Video messages	10,168	9,515	9,142	9,508	9,193	12,344

3.2 Network Model and Backbone Extraction

The network model used in this work creates an abstraction for user interactions in WhatsApp groups as a vehicle for information dissemination. Given that goal, the model focuses on the *content* shared, by connecting users who shared the "same message" at least once, regardless of whether they shared it on the same group or on different groups. By "same message" we mean messages that were identified as carrying near duplicate content, as described in Sect. 3.1. We refer to such networks as *media-centric* networks.

Specifically, given our dataset, we created a set of graphs $\mathcal{G} = \{G^1, G^2, \dots, G^{\Delta T}\}$, in which each G^w models user interactions during week w (i.e., $\Delta T = 6$ in our case). Each graph $G^w(V, E)$ is structured as follows. Each vertex in V refers to a user who posted a message during week w in one of the groups. An undirected edge $e = (v_i, v_j)$ exists in E if users corresponding to vertices v_i and v_j shared at least one message in common during week w. The weight of e corresponds to the number of messages both users shared in common during w.

As defined, our network model captures *many-to-many* interactions, i.e., interactions that occur among multiple (possibly more than two) users at once - in our case, co-sharing the same media content. This kind of interaction occurs in a range of other environments, such as the networks that emerge from relationships based on co-authorship, emails sent to a group of people and congressmen voting sessions [5,11,30], and raises different modeling challenges if compared to traditionally studied *one-to-one* interactions (e.g., friendship links) [4,14].

In particular, modeling sequences of many-to-many interactions into a network may lead to the emergence of a (potentially large) number of spurious edges, reflecting random or sporadic user activities. Such spurious edges may pollute the network, obfuscating the real underlying structure that better represents the phenomenon under study. We illustrate this problem by taking a fictitious example in our context. Suppose two different scenarios: (1) one particular viral content is massively disseminated through the WhatsApp network as many users shared it in different groups and (2) a smaller set of users repeatedly spread the same content, reaching different audience which ultimately leads to a large spread. By simply looking at the topology of the network that emerges from these two scenarios, one may consider both groups of users in (1) and (2) as communities. However, we are here interested in identifying strong and consistent co-sharing behavior, as in (2), as opposed to sporadic connections, as possibly in (1). As more many-to-many interactions are added to the network, more edges of different natures are added, resulting in a richer but quite noisier topology. We want to remove this noise to be able to focus on what is fundamental to the large scale dissemination of content in the network.

In other words, we want to identify pairs of users (i.e., edges) who disproportionately shared messages in common, filtering out edges resulted from randomness and sporadic co-sharing, thus revealing the underlying *network backbone* [9]. By definition, the backbone contains only the *salient* edges more fundamentally related to the phenomenon under study (information dissemination, in our case).

There is a rich and vast literature on methods to extract the backbone from networks [9,30,37]. Here, we aimed to identify when an edge connecting two vertices reflects a strong connection between them when compared to the other spurious connections they both may have with other peers. We experimented with two state-of-the-art backbone extraction methods which are driven by that goal, namely Noise Corrected Method [9] and Disparity Filter Method [37], selecting the latter. Our choice follows the approach in [9]: we selected the method that, according to our experiments, delivers the best trade-off between the number of edges with lower weight removed and the structural connectivity of the resulted backbone. The latter was measured in terms of clustering coefficient and modularity metrics (discussed in the next section), preserving those edges with higher weights. The Disparity Filter Method was able to remove a larger number of spurious/sporadic edges while still maintaining modularity and clustering coefficient measures comparable to the complete network.

The Disparity Filter algorithm works as follows. Let's define the strength of vertex v_i as the sum of all edge weights attached to v_i. The algorithm considers an edge attached to v_i as salient if it represents a "large fraction" of v_i's strength. Specifically, each edge attached to v_i is tested against the null hypothesis that the weights of all edges of v_i are uniformly distributed. Salient edges are those whose weights deviate significantly from this hypothesis. Notice that an edge is tested twice, once for each vertex it is incident to, and it is considered salient if it is statistically significant for both vertices when compared to a p-value. Edges that are not considered salient are removed from the graph. In our experiments we adopted a p-value of 0.1. This value was selected in preliminary experiments, by running the algorithm with various options (ranging from 0.01 to 0.32) and choosing the one that led to the best tradeoff between statistical significance, number of remaining vertices in the backbone and backbone connectivity. This choice of p-value is consistent with prior studies on backbone extraction, which report that very small p-values lead to a large number of nodes removed, ultimately breaking the original network and turning the analysis unfeasible [37]. Thus the need for a choice that meets the aforementioned tradeoff.

3.3 Community Identification and Characterization

As a final step, we identify groupings of users who impact information spread in the network by, intentionally or not, sharing common content in a disproportionately high frequency. Studying these groupings, here referred to as *communities* (to avoid confusion with the original WhatsApp groups), reveals how they are structured, how such structure relates to the Whatsapp groups and how they evolve over time. Ultimately, we aim at bringing novel insights into how information virality may occur [34], despite the restrictions in group membership.

We identify user communities in the backbone extracted from each graph G^w ($w = 1...6$) using the Louvain algorithm [6]. It is a widely used community detection algorithm [32] that relies on a heuristic approach to iteratively build hierarchical partitions of the backbone. Specifically, it is based on a greedy optimization of the modularity, which is a metric of quality of these partitions. The modularity Q is defined [6] as $Q = \frac{1}{2M} \sum_{ij} \left[A_{ij} - \frac{k_i k_j}{2m} \right] \delta(c_i, c_j)$, where A_{ij} is the weight of edge (v_i,v_j); k_i (k_j) is the sum of the weights of the edges attached to v_i (v_j); m is the sum of all of the edge weights in the graph; c_i (c_j) is the community assigned to v_i (v_j); and $\delta(c_i, c_j) = 1$ if c_i=c_j or 0 otherwise.

Note that the communities are built from pairs of users who share similar content more often than expected by chance. Such groupings could be driven by intentional behavior (i.e., by orchestration), by coincidence (as side effect of the general information diffusion process) or by a mix of both. We here do not distinguish between these effects, although the stronger the edge weights the greater the chance that some sort of coordination is in place. Characterizing the effects behind community formation is an interesting line of future research.

Our community identification process revealed a number of very small communities (e.g., three-four nodes), often organized as small trees. We chose to disregard small groupings (fewer than 5 vertices), focusing our analyses instead on the larger and more impactful communities. These communities were analyzed according to two dimensions: structural properties and temporal dynamics.

The analysis of structural properties aimed to quantify the quality of each grouping and what it represents to the diffusion of information during the period under analysis. For the former, we make use of the clustering coefficient metric computed for each community [31], which measures the degree to which vertices in the community tend to cluster together. For the latter, we compute the group coverage and the content coverage of each community. Group coverage is the fraction of all monitored groups that were reached by posts from community members. Content coverage is the fraction of all contents shared during the period that were shared by the community members. Larger values of either metric reflect greater importance of the community to the information spread.

The analysis of temporal dynamics aimed to characterize the evolution of communities in the backbone and quantify changes in community membership over time. To that end, we used two metrics (as in [11]), always computed for snapshot w with respect to snapshot $w-1$. The first metric, called *persistence*, captures the continuous presence of the same users in the network backbone over time. It is computed as the fraction of users in the backbone at snapshot $w-1$ that remain in the backbone in snapshot w. This metric quantifies the presence of users who remain important for content dissemination over time.

Note that persistence does not distinguish between users who, despite remaining in the backbone over time, often change community from those who remain in the same community. The latter might reflect a potential coordinated effort to boost information spread. To capture the permanence of members in the same community, we adopted the Normalized Mutual Information (NMI) [26].

Let X and Y be the sets of communities identified in snapshots $w-1$ and w, respectively. Let also $P(x)$ be the probability of a randomly selected user being assigned to community x in X, and $P(x, y)$ be the probability of a randomly selected user being assigned to communities x in X and y in Y. Finally, let $H(x)$ be the Shannon entropy for X defined as $H(X) = -\sum_x P(x) \log P(x)$. The NMI X and Y is defined as $NMI(X,Y) = \frac{\sum_x \sum_y P(x,y) \log \frac{P(x,y)}{P(x)P(y)}}{\sqrt{H(X)H(Y)}}$. It can be thought as the information "overlap" between X and Y, or how much we learn about Y from X (and vice-versa). Its value ranges from 0 (all members changed their communities) to 1 (all members remained in the same communities).

In addition to characterizing communities, we also analyzed the importance of users to the information dissemination process. We did so by computing the impact on the (group and content) coverage metrics as vertices are removed from the backbone according to a ranking of importance. We experimented with different user/vertex rankings built based on metrics of activity level (number of messages shared and number of groups the user participates in) and metrics of centrality in the backbone. Our goal by comparing those rankings is to assess the extent to which the backbone and its communities are able to reveal important

Table 2. Structural properties of the extracted backbones

Metrics	Weeks					
	1	2	**3**	4	5	**6**
# Users	114	162	132	216	143	338
# Edges	346	767	500	1,676	617	3,499
# Connected components	1	1	1	1	2	2
Average clustering coefficient	0.49	0.61	0.58	0.63	0.57	0.63
Average degree	6.07	9.47	7.58	15.52	8.63	20.70
Average edge weight	6.95	9.10	8.05	12.66	7.36	8.91
# Communities	5	6	6	5	7	8
Modularity	0.70	0.61	0.55	0.54	0.59	0.57

users to information dissemination in the system (as captured by the coverage metrics), compared to the activity metrics, priorly analyzed in [8,28,34].

We built four vertex rankings based on degree centrality and closeness (which relates to the average distance of the vertex to all other vertices) [3]. In each case, we use the standard metric, computed over the complete backbone, along with a tuned variation, referred to as community centrality. The latter, defined in [15], considers the embeddedness of the vertex in its community and the relations between this community and the others, and quantifies the vertex's ability to disseminate information on its own community and on the overall network.

Specifically, the community centrality of a vertex v_i assigned to community c is computed by combining two components, a local one and a global one [15]. The local component α_i^L quantifies the (degree or closeness) centrality of v_i in its own community c; whereas the global component α_i^G quantifies the centrality of v_i in the backbone by considering only edges connecting vertices of different communities (i.e., inter-community edges). These two components are combined by a weighting factor μ_c that is the fraction of all inter-community edges that are incident to community c. In other words, the community (degree or closeness) centrality of vertex v_i, α_i, is defined as $\alpha_i = (1 - \mu_c) * \alpha_i^L + \mu_c * \alpha_i^G$.

4 Characterization Results

We now discuss the results of the characterization of the communities that emerged from the networks generated according to methodology described in Sect. 3. Table 2 presents a summary of the structural properties of the backbones extracted from the networks in set $\mathcal{G}$. Once again, we show the snapshots containing 1st and 2nd rounds of the general elections (weeks 3 and 6) in bold.

We first note that the size of the backbone, in number of users and edges, varies greatly over the weeks. This variation is consistent, though in higher degree, with the variations in the amounts of participation and sharing activity in the monitored groups over the period, presented in Table 1. We also note that the backbones are formed by at most 2 connected components (often only one), with a reasonably strong average clustering coefficient (ranging from 0.49 to 0.63) and large average degree (6.07 to 20.7). These measures suggest well connected topologies and also hint at the formation of communities.

The table also shows the average edge weight in each backbone. These numbers should be analyzed in light of the average number of messages per user in a group, shown in Table 1. Note that the backbone edges represent a large fraction of all messages shared by the users, on average. Moreover, by combining average degree and average edge weight, we observe that each user in the backbone simultaneously shared multiple contents with many other users. All these results show the multiplicity of media co-sharing among users and highlight the need for investigating higher level user structures, notably communities. Indeed, as shown in the last two rows of Table 2, we identified from 5 to 8 communities in the backbones, with an overall quality in terms of structural connectivity rather high (modularity between 0.54 and 0.70). Next, we delve deeper into these communities.

Figure 2 provides an overview of different community properties. Figure 2a shows the sizes (in number of nodes) of the communities identified in each snapshot (week), with each point representing a different community. We observe a great diversity of community sizes, normally constrained to fewer than 40 users, but it is also noticeable communities with more than 50 members. As an example, we identified a community with 26 members who shared political-driven content about presidential candidates and online campaigns on over 20 distinct WhatsApp groups in the week preceding the 1st round of election. This result suggests that the communities are dynamically built over time with variable number of users. We also correlate community size with average clustering coefficient, computed for community nodes, which is a measure of internal connectivity. Figure 2b shows these results for communities in all snapshots. We observe that most communities are strongly connected (even the larger ones) as the vast majority of them have average clustering coefficient above 0.50. Thus, in essence, the identified user communities are well structured, and, despite some size diversity over the snapshots, offer clear indications of consistent user co-sharing activity.

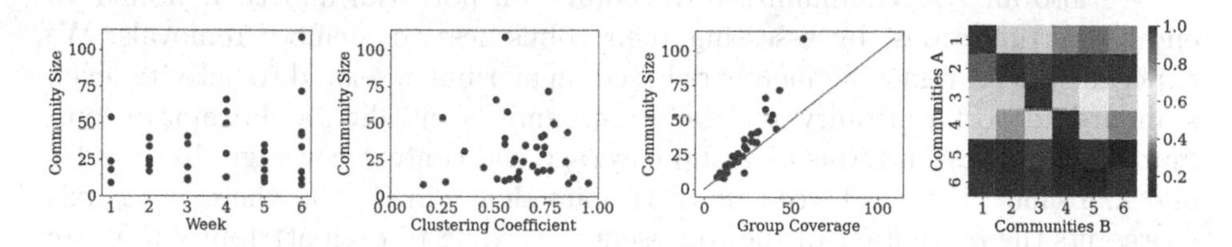

Fig. 2. Community properties: (a) Size; (b) Average clustering coefficient; (c) Group coverage and (d) Similarity of group coverage.

We now analyze the reach of these communities. Figure 2c shows a scatter plot with community size versus group coverage, for all communities in all snapshots. Recall that the latter is the number of groups the members of the communities participate in. There is a strong one-to-one relationship, in which the community size is strongly correlated with the number of groups it reaches. For larger communities, the sizes often are greater than the number of groups covered. These results suggest a broad reach in the ability to disseminate information, since communities often have members participating in multiple groups during the same time period. Moreover, the redundancy in some larger communities suggests some degree of robustness as well.

Taking a step further, we analyze the intersection of group coverage for different pairs of communities. Given two communities A and B, the intersection of B with A is the fraction of all groups covered by A that are also covered by B. Figure 2d shows these fractions for one snapshot – the third week ($w = 3$), which contains the 1st election round. Results for the other snapshots are similar, being thus omitted. In the figure, each cell in the upper triangle shows the intersection of community B (column) with community A (row), and each cell in the lower triangle shows the intersection of A with B. In general, around half of the communities are built over distinct groups, due to the small similarity values (around 20%). On the other hand, some communities are built over similar groups, reaching up to 70% of coverage similarity. Such communities contribute to robust, distinct and parallel content dissemination on the WhatsApp network.

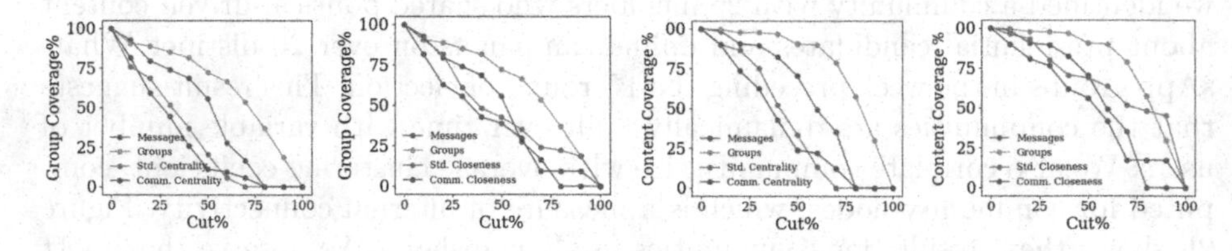

Fig. 3. Impact of removing users in decreasing centrality degree on group (a) and Content coverage (c) and in Decreasing closeness (b and d, respectively). (Color figure online)

We also analyze communities' dissemination potential and their ability to endure perturbations, by assessing their robustness to member removals. We experiment with removing members based on attributes related to activity level, standard network centrality and node community centrality, and evaluate community robustness in terms of group coverage and content coverage. Results for one snapshot – the third week ($w = 3$) – are shown in Fig. 3, where the x-axis represents the percentage of the top users, according to each attribute, that are removed ($n\%$ cut). Results for the other snapshots are quite similar.

Let's start by looking into group coverage. Figures 4a and 4b show that, by comparing the results for the attributes related to activity (number of messages shared and number of groups the user participates in) with those for standard (std.) network centrality metrics, we see that the latter are more relevant to detect the most important users for group coverage. This is true for both centrality metrics considered, i.e., centrality degree, which captures the node's ability to retrieve information from the network, and closeness, which relates to the node's efficiency to spread information in the network. Yet, the results for the node community centrality metrics (red lines in both figures) reveal that the communities are very important to the information dissemination through the network. This is due to the successful identification of the most important users and the positions they occupy both within its community and in relation to the whole network. Analogously, the same trend is observed for content coverage, as shown in Figs. 4c and 4d. Thus, using the community structure for identifying the most important users to information dissemination is relatively more effective than simply checking the activity level or the standard centrality metrics.

Finally, we also analyze the temporal evolution of the backbone and its communities. First, we quantify the persistence of users in the backbone over consecutive weeks. As shown in Fig. 4, a considerable amount of users (from 20% to 42%) remain in the backbone, repeatedly engaging on the weekly sharing activities. Yet, there is a large fraction of newcomers week after week. It could be due to new users who joined the groups (see Table 1) or simply reflect the replacement of non-persistent users by others as result of a natural diversity of user behavior over time. Regardless, the results indicate a highly dynamic backbone.

Focusing on the persistent users, we analyze how community members change over time based on NMI. As shown in Fig. 5, we compute NMI considering all persistent users (blue line) and only the top 10 users who shared more content in the week (green line). Considering all persistent users, there is a high mobility of users across communities (low NMI). This is illustrated in Fig. 6, which shows, from one week to the next, events associated with community membership such as splits, merges, births and deaths. In this figure, the wider the line, the greater the number of members migrating between communities, and the larger the diameter of the circle, the greater the number of members of that community. Once again, we see that persistent users in general often change community, engaging in different but strong co-sharing activities over time. However, if we zoom in the top 10 most active users (green line in Fig. 5), we observe a stronger tendency to continue in the same community (higher NMI), suggesting that these most active users regularly share common content over time.

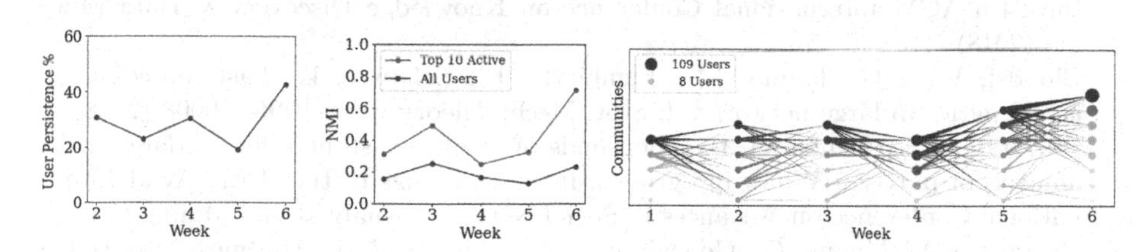

Fig. 4. Persistence (Color figure online)

Fig. 5. NMI (Color figure online)

Fig. 6. Community evolution

5 Conclusions and Future Work

This work analyzed the underlying network structure of information dissemination on WhatsApp publicly accessible groups in Brazil. By monitoring over 150 groups, our study revealed the formation of strongly connected user communities that cross the boundaries of traditional groups, fostering content spread at large. We found that these communities co-exist in the same groups constantly sharing broad content. By analyzing backbone and community centrality metrics, we were able to uncover users who are very important to the information dissemination, and those users would not be found if we looked only at their activity levels (numbers of messages and groups), as done in prior work. Moreover, by analyzing the temporal dynamics of the communities, we found that, despite constant changes in community membership, there is a number of users who persist in the network backbone over time, some of whom even remain tightly connected in the same community, suggesting coordinated efforts to boost information spread.

Future work includes analyzing the relations between content and community properties, zooming into spread of particular types of content (e.g., misinformation), and assessing user engagement and coordinated efforts in the communities.

Acknowledgements. This work was partially supported by grants from FAPEMIG, CNPQ and CAPES.

References

1. Abbe, E.: Community detection and stochastic block models: recent developments. J. Mach. Learn. Res. **18**, 6446–6531 (2017)
2. Agrawal, K., Garg, S., Patel, P.: Spatio-temporal outlier detection technique. Int. J. Comput. Sci. Commun. **6**, 330–337 (2015)
3. Barabási, A.L., et al.: Network Science. Cambridge University Press, Cambridge (2016)
4. Benson, A.R., Abebe, R., Schaub, M.T., Jadbabaie, A., Kleinberg, J.: Simplicial closure and higher-order link prediction. In: Proceedings of the National Academy of Sciences, pp. 11221–11230 (2018)
5. Benson, A.R., Kumar, R., Tomkins, A.: Sequences of sets. In: Proceedings of the 24th ACM International Conference on Knowledge Discovery & Data Mining (2018)
6. Blondel, V.D., Guillaume, J.L., Lambiotte, R., Lefebvre, E.: Fast unfolding of communities in large networks. J. Stat. Mech: Theory Exp. **2008**, 10008 (2008)
7. Bursztyn, V.S., Birnbaum, L.: Thousands of small, constant rallies: a large-scale analysis of partisan Whatsapp groups. In: Proceedings of the IEEE/ACM International Conference on Advances in Social Networks Analysis and Mining (2019)
8. Caetano, J.A., Magno, G., Gonçalves, M.A., Almeida, J.M., Marques-Neto, H.T., Almeida, V.A.F.: Characterizing attention cascades in whatsapp groups. In: Boldi, P., Welles, B.F., Kinder-Kurlanda, K., Wilson, C., Peters, I., Jr., W.M. (eds.) Proceedings of the 10th ACM Conference on Web Science, pp. 27–36 (2019)
9. Coscia, M., Neffke, F.M.: Network backboning with noisy data. In: 2017 IEEE 33rd International Conference on Data Engineering (ICDE) (2018)

10. Gomes Ferreira, C.H., de Sousa Matos, B., Almeira, J.M.: Analyzing dynamic ideological communities in congressional voting networks. In: Staab, S., Koltsova, O., Ignatov, D.I. (eds.) SocInfo 2018. LNCS, vol. 11185, pp. 257–273. Springer, Cham (2018). https://doi.org/10.1007/978-3-030-01129-1_16
11. Ferreira, C.H., Murai, F., de Souza Matos, B., de Almeida, J.M.: Modeling dynamic ideological behavior in political networks. J. Web Sci. **1**, 1–14 (2019)
12. Fortunato, S., Hric, D.: Community detection in networks: a user guide. Phys. Rep. **659**, 1–44 (2016)
13. de Freitas Melo, P., Vieira, C.C., Garimella, K., de Melo, P.O.S.V., Benevenuto, F.: Can Whatsapp counter misinformation by limiting message forwarding? In: Cherifi, H., Gaito, S., Mendes, J.F., Moro, E., Rocha, L.M. (eds.) COMPLEX NETWORKS 2019. SCI, vol. 881, pp. 372–384. Springer, Cham (2020). https://doi.org/10.1007/978-3-030-36687-2_31
14. Fu, X., Yu, S., Benson, A.R.: Modelling and analysis of tagging networks in stack exchange communities. J. Complex Netw. (2019)
15. Ghalmane, Z., Cherifi, C., Cherifi, H., El Hassouni, M.: Centrality in complex networks with overlapping community structure. Sci. Rep. **9**, 1–29 (2019)
16. Gilbert, F., Simonetto, P., Zaidi, F., Jourdan, F., Bourqui, R.: Communities and hierarchical structures in dynamic social networks: analysis and visualization. Soc. Netw. Anal. Min **1**, 83–95 (2011). https://doi.org/10.1007/s13278-010-0002-8
17. Girvan, M., Newman, M.E.: Community structure in social and biological networks. Proceedings of the National Academy of Sciences, pp. 7821–7826 (2002)
18. Goel, V.: How whatsapp leads mobs to murder in India, July 2018. https://www.nytimes.com/interactive/2018/07/18/technology/whatsapp-india-killings.html. Posted on 18 July 2018
19. Hoffmann, T., Peel, L., Lambiotte, R., Jones, N.S.: Community detection in networks without observing edges. Sci. Adv. **6**(4), eaav1478 (2020)
20. Jaccard, P.: Etude de la distribution florale dans une portion des alpes et du jura. Bulletin de la Societe Vaudoise des Sciences Naturelles, pp. 547–579 (1901)
21. Jégou, H., Douze, M., Schmid, C.: Product quantization for nearest neighbor search. IEEE Trans. Pattern Anal. Mach. Intell. **33**, 117–128 (2011)
22. Kietzmann, J.H., Hermkens, K., McCarthy, I.P., Silvestre, B.S.: Social media? Get serious! understanding the functional building blocks of social media. Bus. Horiz. **54**, 241–251 (2011)
23. Kordopatis-Zilos, G., Papadopoulos, S., Patras, I., Kompatsiaris, I.: FIVR: fine-grained incident video retrieval. IEEE Trans. Multimedia **21**, 2638–2652 (2019)
24. Kumar, R., Novak, J., Tomkins, A.: Structure and evolution of online social networks. In: Yu, P., Han, J., Faloutsos, C. (eds.) Link Mining: Models, Algorithms, and Applications. Springer, New York (2010). https://doi.org/10.1007/978-1-4419-6515-8_13
25. Kumar, S., Shah, N.: False information on web and social media: a survey. In: Advances and Applications, Social Media Analytics (2018)
26. Lancichinetti, A., Fortunato, S.: Detecting the overlapping and hierarchical community structure in complex networks. New J. Phys. **11**, 033015 (2009)
27. Magenta, M., Gragnani, J., Souza, F.: How Whatsapp is being abused in Brazil's election, October 2018. https://www.bbc.com/news/technology-45956557. Acessed on 24 May 2020
28. Maros, A., Almeida, J., Benevenuto, F., Vasconcelos, M.: Analyzing the use of audio messages in Whatsapp groups. In: Proceedings of The Web Conference (2020)

29. Mikolov, T., Sutskever, I., Chen, K., Corrado, G.S., Dean, J.: Distributed representations of words and phrases and their compositionality. In: Advances in Neural Information Processing Systems, vol. 26, pp. 3111–3119. Curran Associates, Inc. (2013)
30. Newman, M.: Network structure from rich but noisy data. Nat. Phys. **14**, 542–545 (2018)
31. Onnela, J.P., Saramäki, J., Kertész, J., Kaski, K.: Intensity and coherence of motifs in weighted complex networks. Phys. Rev. E **71**, 065103 (2005)
32. Papadopoulos, S., Kompatsiaris, I., Vakali, A., Spyridonos, P.: Community detection in social media. Data Min. Knowl. Disc. **24**, 515–554 (2012). https://doi.org/10.1007/s10618-011-0224-z
33. Resende, G., Melo, P., C.dS. Reis, J., Vasconcelos, M., Almeida, J.M., Benevenuto, F.: Analyzing textual (mis)information shared in Whatsapp groups. In: Proceedings of the 10th ACM Conference on Web Science. WebSci 2019. Association for Computing Machinery (2019)
34. Resende, G., Melo, P.F., Sousa, H., Messias, J., Vasconcelos, M., Almeida, J.M., Benevenuto, F.: (Mis)information dissemination in whatsapp: Gathering, analyzing and countermeasures. In: The World Wide Web Conference, WWW 2019, San Francisco, CA, USA, May 13–17, 2019, pp. 818–828. ACM (2019)
35. Rossetti, G., Cazabet, R.: Community discovery in dynamic networks: a survey. ACM Comput. Surv. (CSUR) **51**, 1–37 (2018)
36. Rossetti, G., Pappalardo, L., Pedreschi, D., Giannotti, F.: Tiles: an online algorithm for community discovery in dynamic social networks. Mach. Learn. **106**(8), 1213–1241 (2016). https://doi.org/10.1007/s10994-016-5582-8
37. Serrano, M.A., Boguna, M., Vespignani, A.: Extracting the multiscale backbone of complex weighted networks. Proc. Natl. Acad. Sci. **106**, 6483–6488 (2009)
38. Song, J., Yang, Y., Huang, Z., Shen, H.T., Hong, R.: Multiple feature hashing for real-time large scale near-duplicate video retrieval. In: Proceedings of the 19th ACM International Conference on Multimedia, pp. 423–432 (2011)
39. Tang, L., Liu, H., Zhang, J., Nazeri, Z.: Community evolution in dynamic multimode networks. In: Proceedings of International Conference on Knowledge Discovery and Data Mining, pp. 677–685 (2008)
40. WhatsApp: two billion users on WhatsApp. https://blog.whatsapp.com/10000666/Two-Billion-Users-Connecting-the-World-Privately. Accessed 19 May 2020
41. WhatsApp: about WhatsApp (2020). https://www.whatsapp.com/about/. Accessed 19 May 2020
42. WhatsApp: keeping WhatsApp personal and private. https://blog.whatsapp.com/Keeping-WhatsApp-Personal-and-Private (2020). Accessed 19 May 2020
43. Zauner, C., Steinebach, M., Hermann, E.: Rihamark: perceptual image hash benchmarking. In: Memon, N.D., Dittmann, J., Alattar, A.M., Delp III, E.J. (eds.) Media Watermarking, Security, and Forensics III, vol. 7880. SPIE, Bellingham (2011)
44. Zhang, X.S., et al.: Modularity optimization in community detection of complex networks. EPL (Europhys. Lett.) **87**, 38002 (2009)

Structural Invariants in Individuals Language Use: The "Ego Network" of Words

Kilian Ollivier(✉), Chiara Boldrini, Andrea Passarella, and Marco Conti

CNR-IIT, Via G. Moruzzi 1, 56124 Pisa, Italy
{kilian.ollivier,chiara.boldrini,andrea.passarella,
marco.conti}@iit.cnr.it

Abstract. The cognitive constraints that humans exhibit in their social interactions have been extensively studied by anthropologists, who have highlighted their regularities across different types of social networks. We postulate that similar regularities can be found in other cognitive processes, such as those involving language production. In order to provide preliminary evidence for this claim, we analyse a dataset containing tweets of a heterogeneous group of Twitter users (regular users and professional writers). Leveraging a methodology similar to the one used to uncover the well-established social cognitive constraints, we find that a concentric layered structure (which we call *ego network of words*, in analogy to the ego network of social relationships) very well captures how individuals organise the words they use. The size of the layers in this structure regularly grows (approximately 2–3 times with respect to the previous one) when moving outwards, and the two penultimate external layers consistently account for approximately 60% and 30% of the used words (the outermost layer contains 100% of the words), irrespective of the number of the total number of layers of the user.

Keywords: Language · Cognitive constraints · Structural invariants

1 Introduction

Language production relies on many cognitive processes that unfold at every sequential step of word retrieval, essentially during the selection of the lexical element that will symbolize the concept that has to be expressed in the sentence [19]. The brain has acquired strategies to process efficiently the mental lexicon that contains a large number of words (e.g., for retrieving a single word among 40K others in less than 250 ms). These strategies are executed unconsciously and they take advantage of language properties such as word frequencies [4,23] (the brain takes less time to retrieve words that are commonly used) to spend the least amount of time and effort in this task. In this work, we are interested in finding indirect traces of these cognitive processes in the written

S. Aref et al. (Eds.): SocInfo 2020, LNCS 12467, pp. 267–282, 2020.
https://doi.org/10.1007/978-3-030-60975-7_20

production. Specifically, we aim at investigating, through a data-driven approach, whether a regular structure can be found in the way people use words, as a "symptom" of cognitive constraints in this mental process. We argue that words usage might present similar properties to other mental processes which are known to be driven by cognitive constraints, specifically the way how humans allocate cognitive capacity to maintaining social relationships.

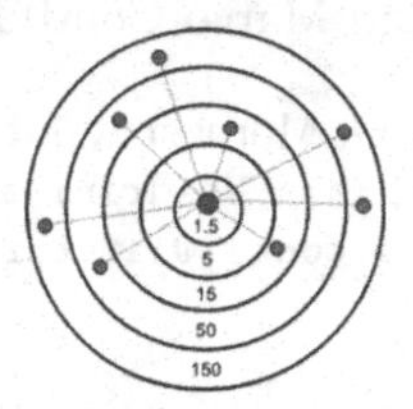

Fig. 1. The *ego network* of social relationships. The red dot symbolizes the ego and the blue dots the alters with whom the ego maintains an active social relationship. The numbers correspond to the layers sizes. (Color figure online)

The cognitive efforts that we allocate to socialization have been extensively studied by anthropologists, and their findings [10] show that the social life of humans is constrained, through time and cognitive capacity, to 150 meaningful relationships per person (a limit that goes under the name of *Dunbar's number*, from the scientist who first postulated its existence). This limit is observable in primates as well, where it is related to how many peers can be effectively groomed by the animals to reinforce social bonds. In humans, these 150 social relationships can be grouped into classes of different intimacy. Specifically, anthropologists have found that the social relations around the average individual can be grouped into at least 4 concentric layers [17,26], starting from the innermost one which typically includes our closest family members. The typical sizes of these layers are 5, 15, 50, 150. Many people also feature an additional layer, whose size is around 1.5 people, included in the first layer, which comprises relationships with extremely high social closeness [12]. This structure of social relationships, illustrated in Fig. 1, is typically referred to as *ego network*. A characteristic fingerprint of these social circles is their scaling ratio (i.e. the ratio between the sizes of consecutive layers), which has been found to be approximately around 3, regardless of the specific social network considered. Interestingly, both real-life and online social networks follow this social organization [12,15,16,21]. The discovery of this social structure (stratified in concentric layers) and its invariants (in terms of number of layers and scaling ratio across different and heterogeneous social networks) has represented a breakthrough moment in this research area. Many subsequent studies have leveraged this aggregate representation through social circles to better understand social-dependent human behaviour, such as how humans trust each other [24] or how they share resources and information [1,2].

Building upon the above considerations, in this work we set out to investigate the presence of an analogous structure and structural invariants in cognitive processes beyond the well-established social ones. Indeed, socialization is just one of the many cognitive processes that we entertain in our daily life. Thus, it is reasonable to expect that similar limitations in our cognitive capacity yield characteristic structural properties in other domains as well. Here, we focus on the cognitive process associated with language production, which, as described above, is tightly related to our cognitive capacity. Moreover, language is intimately linked to sociality, as there are hypotheses (that go under the name of *social gossip theory of language evolution* [11]) postulating that language has been developed as a more efficient way for grooming social relationships: with *vocal grooming*, we can reach more peers at the same time. We already had pieces of evidence suggesting the existence of cognitive limits in language production. One of the most prominent examples, which has been reported by G. Zipf [27] in 1932, is the empirical observation that the frequency of words in a corpus is inversely proportional to its position in the frequency table. It is also well-known that our vocabulary size is limited: e.g., an average 20-year-old native speaker of American English knows 42,000 words [6]. In this work, we try to go one step further and investigate more complex structural properties, leveraging the approach used to uncover cognitive constraints in the social domain. To the best of our knowledge, this research perspective has never been tackled before in the related literature. Please note that when we refer to the structural properties of language use, we do not refer to grammar but to the language-agnostic way in which lemmas are assigned a cognitive effort by their users.

An advantage of studying language production is that myriad textual datasets are available online. In trying to find out which cognitive constraints affect the production of language, our intuition is that the more spontaneous this production is, the greater the time constraint, and the more visible the cognitive limit. We argue that Twitter is a platform that facilitates a spontaneous writing style, much more so than newspaper articles or speech transcripts (just to mention a few other textual formats readily available online). For this reason, we choose Twitter for this initial investigation of the cognitive constraints in language production, leaving the analysis of other textual dataset as future work. We have collected a diverse dataset of tweets for our analysis, including tweets from regular Twitter users and professional writers (Sect. 2). Then, leveraging a methodology similar to the one used to uncover social constraints, we study the structural properties of language production on Twitter as a function of the individual word usage frequency, and we provide preliminary evidence for a set of cognitive constraints that naturally determine the way we communicate (Sect. 3). Specifically, our main findings are the following:

- Similarly to the social case, we found that a *regular concentric, layered structure* (which we call *ego network of words* in analogy to the ego networks of the social domain) very well captures how an individual organizes their cognitive effort in language production. Specifically, words can be typically grouped in between 5 and 7 layers of decreasing usage frequency moving outwards,

regardless of the specific class of users (regular vs professional) and of the specific time window considered.
- One structural invariant is observed for the *size of the layers*, which approximately doubles when moving from layer i to layer $i+1$. The only exception is the innermost layer, which tends to be approximately 5 five times smaller than the next one. This suggests that the innermost layer, the one containing the most used words, may be drastically different from the others.
- A second structural invariant emerges for the *external layers*. Users with more clusters organise differently their innermost layers, without modifying significantly the size of the most external ones. In fact, while the size of all layers beyond the first one linearly increases with the most external layer size, the second-last and third-last layer consistently account for approximately 60% and 30% of the used words, irrespective of the number of clusters of the user.

2 The Dataset

The analysis is built upon four datasets extracted from Twitter, using the official Search and Streaming APIs (note that the number of downloadable tweets is limited to 3200 per user). Each of them is based on the tweets issued by users in four distinct groups:

Journalists Extracted from a Twitter list containing New York Times journalists[1], created by the New York Times itself. It includes 678 accounts, whose timelines have been downloaded on February 16th 2018.

Science writers Extracted from a Twitter list created by Jennifer Frazer[2], a science writer at *Scientific American*. The group is composed of 497 accounts and has been downloaded on June 20th 2018.

Random users #1 This group has been collected by sampling among the accounts that issued a tweet or a retweet in English with the hashtag *#MondayMotivation* (at the download time, on January 16th 2020). This hashtag is chosen in order to obtain a diversified sample of users: it is broadly used and does not refer to a specific event or a political issue. As the accounts are not handpicked as in the two first groups, we need to make sure that they represent real humans. The probability that an account is a bot is calculated with the Botometer service [8], which is based not only on language-agnostic features like the number of followers or the tweeting frequency, but also on linguistic features such as grammatical tags, or the number of words in a tweet [25]. The algorithm detects 29% of bot accounts, such that this dataset is composed of 5183 users.

Random users #2 This group has been collected by sampling among the accounts which issued a tweet or a retweet in English, from the United Kingdom (we set up a filter based on the language and country), at download time

[1] https://twitter.com/i/lists/54340435.
[2] https://twitter.com/i/lists/52528869.

on February 11th 2020. 23% of the accounts are detected as bot, such that this group contains 2733 accounts.

These groups are chosen to cover different types of users: the first two contain accounts that use language professionally (journalists and science writers), the other two contain regular users, which are expected to be more colloquial and less controlled in their language use. Please note that we discard retweets with no associated comments, as they do not include any text written by the target user, and tweets written in a language other than English (since most of the NLP tools needed for our analysis are optimised for the English language). In our analysis, we only consider active Twitter accounts, which we define as an account not abandoned by its user and that tweets regularly. Further details on this preprocessing step are provided in Appendix A.1.

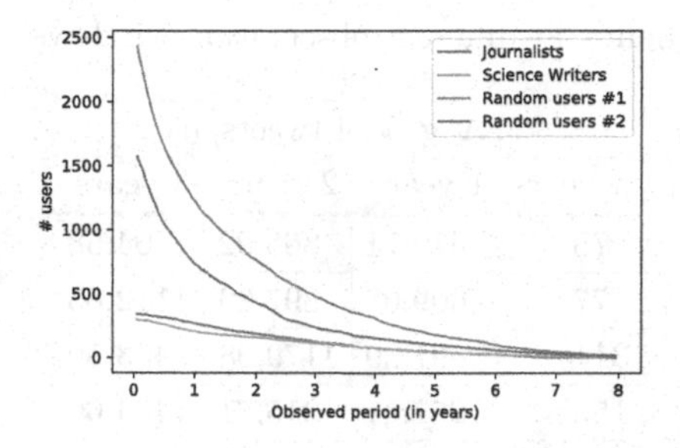

Fig. 2. Number of selected timelines depending on the observation window.

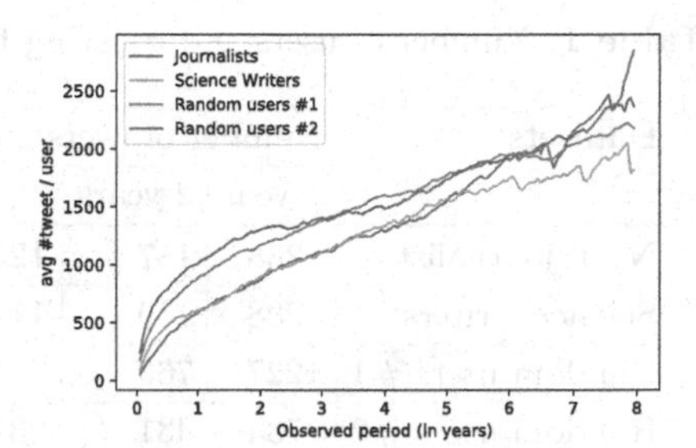

Fig. 3. Average number of tweets depending on the observation window.

2.1 Extracting User Timelines with the Same Observation Period

The observed timeline size is only constrained by the number of tweets (limited by API), thus the observation period varies according to the frequency with which the account is tweeting: for very active users, the last 3200 tweets will only cover a short time span. This raises the following problem: as random users are generally more active, their observation period is shorter, and this may create a significant sampling bias. In fact, the length of the observation period affects the measured word usage frequencies discussed in Sect. 3.1 (specifically, we cannot observe frequencies lower than the inverse of the observation period). In order to guarantee a fair comparison across user categories and to be able to compare users with different tweeting activities without introducing biases, we choose to work on timelines with the same duration, by restricting to an observation window T. To obtain timelines that have the same observation window T (in years), we delete all those with a duration shorter than T and remove tweets written more than T years ago from the remaining ones. Increasing T therefore reduces the number of profiles we can keep (see Fig. 2): for a T larger than 2 years, that number is divided by two, and for a T larger than 3 years, it falls below 500 for all datasets. On the contrary, the average number of tweets per timeline increases linearly with T (Fig. 3). The choice of an observation window will then result from a trade-off between a high number of timelines per dataset and a large average number of tweets per timeline. To simplify the choice of T, we only select round numbers of years. We can read in Table 1 that, beyond 3 years, the number of users falls below 100 for some datasets. On the other hand, the number of tweets for $T = 1$ year remains acceptable (> 500). We, therefore, decided to carry on the analysis with $T \in \{1 \text{ year}, 2 \text{ years}, 3 \text{ years}\}$. Please note that random users have a higher frequency of tweeting than others. This difference tends to smooth out when the observation period is longer (Table 1). This can be explained by the fact that the timelines with the highest tweet frequency are excluded in that case because their observation period is too small.

Table 1. Number of users and tweeting frequency at different observation windows.

Datasets	Number of users				Avg # of tweets/user		
	1 year	2 years	3 years	4 years	1 year	2 years	3 years
NYT journalists	268	187	125	75	579.71	865.02	1104.58
Science writers	208	159	117	77	609.08	897.29	1112.63
Random users #1	1227	765	481	311	897.29	1179.98	1403.50
Random users #2	734	431	237	153	1057.41	1315.71	1404.60

2.2 Word Extraction

Since the analysis has a focus on words and their frequency of use, we take advantage of NLP techniques for extracting them. As first step, all the syntactic marks that are specific to communication in online social networks (mentions with @, hashtags with #, links, emojis) are discarded (see Table 3 in Appendix A.2 for a summary). Once the remaining words are tokenized (i.e., identified as words), those that are used to articulate the sentence (e.g., "with", "a", "but") are dropped. This type of words is called a functional word as opposed (in linguistics) to lexical words, which have a meaning independent of the context. These two categories involve different cognitive processes (syntactic for functional words and semantic for lexical words), different parts of the brain [9], and probably different neurological organizations [13]. We are more interested in lexical words because their frequency in written production depends on the author's intentions, as opposed to functional words frequencies that depend on the language characteristics[3]. Moreover, lexical words represent the biggest part of the vocabulary. Functional words are generally called stop-words in the NLP domain and many libraries provide tools to filter them out.

[3] Functional words may also depend on the style of an author (and due to this they are often used in stylometry). Still, whether their usage require a significant cognitive effort is arguable, hence in this work we opted for their removal.

As this work will leverage word frequencies as a proxy for discovering cognitive properties, we need to group words derived from the same root (e.g. "work" and "worked") in order to calculate their number of occurrences. This operation can be achieved with two methods: stemming and lemmatization. Stemming algorithms generally remove the last letters thanks to complex heuristics, whereas lemmatization uses the dictionary and a real morphological analysis of the word to find its normalized form. Stemming is faster, but it may cause some mistakes of overstemming and understemming. For this reason, we choose to perform lemmatization. Once we have obtained the number of occurrences for each word base, we remove all those that appear only once to leave out the majority of misspelled words. Table 4 in Appendix A.2 contains examples of the entire preprocessing part.

3 From Word Usage to Cognitive Constraints

Recalling that our goal is to investigate the structure and structural invariants in language production, in this section we present the main findings of our study. The methodology of our analysis is as follows. First, we analyse the frequency of words usage and the richness in vocabulary across the datasets, as preliminary characterisation of the different types of users we consider (Sect. 3.1). We then perform a clustering analysis to investigate whether regular groups of words used at different frequencies can be identified. This is the same method used to analyse the structural properties of human social networks, as explained in Sect. 1. Based on this analysis, we describe the structural invariants that we found in Sects. 3.2 and 3.3.

3.1 Preliminaries

Let us focus on a tagged user j. When studying the social cognitive constraints, the contact frequency between two people was taken as proxy for their intimacy and, as a result, for their cognitive effort in nurturing the relationship. Similarly, the frequency f_i at which user j uses word i is considered here as a proxy of their "relationship". Frequency f_i is given by $\frac{n_{ij}}{t_i}$, where n_{ij} denotes the number of occurrences of word i in user j's timeline, and t_i denotes the observation window of j's account in years.

Figure 4 shows the frequency distribution for the different categories of users (regular users vs professional writers) and for the different observation windows (1, 2, 3 years). We can make two observations. First, the distributions exhibit a heavy-tailed behaviour (see Table 5 in Appendix A.2 for the fitting of individual users word frequency). Second, the distributions are very similar two by two: specialized users (journalists and science writers) who fulfill a particular role of information in the social network, and randoms users who are samples of more regular users. The first group seems to use more low-frequency words, while the second group uses a larger proportion of high-frequency words. Based on the second observation, we can compare the datasets based on two criteria:

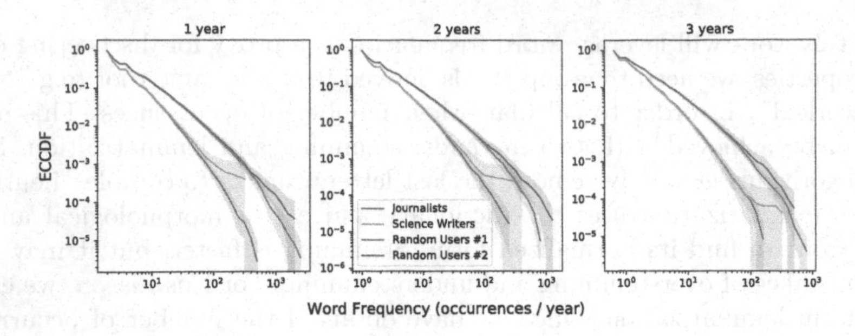

Fig. 4. Aggregate visualization of user-wise empirical CCDF of word usage frequencies, in log-log scale. The solid line corresponds to the average and the shaded area to the 95% confidence intervals.

verbosity, which counts the total number of words per tweet, and *lexical richness*, which counts the number of distinct words per tweet (Figs. 5-6). Despite a lower verbosity (Fig. 5), the vocabulary of specialized users seems richer than those of random users (Fig. 6). This is intuitive, but not necessarily expected. Professional users certainly have a higher diversity in the use of words. However, this manifests also in Twitter, i.e., in an environment where they do not write necessarily for professional reasons, but where they are (supposedly) writing in a more immediate and informal way.

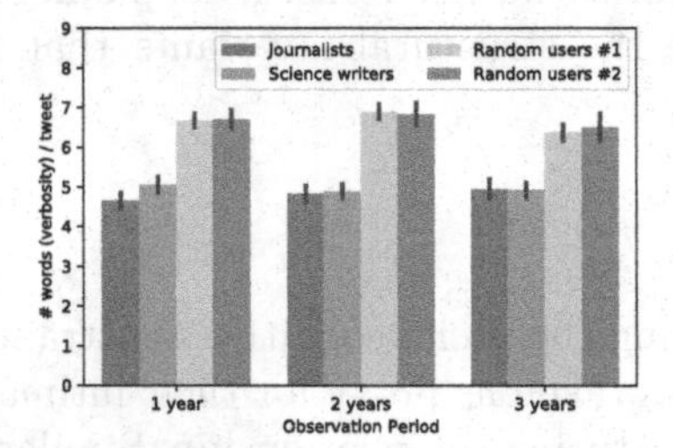

Fig. 5. Average verbosity, with 95% confidence intervals.

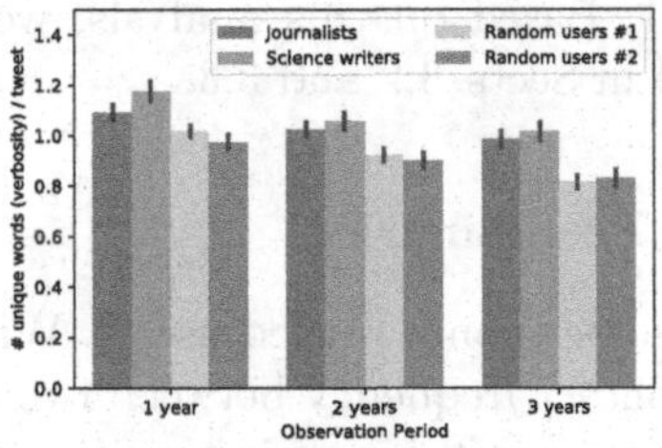

Fig. 6. Average lexical richness, with 95% confidence intervals.

3.2 Many Words, Just a Few Groups

Using the frequencies described in the previous section, we now investigate whether the words of a user can be grouped into homogeneous classes, and whether different users feature a similar number of classes or not. To this aim, for each user, we leverage a clustering algorithm to group words with a similar frequency. The selected algorithm is Mean Shift [14], because as opposed to Jenks [18] or KMeans [20], it is able to find the optimal number of clusters without fixed parameters. The original Mean Shift algorithm has nevertheless a drawback: it is only able to find the estimated density peaks with a fixed bandwidth kernel. The bandwidth is estimated based on the distribution of the pairwise distances between all the frequency values. However, in our case the distance between frequencies is not homogeneous: most of them are concentrated in the lowest values, close to each other. Hence, the selected bandwidth is fitted for estimating the density in that area, but not in the tail of the distribution. For that reason, a log-transformation is applied to the frequency values prior to the Mean Shift run: it still allows a fine mode detection in low-frequency part and compresses high values to allow detection of modes with a larger width. The use of a logarithmic scale is also used by psychological researchers to explain the impact of word frequency on their cognitive processing [5].

The histograms of the obtained optimal number of clusters are shown in Fig. 7. It is interesting to note that, despite the heterogeneity of users (in terms of tweeting frequency, verbosity, and lexical richness), the distributions are always quite narrow, with peaks consistently between 5 and 7 clusters. The observation period seems to have a very limited effect on the resulting cluster structure. This means that, after one year, the different groups of words can be already identified reliably. In addition, this limited effect actually reinforces the idea of a natural grouping: when more words are added (longer observation period) the clusters become slightly fewer, not slightly more. Hence, new words tend to reinforce existing clusters. Thus, similarly to the social constraints case, also for language production we observe a fairly regular and consistent structure. This is the first important result of the paper, hinting at the existence of structural invariants in cognitive processes, which we summarise below.

Cognitive Constraint 1: Individual distributions of word frequencies are divided into a consistent number of groups. Since word frequencies impact the cognitive processes underlying word learning and retrieval in the mental lexicon [22], these groups can be an indirect trace of these processes' properties. The number of groups is only marginally affected by the class (specialized or generic) the users belong to or by the observation window. This regularity might also suggest that these groups of words correspond to linguistic functional groups, and we plan to investigate this as future work.

3.3 Exploring the Group Sizes

We now study the size of the clusters identified in the previous section. For the sake of statistical reliability, we only consider those users whose optimal number of clusters (as identified by Mean Shift) corresponds to the most popular number of clusters (red bars) in Fig. 7. This allows us to have a sufficient number of samples in each class. We rank each cluster by its position in the frequency distribution: cluster #1 is the one that contains the most frequent words, and the last cluster is the one that contains the least used. Following the convention of the Dunbar's model discussed in Sect. 1, these clusters can be mapped into concentric layers (or circles), which provide a cumulative view of word usage.

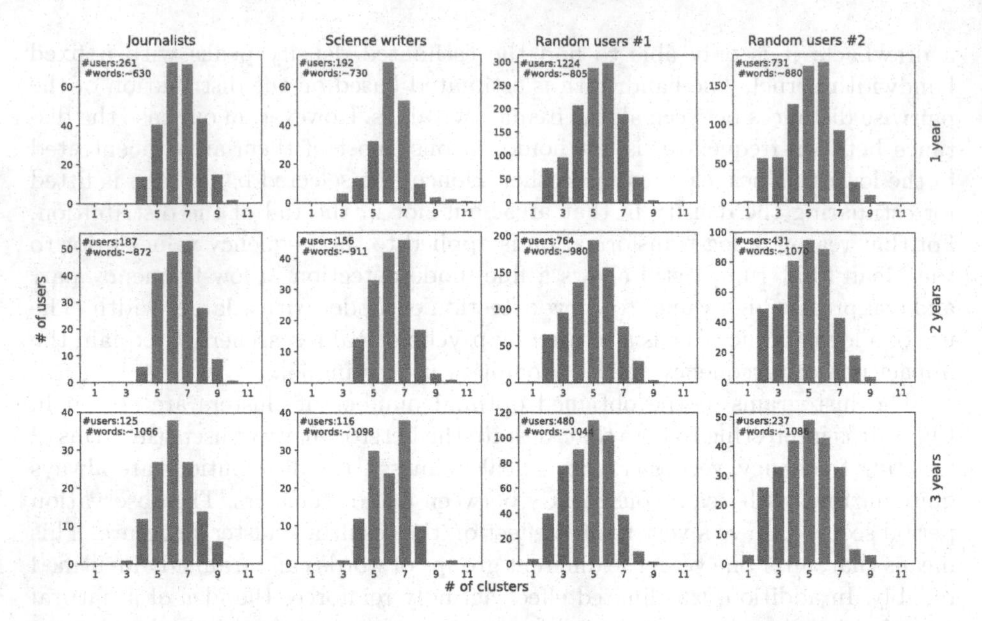

Fig. 7. Number of clusters obtained applying Mean Shift to log-transformed frequencies. The most frequent number of clusters are highlighted in red. (Color figure online)

Specifically, layer i includes all clusters from the first to the i-th. Layers provide a convenient grouping of words used *at least* at a certain frequency. We refer to this layered structure as *ego network of words.*

Figure 8 shows the average layer sizes for every dataset and different observation periods. As expected, for a given number of clusters, the layer size increases as we expand the observation period, because more words are brought in. For a given number of clusters we also observe a striking regularity across the datasets, with confidence intervals overlapping in practically all settings. Typically, the layer sizes are slightly higher for journalists and science writers ($T = 2$ years and $T = 3$ years). The main reason is that their lexicon is generally richer than those of regular users (as discussed in Sect. 3.1) and this is reflected in their layer size.

Another typical metric that is analysed in the context of social cognitive constraints is the scaling ratio between layers, which, as discussed earlier, corresponds to the ratio between the size of consecutive layers. The scaling ratio is an important measure of regularity, as it captures a relative pattern across layers, beyond the absolute values of their size. Figure 9 shows the scaling ratio of the layers in language production. We can observe the following general behavior: the scaling ratio starts with a high value between layers #1 and #2, but always gets closer to 2–3 as we move outwards. This empirical rule is valid whatever the dataset and whatever the observation period. This is another significant structural regularity, quite similar to the one found for social ego networks, as a further hint of cognitive constraints behind the way humans organise word use.

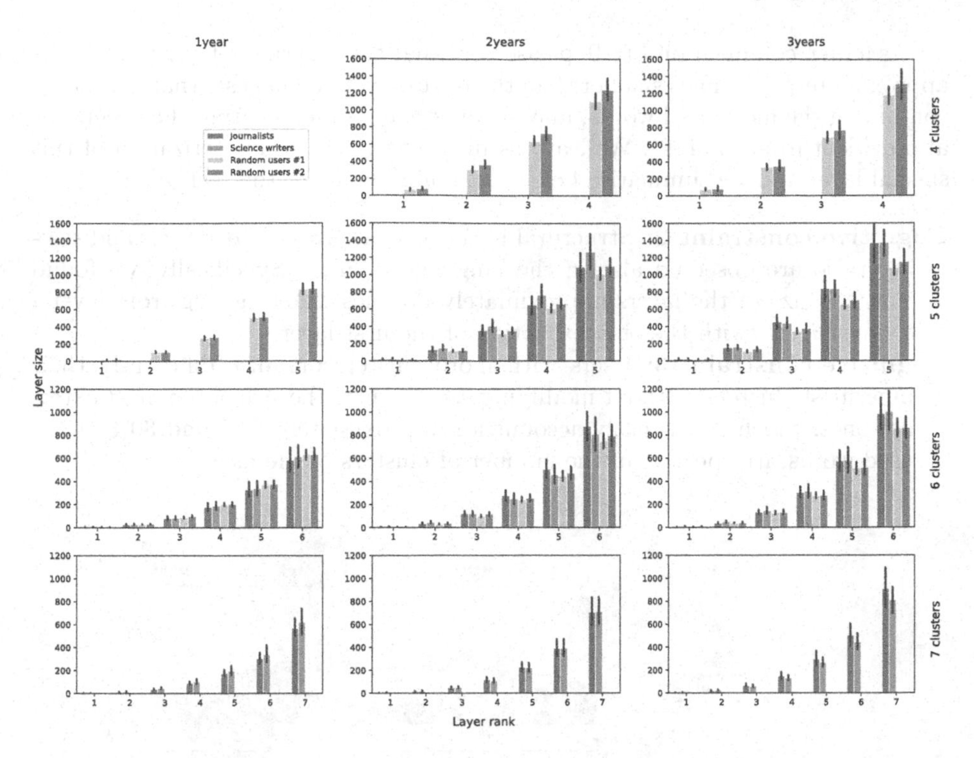

Fig. 8. Average layer size (with 95% confidence intervals) for the various datasets, different number of clusters (rows), and different observation periods (columns).

In order to further investigate the structure of the word clusters, we compute the linear regression coefficients between the total number of unique words used by each user (corresponding to the size of the outermost layer) and the individual layer sizes. Due to space limits, in Table 2 we only report the exact coefficients for the journalists dataset with T=1 year (but analogous results are obtained for the other categories and observation windows) and in Fig. 10 we plot the linear regression for all the user categories with $T = 1$ year. Note that the size of the most external cluster is basically the total number of words used by an individual in the observation window. It is thus interesting to see what happens when this number increases, i.e., if users who use more words distribute them uniformly across the clusters, or not. Table 2 shows two interesting features. First, it shows another regularity, as the size of all layers linearly increases with the most external cluster size, with the exception of the first one (Fig. 10). Moreover, it is quite interesting to observe that the second-last and third-last layer consistently account for approximately 60% and 30% of the used words, irrespective of the number of clusters. This indicates that users with more clusters split at a finer granularity words used at highest frequencies, i.e., they organise differently their innermost clusters, without modifying significantly the size of the most external ones.

As a final comment on Fig. 9, please note that the innermost layer tends to be approximately five times smaller than the next one. This suggests that this layer, containing the most used words, may be drastically different from the others (as also evident from Table 2). We leave as future work the characterization of this special layer and we summarise below the main results of the section.

Cognitive constraint 2: Structural invariants in terms of layer sizes and scaling ratio are observed also in the language domain. Specifically, we found that the size of the layers approximately doubles when moving from layer i to layer $i+1$, with the only exception of the first layer.

Cognitive constraint 3: Users with more clusters organise differently their innermost clusters, without modifying significantly the size of the most external ones, which consistently account for approximately 60% and 30% of the used words, irrespective of the number of clusters of the user.

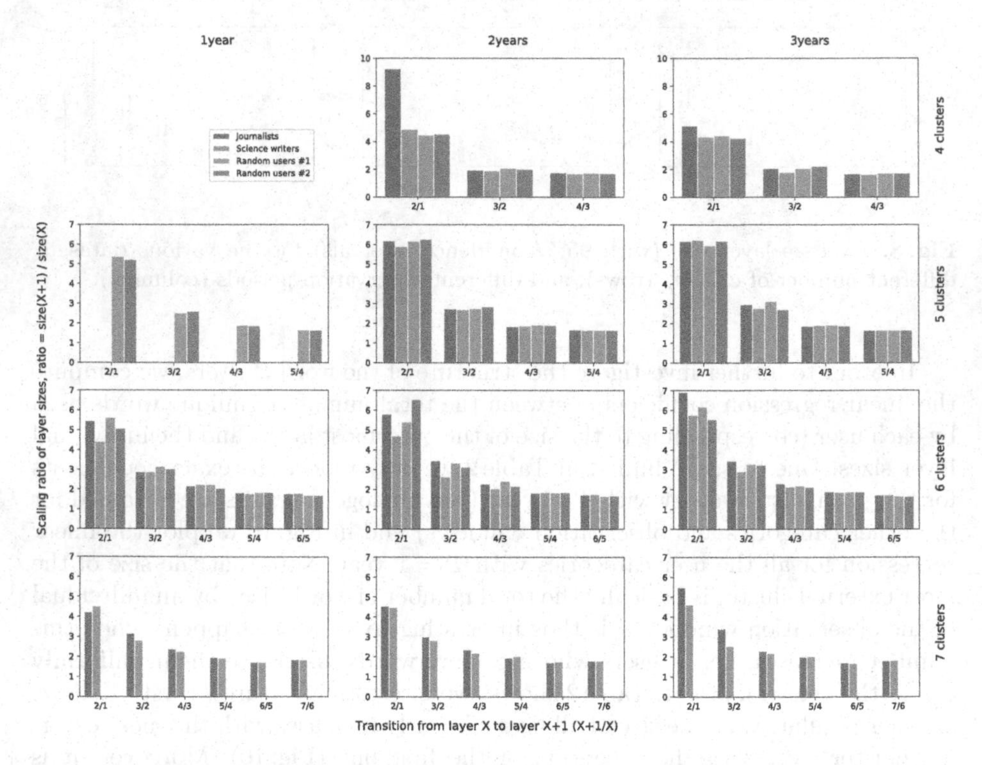

Fig. 9. Scaling ratio for the various datasets, different number of clusters (rows), and different observation periods (columns).

Table 2. Linear coefficients obtained for the journalists dataset with $T = 1$ year.

Opt. # of clusters	Cluster Rank						
	1	2	3	4	5	6	7
4 clusters	0.04	0.33	0.61	1.00			
5 clusters	0.02	0.13	0.33	0.62	1.00		
6 clusters	0.01	0.04	0.14	0.32	0.59	1.00	
7 clusters	0.00	0.02	0.06	0.16	0.32	0.56	1.00

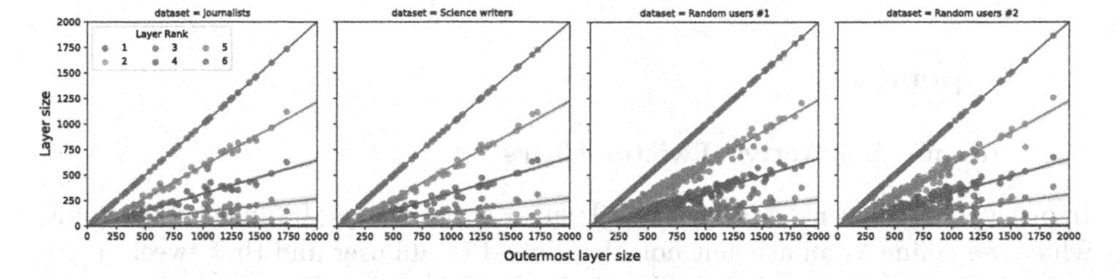

Fig. 10. Linear regression between the total number of unique words used by each user (corresponding to the size of the outermost layer) and the individual layer sizes.

4 Conclusion

In this paper, we investigated, through a data-driven approach, whether a regular structure can be found in the way people use words, as a symptom of cognitive constraints in their mental process. This is motivated by the fact that other mental processes are known to be driven by cognitive constraints, such as the way how humans allocate cognitive capacity to social relationships. To this aim, we collected a diverse dataset from Twitter (identified as one of the major sources of informal and spontaneous language online), including tweets from regular Twitter users and from professional writers. Then, leveraging a methodology similar to the one used to uncover social constraints, we have analysed the structural properties of language production on Twitter, uncovering regularities that constitute preliminary evidence of the aforementioned cognitive constraints. Specifically, we have found that, similarly to the social case, a concentric layered structure (ego network of words) very well captures how an individual organizes their cognitive effort in language production. Words can be grouped typically in between 5 and 7 layers, regardless of the specific class of users. We also observe a structural invariant in the size of the layers, which grow approximately 2–3 times when moving from a layer to the next one. A second structural invariant emerges for the external layers, which, regardless of the number of clusters of the user, consistently account for approximately 60% and 30% of the used words. While these findings are restricted to language production on Twitter, as future work we plan to generalize these results to other types of written communication.

Acknowledgements. This work was partially funded by the SoBigData++, HumaneAI-Net, MARVEL, and OK-INSAID projects. The SoBigData++ project has received funding from the European Union's Horizon 2020 research and innovation programme under grant agreement No 871042. The HumaneAI-Net project has received funding from the European Union's Horizon 2020 research and innovation programme under grant agreement No 952026. The MARVEL project has received funding from the European Union's Horizon 2020 research and innovation programme under grant agreement No 957337. The OK-INSAID project has received funding from the Italian PON-MISE program under grant agreement ARS01 00917.

A Appendix

A.1 Identifying Active Twitter Users

In order to be relevant to our work, a Twitter account must be an active account, which we define as an account not abandoned by its user and that tweets regularly. A Twitter account is considered abandoned, and we discard it, if the time since the last tweet is significantly bigger (we set this threshold at 6 months, as previously done also in [3]) than the largest period of inactivity for the account. We also consider the tweeting regularity, measured by counting the number of months where the user has been inactive. The account is tagged as sporadic, and discarded, if this number of months represents more than 50% of the observation period (defined as the time between the first tweet of a user in our dataset and the download time). We also discard accounts whose entire timeline is covered by the 3200 tweets that we are able to download, because their Twitter behaviour might have yet to stabilise (it is known that the tweeting activity needs a few months after an account is created to stabilise).

A.2 Additional Tables

Table 3. In the process of word extraction, the tweet is decomposed in tokens which are usually separated by spaces. These tokens generally corresponds to words, but they can also be links, emojis and others markers that are specific to the online language such as hashtags. The table gives the percentage of hashtags, links and emojis, which are tokens filtered out from the datasets.

	Percentage of hashtags	Percentage of links	Percentage of emojis
Journalists	1.34 %	7.27 %	0.20 %
Science writers	3.47 %	8.02 %	0.55 %
Random users #1	16.84 %	6.97 %	5.21 %
Random users #2	7.20 %	6.42 %	4.60 %

Table 4. Example of word extraction results.

Original tweet content	List of words after pre-processing
The @Patriots say they don't spy anymore. The @Eagles weren't taking any chances. They ran a "fake" practice before the #SuperBowl	Spy, anymore, chance, run, fake, practice
#Paris attacks come 2 days before world leaders will meet in #Turkey for the G20. Will be a huge test for Turkey	Attack, come, day, world, leader, meet, huge, test, turkey
Latest garden species - the beautiful but destructive rosemary beetle, and a leafhopper (anyone know if this can be identified to species level from photo? Happy to give it a go) #30DaysWild #MyWildCity #gardening	Late, garden, specie, beautiful, destructive, rosemary, beetle, leafhopper, know, identify, specie, level, photo, happy

Table 5. Percentage of users for which the hypothesis that the word frequency distribution is a power-law is rejected with a p-value below 0.1, 0.05 and 0.01. The p-value is obtained with the Kolmogorov-Smirnov test, using the fitting technique described in [7].

	Journalists			Science writers			Random dataset #1			Random dataset #2		
	1 year	2 years	3 years	1 year	2 years	3 years	1 year	2 years	3 years	1 year	2 years	3 years
p<0.1	22.1 %	33.2 %	49.6 %	24.8 %	37.3 %	47.9 %	41 %	49.7 %	52.6 %	50.3 %	62 %	61.4 %
p<0.05	17.6 %	25.1 %	39.2 %	19.4 %	29.1 %	39.3 %	32.8 %	43.5 %	43 %	43.9 %	52.1 %	52.4 %
p<0.01	8.6 %	11.2 %	18.4 %	12.1 %	15.8 %	17.1 %	21.4 %	27.8 %	27.8 %	30.4 %	37.3 %	35.2 %

References

1. Aral, S., Van Alstyne, M.: The diversity-bandwidth trade-off. Am. J. Sociol. **117**(1), 90–171 (2011)
2. Arnaboldi, V., Conti, M., La Gala, M., Passarella, A., Pezzoni, F.: Information diffusion in OSNs: the impact of nodes' sociality. In: Proceedings of the 29th Annual ACM Symposium on Applied Computing, pp. 616–621. ACM (2014)
3. Boldrini, C., Toprak, M., Conti, M., Passarella, A.: Twitter and the press: an ego-centred analysis. In: Companion Proceedings of the The Web Conference 2018, pp. 1471–1478 (2018)
4. Broadbent, D.E.: Word-frequency effect and response bias. Psychol. Rev. **74**(1), 1 (1967)
5. Brysbaert, M., Mandera, P., Keuleers, E.: The word frequency effect in word processing: an updated review. Curr. Direct. Psychol. Sci. **27**(1), 45–50 (2018)
6. Brysbaert, M., Stevens, M., Mandera, P., Keuleers, E.: How many words do we know? Practical estimates of vocabulary size dependent on word definition, the degree of language input and the participant's age. Front. Psychol. **7**(Jul), 1116 (2016)
7. Clauset, A., Shalizi, C.R., Newman, M.E.: Power-law distributions in empirical data. SIAM Rev. **51**(4), 661–703 (2009)
8. Davis, C.A., Varol, O., Ferrara, E., Flammini, A., Menczer, F.: BotOrNot: a system to evaluate social bots. In: Proceedings of the 25th International Conference Companion on World Wide Web, pp. 273–274 (2016)

9. Diaz, M.T., McCarthy, G.: A comparison of brain activity evoked by single content and function words: an FMRI investigation of implicit word processing. Brain Res. **1282**, 38–49 (2009)
10. Dunbar, R.: The social brain hypothesis. Evol. Anthropol. **9**(10), 178–190 (1998)
11. Dunbar, R.: Theory of Mind and the Evolution of Language. Approaches to the Evolution of Language (1998)
12. Dunbar, R.I., Arnaboldi, V., Conti, M., Passarella, A.: The structure of online social networks mirrors those in the offline world. Soc. Netw. **43**, 39–47 (2015)
13. Friederici, A.D., Opitz, B., Von Cramon, D.Y.: Segregating semantic and syntactic aspects of processing in the human brain: an FMRI investigation of different word types. Cerebr. Cortex **10**(7), 698–705 (2000)
14. Fukunaga, K., Hostetler, L.: The estimation of the gradient of a density function, with applications in pattern recognition. IEEE Trans. Inf. Theor. **21**(1), 32–40 (1975)
15. Gonçalves, B., Perra, N., Vespignani, A.: Modeling users' activity on twitter networks: validation of Dunbar's number. PloS ONE **6**(8), e22656 (2011)
16. Haerter, J.O., Jamtveit, B., Mathiesen, J.: Communication dynamics in finite capacity social networks. Phys. Rev. Lett. **109**(16), 168701 (2012)
17. Hill, R.A., Dunbar, R.I.: Social network size in humans. Hum. Nat. **14**(1), 53–72 (2003)
18. Jenks, G.F.: Optimal data classification for choropleth maps. Department of Geography, University of Kansas Occasional Paper (1977)
19. Levelt, W.J., Roelofs, A., Meyer, A.S.: A theory of lexical access in speech production. Behav. Brain Sci. **22**(1), 1–38 (1999)
20. MacQueen, J., et al.: Some methods for classification and analysis of multivariate observations. In: Proceedings of the Fifth Berkeley Symposium on Mathematical Statistics and Probability, Oakland, CA, USA, vol. 1, pp. 281–297 (1967)
21. Miritello, G., et al.: Time as a limited resource: communication strategy in mobile phone networks. Soc. Netw. **35**(1), 89–95 (2013)
22. Perfetti, C.A., Wlotko, E.W., Hart, L.A.: Word learning and individual differences in word learning reflected in event-related potentials. J. Exp. Psychol. Learn. Memory Cogn. **31**(6), 1281 (2005)
23. Qu, Q., Zhang, Q., Damian, M.F.: Tracking the time course of lexical access in orthographic production: an event-related potential study of word frequency effects in written picture naming. Brain Lang. **159**, 118–126 (2016)
24. Sutcliffe, A.G., Wang, D., Dunbar, R.I.: Modelling the role of trust in social relationships. ACM Trans. Internet Technol. (TOIT) **15**(4), 16 (2015)
25. Varol, O., Davis, C.A., Menczer, F., Flammini, A.: Feature engineering for social bot detection. In: Feature Engineering for Machine Learning and Data Analytics, pp. 311–334. CRC Press (2018)
26. Zhou, W.X., Sornette, D., Hill, R.a., Dunbar, R.I.M.: Discrete hierarchical organization of social group sizes. Proc. Biol. Sci. Roy. Soc. **272**(1561), 439–444 (2005)
27. Zipf, G.K.: Human Behavior and the Principle of Least Effort (1949)

Dynamics of Scientific Collaboration Networks Due to Academic Migrations

Pavlos Paraskevopoulos(✉), Chiara Boldrini, Andrea Passarella, and Marco Conti

CNR-IIT, Via G. Moruzzi 1, 56124 Pisa, Italy
{p.paraskevopoulos,c.boldrini,a.passarella,m.conti}@iit.cnr.it

Abstract. Academic migration is the change of host institution by a researcher, typically aimed at achieving a stronger research profile. Scientific features such as the number of collaborations, the productivity and its research impact tend to be directly affected by such movements. In this paper we analyse the dynamics of the collaboration network of researchers as they move from an institution to the next one. We specifically highlight cases where they increase and when they shrink, and quantify the dependency between the collaboration networks before and after such a movement. Finally, we drill down the analysis by dividing movements depending on the career stage of the researchers. The analysis shows a remarkable dynamism of collaboration network across migrations. Interestingly, not always movements result in larger collaboration networks, while the overall similarity between networks across movements is quite limited on average. Qualitatively, the same effects can be found at all career stages, while clearly the magnitude of them might vary. These results are based on a dataset extracted from Scopus, containing detailed scientific information for the publications of 84,141 researchers.

Keywords: Academic mobility · Collaboration networks · Scopus dataset

1 Introduction

Mobility is an important aspect of a researcher's life, affecting the career of the scientist in many ways [3,4,7–9,12,13]. Through the change of host institutions, new career opportunities are chased, positions with higher prestige acquired, stronger collaborations can be created, and novel projects are started. As a result, the collaboration network, the productivity and the research impact of the studies of the researcher are possibly affected every time a researcher changes host institution. Thus, mobility, both international and domestic, is rightly considered as a very important part of the research career. At first glance, the outcome of a movement may seem only beneficial. However, this is not always true. The downside to the creation of new opportunities is that old collaborations may collapse and old projects may be abandoned. The goal of this work

S. Aref et al. (Eds.): SocInfo 2020, LNCS 12467, pp. 283–296, 2020.
https://doi.org/10.1007/978-3-030-60975-7_21

is to assess the impact of academic migrations on the collaboration network of a researcher, also considering whether such features vary across different career stages. Moving towards this direction, we analyse how the size and composition of the collaboration network vary before and after an institution change. Finally, we investigate how the career duration affects the outcome of a movement in terms of collaboration network.

For our analysis, we use a dataset containing details for the scientific publications (extracted from the Scopus repository) of 84,141 mobile authors that have moved 561,389 times. The key element of our analysis is the author's ego network, which is defined as the set of co-authors (alters) of the given author (ego) while being at a certain institution. For studying the effects of a host institution change on the ego network, we focus on the analysis of a set of characteristics such as the change of the ego network size, and the similarity between the ego networks before and after the movement. We first consider an overall exploratory analysis, where we study all movements together, and analyse (i) the percentage change of the ego network size, (ii) the overlap between the ego networks across each movement, (iii) how much of the old network "survives" in the new one and (iv) to what extent the new network is built on old collaborations. We then drill down the analysis by separating movements that result in an increase of ego network size from those that result in a reduction. Finally, we separately analyse the same figures based on the different career stages of researchers. The key results we highlight from our analysis are as follows:

- On average, after a movement, the collaboration network tends to expand. However, the overlap between the old and new networks is not particularly high (Jaccard similarity between 10%–20%) and this result holds true both when the network size increases and when it decreases.
- On average, approximately 30% of the old collaborations carry over to the new network, and they amount to ~20% of the latter. This turnover is related to the cognitive effort required to nurture collaborations: one cannot just add new relationships because time needs to be invested for their maintenance. In order to bring in new collaborations one needs to replace old ones.
- When the network size increases after a movement, still not all old collaborations are maintained. On average, only 36% of old collaborations are carried over to the new network. When the network shrinks, this fraction is even smaller (around 14%) but the old collaborations relatively weight much more in the new network (around 36%).
- Movements tend, on average, to be more disruptive as the career progresses, partly because the networks after a movement are larger for more senior people but also because former collaborations tend to be curtailed more.

The rest of the document is organized as follows. In Sect. 2 we present the related work. We present our dataset and its features in Sect. 3. Section 4 describes the metrics used for our analysis. We present our findings in Sect. 5, and draw conclusions in Sect. 6.

2 Related Work

Given the omnipresence of academic migrations in the careers of researchers, the movements of scientists have received a lot of attention in recent years. The migration flows from a country to another and how the language similarities affect them have been studied in [6]. Their finding is that language similarity is one of the most important factors in relocation decisions. Franzoni *et al.* in [4] focus on the effects of academic mobility on the impact factor of the published papers, finding positive effects on the career of researchers that migrate. Movements to a higher- or a lower-ranked institution are considered in [2]. The main finding is that movement to a lower-ranked institution is affecting negatively the profile of a researcher, but the impact of authors that move to highly-ranked institutions remains the same. Sugimoto *et al.* [9] study the benefits of global researchers mobility. They found that mobile scholars always achieve higher citation rates than non-mobile ones. Mobile researchers do not cut their ties with their country of origin but instead work as bridges between different countries. Mobility of researchers is also considered in [5,10,11], but from a different perspective, hence we do not discuss them further.

Closest to our work are the contributions by Petersen [7] and Arnaboldi *et al.* [1]. Petersen [7] focuses on geographic displacements and their effect on the number of citations and collaboration network. His findings are threefold: (i) migration is associated with a significant churning in the collaboration network, (ii) the professional ties created after a migration event are less strong, (iii) career benefits of mobility are common to all ranks, not just to elite scientists. Despite considering the collaboration network, Petersen is mostly interested in the country to which the collaborators belong to, specifically focusing, e.g., on the effect of the country of previous collaborators on relocation decisions. Arnaboldi *et al.* [1] also study the collaboration network of scientists, but they are mostly interested in uncovering the cognitive efforts behind collaboration maintenance. In addition, they do not consider the effects of mobility at all.

3 Dataset

The dataset that we use for our study is extracted from Scopus[1], a popular academic knowledge repository managed by Elsevier, containing detailed records for the scientific publications of researchers from various disciplines. These records include the publications of a researcher, the authors of each publication and the affiliation of each of the authors at the time each study was published. Our dataset covers 84,141 authors. The average number of co-authors per paper in our dataset is 26, with a standard deviation of 219. The maximum number of co-authors in a paper is 5,563, while 96.8% of the papers have less than 26

[1] The APIs used are the following: Author Search API, the Author Retrieval API, and the Affiliation Retrieval API.

co-authors. Setting the latter as the threshold for assuming a meaningful collaboration, we dropped from the analysis the papers that have more than 26 authors.

The affiliations specified in the publications by each author are used for the creation of affiliation time series, where each timeslot contains information related to the affiliation and the date. The available timeslots of the time series are defined by the month and the year a paper was published and for every timeslot the author is considered to have only one available affiliation. In case a researcher has more than one affiliation declared in a timeslot's publications, we keep the one that appears more times in the timeslot. If we have affiliations with equal number of appearances, we keep the one that is declared as the author's affiliation in the closest in time (either in the past or the future) to the investigated timeslot.

Every affiliation change from an institution $inst_a$ to a different institution $inst_b$ for two consecutive timeslots defines a *movement*[2]. The duration of the *institution stay* of a researcher at $inst_a$ is defined as the period between the first paper the researcher published under the affiliation $inst_a$ and the first paper published under the affiliation $inst_b$. As a result, we have *institution stays* that have a minimum duration of 1 month. For the rest of the analysis, taking into consideration the example of the movement from $inst_a$ to $inst_b$, we define two networks for a given author. The first one is the old network of the ego (net_{old}), containing the co-authors of the ego during the stay at $inst_a$. The second one contains the co-authors of the ego during the stay at $inst_b$ (net_{new}). Finally, we define as *research career duration* the period between the first and the last publication of the author. Since different career stages may be affected differently by academic relocations, we extract the current career stage from the career duration using the following mapping, derived by considering the typical duration of the respective career stages: PhD students (PhD, 0 to 3 years), Young Researchers (YR, 3 to 6 years), Assistant Professors (AP, 6 to 10 years), Associate Professors (AsoP, 10 to 28 years), Full Professors (FP, 28 to 38 years) and Distinguished Professors (DP, 38+ years). Our dataset consists of records that depict the activity of 84,141 authors that have at least one movement during their research career. Based on the career stage classification, our dataset consists of 26 PhD, 4,026 YR, 18,157 AP, 51,129 AsoP, 7,846 FP and 2,957 DP. The total number of movements that these authors have done is 561,389.

After analyzing the distribution of the *institution stay* duration, we found that 343,312 movements have been done either before or after an institution stay of length less than 6 months (180 days). The occurrence of short institution stays could be caused either due to virtual migration of authors that have many

[2] We are well aware that changes of affiliations as proxies for academic movements suffer from several limitations (e.g., authors with multiple affiliations could create spurious short detected movements). However, at the moment, it is the only approximation that allows researchers to study scientists' mobility at a large scale. We have in place some preprocessing aimed at mitigating the impact of such limitations (such as the removal of short movements, discussed later on).

affiliations or due to delays for the publication of accepted papers that contain the affiliation of the author at the time the paper was submitted. The remaining 218,077 movements (corresponding to stays above 6 months) have been done by 77,713 authors. These 77,713 authors are classified as: 13 PhD, 3,213 YR, 16,220 AP, 47,857 AsoP, 7,544 FP and 2,866 DP. In the following of the paper, most of the times we do not consider such short stays, hence the impact of, e.g., spurious short movements caused by multiple affiliations for the same author, is very limited. We explicitly mention when we consider also them, which is done primarily to assess that including or excluding such short movements does not change the essence of our results.

4 Methodology

The movement of a researcher to a new institution may affect the collaborations of the researchers. While it might be intuitively assumed that movements result in a higher number of collaboration, this may not be always the case in reality. The main purpose of this paper is to analyse this aspect, and the resulting overlap between collaborations across movements. Specifically, we define a *positive* movement as one that is associated with an increase of the ego network, while a *negative* movement denotes one associated with a decrease. Below, we define a set of metrics that capture the different ways a collaboration network might be affected by a movement. To this aim, we focus on a tagged movement j of researcher i. Based on the notation introduced in the previous section, we denote with $net_{old}^{(ij)}$ the set of collaborators of author i before movement j and with $net_{new}^{(ij)}$ the set of collaborators of author i after movement j. Since in the following we unambiguously refer to a tagged pair (author i, movement j), we will drop the corresponding superscript.

First, we define a set of metrics focused on the size difference between the old and new collaboration network. We start with the "difference of the network sizes" ($size_{diff}$), defined as the difference between the sizes of the new and the old ego networks. We say that the effect of the movement is neutral when the $size_{diff}$ is equal to zero.

Definition 1. *The "difference of the network size" ($size_{diff}$) captures the difference between the size of the old and new network after a movement and is given by the following:*

$$size_{diff} = |net_{new}| - |net_{old}|. \tag{1}$$

In order to further characterise this difference, we also calculate the "difference ratio" ($diff_{ratio}$), which captures the relative change between the size of the old and new network.

Definition 2. *The "difference ratio" ($diff_{ratio}$) measures the relative change between the size of the old and new network, and it is given by the following:*

$$diff_{ratio} = \frac{|net_{new}| - |net_{old}|}{|net_{old}|}. \tag{2}$$

Distinguishing between a positive movement (i.e., one that grows the ego network) and a negative one (i.e., one that shrinks), the third metric $size_{ratio}$ measures how much larger is the largest network with respect to the smaller one.

Definition 3. *The "Ratio of the network size"($size_{ratio}$) can be computed as follows:*

$$size_{ratio} = \begin{cases} \frac{|net_{new}|}{|net_{old}|} & |net_{new}| \geq |net_{old}| \\ -\frac{|net_{old}|}{|net_{new}|} & |net_{new}| < |net_{old}| \end{cases} \tag{3}$$

Notice that, due to this definition, in the case of positive movements the $size_{ratio}$ is the multiplicative factor of increase of the new network size with respect to the old, while in the case of negative movements it is the division factor of decrease of the old network size with respect to the new.

While the previous group of metrics capture changes in network size, they do not shed light on how the composition of the network is modified after a movement. In order to assess the latter, we introduce the following additional metrics, that, taken together, fully characterise the compositional changes in the collaboration network.

Definition 4. *The similarity between the old and new collaboration network is measured with the Jaccard similarity, which can be obtained as follows:*

$$net_{sim} = \frac{|net_{old} \cap net_{new}|}{|net_{old} \cup net_{new}|}. \tag{4}$$

Definition 5. *The "Network kept" (net_{kept}) measures the fraction of old collaborators that keep collaborating with the tagged author after the movement, and it is given by the following:*

$$net_{kept} = \frac{|net_{old} \cap net_{new}|}{|net_{old}|}. \tag{5}$$

Definition 6. *The "Network dependency" (net_{dep}) measures the weight of old collaborations in the new network, and it is given by the following:*

$$net_{dep} = \frac{|net_{old} \cap net_{new}|}{|net_{new}|}. \tag{6}$$

In short, net_{sim} is a general measure of similarity between the two networks, while the other two capture the fraction of old collaboration that carry over to the new network and their relative weight in it. When needed, we will also denote with $|net_{inter}|$ the size of the intersection between the old and new network.

5 Results

In this section we investigate how academic movements affect the collaboration network of scientists. We present our general findings in Sect. 5.1. In Sect. 5.2, we focus our attention on the impact of positive or negative movements (i.e., movements that are associated with a growth or decrease in network size). Then, in Sect. 5.3 we study how the above results are affected by the career stage of researchers. Please note that the methodology used in the paper does not allow to establish clear causal relationships, but only co-occurrences, between academic movements and the observed changes in the collaboration network. The assessment of cause-effect patterns is left as future work.

5.1 Sizes of Networks and Changes in Collaborations Across Movements

We start by studying what happens to the size of an author's collaboration network after they move from an institution to another one. For the moment, we do not filter out short inter-movement periods. Taking into consideration all the 561,389 movements in our dataset, we calculate the parameter $diff_{ratio}$ for every movement each author has recorded and we get the average $diff_{ratio}$ for each author. Averaging among all the authors, we find (Table 1) that the average $diff_{ratio}$ per author is 2.84, indicating that on average a movement is associated with an expansion of the network by 284%. This large average is due to some researchers whose network significantly grows after a movement (as also highlighted by the skewness in the $diff_{ratio}$ distribution in Fig. 1, which we will discuss later on). In Table 1 we also report the median $diff_{ratio}$ value, which is around 8%. Thus, as expected, movements are generally associated with a positive impact on the collaboration network.

The fact that the number of collaborations increase after a movement tells us nothing about the composition of such network. To better understand this point, first we calculate the average intersection (net_{inter}) of the networks before and after a movement, for each author. From Table 1, we see that the average intersection of the authors is 1.93, indicating that (on average) a researcher keeps approximately 2 collaborations when moving to a new institution. In order to assess more precisely the similarity of the collaboration network before and after a movement, we compute the Jaccard similarity (Definition 4) at each movement, then we average these values across authors, obtaining an average Jaccard similarity of 0.11 (Table 1). Please note that a low similarity value was expected, since the Jaccard similarity also takes into account the difference in network size (and we have shown that the new network is approximately 4 times the old one, on average). However, if the old collaborations were fully retained, the similarity should be around $\frac{1}{4}$. The fact that it is lower implies that some old collaborations are removed. We will analyse this aspect more in detail later on in the section.

The second column of Table 1 shows the values of the metrics described above but obtained discarding movements associated with short stays (smaller than 6 months). All the values change only slightly with respect to the first column. As short stays might also include non real migration events, and since including them or not does not make significant differences, in the following we consider the dataset where those short stays have been removed.

Table 1. Comparison of old and new collaboration networks with and without short stays. The first column contains the values obtained from both short and long stays, the second column corresponds to the values obtained discarding inter-movement periods shorter than 6 months.

	No threshold	**180 days**
# of Movements	561,389	218,077
# of Authors	84,141	77,713
	Difference ratio ($diff_{ratio}$)	
Avg	2.84	3.09
Std	11.25	10.4
Median	0.08	0.36
	Jaccard similarity (net_{sim})	
Avg	0.11	0.12
Std	0.16	0.16
	Intersection (net_{inter})	
Avg	1.93	2.15
Std	3	3.3

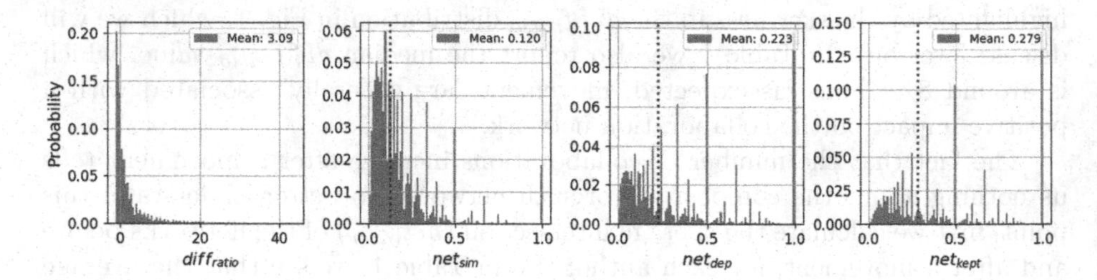

Fig. 1. Distribution of the changes in size ($diff_{ratio}$) and composition of collaboration networks (net_{sim}, net_{dep}, net_{kept}) after a movement.

Figure 1 shows, in the first two charts, the distribution of the $diff_{ratio}$ metric and the Jaccard similarity net_{sim}, respectively. It is interesting to note that both distributions are quite spread: the size difference ratio could increase up to 20–40 times, while there are cases of complete overlap between the old and new network (corresponding to the peak at 1 for net_{sim}), even though the average overlap is in the order of 10% only. As discussed above, the Jaccard similarity values that we obtain hint at the fact that collaboration networks before and after a movement are not just different because of their size. In order to better investigate this aspect, we now focus on the analysis of net_{kept} and net_{dep}, which capture specific variations in the composition of the collaboration network. The distributions of these metrics are shown in the third and fourth plot in Fig. 1, respectively. We start with the analysis of the percentage of the old network that is retained after a movement (corresponding to the net_{kept} metric). We find that on average, 27.9%

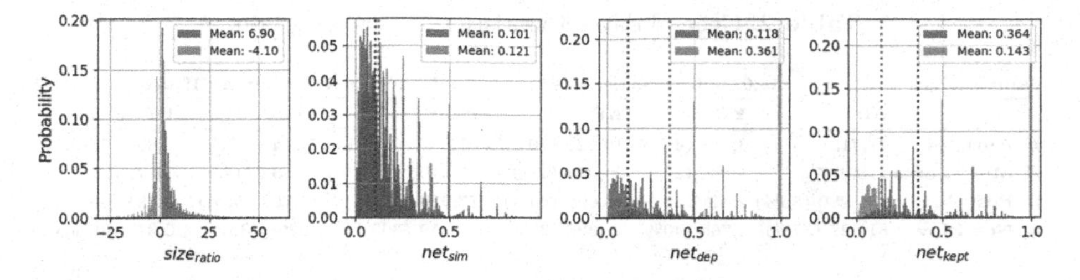

Fig. 2. Distribution of the changes in size ($size_{ratio}$) and composition of collaboration networks (net_{sim}, net_{dep}, net_{kept}) after positive (blue) and negative (red) movements. (Color figure online)

of the old network keeps collaborating with the ego also after the movement. The fact that more than 72% of the old collaborations are dropped could probably be related to the cognitive effort required to nurture collaborations, restricting the ability to maintain the old ones. On the other hand, looking at the net_{dep} distribution, we observe that the collaborations overlapping between the old and new network are, on average 22.3% of the new network. These two metrics allow us to better understand that the old collaborations can be seen as a good starting point after a movement, however, there is on average a very significant turnover associated with every movement. Interestingly, as we can see in Fig. 1, there are some extreme cases that keep active all of their old collaborations (net_{kept} equal to 1), while some others form a new network that completely consists of old collaborations (net_{kept} equal to 1). These are cases quite different from the ones described by the "average" effect of movements.

5.2 Positive vs. Negative Movements

The first results of our analysis indicate that, on average, after a movement the overall number of collaborations increases, even if 72% of old collaborations are curtailed in the process. However, from the existence of negative values in the $diff_{ratio}$ (Fig. 1), it was clear that a movement would not always be associated with an increase in the number of collaborations a researcher has. More precisely, only 121,980 (56%) movements are associated with a positive effect on the network size, while 81,027 (37%) movements are linked to its decrease (Table 2, column "All"). The remaining 7% of the movements results in equal size networks for both institution stays before and after a movement.

Figure 2 and Table 3 (column "All") show the analysis of positive and negative movements separately. Specifically, in Fig. 2 blue histograms correspond to positive movements, while red histograms correspond to negative movements. Analyzing, in Table 3 column "All", the ratio of the network size ($size_{ratio}$) for the positive and the negative movements (Definition 3), we found that the positive movements correspond to new collaboration networks that are on average 6.9 times the size of the old network. On the other hand, the negative movements would result into networks of a size 4.1 times smaller than the network size prior

Table 2. General dataset statistics per career group.

Sample size	218,077 Movements—**Pos**: 121,980 / **Neg**: 81,027 / **Neutral**: 15,070				
	All	**YR**	**AP**	**AsoP**	**FP**	**DP**
# Authors	77,713	3,213 (4.1%)	16,220 (20.9%)	47,857 (61.6%)	7,544 (9.7%)	2,866 (3.7%)
# Movements	218,077	4,360 (2%)	29,093 (13.3%)	136,105 (62.4%)	32,392 (14.9%)	16,113 (7.4%)
# Pos Mov	121,980 (56%)	2,111 (48%)	15,696 (54%)	77,923 (57%)	17,754 (55%)	8,491 (53%)
# Neg Mov	81,027 (37%)	1,765 (40%)	11,326 (39%)	49,359 (36%)	12298 (38%)	6,275 (39%)

the movement. As far as it concerns the similarity net_{sim}, it was similar for both positive and negative movements with values around 0.1. Interestingly, 40,412 positive movements and 27,190 negative movements have *0* Jaccard similarity, indicating that after these movements no prior collaboration remained active.

The *dependency* of the new network on the old network (i.e., the ratio between the intersection and the size of the new network – Definition 6), after a positive movement, is 11,8%, while it jumps to the 36,1% for the negative movements (Table 3). Remember that this metric captures how much the new network is based on old collaborations. While this increase is expected (since in the case of negative movements the new network size is smaller), it is, however, interesting to note that also in this case not all old relationships are kept. As it is clear from Fig. 2, the percentage of times where this is the case (i.e., the net_{dep} metric is equal to 1) is only about 20%.

The net_{kept} metric allows us to analyse the dual phenomenon, i.e., what is the proportion of old collaborations that are kept in the new network. In case of positive movements, 36.4% of old collaborations are kept, while this metric for the negative movements is 14.3% (Table 3). The fact that this metric is higher in case of positive movements could also be expected, as the new network is bigger, and therefore there is "more space" to accommodate also old collaborations. It is quite interesting to note, however, that despite such additional "space" only about 20% of authors keep all their old collaborations when moving (depicted by a value of net_{kept} equal to 1 in Fig. 2).

5.3 Impact of Migrations at the Different Career Stages

In this section, we take into account how the career stage of a researcher affects the results discussed so far. Indeed, it is reasonable to expect that an academic movement has a different effect on the network of collaborations if you are a PhD student or a full professor. It is not immediately clear in which direction this effect applies, though. On the one hand, for junior scientists, establishing new collaborations might be more difficult. On the other hand, the incentives for them to establish a widespread professional network are definitely stronger. Taking this into consideration, we split our dataset into the 6 career groups defined in Sect. 3 (we drop the PhD students from the analysis because they are too few). As shown in Table 2, the bulk of the authors in our dataset can be classified as associated professors, and they also account for the majority of movements. The proportion between positive and negative movements is very

Table 3. Average network statistics per career group.

	All	YR	AP	AsoP	FP	DP
	Difference ratio ($diff_{ratio}$)					
All	3.09	0.99	1.67	3.15	4.13	3.65
	Ratio of the network size ($size_{ratio}$)					
Pos	6.9	3.5	4.5	6.7	9	8.4
Neg	-4.1	-3.2	-3.35	-4	-4.8	-5.3
	Jaccard similarity (net_{sim})					
All	0.12	0.18	0.14	0.12	0.11	0.1
Pos	0.1	0.14	0.12	0.1	0.09	0.08
Neg	0.12	0.16	0.13	0.12	0.11	0.1
	Network kept (net_{kept})					
All	0.27	0.32	0.3	0.28	0.27	0.25
Pos	0.36	0.4	0.37	0.37	0.36	0.34
Neg	0.14	0.19	0.16	0.14	0.13	0.12
	Network dependency (net_{dep})					
All	0.22	0.3	0.25	0.22	0.22	0.21
Pos	0.12	0.17	0.14	0.11	0.1	0.1
Neg	0.36	0.41	0.37	0.36	0.36	0.35

similar, regardless of the career stage, with the positive movements being around 55% and the negative movements being between 36% and 40%. The only career group that slightly differs is the group of Young Researcher, who have a 48% of movements resulting into increased network size. This slightly smaller percentage could be due to the short career duration of the authors that results into short institution stays. The latter, reinforces our idea that short institution stays are not ideal for forming an adequate and representative network.

From the network size standpoint, we see in Table 3 that a movement would (on average) be associated with increase in the network size of the author, regardless of the career group the authors belongs to. This increase is around 100% for YR, 167% for AP, 315% for AsoP, 413% for FP, and 365% for DP. Given the skewness of the $diff_{ratio}$ distribution, the median values are smaller but they substantially confirm this growth trend. Thus, it seems that the longer the career duration, the higher the expansion of the network. The fact the DP grow their networks slightly less than FP may be due to the fact that at the career stage of distinguished professor expanding collaborations might be less relevant. If we distinguish between positive and negative movements, the $size_{ratio}$ metric tells us that, for positive movements, the trend described above is substantially confirmed: the collaboration network increasingly grows from YR to FP. Then, it faces a minor setback for DP, which are, as discussed above, a corner case. Quite interestingly, the trend in the negative movements is the same in terms of absolute values of $size_{ratio}$ but the implications are opposite: when authors curtail collaborations after a movement, they do so, much more when they are senior than junior.

Let us now focus on the variations of the collaboration network composition, at the different career stages. The variation of the average Jaccard similarity values across the career stages is less pronounced than the variation in network size. The overall low values are substantially expected based on the considerations in Sect. 5.1 about the network size difference. It is also interesting to note that the difference between the case of positive and negative movements is barely noticeable. The net_{kept} values in Table 3 tell us that the different career stages do not impact significantly on the fraction of the old network that is retained after a movement: indeed, the average net_{kept} is always around 0.3. Slightly more variation is observed for the average net_{dep}, corresponding to the fraction of old collaborations in the new network, but the difference is still limited. We now consider the combined effect of net_{kept} and net_{dep} for positive and negative movements. We substantially confirm the findings discussed in Sect. 5.2: when authors grow their network after a movement, they retain an important fraction of the old network, which, obviously, does not weight as much in the new network (which is larger). When the collaboration network shrinks, there is anyhow a significant collaborations turnover, with ~15% of old relationships retained, which now account for ~35% of collaborations in the new network.

6 Conclusions

The changing of host institutions is a fundamental part of the research career, yet the impact that such movements have on a scientist's network of collaborations is not completely understood. In this study, we have presented a preliminary investigation of the potential effects a movement has on the network of a researcher from the standpoint of network size and composition of the collaboration network. For our study, we have used records of scientific publications published by 84,141 researchers, extracted from the well-known platform Scopus. Our findings highlight that on average, the collaboration network after a movement tends to be larger. Yet, the overlap between the old and new networks is not particularly high (Jaccard similarity between 10%–20%), regardless of the movement being positive or negative. In terms of how much the old network "survives" in the new one, we found that, on average, approximately 30% of the old collaborations carry over to the new network. Old collaborations amount typically to ~ 20% of the new network. Distinguishing between positive and negative movements, when the network size increases after a movement, still, not all old collaborations are maintained. On average, only 36% of old collaborations are carried over to the new network. When the network shrinks, this fraction is even smaller (around 14%) but the old collaborations relatively weight much more in the new network (around 36%). The career stage of scientists does not affect significantly the above results. Movements tend, on average, to be slightly more disruptive as the career progresses, partly because the networks after a movement are larger for more senior people but also because former collaborations tend to be curtailed more.

Acknowledgements. This work was partially funded by the SoBigData++, HumaneAI-Net, MARVEL, and OK-INSAID projects. The SoBigData++ project has received funding from the European Union's Horizon 2020 research and innovation programme under grant agreement No 871042. The HumaneAI-Net project has received funding from the European Union's Horizon 2020 research and innovation programme under grant agreement No 952026. The MARVEL project has received funding from the European Union's Horizon 2020 research and innovation programme under grant agreement No 957337. The OK-INSAID project has received funding from the Italian PON-MISE program under grant agreement ARS01 00917.

The work of Pavlos Paraskevopoulos was supported by the ERCIM Alain Bensoussan Fellowship Program.

References

1. Arnaboldi, V., Dunbar, R.I.M., Passarella, A., Conti, M.: Analysis of co-authorship ego networks. In: Wierzbicki, A., Brandes, U., Schweitzer, F., Pedreschi, D. (eds.) NetSci-X 2016. LNCS, vol. 9564, pp. 82–96. Springer, Cham (2016). https://doi.org/10.1007/978-3-319-28361-6_7
2. Deville, P., Wang, D., Sinatra, R., Song, C., Blondel, V.D., Barabási, A.L.: Career on the move: geography, stratification, and scientific impact. Sci. Rep. **4**, 4770 (2014)
3. Fortunato, S., et al.: Science of science. Science (New York, N.Y.) **359**(6379), eaao0185 (2018). https://doi.org/10.1126/science.aao0185
4. Franzoni, C., Scellato, G., Stephan, P.: The mover's advantage: the superior performance of migrant scientists. Econ. Lett. **122**(1), 89–93 (2014)
5. James, C., Pappalardo, L., Sîrbu, A., Simini, F.: Prediction of next career moves from scientific profiles. arXiv preprint arXiv:1802.04830 (2018)
6. Moed, H.F., Aisati, M., Plume, A.: Studying scientific migration in scopus. Scientometrics **94**(3), 929–942 (2013)
7. Petersen, A.M.: Multiscale impact of researcher mobility. J. R. Soc. Interface **15**(146), 20180580 (2018)
8. Robinson-Garcia, N., Sugimoto, C.R., Murray, D., Yegros-Yegros, A., Larivière, V., Costas, R.: The many faces of mobility: using bibliometric data to measure the movement of scientists. J. Informetr. **13**(1), 50–63 (2019). https://doi.org/10.1016/J.JOI.2018.11.002. https://www.sciencedirect.com/science/article/pii/S1751157718300865
9. Sugimoto, C.R., Robinson-García, N., Murray, D.S., Yegros-Yegros, A., Costas, R., Larivière, V.: Scientists have most impact when they're free to move. Nature News **550**(7674), 29 (2017)
10. Urbinati, A., Galimberti, E., Ruffo, G.: Hubs and authorities of the scientific migration network. arXiv preprint arXiv:1907.07175 (2019)

11. Vaccario, G., Verginer, L., Schweitzer, F.: The mobility network of scientists: Analyzing temporal correlations in scientific careers. arXiv preprint arXiv:1905.06142 (2019)
12. Verginer, L., Riccaboni, M.: Brain-Circulation Network: The Global Mobility of the Life Scientists. arXiv, October 2018. http://eprints.imtlucca.it/4072/1/EIC_WP_4_2018.pdf
13. Wang, J., Hooi, R., Li, A.X., Chou, M.H.: Collaboration patterns of mobile academics: the impact of international mobility. Sci. Public Policy **46**(3), 450–462 (2019). https://doi.org/10.1093/scipol/scy073. https://academic.oup.com/spp/article/46/3/450/5281269

Social Search and Task-Related Relevance Dimensions in Microblogging Sites

Divi Galih Prasetyo Putri, Marco Viviani(✉), and Gabriella Pasi

Department of Informatics, Systems, and Communication (DISCo), Information and Knowledge Repersentation, Retrieval, and Reasoning (IKR3) Lab, University of Milano-Bicocca, Edificio U14, Viale Sarca 336, 20126 Milan, Italy
d.putri@campus.unimib.it, {marco.viviani,gabriella.pasi}@unimib.it
http://www.ir.disco.unimib.it/

Abstract. Social media, and in particular microblogging sites, allow users to post multiple kinds of content for different purposes. Content may be purely conversational, or news-related, or event-related. To find information relevant to users in this heterogeneous mass of content, it would be important to consider the task for which search is carried out, and the most suitable relevance dimensions. In the last years, despite the social search problem has been increasingly investigated, this aspect has not been sufficiently analyzed. For this reason, in this paper, we focus on different search tasks in the microblog search context, and we identify some related relevance dimensions. We also report some experiments we have made to verify the impact of the identified relevance dimensions on the system effectiveness, with respect to the considered search tasks.

Keywords: Social search · Microblog search · Multidimensional relevance · Relevance dimensions · Interestingness · Informativeness · Opinionatedness · Credibility

1 Introduction

The spread of social media has made it possible for anyone to generate *User-Generated Content* (UGC), and this has led to the availability of a huge amount of online information (which is, in many cases, misinformation [21,46]). Among the several social platforms available, microblogging services play a fundamental role in content generation. Twitter, in particular, is incredibly popular, with more than 300 million monthly active users in the first quarter of 2019.[1] Microblogs allow users to freely share content in the form of short texts. Microblog posts can be constituted of personal updates or discussions, the so-called *conversational* content, but also by *newsworthy* content, such as news and public events [8]. This second category of content is of fundamental importance to users if we consider that many people turn to microblogging platforms as their main

[1] https://www.statista.com/statistics/282087/number-of-monthly-active-twitter-users/.

S. Aref et al. (Eds.): SocInfo 2020, LNCS 12467, pp. 297–311, 2020.
https://doi.org/10.1007/978-3-030-60975-7_22

source of news. For example, from 2013 to 2017, the percentage of Twitter users employing the platform to fulfill their information needs has grown by more than twenty percent.[2] In this scenario, characterized by huge amounts of heterogeneous information, implementing *social search* services to effectively provide users with social media contents satisfying their information needs becomes crucial.

In general, when developing a *search engine*, the key issue is how to assess the *relevance* of a document (e.g., a textual document, a Web page, a social media post, etc.) to a user *query* that expresses the user's information needs. Relevance is a complex notion, which relies on several dimensions [2,6,27]. *Topicality* refers to the 'aboutness' of a document to the *topic* suggested by a query, and it constitutes the core relevance dimension. Additional relevance dimensions such as *popularity* [33] and *novelty* [5] have been identified in the literature to improve the search process. Relevance depends on the *search task* and the *context* in which search is performed, which concur to define the so-called *situational relevance* [2]. In this scenario, the relevance dimensions that are effective to the retrieval process are those able to capture the usefulness of the documents in helping the user to understand, make decisions, and solve the problems related to the search task or context underlying the user's information needs. The impact of relevance dimensions can dynamically change or shift during the search process [39]. For example, at the initial stage of the search process, the user could be more focused on the topical relevance of documents. Later, the user could also be interested in considering different properties of the retrieved documents w.r.t. to the considered search context.

The complex process of relevance assessment also concerns social search, which includes *microblog search* that is the focus of this paper. Retrieving useful information in microblogs is not an easy task, and some peculiar characteristics of the UGC diffused in these platforms make information processing and analysis a challenging problem. In fact, as previously pointed out, microblog posts are short texts that do not have a homogeneous linguistic quality, as users tend to employ abbreviations and acronyms, they remove vowels and articles [14], and even shorten URLs [41]. Moreover, a lot of users spread rumour, misinformation, hoaxes, and spam. Hence, also the assessment of the credibility of such content has become an issue [46]. Finally, given that these platforms are used to post content that covers various information needs in different contexts, the purpose for which the search is performed is another fundamental aspect to be taken into consideration [10,42]. For the above reasons, researchers have recently paid attention to some microblog aspects that could be useful in the context of social search, such as *interestingness* [1,30], *informativeness* [38], *credibility* [25], and *opinionatedness* [13]. Each of the above-mentioned properties has been investigated w.r.t. microblog search individually, and not fully considering the concepts of multidimensional and situational relevance.

[2] https://www.journalism.org/2017/09/07/news-use-across-social-media-platforms-2017/.

For this reason, in this article, our intent is to consider these properties as multiple relevance dimensions, and investigate which of them may have an impact on the effectiveness of microblog search in relation to specific search tasks. To the aim of conducting an exploratory study, we consider two microblog search tasks: (*i*) the *disaster-related retrieval* task, and (*ii*) the *opinion retrieval* task. For each of the above-mentioned tasks, the previously discussed four relevance dimensions are considered beyond topicality. Our hypothesis is that specific search tasks in a microblogging context are impacted differently by distinct relevance dimensions. For example, we can hypotesize that in the disaster-related retrieval task both informativeness and credibility should play a major role, and that in opinion retrieval, both opinionatedness and credibility should strongly impact the process of tweet retrieval. In order to verify this hypothesis, in the performed experiments we have employed two publicly available task-related data sets, and as baseline we have considered a system that employs topicality as the only relevance dimension. The results of the experiments confirm that the effectiveness of a search process guided by a specific search task is affected by different relevance dimensions in providing useful documents to users.

2 Related Work

In this section, we present a synthetic review of the research contributions addressing the task of microblog search, by specifically focusing on distinct properties (of both documents and microblogging sites) that can be interpreted as relevance dimensions in this context.

Early work on microblog search simply tried to exploit some *features* specific to microblogging platforms in the ranking process. In [9] and [28], the authors employ *learning-to-rank* [20] to inject in the search process some features related to the 'authoritativeness' of an account and to the network structure, such as the number of times a user is mentioned in tweets, the number of her/his followers, the number of lists or groups a user belongs to, a popularity score estimated by the PageRank algorithm based on the retweet network, the presence of URLs in the tweets, etc. However, specific relevance dimensions and the role played by multidimensional relevance in microblog search are not analysed in the above-mentioned works.

Some other works have investigated different properties that can be useful in ranking microblog posts, beyond topicality; they have been presented as quality indicators, and include *interestingness* [1,30,40], *informativeness* [4,17], and *credibility* [17,25,47]. *Interestingness* has been discussed as a *static* quality measure of a microblog post. In Twitter, in particular, it has been considered as related to the number of *retweets* obtained by a tweet. In [1,30], the authors predict the probability of a tweet being retweeted, and use this probability score as the interestingness score. In [1], in particular, the authors employ a Naïve Bayes model to compute such probability. Furthermore, they explore two different methods to incorporate interestingness in the ranking process, by *reranking* and *filtering* the retrieved tweets based on the obtained score.

Information about the retweet rate is also used as a signal of *informativeness*. In [4], it has been observed that informative tweets are more likely to be relevant, and informativeness is highly correlated with a high number of retweets. Huang *et al.* in [17] develop a heterogeneous network, based on information from tweets, users, and Web documents to compute tweet informativeness, and tweets are ranked based on this dimension. The authors argue that an informative tweet is usually posted by a credible user, and shared (i.e., retweeted) by many users. In addition, a tweet that can be linked to a Web document is more likely to be informative.

Also, the concept of *credibility* has been considered in the context of microblog search. In [25], the authors employ some credibility indicators of blog posts, introduced in [49], to assess the credibility of tweets. The authors also consider some additional microblogging-related indicators, such as the frequency of repost of a tweet, its recency, and the number of followers of its author. The values associated with credibility indicators are computed as probabilities, and range in the $[0, 1]$ interval. Later, these values are linearly combined to get an overall credibility score, which is incorporated in the retrieval model. Vosecky *et al.* in [47] consider a network composed by tweets and links to specific Web sites mentioned in the tweets. Based on this network, they compute features including the *reputation* of URLs, retweet sources, and hashtags. The authors argue that tweets that link to a source with a good reputation (e.g., well-known news Web sites or trustworthy users) carry high-credible information. In the paper, reputation features are employed in a *learning-to-rank* algorithm.

In addition to the above-mentioned properties, the so-called *opinionatedness* has been proposed in the *opinion retrieval* context [13,22]. Opinionatedness can be defined as the likelihood of a document to express an opinion about a topic.

Although several properties of microblogs and their impact on retrieval have been investigated in the literature, the concept of task-related multidimensional relevance has not been investigated yet. For this reason, in this work, we propose an approach to microblog search that relies on the identification of distinct relevance dimensions for different *search tasks*, and we evaluate their impact on the search effectiveness.

3 A Task-Based Model for Assessing Relevance in Microblog Search

In Information Retrieval, a search task is defined as "an activity to be performed in order to accomplish a goal" [43]. In microblog search, different search tasks can be performed, due to the characteristics of social media platforms. In this paper, we investigate the hypothesis that distinct relevance dimensions can have a different impact on different search tasks; to this aim, we consider two specific search tasks, i.e., (*i*) *disaster-related retrieval*, and (*ii*) *opinion retrieval*. Disaster-related retrieval refers to the identification of specific tweets related to a disaster situation, while opinion retrieval refers to the identification of opinionated tweets about a specific topic. Different search tasks are related to different *objectives.* The disaster-related retrieval task has the main objective of helping emergency relief operations and emergency preparedness initiatives [12]. The opinion retrieval task has the main objective of understanding the user's opinion about the query topic. In this context, besides being topically relevant, tweets should also contain affective content, related to the user's opinions, either being positive or negative. The choice of the above specific search tasks is due to the interest of the tasks, and to the fact that both of them can be evaluated on available public data collections, as explained in Sect. 4.1.

3.1 Relevance Dimensions

In this paper, we consider four *relevance dimensions* that can have a possible impact on the effectiveness of the search tasks, as illustrated in Sect. 2. The way in which the four dimensions are modeled is detailed in the following subsections.

Interestingness. In the literature, *interestingness* has been defined as the extent to which a content significantly arouses reactions [48]. Microblogging services enable users to republish content that could be of interest to other users. Thus, a tweet that received a high number of retweets can be considered as of general interest [1,29,30]. In this work, we follow the same approach adopted in [29] to compute the *interestingness score* of a single tweet. This score is computed as the probability of a tweet being retweeted.

To do so, *logistic regression* [16], as implemented in the `Python` library `scikit-learn`,[3] is used to train a model on several features related to tweets, such as the presence of URLs, usernames, hashtags, exclamation marks, question marks, emoticons, sentiment. The logistic regression model learns the influence that features have on the probability of a message being retweeted; this way, after this training phase, we are able to compute the probability of a new tweet being retweeted. To train the model, a corpus of tweets is required. Because the datasets used in [29,30] are not publicly available, we have used the data collection introduced in [53], in which tweets have been collected by using a set of relevant keywords and hashtags associated with 30 real-world events. From this collection, we have extracted around 20 thousand tweets. Specifically, we have selected ten thousand tweets with zero retweets and ten thousand tweets with more than 10 retweets. The tweets with zero retweets are considered as *not interesting*, while the rest as *interesting*.

Informativeness. In [4], *informativeness* has been defined as "specific information that people might care about". In this work, to assess the informativeness of a tweet, we follow the approach proposed in [23] in the context of crisis events. This approach has proven to be effective not only to assess the informativeness of crisis-related content, but also to other kinds of content [18]. To calculate

[3] https://scikit-learn.org/stable/modules/generated/sklearn.linear_model.Logistic Regression.html.

an *informativeness score* for each tweet, we have employed the data collection introduced in [32]. The dataset contains tweets gathered during 26 crisis events in the period 2012–2013, which are labeled based on their level of informativeness. Labels assume binary values; the 1 value is employed if a tweets is *related and informative*,[4] and the 0 value is associated with tweets that are *related but not informative*, or *not related*. As for the case of interestingness, based on the above-mentioned training data, we have used *logistic regression* to train a model to compute the probability of informativeness of a new tweet.[5] In this paper, we consider the same features proposed in [23] to train the model, such as the presence of URLs, the number of words, stopwords, sentiment-related words, hashtags, user's mentions, the tweet length, and some other features identifying the content style.

Opinionatedness. The concept of *opinionatedness* has been discussed in the context of opinion retrieval, as briefly illustrated in Sect. 2. A way to assess opinionatedness for a document is to compute both a *term-based opinion score*, based on the presence of opinionated terms in a document, and a *stylistic-based opinion score*, which focuses on some stylistic aspects of the document under consideration. In this work, we have used the definition and the approach proposed in [13] to compute an *opinionatedness score* associated with a tweet. The authors propose to compute this score as a linear combination of a term-based opinion score and a stylistic-based opinion score. The former is computed as the average opinion score over all terms in the document, as illustrated in [13], and by employing the AFINN Lexicon to identify opinionated terms [31]. The latter is computed by considering the stylistic variations that a tweet contains, as illustrated in [13]. In this paper, the following stylistic features have been considered to capture stylistic variations: the presence of emoticons, exclamation marks, emphatic lengthening, and opinionated hashtags.

Since stylistic variations are topic-dependent in capturing opinionatedness, we have performed *topic modelling* by using *Latent Dirichlet Allocation* (LDA), as implemented in `gensim`,[6] to identify topics and related tweets. The importance of each stylistic variation is calculated for each topic based on the tweets assigned to the topic.

Credibility. In the literature, *credibility* has been referred to as a perceived quality of the information receiver; it involves several factors related to the source of information, the information itself, and the media employed to diffuse information [11]. In this paper, to assess credibility of microblog contents, we adopt the approach described in [35], which is a model-driven approach based *Multi-Criteria Decision Making* (MCDM). A microblog post (i.e., a tweet) is considered

[4] In this case, being a *related* tweet means that the tweet discusses the crisis event.
[5] Also for informativeness, logistic regression has been performed by employing the model implemented by the `scikit-learn` library [36], using the default parameters.
[6] https://radimrehurek.com/gensim/models/ldamodel.html.

credible if it 'satisfies' several criteria, i.e., some features in the microblogging context, which can be interpreted from a credibility point of view (e.g., a tweet written by a user who has many friends can be considered as more credible w.r.t. a tweet written by a user with no friends). To compute the *credibility score* of a tweet, we exploited some features taken from the literature [15,26], employed in association with the model presented in [45]. The considered features include, for each tweet: the number of retweets, the number of followers, friends, tweets of the author, the presence of URLs, and the author's account age w.r.t. the date in which the tweet has been posted. Formally, with each tweet, distinct credibility scores $x_1, \ldots, x_n$ are associated, one for each feature, as illustrated in [45]; each score expresses how much the tweet is credible based on the related feature. In order to obtain an overall credibility score $\mathcal{A}(x_1, \ldots, x_n)$ for the tweet, the feature-related credibility scores must be *aggregated* through a suitable function $\mathcal{A}$. As suggested in [34,45], *Ordered Weighted Averaging* (OWA) [50] aggregation operators are employed.

3.2 Combining Relevance Dimensions

To evaluate the impact of each relevance dimension on each of the two considered search tasks, we have implemented a simple *retrieval model*, which implements a *linear combination* of the two relevance scores related to *topicality* and an *additional relevance dimension* respectively. For each tweet, the combined relevance score (named *Retrieval Status Value* - RSV) is then computed as follows:

$$\text{RSV} = \alpha\text{RSV}_t + (1 - \alpha)\text{RSV}_i \quad (1)$$

where RSV_t is the topicality score, and RSV_i is the score associated with the additional relevance dimension.

To compute the RSV_t score, we have employed the *query likelihood model* [24], which builds a probabilistic language model from each document d, and ranks documents based on the probability of the model to generate the query q. This is interpreted as the likelihood that the document is relevant to the query. Formally:

$$P(d|q) = \frac{P(q|d)P(d)}{P(q)} \quad (2)$$

where $P(q|d)$ is the probability of the query q under the language model derived from d, $P(q)$ is the same for all documents, and so can be ignored, and $P(d)$ is the prior probability of a document often treated as uniform across all d and so it can also be ignored. In the paper, we have used the *Dirichlet prior smoothing* in association with the query likelihood model to compute $P(d|q)$, as illustrated in detail in [52].

The RSV_i score is obtained by applying the approaches described in Sect. 3.1 for each relevance dimension i. The weight α takes values in the $[0, 1]$ interval, and, in the evaluation process, we tune its value by considering steps of 0.1.

The obtained overall RSV score represents the *estimated relevance* of the tweet to a query (based on multiple relevance dimensions), based on which the tweets are ranked.

As it will be illustrated in detail in Sect. 4, we made a comparative evaluation of a baseline microblog search engine computing relevance as topicality only, according to Eq. (2), and microblog search engines assessing relevance by combining topicality with just an additional dimension at a time, according to Eq. (1).

4 Evaluating the Impact of Relevance Dimensions

This section presents the experiments that have been carried out to evaluate the impact of the relevance dimensions presented in Sect. 3.1 on the disaster-related retrieval and the opinion retrieval search tasks. To perform the evaluations, two publicly available datasets have been employed, which are described in Sect. 4.1; each dataset is related to one of the two considered search tasks.

4.1 Description of the Datasets

In Information Retrieval, in order to evaluate the effectiveness of search engines, large test collections are generated and made publicly available. Such collections are usually constructed so that, besides the document collection, also a set of queries is provided (generally topics); for each query, the documents deemed relevant by a pool of assessors are also provided.

To evaluate the impact of the considered relevance dimensions on the disaster-related task, a dataset has been used that was introduced at the ECIR 2017 Workshop on Exploitation of Social Media for Emergency Relief and Preparedness (SMERP 2017) [12]. This dataset is composed of around 42,000 tweets and is split into two parts. The SMERP level-1 dataset contains tweets collected on the first day of the disaster (around 28,000 tweets), while the SMERP level-2 dataset contains the tweets that are collected during the second day of the disaster (around 14,000 tweets). In both datasets, four topics are discussed, i.e., available resources, required resources, infrastructure damages, and rescue activities.[7] For the opinion retrieval task, we used the dataset introduced in [22]. The original dataset refers to 5,000 tweets and 50 topics collected in 2011. Some of the original tweets could not be crawled anymore because, in the meantime, they have been set as private or permanently deleted. So, the considered dataset contains around 3,100 tweets. In both the considered data collections, tweets are simply labelled as *relevant* or *not relevant*.

All the considered data have been preprocessed by using the Anserini IR open-source toolkit [51], and indexed using Apache Lucene.[8] In particular, during the preprocessing phase, the tweets have been tokenized and normalized. Then, stemming has been applied to all normalized tokens by using the Porter Stemmer [37] that is implemented in Lucene.

---

[7] https://www.computing.dcu.ie/~dganguly/smerp2017/.
[8] https://lucene.apache.org/.

4.2 Experimental Setting

For each search task, we have conducted several experiments by implementing multiple microblog search engines; each search engine implements a model that combines topicality with just an additional relevance dimension, based on Eq. (1). Hence, we have the following 4 search engine configurations, which we denote as: *Topicality + Informativeness*, *Topicality + Interestingness*, *Topicality + Opinionatedness*, and *Topicality + Credibility*. We compare the results produced by each search engine with a *Topicality-based Baseline*, i.e., a search engine that estimates relevance based on topicality only. In order to compute the topicality score, as discussed in Sect. 3.2, we have employed the *query likelihood model* together with the *Dirichlet prior smoothing*, as implemented in Lucene. Since for the disaster-related retrieval task we have considered both datasets illustrated in Sect. 4.1, we have performed 15 evaluations, 5 for the disaster-related retrieval task w.r.t. the SMERP level-1 dataset (i.e., the results produced by the Topicality-based Baseline and by the 4 search engines combining topicality with an additional relevance dimension), 5 for the disaster-related retrieval task w.r.t. the SMERP level-2 dataset, and the remaining 5 for the opinionated tweets dataset, related to the opinion retrieval task.

To comparatively evaluate the effectiveness of the results produced by each search engine for each considered search task, we have employed the *bpref*, *Precision at k* (P@k), and *Mean Average Precision* (MAP) measures, which are the same suggested and employed in [12] as evaluation metrics.

Bpref. The *bpref* measures considers whether relevant documents are ranked above irrelevant ones [3,7]. It is designed to be robust to missing relevance judgments, since in large collections not all the documents can be judged due to the onerousness of this operation. It is formally defined as:

$$bpref = \frac{1}{R}\sum_{r}\left(1 - \frac{|n \text{ ranked higher than } r|}{\min(R,N)}\right) \tag{3}$$

where R is the total number of truly relevant documents, N is the number of known irrelevant documents, r is a relevant document in the ranked result list, and n is a non-relevant document retrieved by the system.

Precision@*k*. *Precision* is a way of measuring the accuracy of a system. Specifically, it is the fraction of the documents retrieved that are relevant to the user's information needs. In this paper, in particular, we consider *Precision at* k (P@k), which is defined as the total number R of relevant documents from the top-k retrieved documents. It is formally defined as:

$$\text{P@}k = \frac{R}{k} \tag{4}$$

Specifically, in this article we have considered $k = 20$, thus obtaining values of P@20. This choice was made for comparability reasons, as the P@20 measure is the same used in [12] in the context of the disaster-related retrieval.

Mean Average Precision. To compute *Mean Average Precision* (MAP), it is first necessary to compute *Average Precision* (AP), which is the mean of the precision scores after each relevant document is retrieved. AP is formally defined as:

$$\text{AP} = \frac{1}{R} \sum_{k \in \mathcal{R}} \text{P@}k \tag{5}$$

where R is the total number of relevant documents, and $\mathcal{R}$ is the set of the ranks of the relevant documents. MAP is computed as the mean of AP on the entire query set. Formally:

$$\text{MAP} = \frac{1}{Q} \sum_{q=1}^{Q} \text{AP}_q \tag{6}$$

where Q is the total number of queries in the query set, and AP_q is the average precision for the query q.

5 Analysis of the Results

In this section, we present the results obtained by performing the evaluation strategy illustrated in Sect. 4. In Table 1, the results obtained in terms of *bpref*, P@20, and MAP are illustrated for each implemented search engine for each task. This allows to evaluate the different impact that each relevance dimension has on the different search tasks.

As explained in Sect. 3.2, we have used a linear combination of the considered relevance dimensions to get the overall relevance score of a tweet. In the performed experiments, we have set different values of α in the linear combination (i.e., Eq. (1)), and we have selected, as α value, the one providing the best search engine performance with respect to each specific data collection (i.e., with respect to each search task). For each data collection, and for each search engine, the scores that are indicated in bold in Table 1 refer to the results that outperform the *Topicality-based Baseline* for the considered search task.

5.1 Impact of Distinct Relevance Dimensions

The obtained results show that despite the method we employ is quite simple, we obtain an indication of the possible impact that each of the considered relevance dimensions may have on the different search tasks. In fact, we observe that the search engines whose retrieval model is based on the combination of *Topicality* + *Interestingness*, *Topicality* + *Informativeness*, and *Topicality* + *Credibility*, perform better than the *Topicality-based Baseline* in the disaster-related retrieval task, for both SMERP level-1 and level-2 datasets. When considering the opinion retrieval task, it can be noticed that neither informativeness nor interestingness as relevance dimensions have a significant impact on the results. On the contrary, we can observe that using credibility and opinionatedness as relevance dimensions is useful for the opinion retrieval task.

Table 1. Results obtained for each microblog search engine w.r.t. different search tasks.

Microblog search engine	α value	Evaluation measure		
		bpref	P@20	MAP
Disaster-related retrieval task				
Dataset: SMERP level-1				
Topicality-based baseline	-	0.0300	0.0875	0.0213
Topicality + Interestingness	0.8	**0.0409**	**0.1250**	**0.0251**
Topicality + Informativeness	0.6	**0.0564**	**0.1250**	**0.0247**
Topicality + Opinionatedness	0.9	0.0290	0.0875	0.0200
Topicality + Credibility	0.8	**0.0390**	**0.1000**	0.0138
Dataset: SMERP level-2				
Topicality-based baseline	-	0.0459	0.1875	0.0215
Topicality + Interestingness	0.8	**0.0562**	0.1250	**0.0275**
Topicality + Informativeness	0.7	**0.0589**	**0.2000**	**0.0290**
Topicality + Opinionatedness	0.9	**0.0479**	0.1375	0.0213
Topicality + Credibility	0.9	**0.0488**	0.1500	**0.0222**
Opinion retrieval task				
Topicality-based baseline	-	0.1398	0.1470	0.2126
Topicality + Interestingness	0.9	0.1250	0.1460	0.2041
Topicality + Informativeness	0.9	0.1227	0.1310	0.1962
Topicality + Opinionatedness	0.4	**0.2049**	**0.1910**	**0.2517**
Topicality + Credibility	0.7	**0.1492**	**0.1670**	**0.2186**

These preliminary results confirm our hypothesis that for a given search task some specific dimensions may have an impact while others could be not considered. Thus, the impactful dimensions should be emphasised by the retrieval process to improve the effectiveness of search for a given task.

5.2 Evaluation Issues

An important problem that we have encountered during the development of this work is that the available relevance judgements of the considered data collections that we employed as ground truth might be not totally adequate to support our experiments. In fact, at the current stage of research in this field, the retrieved results are mostly judged by external annotators without explicitly considering the search context or multiple relevance dimensions, but mostly based on topicality [19,44]. For example, in disaster-related retrieval, the tweet "`Go to blood donations! #terremoto #earthquake`" is assessed to be relevant to the query "`What resources are required`". The tweet is talking about the resource *blood*, but it is not useful when considering the goal of the task.

The tweet does not contain any practical information to support the activity of emergency response operations.

Despite this problem (which, however, concerns most of the works that have to face the same research issue) the results produced by our preliminary experiments are still able to outline that our hypothesis deserves to be taken into consideration in future research on microblog search.

6 Conclusions

In this work, we have tackled the problem of assessing the impact of multidimensional relevance in task-based microblog search. In particular, we have analyzed the impact of different relevance dimensions on different search tasks, by combining a single relevance dimension in addition to topicality in the retrieval model of different search engines. Although in this preliminary work we have only used a simple linear combination of relevance dimensions, the obtained results show that there is indeed a relationship between a search task and specific relevance dimensions. This suggests that, for different search tasks, some relevance dimensions should be prioritized in the computation of an overall Retrieval Status Value based on which documents are ranked.

As a future development of the proposed study, we plan to incorporate more than one relevance dimension at a time together with topicality per search task, also considering more complex aggregation strategies. Further experiments on other search tasks will be performed. Moreover, we intend to construct ad hoc datasets via crowdsourcing, in order to be release labelled data where relevance is assessed w.r.t. multiple aspects.

References

1. Alhadi, A.C., Gottron, T., Kunegis, J., Naveed, N.: LiveTweet: microblog retrieval based on interestingness and an adaptation of the vector space model. In: TREC (2011)
2. Borlund, P.: The concept of relevance in IR. J. Am. Soc. Inform. Sci. Technol. **54**(10), 913–925 (2003)
3. Buckley, C., Voorhees, E.M.: Retrieval evaluation with incomplete information. In: Proceedings of the 27th Annual International ACM SIGIR Conference on Research and Development in Information Retrieval, pp. 25–32 (2004)
4. Choi, J., Croft, W.B., Kim, J.Y.: Quality models for microblog retrieval. In: Proceedings of the 21st ACM International Conference on Information and Knowledge Management, pp. 1834–1838. ACM (2012)
5. Cooper, W.S.: On selecting a measure of retrieval effectiveness. J. Am. Soc. Inf. Sci. **24**(2), 87–100 (1973)
6. da Costa Pereira, C., Dragoni, M., Pasi, G.: Multidimensional relevance: prioritized aggregation in a personalized information retrieval setting. Inf. Process. Manag. **48**(2), 340–357 (2012)
7. Craswell, N.: Bpref. In: Liu, L., Ozsu, M.T. (eds.) Encyclopedia of Database Systems, pp. 266–267. Springer, Boston (2009). https://doi.org/10.1007/978-0-387-39940-9_489

8. De Grandis, M., Pasi, G., Viviani, M.: Fake news detection in microblogging through quantifier-guided aggregation. In: Torra, V., Narukawa, Y., Pasi, G., Viviani, M. (eds.) MDAI 2019. LNCS (LNAI), vol. 11676, pp. 64–76. Springer, Cham (2019). https://doi.org/10.1007/978-3-030-26773-5_6
9. Duan, Y., Jiang, L., Qin, T., Zhou, M., Shum, H.Y.: An empirical study on learning to rank of tweets. In: Proceedings of the 23rd International Conference on Computational Linguistics, pp. 295–303. Association for Computational Linguistics (2010)
10. Efron, M.: Information search and retrieval in microblogs. J. Am. Soc. Inform. Sci. Technol. **62**(6), 996–1008 (2011)
11. Fogg, B., Tseng, H.: The elements of computer credibility. In: Proceedings of the SIGCHI Conference on Human Factors in Computing Systems, pp. 80–87. ACM (1999)
12. Ghosh, S., Ghosh, K., Ganguly, D., Chakraborty, T., Jones, G.J., Moens, M.F.: ECIR 2017 workshop on exploitation of social media for emergency relief and preparedness (SMERP 2017). In: ACM SIGIR Forum, vol. 51, pp. 36–41. ACM (2017)
13. Giachanou, A., Harvey, M., Crestani, F.: Topic-specific stylistic variations for opinion retrieval on Twitter. In: Ferro, N., et al. (eds.) ECIR 2016. LNCS, vol. 9626, pp. 466–478. Springer, Cham (2016). https://doi.org/10.1007/978-3-319-30671-1_34
14. Gouws, S., Metzler, D., Cai, C., Hovy, E.: Contextual bearing on linguistic variation in social media. In: Proceedings of the Workshop on Languages in Social Media, pp. 20–29. Association for Computational Linguistics (2011)
15. Gupta, A., Kumaraguru, P., Castillo, C., Meier, P.: TweetCred: real-time credibility assessment of content on Twitter. In: Aiello, L.M., McFarland, D. (eds.) SocInfo 2014. LNCS, vol. 8851, pp. 228–243. Springer, Cham (2014). https://doi.org/10.1007/978-3-319-13734-6_16
16. Hosmer Jr., D.W., Lemeshow, S., Sturdivant, R.X.: Applied Logistic Regression, vol. 398. Wiley, Hoboken (2013)
17. Huang, H., et al.: Tweet ranking based on heterogeneous networks. In: Proceedings of COLING 2012, pp. 1239–1256 (2012)
18. Imran, M., Elbassuoni, S., Castillo, C., Diaz, F., Meier, P.: Practical extraction of disaster-relevant information from social media. In: Proceedings of the 22nd International Conference on World Wide Web, pp. 1021–1024. ACM (2013)
19. Jiang, J., He, D., Kelly, D., Allan, J.: Understanding ephemeral state of relevance. In: Proceedings of the 2017 Conference on Conference Human Information Interaction and Retrieval, pp. 137–146. ACM (2017)
20. Liu, T.Y.: Learning to rank for information retrieval. Found. Trends Inf. Retrieval **3**(3), 225–331 (2009)
21. Livraga, G., Viviani, M.: Data confidentiality and information credibility in on-line ecosystems. In: Proceedings of the 11th International Conference on Management of Digital EcoSystems, pp. 191–198 (2019)
22. Luo, Z., Osborne, M., Wang, T.: An effective approach to tweets opinion retrieval. World Wide Web **18**(3), 545–566 (2015)
23. Mahata, D., Talburt, J.R., Singh, V.K.: From chirps to whistles: discovering event-specific informative content from Twitter. In: Proceedings of the ACM Web Science Conference, p. 17. ACM (2015)
24. Manning, C.D., Raghavan, P., Schütze, H.: Introduction to Information Retrieval. Cambridge University Press, Cambridge (2008)

25. Massoudi, K., Tsagkias, M., de Rijke, M., Weerkamp, W.: Incorporating query expansion and quality indicators in searching microblog posts. In: Clough, P., et al. (eds.) ECIR 2011. LNCS, vol. 6611, pp. 362–367. Springer, Heidelberg (2011). https://doi.org/10.1007/978-3-642-20161-5_36
26. Mitra, T., Gilbert, E.: Credbank: A large-scale social media corpus with associated credibility annotations. In: Ninth International AAAI Conference on Web and Social Media (2015)
27. Mizzaro, S.: How many relevances in information retrieval? Interact. Comput. **10**(3), 303–320 (1998)
28. Nagmoti, R., Teredesai, A., De Cock, M.: Ranking approaches for microblog search. In: Proceedings of the 2010 IEEE/WIC/ACM International Conference on Web Intelligence and Intelligent Agent Technology, vol. 1, pp. 153–157. IEEE Computer Society (2010)
29. Naveed, N., Gottron, T., Kunegis, J., Alhadi, A.C.: Bad news travel fast: a content-based analysis of interestingness on Twitter. In: Proceedings of the 3rd International Web Science Conference, pp. 1–7 (2011)
30. Naveed, N., Gottron, T., Kunegis, J., Alhadi, A.C.: Searching microblogs: coping with sparsity and document quality. In: Proceedings of the 20th ACM International Conference on Information and Knowledge Management, pp. 183–188. ACM (2011)
31. Nielsen, F.: A new ANEW: evaluation of a word list for sentiment analysis in microblogs. In: Proceedings of the ESWC2011 Workshop on Making Sense of Microposts: Big things come in small packages 718 in CEUR Workshop Proceedings, Heraklion (2011)
32. Olteanu, A., Vieweg, S., Castillo, C.: What to expect when the unexpected happens: social media communications across crises. In: Proceedings of the 18th ACM Conference on Computer Supported Cooperative Work & Social Computing, pp. 994–1009. ACM (2015)
33. Page, L., Brin, S., Motwani, R., Winograd, T.: The pagerank citation ranking: bringing order to the web. Technical report, Stanford InfoLab (1999)
34. Pasi, G., De Grandis, M., Viviani, M.: Decision making over multiple criteria to assess news credibility in microblogging sites. In: Proceedings of IEEE World Congress on Computational Intelligence (WCCI) 2020. IEEE (2020)
35. Pasi, G., Viviani, M.: Application of aggregation operators to assess the credibility of user-generated content in social media. In: Medina, J., et al. (eds.) IPMU 2018. CCIS, vol. 853, pp. 342–353. Springer, Cham (2018). https://doi.org/10.1007/978-3-319-91473-2_30
36. Pedregosa, F., et al.: Scikit-learn: machine learning in Python. J. Mach. Learn. Res. **12**, 2825–2830 (2011)
37. Porter, M.: The Porter stemming algorithm, 2005 (2008). http://www.tartarus.org/martin/PorterStemmer/index.html
38. Surdeanu, M., Ciaramita, M., Zaragoza, H.: Learning to rank answers to non-factoid questions from web collections. Comput. Linguist. **37**(2), 351–383 (2011)
39. Tang, R., Solomon, P.: Toward an understanding of the dynamics of relevance judgment: an analysis of one person's search behavior. Inf. Process. Manag. **34**(2–3), 237–256 (1998)
40. Tao, K., Abel, F., Hauff, C., Houben, G.-J.: Twinder: a search engine for Twitter streams. In: Brambilla, M., Tokuda, T., Tolksdorf, R. (eds.) ICWE 2012. LNCS, vol. 7387, pp. 153–168. Springer, Heidelberg (2012). https://doi.org/10.1007/978-3-642-31753-8_11
41. Tao, K., Hauff, C., Abel, F., Houben, G.J.: Information retrieval for Twitter data, pp. 195–206. Digital Formations, Peter Lang (2013)

42. Teevan, J., Ramage, D., Morris, M.R.: # twittersearch: a comparison of microblog search and web search. In: Proceedings of the Fourth ACM International Conference on Web Search and Data Mining, pp. 35–44. ACM (2011)
43. Vakkari, P.: Task-based information searching. Ann. Rev. Inf. Sci. Technol. **37**(1), 413–464 (2003)
44. Verma, M., Yilmaz, E., Craswell, N.: On obtaining effort based judgements for information retrieval. In: Proceedings of the Ninth ACM International Conference on Web Search and Data Mining, pp. 277–286. ACM (2016)
45. Viviani, M., Pasi, G.: A multi-criteria decision making approach for the assessment of information credibility in social media. In: Petrosino, A., Loia, V., Pedrycz, W. (eds.) WILF 2016. LNCS (LNAI), vol. 10147, pp. 197–207. Springer, Cham (2017). https://doi.org/10.1007/978-3-319-52962-2_17
46. Viviani, M., Pasi, G.: Credibility in social media: opinions, news, and health information—a survey. Wiley Interdisc. Rev. Data Mining Knowl. Discov. **7**(5), e1209 (2017)
47. Vosecky, J., Leung, K.W.-T., Ng, W.: Searching for quality microblog posts: filtering and ranking based on content analysis and implicit links. In: Lee, S., Peng, Z., Zhou, X., Moon, Y.-S., Unland, R., Yoo, J. (eds.) DASFAA 2012. LNCS, vol. 7238, pp. 397–413. Springer, Heidelberg (2012). https://doi.org/10.1007/978-3-642-29038-1_29
48. Webberley, W.M., Allen, S.M., Whitaker, R.M.: Retweeting beyond expectation: inferring interestingness in Twitter. Comput. Commun. **73**, 229–235 (2016)
49. Weerkamp, W., De Rijke, M.: Credibility improves topical blog post retrieval. In: Proceedings of ACL 2008: HLT, pp. 923–931 (2008)
50. Yager, R.R.: On ordered weighted averaging aggregation operators in multicriteria decision making. IEEE Trans. Syst. Man Cybern. **18**(1), 183–190 (1988)
51. Yang, P., Fang, H., Lin, J.: Anserini: enabling the use of lucene for information retrieval research. In: Proceedings of the 40th International ACM SIGIR Conference on Research and Development in Information Retrieval, pp. 1253–1256. ACM (2017)
52. Zhai, C., Lafferty, J.: A study of smoothing methods for language models applied to information retrieval. ACM Trans. Inf. Syst. (TOIS) **22**(2), 179–214 (2004)
53. Zubiaga, A.: A longitudinal assessment of the persistence of Twitter datasets. J. Assoc. Inf. Sci. Technol. **69**(8), 974–984 (2018)

Regional Influences on Tourists Mobility Through the Lens of Social Sensing

Helen Senefonte[1,2(✉)], Gabriel Frizzo[1], Myriam Delgado[1], Ricardo Lüders[1], Daniel Silver[3], and Thiago Silva[1,3]

[1] Universidade Tecnológica Federal do Paraná (UTFPR), Curitiba, Brazil
prof.helen.mattos@gmail.com
[2] Universidade Estadual de Londrina, Londrina, Brazil
[3] University of Toronto, Toronto, Canada

Abstract. This study aims at exploring social media data to evaluate how regional and cultural characteristics influence the mobility behavior of tourists and residents. By considering information taken from the mobility graphs of users from different countries, we observe that users' origins can influence their choices. Additionally, the analysis performed in the experiments shows that a regression model could enable the prediction of the behavior of a tourist from a specific country when visiting another country, based on their cultural distances (obtained offline). The ability to explore the cultural characteristics of each nationality in different destinations shows a promising way to improve recommendation systems for points of interest and other services to particular groups of tourists.

Keywords: Cross-cultural study · Tourists · LBSNs · Mobility

1 Introduction

Studies based on traditional data sources tend to suffer from poor scalability. Consequently, the experiments are limited, and results are restricted to small regions, like a city or state. The use of location-based social network (LBSN) data can mitigate the scalability problem enabling the study of social behavior for larger populations. These studies contribute to solving problems in many areas ranging from socio-economic issues [2,3] to assisting tourism-related activities [1,13].

Planning for urban tourism is important in many aspects, not least economically. Considering data from foreign tourists, world tourism generated revenues of more than one billion US dollars in 2017 alone [6]. Tourism is also a crucial source of jobs in the labor market: according to the World Tourism Organization, it accounts for 1 in 10 jobs [6]. To keep tourist activity in a particular location attractive, it is essential to understand tourists' preferences in order to offer better and smarter services.

S. Aref et al. (Eds.): SocInfo 2020, LNCS 12467, pp. 312–319, 2020.
https://doi.org/10.1007/978-3-030-60975-7_23

With this goal in mind, this research aims at exploring social media data to study regional influences on the behavior of tourists and residents in the context of mobility. The central questions that guide this research are: (i) In terms of mobility, how do tourists behave in countries where they travel? Do they move similarly to residents of destination countries, or do they have similar mobility patterns to residents of their own countries of origin? (ii) Do tourists with the same cultural habits perform similar tourist activities?

To answer the previous questions, we first define the dataset, which encompasses data of eight countries in different regions worldwide, aiming to establish two main categories of users for each country: tourists and residents. The mobility of tourists and residents in distinct countries is analyzed in this study with Foursquare-Swarm check-ins. We formalize the mobility graph to represent residents and tourists' movement at different times throughout the day. Using the type of location visited and the geographic coordinates, this representation captures, in a certain way, the semantic of mobility. After that, we present the concept of *behavioral distance*, which is based on the vectors obtained from the linearization of adjacency matrices representing the mobility graphs of residents and tourists. The behavioral distance allows us to perform two main analyses: (i) the influence of origin and destination on tourists of different countries; (ii) the inference, from a linear regression model, of the behavioral distance of a specific country considering its cultural distance from the other countries.

2 Related Work

Several works on tourist behavior use data obtained from traditional approaches such as questionnaires and interviews. For example, by exploring questionnaire data, Zieba [14] studied how individual features of Austrian tourists can influence their travel motivation. Additionally, the author investigated if tourists' habits are different in other countries. Scuderi and Chiara [8] analyzed expense patterns using a tourist card (which provides discounts in the city of Trentino).

Given the vast amount of data available from LBSN, there is a broad spectrum of possibilities for carrying on studies about urban societies in an unprecedented scale [10], including those focused on cultural aspects. For example, Silva et al. [9] explored Foursquare check-ins on food venues to automatically identify cultural boundaries, whose results are compatible with those obtained using traditional data.

In the context of tourists' behavioral pattern analysis exploring LBSNs data, Vu et al. [12] explore Foursquare check-ins to study the preferences of Malaysian tourists in various countries and their differences from Thai tourists. The study suggests that despite the regional proximity of Malaysia and Thailand, it is possible to find relevant differences in the users' preferences of each country.

Long et al. [5] explore temporal characteristics of tourist check-ins to analyze their movements, finding that most tourists tend to exhibit higher diversity in their activities. Ferreira et al. [1] consider the preferences of tourists and residents represented by Foursquare check-ins and use a spatiotemporal graph model to suggest that each group has distinct preferences in different times for specific places of the studied cities. Veiga et al. [11] extract tourist mobility from two different Foursquare datasets. The authors tested two different ways of identifying tourists and concluded that the results are not significantly influenced by the identification process. In addition, big cities tend to dominate the mobility pattern for the whole country.

The present work differs from the previous ones due to its focus on the analysis of the users' origins influence on their mobility decisions. A comparison between tourists and residents has been carried out, obtaining different groups in terms of mobility patterns. In addition, we quantified regional influence on mobility based on the concept of *behavioral distance*. We found evidence that the user's origin can influence the activities performed. Moreover, we evaluated a model to predict the behavioral distance using as a predictor the cultural dimensions defined by Inglehart and Welzel [4].

3 Data and Method

The considered dataset is publicly available and composed by Foursquare-Swarm check-ins shared on Twitter between January and June 2014 (≈20 million check-ins from ≈1.2 million unique users). These data are obtained by a web crawler developed especially for this purpose. Each record of this dataset identifies (by a single ID) the user who performed the activity, the GPS location of the venue visited, the time and category of the venue (names and organization are established by Foursquare-Swarm[1]). Aiming at getting a large data volume and diversity of world regions, the following countries are considered: Brazil (BR), United States (US), Indonesia (ID), Japan (JP), Malaysia (MY), Mexico (MX), the United Kingdom (UK), and Turkey (TR).

It is essential to identify tourists and residents to reach the goals of this study. Our dataset allows us to identify the country of users' residence with minor inconsistencies. We only consider users with a minimum of 2 check-ins, totaling 1010953 unique users. The country of users' residence is chosen as the country where they spent most of the time. Elsewhere, they are considered tourists. In the case of two subsequent *check-ins* are made in different countries, we suppose that the user spends this time window in the first country. Although these premises bring possible misinterpretation, we believe it could have at most a minor impact on the results due to the large volume of data. Previous works in the literature have successfully used similar strategies [1,7].

Subsequent *check-ins* are modeled in a *mobility graph* which is a directed graph $G_m(V, E)$ obtained for each country m available in our dataset. Vertices $v_i \in V$ represent ten main categories[2] of Foursquare-Swarm. If a user makes two subsequent *check-ins* at v_i and v_j within no more than two periods of the day,

[1] https://developer.foursquare.com/docs/resources/categories.

[2] Arts & Entertainment; College & University; Professional & Other Places; Residences; Outdoors & Recreation; Shops & Services; Nightlife Spots; Food; Travel & Transport; and Event.

a transition or edge $e(i, j) \in E$ connects v_i to v_j. A period of the day includes: (i) morning between 6:00 am and 9:59 am; (ii) noon between 10:00 am and 2:59 pm; (iii) afternoon between 3:00 pm and 6:59 pm; (iv) night between 7:00 pm and 11:59 pm; and (v) dawn between 00:00 am and 05:59 am. For each country, there is one mobility graph for residents and several mobility graphs for tourists of a certain country visiting other countries.

Each mobility graph is represented by an adjacency matrix 10×10 corresponding to ten main categories mentioned above. Each element (i, j) of the matrix corresponds to the number of transitions made by all users between categories i and j. The matrix lines are then concatenated in a 100 position vector called *mobility vector*. Comparisons between mobility vectors are made by *Canberra* distance combined with *Ward* linkage criteria[3].

We are interested in evaluating how tourists are influenced by the countries of their origin and destination. Let's assume two countries O (origin) and D (destination) with mobility vectors defined for Residents O, Residents D, and tourists of O in D denoted by Tourists O/D. We compare these mobility vectors for two cases: (i) Tourists $O/D\times$ Residents O; (ii) Tourists $O/D\times$ Residents D. The first case shows how far (or close) tourists of O are from residents of O when visiting D, whereas the second case shows how tourists of O are far (or close) from residents of D when visiting it. Taking a country O as a reference, mobility vectors for Tourists O/D_i can be built for different destination countries D_i. For example, taking USA as country O, the distance between mobility vectors of USA tourists in country D_i and residents of USA is given by $dist_i(O/D_i, O)$ (similarly for $dist_i(O/D_i, D_i)$ when residents of D_i are considered). The *behavioral distance* (Bh) is then defined in this work as the RMS (*Root Mean Square*) over $dist_i$ calculated with the Canberra distance. Therefore, there are two different behavioral distances: Bh_O and Bh_D depending on whether tourists are compared with their origin O or their destination D_i, respectively.

According to Inglehart and Welzel [4], the *cultural distance* (Ct) is based on the space defined by the two major dimensions of cross-cultural variation: f_{TS} (traditional values versus secular-rational values), and f_{SS} (survival values versus self-expression values). Considering the World Value Survey data from wave 6, 2010–2014, every country is positioned in this two-dimensional space. A particular country O has a cultural distance $Ct(O) = \frac{1}{N}\sum_{i=1}^{N}(euc(O, D_i))$ to a set of countries $\{D_1, D_2, \ldots, D_N\}$, where euc is the euclidean distance. In other words, it is the average Euclidean distance between O and all other countries in the space defined by their cultural dimensions. Finally, we investigate the relationship between cultural and behavioral distances in a set of countries using a linear regression model (ordinary least squares). The behavioral distance (dependent variable) is evaluated from the cultural distance (independent one) by considering two different regression models for Bh_O and Bh_D in the experiments.

[3] Other distances and linkage criteria were evaluated, but this combination provided the most consistent results.

4 Results

4.1 Measuring Regional Influence on Behavior

The behavioral distance mentioned in Sect. 3 is calculated for each analyzed country. Figure 1 presents these results considering the origin of some countries and the destination of all the studied countries. In this figure, the countries mentioned in the x axis represent the variable O (that is, tourist origin country).

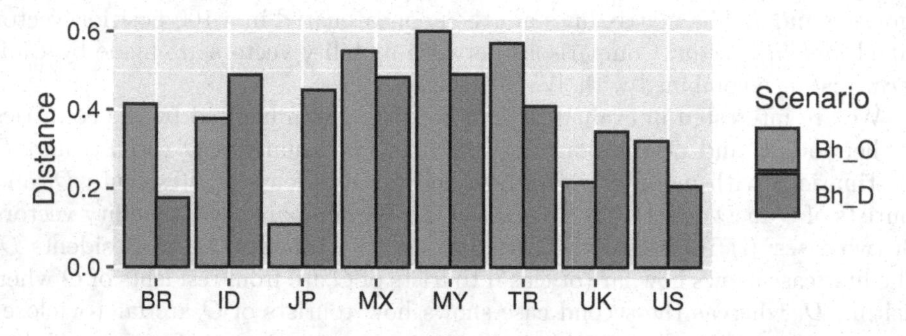

Fig. 1. Behavioral distances Bh_O and Bh_D for tourists of O (BR, ID, JP, MX, MY, TR, UK, US) visiting all other considered countries.

According to this result, there is an indication that Brazilian, Mexican, Turkish and Malaysian tourists distance considerably from their original behavior (from residents of their country of origin). Brazilian tourists tend to absorb more intensively the local habits of countries visited in contrast to the other tourists. On the contrary, Japanese tourists tend to maintain their original behavior when leaving the country (close to Japanese residents).

The relative differences between behavioral distances (Bh_O and Bh_D) of Fig. 1 is presented in Fig. 2. As we can see, Brazil and Japan are the countries with the highest discrepancy, with Brazil being more influenced by the destination and Japan in the opposite way. For the other countries, the discrepancies are lower, especially for Mexico.

4.2 Regressions on Cultural and Behavioral Dimensions

Here we present the results for the linear regression models explained in Sect. 3. Table 1 summarizes the results. As we can see, the preliminary results suggest that there is a considerable potential of using the cultural distance based on the World Value Survey to explain the behavioral distance, especially in the case Bh_O with $R^2 = 0.42$ (considering all countries) and p-value < 0.1 for all coefficients. This is not the case for Bh_D whose regression models are not relevant with $R^2 = 0.2$ (all countries).

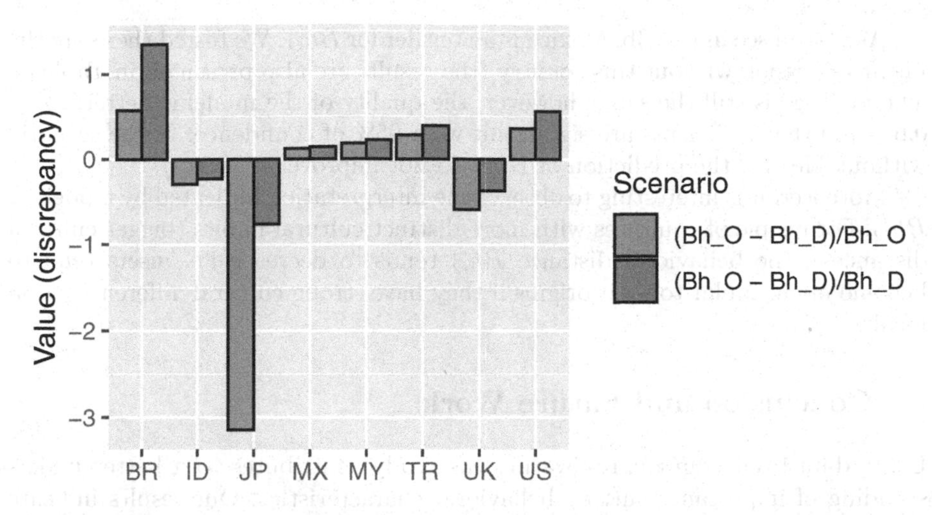

Fig. 2. Relative differences between behavioral distances Bh_O and Bh_D for all countries.

Table 1. Linear regression parameters and quality index R^2, for models predicting the behavioral distances Bh_O and Bh_D for all countries and without Mexico. Significance values codes: ** $p < 0.01$, * $p < 0.05$, .$p < 0.1$.

	All countries		Without Mexico	
	Bh_O	Bh_D	Bh_O	Bh_D
(Intercept)	0.832**	0.174	0.876**	0.19
Cultural distance	−0.251	0.119	−0.294*	0.104
R^2	0.42	0.2	0.64	0.15

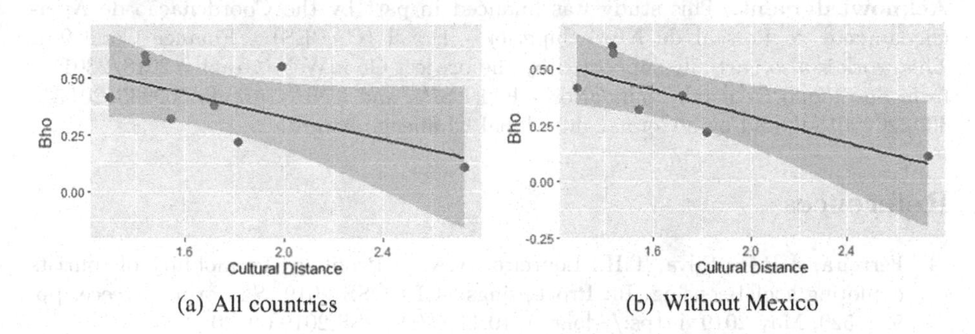

(a) All countries

(b) Without Mexico

Fig. 3. Linear regressions for Bh_O considering all countries and removing Mexico as an outlier.

As we can see in Fig. 3b, Mexico is an outlier for Bh_O. We tested therefore the linear regression without this country (the results are also presented in Table 1). The message is still the same; however, the quality of the model is better: $R^2 = 0.64$, and the coefficients are significant with 95% of confidence. In the scenario without Mexico, the prediction of Bh_D do not improve.

Moreover, it is interesting to observe the interpretation indicated by models of Bh_O. For people of countries with more distinct cultural habits (bigger cultural distances), the behavioral distance Bh_O tends to decrease i.e., users tend to become more similar to their origins if they have strong cultural differences with locals.

5 Conclusion and Future Work

Using data from Foursquare-Swarm, this study contributes to a better understanding of important tourists' behavioral characteristics. Our results indicate that tourists typically have behavior influenced by their origin (especially when their cultural distance from the locals is large). We also obtained an indication that offline measures of cultural distances, such as those measured by the World Value Surveys [4], could be good predictors for these behavioral differences. This is interesting because those measures are publicly available for many countries. Our results indicate that place recommendation systems for tourists could be leveraged by a tourist behavior estimation which could take into account additional information such as users' origin, destination, and their cultural characteristics.

This study could be expanded in numerous ways. For example, the availability of places that are part of tourists' routine is one factor that can influence their choices and should be studied in the future. An Indonesian tourist may have difficulty finding his cultural preferences in a western country due to religious and gastronomic differences; however, he will find it more comfortable in countries with a similar profile. We are also working to expand the dataset and evaluate other statistical models to better understand the phenomena under study.

Acknowledgment. This study was financed in part by the Coordenação de Aperfeiçoamento de Pessoal de Nível Superior - Brasil (CAPES) - Finance Code 001. This work is also partially supported by the project GoodWeb (Grant #2018/23011-1 from São Paulo Research Foundation - FAPESP), and CNPq, grants 309935/2017-2, 439226/2018-0, and by a Connaught Global Challenge Award.

References

1. Ferreira, A.P.G., Silva, T.H., Loureiro, A.A.F.: Profiling the mobility of tourists exploring social sensing. In: Proceedings of DCOSS 2019, Santorini, Greece, pp. 522–529, May 2019. https://doi.org/10.1109/DCOSS.2019.00100
2. Hristova, D., Williams, M.J., Musolesi, M., Panzarasa, P., Mascolo, C.: Measuring urban social diversity using interconnected geo-social networks. In: Proceedings of WWW 2019, Montreal, Canada (2016)

3. Huang, C., Wang, D.: Unsupervised interesting places discovery in location-based social sensing. In: Proceedings of DCOSS 2016, Washington, USA, pp. 67–74 (2016)
4. Inglehart, R., Welzel, C.: Changing mass priorities: the link between modernization and democracy. Perspect. Polit. **8**(2), 551–567 (2010). https://doi.org/10.1017/s1537592710001258
5. Long, X., Jin, L., Joshi, J.: Towards understanding traveler behavior in location-based social networks. In: Proceedings of GLOBECOM 2013, pp. 3182–3187, December 2013
6. Organization, W.T.: UNWTO Tourism highlights. 2018 edn. (2018). https://doi.org/10.18111/9789284419876
7. Paldino, S., Bojic, I., Sobolevsky, S., Ratti, C., González, M.C.: Urban magnetism through the lens of geo-tagged photography. EPJ Data Sci. **4**(1), 1–17 (2015). https://doi.org/10.1140/epjds/s13688-015-0043-3
8. Scuderi, R., Dalle Nogare, C.: Mapping tourist consumption behaviour from destination card data: what do sequences of activities reveal? Int. J. Tour. Res. **20**(5), 554–565 (2018)
9. Silva, T.H., de Melo, P.O.V., Almeida, J.M., Musolesi, M., Loureiro, A.A.: A large-scale study of cultural differences using urban data about eating and drinking preferences. Inf. Syst. **72**(Supplement C), 95–116 (2017). https://doi.org/10.1016/j.is.2017.10.002
10. Silva, T.H., et al.: Urban computing leveraging location-based social network data: a survey. ACM Comput. Surv. **52**(1), 17:1–17:39 (2019). https://doi.org/10.1145/3301284
11. Veiga, D.A., Frizzo, G.B., Silva, T.H.: Cross-cultural study of tourists mobility using social media. In: Proceedings of WebMedia 2019, pp. 313–316. ACM, Rio de Janeiro (2019)
12. Vu, H.Q., Li, G., Law, R.: Cross-country analysis of tourist activities based on venue-referenced social media data. J. Travel Res. 0047287518820194 (2019). https://doi.org/10.1177/0047287518820194
13. Yang, D., Zhang, D., Qu, B.: Participatory cultural mapping based on collective behavior data in location-based social networks. ACM Trans. Intell. Syst. Technol. **7**(3), 30:1–30:23 (2016). https://doi.org/10.1145/2814575
14. Zieba, M.: Cultural participation of tourists-evidence from travel habits of austrian residents. Tour. Econ. **23**(2), 295–315 (2017)

Genuine Personal Identifiers and Mutual Sureties for Sybil-Resilient Community Growth

Gal Shahaf[1], Ehud Shapiro[1], and Nimrod Talmon[2](✉)

[1] Weizmann Institute, Rehovot, Israel
{gal.shahaf,ehud.shapiro}@weizmann.ac.il
[2] Ben-Gurion Univesity, Beersheba, Israel
talmonn@bgu.ac.il

Abstract. Observing its lack, we introduce the notion of a *genuine personal identifier*—a globally unique and singular identifier of a person—and present a foundation for a decentralized, grassroots, bottom-up process by which every human being may create, own, and protect the privacy of a genuine personal identifier. Our solution is based on a web-of-trust and is designed for a distributed realization; we apply graph-theoretic notions to show that digital communities can grow indefinitely while ensuring bounded sybil penetration.

Keywords: Social networks · Digital communities · Sybil attacks

1 Introduction

Providing credible identities to all by 2030 is a UN Sustainable Development Goal. Yet, current top-down digital identity-granting solutions are unlikely to close the 1Bn-people gap [3] in time, as they are not working for citizens of failed states nor for people fleeing physical or political harshness [2,6]. Concurrently, humanity is going online at an astonishing rate, with more than half the world population now being connected. Still, online accounts do not provide a solution for credible digital identity either, as they may easily be fake, resulting in lack of accountability and trust. For example, Facebook reports the removal of 5.4Bn fake accounts in the first 9 months of 2019 [5,9]. The profound penetration of fake accounts on the net greatly hampers its utility for credible human discourse and for deliberations and democratic decision making; it makes the net unsuitable for vulnerable populations, including children and the elderly; it makes the use of the net for person-to-person transactions, notably direct philanthropy, precarious; and in general it turns the net into an inhuman, even dangerous, ecosystem. As an aside, we note that the panacea of cryptocurrencies for the lack of credible

An extended version of this paper that includes proofs and further elaboration appears in [16].

S. Aref et al. (Eds.): SocInfo 2020, LNCS 12467, pp. 320–332, 2020.
https://doi.org/10.1007/978-3-030-60975-7_24

personal identities on the net is the reckless employment of the environmentally-harmful proof-of-work protocol.

Filling the Gap Between the Computational and Social Sciences. Our aim is a conceptual and mathematical foundation for allowing every person to create, own, and protect the privacy of a globally-unique and singular identifier, henceforth referred to as *genuine personal identifier*. In order to provide services for community members, a social institution must first identify the individuals that form the community. In this sense, it is hard to overestimate the necessity of genuine personal identifier for a functioning digital institutions, including: egalitarian democratic governance in digital local communities and global movements; digital cooperatives and digital credit unions; direct philanthropy; child-safe digital communities; preventing unwanted digital solicitation; banishing deep-fake (by marking as spam videos not signed by a genuine personal identifier); credible and durable digital identities for people fleeing political or economic harshness; accountability for criminal activities on the net; and egalitarian cryptocurrencies employing an environmentally-friendly consensus protocol among owners of genuine personal identifiers [4]. Furthermore, genuine personal identifiers may provide the necessary digital foundation for a notion of *global citizenship* and, subsequently, for democratic global governance [17].

A solution that is workable for all must be decentralized, distributed, grassroots, and bottom-up. Solid foundations are being laid out by self-sovereign identities [10], blockchain-based identity management [7,19], and the W3C Decentralized Identifiers [11] and Verifiable Claims [18] emerging standards, which aim to let people freely create and own identifiers and associated credentials. We augment this freedom with the goal that each person declares exactly one identifier as her *genuine personal identifier*. We note that besides the genuine personal identifier, one may create, own and use any number of identifiers of other types.

Challenges. On the technical side, becoming the owner of a genuine personal identifier is simple: (1) Choose a new cryptographic key-pair (v, v^{-1}), with v being the public key and v^{-1} the private key; (2) Claim v to be your genuine personal identifier by publicly posting a declaration that v is a genuine personal identifier, signed with v^{-1}. With a suitable app, this could be done literally with a click of a button.[1] A declaration of a genuine personal identifier, by itself, does not reveal the person making the declaration; it only reveals to all that someone who knows the secret key for the public key v claims v as her genuine personal identifier. Depending on personality and habit, the person may or may not publicly associate oneself with v. E.g., a person with truthful social media accounts may wish to associate these accounts with its newly-minted genuine identifier.

The straightforward procedure of generating an identifier conceals a handful of difficulties and complexities that this paper is nothing but an initial

[1] As the public key may be quite long, one may also associate oneself with a shorter "nickname", a hash of the public key, e.g. a 128-bit hash (as a UUID) or a 256-bit hash (as common in the crypto world).

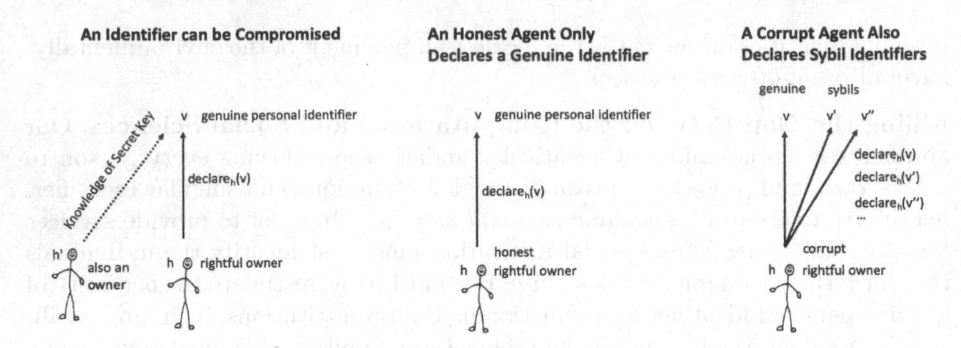

Fig. 1. A compromised identifier (left) and honest/corrupt agents (middle/right).

investigation into describing, analyzing, and preventing them. Some of them are enumerated below; h denotes an agent: (1) The key-pair (v, v^{-1}) is not new, or else someone got hold of it between Step 1 and Step 2 above. Either way, someone else has declared v to be a genuine personal identifier prior to the declaration by h. In which case h cannot declare v; (2) Agent h failed to keep v^{-1} secret so that other people, e.g. h', know v^{-1}, in which case h' is also an *owner* of v and, thus, v is *compromised.* Figure 1 (left) illustrates a compromised personal identifier; (3) The agent h intended to divulge his association with his public key v only on a need-to-know basis, but the association of v and h has become public knowledge, prompting agent h to replace his genuine personal identifier v with a new one; (4) Agent h declared v as his genuine personal identifier, but later also declared another personal identifier v'. Then, v and v' are *duplicates,* v' is a *sybil,* and agent h is *corrupt.* An *honest* agent does not declare sybils. Figure 1 illustrates honest and corrupt agents.

In this paper, we aim to develop the foundation for genuine personal identifiers by employing mutual sureties among owners of personal identifiers, resulting in a mutual-surety graph reminiscent of a web-of-trust. These sureties should be interpreted as mutual recognition of the relation between the identifier and the person who owns it. In the following, we utilize basic concepts of public-key cryptography and graph theory in order to establish a framework for personal identifier update and sybil/byzantine-resilient community growth. For the solution to be complete, additional components are needed, notably a mechanism that encourages honest behavior [14] and a sybil-resilient governance system [15].

Some discussions and proofs are omitted; we refer the reader to [16].

2 Genuine Personal Identifiers

The ingredients needed for a realization of genuine personal identifiers are: (1) A set of agents. Note that, mathematically, the agents form a set (of unique entities) not a multiset (with duplicates). Intuitively, it is best to think of agents as people (or other physical beings with unique personal characteristics, unique personal history, and agency, such as intelligent aliens), which cannot be duplicated, but not as software agents, which can be; (2) A way for agents to create cryptographic key-pairs. This can be realized, e.g., using the RSA standard [12]. Our solution does not require a global standard or a uniform implementation for public key encryption: Different agents can use different technologies for creating and using such key-pairs, as long as the signatures-verification methods are declared; (3) A way for agents to sign strings using their key-pairs. As we assume cryptographic hardness, an agent that does not know a certain key-pair cannot sign strings with it; (4) A bulletin board or public ledger, to which agents may post and observe signed messages, where all agents observe the same order of messages. The weaker requirement that the same order is observed only eventually, as is standard with distributed ledger protocols, could also be accommodated. Considering partial orders is a subject of future work.

Agents and their Personal Identifiers. We assume a set of agents $\mathcal{H}$ that is fixed over time.[2] Agents can create new key-pairs (v, v^{-1}). We assume that an agent that has a key-pair can sign a string, and denote by $v^{-1}(x)$ the string resulting from signing the string x with v^{-1}. Intuitively, each agent corresponds to a human being. Importantly, members of the set $\mathcal{H}$ of agents (e.g., containing all human beings) cannot be referenced explicitly and, in particular, posted signed messages never refer directly to agents $h \in \mathcal{H}$. A key motivation for our work is providing people with digital genuine personal identifiers without accessing any of their intrinsic (e.g. biometric) private properties and without depending upon such properties as identifiers. As we aim personal identifiers to be self-sovereign identities that conform to the W3C Decentralized Identifiers emerging standards, we let agents create and own their personal identifiers. An agent h can publicly declare a personal identifier v for which it knows the private key v^{-1}. A *personal identifier declaration* (we shall use the abbreviation "pid") has the form $v^{-1}(pid(v))$ and can be effected by agent h posting $v^{-1}(pid(v))$ to a public ledger. We denote this action by $declare_h(pid(v))$. We denote by C the sequence of all signed declarations on a public ledger, that corresponds to a finite ordered list of events. Recall that all agents have the same view of the sequence of all declarations made; subsequent work may relax this assumption.

Definition 1 (Personal Identifier). *Let C be a sequence of personal identifier declaration events and $declare_h(pid(v)) \in C$ the first declaration event in which v occurs. Then v is a* personal identifier *and h is the* rightful owner *of v, given C.*

Definition 2. *Let C be a sequence of personal identifier declaration events and h be the rightful owner of personal identifier v in C. Then v is* genuine *if it is the first personal identifier declared in C by h, else v is a* sybil. *An agent h is* corrupt *if it declares any sybils, else h is* honest. *(All notions are relative to C.)*

See Fig. 1 and some remarks: (1) An agent is the rightful owner of its genuine personal identifier as well as of any subsequent sybils that it declares. (2) If h, the rightful owner of v, is corrupt, then its first declared identifier is genuine and the rest of its declared identifiers are all sybils. (3) An honest agent may create

[2] Birth and death of agents will be addressed in future work.

and use many key-pairs for various purposes, yet remain honest as long as it has declared at most one public key as a personal identifier.

Mutual Sureties and Their Graphs. A key element of our approach is pledging of mutual sureties by agents. Intuitively, mutual surety pledges provide a notion of trust between owners of personal identifiers. They allow two agents that know each other and know the personal identifiers declared by each other to vouch for the other regarding the good standing of the personal identifiers. This notion is key to help honest agents fend-off sybils. We consider several types of mutual sureties, of increasing strength, and illustrate the corresponding sybil-resilience each type of mutual sureties obtains. For an agent, say h, to provide surety regarding the personal identifier of another agent, say h', h first has to know h'. How this knowledge is established is not specified in our formal framework, but this is quite an onerous requirement that cannot be taken lightly or satisfied casually. E.g., we may assume that one knows one's family, friends and colleagues, and may diligently get to know new people if one so chooses. We consider several types of sureties of increasing strength, in which an agent h with personal identifier v makes a pledge regarding the personal identifier v' of another agent h'; all assume that the agent h knows the agent h'. We describe four *Surety Types*, which are cumulative as each includes all previous ones, explain when one may choose to pledge them, and present results utilizing them.

Surety of Type 1: *Ownership* of a Personal Identifier. Agent h pledges that agent h' owns personal identifier v'.

Agent h' can prove to agent h that she owns v' without disclosing v'^{-1} to h. This can be done, for example, by h asking h' to sign a novel string x and verifying that $v'(x)$ is signed using v'^{-1}. This surety type is the weakest of all four, it is the one given in "key signing parties", and is implicitly assumed by applications such as PGP and web-of-trust [1]. For a given surety type, we say that the surety is *violated* if its assertion does not hold; in particular, a surety of Type 1 is violated if h' in fact does not know the secret key v'^{-1}. In general, mutual surety between two agents with two personal identifiers is pledged by each of the two agents pledging a surety to the personal identifier of the other agent.[3] We define below three additional surety types, where the format of a surety pledge of Type X by the owner of v to the owner of v' is $v(suretyX(v'))$, $X \in \{1, 2, 3, 4\}$. The corresponding surety event is $pledge_h v(suretyX(v'))$, and the surety enters into effect once both parties have made the mutual pledges. We now take $\mathcal{C}$ to be a record of both declaration events and pledge events.

Definition 3 (Mutual Surety). *The personal identifiers v, v' have* mutual surety of type X, $X \in \{1, 2, 3, 4\}$, *if there are $h, h' \in \mathcal{H}$ for which $pledge_h v(suretyX(v')) \in \mathcal{C} \wedge pledge_{h'} v'(suretyX(v)) \in \mathcal{C}$, in which case h and h' are the* witnesses *for the mutual surety between v and v'.*

[3] We consider undirected graphs, as we require surety to be symmetric. Indeed, one may consider directed sureties.

A sequence of events induces a sequence of surety graphs in which the vertices are personal identifiers that correspond to personal identifier declarations and the edges correspond to mutual surety pledges.

Definition 4 (Surety Graph). *Let $\mathcal{C} = c_1, c_2, \ldots$ be a sequence of events and let $\mathcal{C}_k$ denote its first $k \geq 0$ events. Then, for each $k \geq 0$, $\mathcal{C}_k$ induces a* surety graph of type X, *$GX_k = (V_k, EX_k)$, $X \in \{1, 2, 3, 4\}$, as follows:*[4]

$$V_k = \{v \mid declare_h(pid(v)) \in \mathcal{C}_k \text{ for some } h \in \mathcal{H}\}$$
$$EX_k = \{(v, v') \mid pledge_h v(suretyX(v')) \in \mathcal{C}_k, \; pledge_{h'} v'(suretyX(v)) \in \mathcal{C}_k, \; \text{for some } h, h' \in \mathcal{H}, v, v' \in V_k\}$$

Remark 1. Observe that mutual sureties can be easily pledged by agents, technically. However, we wish agents to be prudent and sincere in their mutual surety pledges. Thus, we expect a mechanism that, on one hand, rewards the pledging of sureties but, on the other hand, punishes for surety violations, for example based on the approach of [14]. While the specifics of such a mechanism is beyond the scope of the current paper, note that with such a mechanism in place, the commissive illocutionary force [13] of a surety pledge will come to bear.

2.1 Updating a Personal Identifier

Once creating a genuine personal identifier is provided for, one shall consider the many circumstances under which a person may wish to update their identifier:

1. **Identifier loss:** The private key was lost.
2. **Identifier theft:** The private key was stolen, robbed, extorted, or otherwise compromised.
3. **Identifier breach of privacy:** The association between the personal identifier and the person was accidentally or maliciously disclosed with unwarranted consequences.
4. **Identifier refresh:** Proactive identifier update to protect against the above.

Update in Case of Loss or Theft. The personal identifier declaration event $declare_h(pid(v))$ establishes v as a personal identifier. To support updating a personal identifier, we add the personal identifier update event $declare_h(pid(v, v'))$, which declares that v' is a new personal identifier that replaces v. A public declaration of identifier update has the form $v'^{-1}(pid(v, v'))$, i.e., it is signed with the new identifier. We refer to declarations of both types as *personal identifier declarations*, and extend the assumption that a new identifier can be declared at most once to this broader definition of identifier declaration. The *validity* of an identifier update declaration is defined inductively, as follows.

[4] We allow surety pledges to be made before the corresponding personal identifier declarations, as we do not see a reason to enforce order.

Definition 5 (Valid identifier Update declaration). *Let $\mathcal{C}$ be a sequence of declarations, V the set of personal identities declared in $\mathcal{C}$, and $h \in \mathcal{H}$. A* personal identifier update event *over V has the form $declare_h(pid(v, v'))$, $v, v' \in V$.*

A personal identifier update event $declare_h(pid(v, v')) \in \mathcal{C}$ is valid *and h is the* rightful owner *of v if it is the first identifier declaration event of v and h is the rightful owner of v'.* □

Valid personal identifier declarations should form linear chains, one per agent, each starting from $pid(v)$ and ending with the currently valid identifier:

Definition 6 (Identifier Provenance Chain). *Let $\mathcal{C}$ be a sequence of declarations and V the declared set of personal identities. A* provenance chain *is a subsequence of $\mathcal{C}$ of the form (starting from the bottom):*

$$\begin{aligned}&declare_{h_k}(pid(v_k, v_{k-1})),\\&declare_{h_{k-1}}(pid(v_{k-1}, v_{k-2})),\\&\ldots\\&declare_{h_1}(pid(v_1)).\end{aligned}$$

Such a provenance chain is valid *if the declarations in it are valid. Such a provenance chain is* maximal *if there is no declaration $declare_h(pid(v, v_k)) \in \mathcal{C}$ for any $v \in V$ and $h \in \mathcal{H}$. A personal identifier v is* current *in $\mathcal{C}$ if it is the last identifier $v = v_k$ in a maximal provenance chain in $\mathcal{C}$.* □

It is very easy for an agent to make an update declaration for its identifier. However, it is just as easy for an adversarial agent wishing to steal the identifier to make such a declaration. Hence, this ability must be coupled with a mechanism that protects the rightful owner of an identifier from identifier theft through invalid identifier update declarations. Here we propose to use a stronger type of mutual sureties to support valid identifier update declarations and help distinguish between them and invalid declarations.

Surety of Type 2: *Rightful* Ownership of a Personal Identifier. Agent h pledges that h' is the rightful owner of personal identifier v'.

In addition to proving to h that it owns v', h' must provide evidence that h' itself, and not some other agent, has declared v'. A selfie video of h' pressing the *declare* button with v', signed with a certified timestamp promptly after the video was taken, and then signed by v', may constitute such evidence. A suitable app may record, timestamp, and sign such a selfie video automatically during the creation of a genuine personal identifier. In particular, this surety is violated if h' in fact did not declared v' as a personal identifier.

Note that immediately following an identifier update declaration, the new identifier may not have any surety edges incident to it. Thus, as a crude measure, we may require that the identifier update would come to bear only after all the Type 2 surety neighbors of the old identifier, or a sufficiently large majority of them, would update their mutual sureties to be with the new identifier. To achieve that, an agent wishing to update its identifier would have to approach its neighbors and to create such updated Type 2 mutual surety pledges.

E.g., consider two friends, agent h and agent h' having a mutual surety pledge between them. If h' would lose her identifier, she would create a new key-pair, make an identifier update declaration, and ask h for a new mutual surety pledge between h's identifier and h''s new identifier. The next observation follows as: (1) a valid provenance chain has a single owner; and (2) whether a Type 2 surety between two identifiers is violated depends on their rightful owners.

Observation 1. *Let $\mathcal{C}$ be a sequence of update declarations and C_1, C_2 be two valid provenance chains in $\mathcal{C}$. If a Type 2 surety pledge between two personal identities $v_1 \in C_1, v_2 \in C_2$ is valid, then any Type 2 surety pledge between two personal identifiers in these provenance chains, $u_1 \in C_1, u_2 \in C_2$ is valid.*

The import of Observation 1 is that a Type 2 mutual surety can be "moved along" valid provenance chains as they grow, without being violated, as it should be. Below we argue that invalid identifier update declarations are quite easy to catch, thus the risk of stealing identities can be managed. In effect, in the full version we show the value of Type 2 surety pledges in defending an identifier against theft via invalid update declarations and consider privacy concerns.

3 Sybil- and Byzantine-Resilient Community Growth

Wishing for sybil-free communities, we acknowledge that one cannot prevent sybils from being declared and, furthermore, perfect detection and eradication of sybils is out of reach. Thus, our aim is to provide the foundation for a digital community of genuine personal identifiers to grow by admitting new identifiers indefinitely, while retaining a bounded sybil penetration. Indeed, democratic governance can be achieved even with bounded sybils penetration [15].

Community History. For simplicity, we assume a single community A and consider elementary transitions obtained by either adding a single member to the community or removing a single community member:

Definition 7 (Elementary Community Transition). *Let $A, A' \subseteq V$ denote two communities in V. We say that A' is obtained from A by an* elementary community transition, *and we denote it by $A \to A'$, if:*

- *$A' = A$, or*
- *$A' = A \cup \{v\}$ for some $v \in V \setminus A$, or*
- *$A' = A \setminus \{v\}$ for some $v \in A$.*

Definition 8 (Community History). *Let $\mathcal{C} = c_1, c_2, \ldots$ be a sequence of events. A* community history *wrt. $\mathcal{C}$ is a sequence of communities $\mathcal{A} = A_1, A_2, \ldots$ such that $A_i \subseteq V_i$ and $A_i \to A_{i+1}$ holds for every $i \geq 1$.*

We do not consider community governance (see [15]) but only the effects of community decisions to add or remove members, thus assume that $\mathcal{C}$ includes $add_A(v)$ and $remove_A(v)$. With this addition, $\mathcal{C} = c_1, c_2, \ldots$ induces a community history $\mathcal{A} = A_1, A_2, ...$, where $A_{i+1} = A_i \cup \{v\}$ if $c_i = add_A(v)$ for $v \in V \setminus A_i$; $A_{i+1} = A_i \setminus \{v\}$ if $c_i = remove_A(v)$ for $v \in A_i$; else $A_{i+1} = A_i$.

Definition 9 (Community, Sybil penetration rate). *Let $\mathcal{C}$ be a sequence of events and let $S \subseteq V$ denote the sybils in V wrt. $\mathcal{C}$. A community in V is a subset of identifiers $A \subseteq V$. The* sybil penetration $\sigma(A)$ *of the community A is given by $\sigma(A) = \frac{|A \cap S|}{|A|}$.*

The following observation is immediate.

Observation 2. *Let $\mathcal{A} = A_1, A_2, ...$ be the community history wrt. a sequence of declarations $\mathcal{C}$. Assume that $A_1 \subseteq H$, and that whenever $A_{i+1} = A_i \cup \{v\}$ for some $v \notin A_i$, it holds that $Pr(v \in S) \leq \sigma$ for some fixed $0 \leq \sigma \leq 1$. Then, the expected sybil penetration rate for every A_i is at most σ.*

I.e., a sybil-free community can keep its sybil penetration rate below σ, as long as the probability of admitting a sybil to it is at most σ. While the simplicity of Observation 2 might seem promising, its premise is naively optimistic. Due to the ease in which sybils can be created and to the benefits of owning sybils in a democratic community, the realistic scenario is of a hoard of sybils and a modest number of genuine personal identifiers hoping to join the community. Furthermore, once a fraction of sybils has already been admitted, it is reasonable to assume that all of them (together with their perpetrators of course) would support the admission of further sybils. Thus, there is no reason to assume neither the independence of candidates being sybils, nor a constant upper bound on the probability of sybil admission to the community. Hence, in the following we explore sybil-resilient community growth under more realistic assumptions.

Sybil-Resilient Community Growth. A far more conservative assumption includes a process employed by the community with the aim of detecting sybils. We shall use the abstract notion of *sybil detector* in order to capture such process, that may take the form of a query, a data-based comparison to other identifiers, or a personal investigation by some other agent. To leverage this detector to sybil-resilient community growth regardless of the sybil distribution among the candidates, we shall utilize a stronger surety type, defined as follows:

Surety of Type 3: Rightful Ownership of a *Genuine* Personal Identifier. Agent h pledges Surety Type 2 and that v' is the genuine personal identifier of h'.

Providing this surety requires a leap of faith: In addition to h obtaining from h' a proof of rightful ownership of v', h must also trust h' not to have declared any other personal identifier prior to declaring v'. There is no reasonable way for h' to prove this to h, hence the leap of faith. Since Type 3 sureties inherently aim to distinguish between genuine identifiers and sybils, sybil-resilient community growth is established upon the underlying $G3$ surety graph. Specifically, we consider a setting where potential candidates to join the community are identifiers with a surety obtained from current community members, so a violation of a surety in one direction as a strong indication that the surety in the other direction is violated. I.e., if $(v, v') \in E$ and v' was shown to be sybil, (i.e., h' has declared some other v'' as a personal identifier before declaring v' as a personal identifier), then v should undergo a thorough investigation in order to determine whether it is sybil as well.

We formalize this intuition in a stochastic process where admissions of new members are interleaved with random sybil detection among community members. We have a tuple (G, A_0, p), where $G = (V, E)$ is an undirected graph, and V is partitioned to genuine and sybil identifiers, $A_0 \subseteq V$ is an initial community, and $0 \leq p \leq 1$ is a probability parameter. The community evolves over time, where A_i indicates the community at time i. The stochastic process is as follows: (1) An identifier is admitted to the community via an elementary community transition $A_{i+1} \to A_i \cup \{v\}$ only if there is some $a \in A_i$ with $(a, v) \in E$; (2) Every admittance of a candidate is followed by a random sybil detection within the community: An identifier $a \in A_i$ is chosen uniformly at random. If a is genuine it is declared as such. If a is sybil, it is successfully detected with probability $0 < p \leq 1$; (3) The detection of a sybil implies the successful detection of its entire connected sybil component (with probability 1). That is, if a is detected as sybil, then the entire connected component of a in the sybil subgraph $G|_S$ is detected and expelled from the community; (4) The sybils are operated from at most k disjoint sybil components in A. Furthermore, we assume that sybils join sybil components uniformly at random, i.e., a new sybil member has a surety to a given sybil component with probability $\frac{1}{k}$, else, it forms a new sybil component.

The main threat in this setting is a hostile takeover of the community by sybils, manifested in this model by the sybil penetration ratio in the steady state of the above stochastic process. In contrast to the probabilistic means employed by the community for self-defense, the sybil operator may act arbitrarily, possibly based upon prior knowledge of the underlying graph and the detection parameter p. Our first result establishes an upper bound on the expected sybil penetration rate, assuming bounded computational resources of the attacker.

Theorem 1. *In the stochastic model described above, obtaining an expected sybil penetration* $E[\sigma(A_i)] \geq \epsilon$ *is NP-hard for every* $\epsilon > 0$.

We establish an upper bound on the sybil penetration rate regardless of the attacker's computational power. The result is formulated in terms of the second eigenvalue of the graph restricted to A_i. ($\lambda(G|_{A_i})$ is defined in [16].

Corollary 1. *Let* $\mathcal{A} = A_1, A_2, \ldots$ *be a community history wrt. a sequence of events* $\mathcal{C}$. *If every community* $A_i \in \mathcal{A}$ *with* $A_i = A_{i-1} \uplus \{v\}$ *satisfies* $\lambda(G|_{A_i}) < \lambda$, *then the expected sybil penetration in every* $A_i \in \mathcal{A}$ *under the stochastic model depicted above, is at most* $\sqrt{\lambda/p}$.

Byzantine-Resilient Community Growth. We consider the challenge of byzantine-resilient community growth. Intuitively, the term *byzantines* aims to

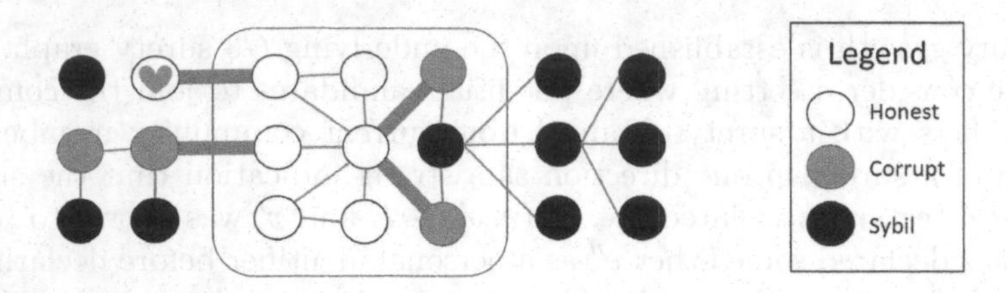

Fig. 2. A sybil-resilient community growth illustration: The white vertices (honest identities) **and** grey vertices (corrupt identities) form the set of genuine identities, while grey vertices (corrupt identities) **and** black vertices (sybil identities) form the set of Byzantines.

capture identifiers owned by agents that are acting maliciously, possibly in collaboration with other malicious agents. Formally, we define byzantines as follows (Fig. 2).

Definition 10. *An identifier is* byzantine *if it is either a sybil or the personal identifier of a corrupt agent. Non-byzantine identifiers are referred to as* harmless. *We denote the byzantine and harmless identifiers in V by $B, H \subseteq V$, respectively. The* byzantine penetration $\beta(A)$ *of a community $A \subseteq V$ is $\beta(A) = \frac{|A \cap B|}{|A|}$.*

Since $S \subseteq B$, it holds that $\sigma(A) \leq \beta(A)$ for every community A, hence an upper bound on the byzantine penetration also provides an upper bound on the sybil penetration. As byzantine identifiers include genuine identifiers, they are inherently harder to detect, and thus the detection-based model described in Sect. 3 is no longer applicable in this setting. Rather, to achieve byzantine-resilient community growth, we rely on a stronger surety type, defined as follows:

Surety of Type 4: Rightful Ownership of a Genuine Personal Identifier by an *Honest* Agent. Agent h pledges Surety Type 3 and, furthermore, that v' is a genuine personal identifier of an honest agent h'.

Here h has to put even greater trust in h': Not only does h has to trust that the past actions of h' resulted in v' being her genuine personal identifier, but she also has to take on faith that h' has not declared any sybils since and, furthermore, that h' will not do so in the future. Note that a Type 4 surety is violated if after h' declares v' it ever declares some other v'' as a personal identifier. See Fig. 3 for illustrations of violations of sureties of Types 3 and 4.

In the following, we provide sufficient conditions for Type 4 sureties to be used for byzantine-resilient community growth. We utilize the notation $A \twoheadrightarrow A'$ to indicate that A' was obtained from A via a finite sequence of elementary community transitions of incremental growth. Formally, $A \twoheadrightarrow A'$ if there exists $k \in \mathbb{N}$ with $A = A_0 \rightarrow A_1 \rightarrow ...A_k = A'$, with $|A_i \setminus A_{i-1}| = 1$ for all $0 \leq i \leq k$.

Theorem 2. *Let $\mathcal{A} = A_1 \twoheadrightarrow A_2 \twoheadrightarrow A_3 \twoheadrightarrow \ldots$ be a community history wrt. a sequence of events $\mathcal{C}$. Set a sequence of degrees $d_1, d_2, \ldots$ and parameters $\alpha, \beta, \gamma, \delta \in [0, 1]$. Assume: (1) The degree of all vertices $v \in A_i$ satisfy*

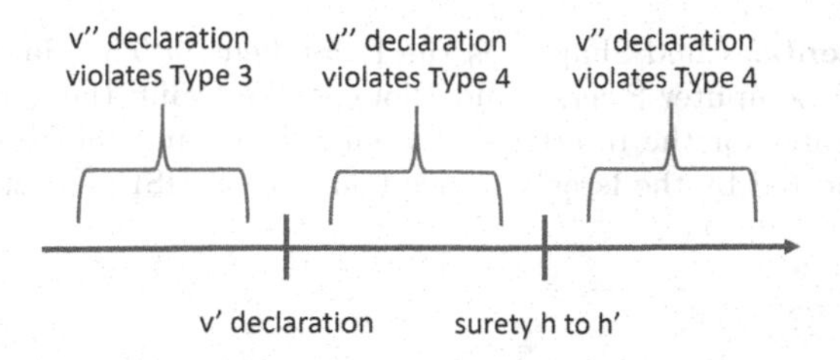

Fig. 3. Violations of Sureties of Type 3 and Type 4. Time goes from left to right, with the points in time in which v' was declared by h' and the surety from h to h' was declared. Then, the braces describe the time regions in which, if another identifier v'' was to be declared by h', would correspond to a violation of a surety of Type 3 or 4.

$deg(v) \leq d_i$ for all $i \in \mathbb{N}$; (2) Every $a \in A_i$ satisfies $\frac{|\{x \in A_i \mid (a,x) \in E\}|}{d_i} \geq \alpha$; (3) $\frac{|A_1 \cap B|}{|A_1|} \leq \beta$; (4) $\frac{e(A_i \cap H, A_i \cap B)}{vol_{A_i}(A_i \cap H)} \leq \gamma$; (5) $|A_i \setminus A_{i-1}| \leq \delta |A_{i-1}|$, with $\beta + \delta \leq \frac{1}{2}$; (6) $\Phi(G|_{A_i}) > \frac{\gamma}{\alpha} \cdot \left(\frac{1-\beta}{\beta}\right)$. Then, every community $A_i \in \mathcal{A}$ has Byzantine penetration $\beta(A_i) \leq \beta$.

Roughly speaking, Theorem 2 suggests that whenever: (1) Each graph $G|_{A_i}$ has a bounded degree d_i; (2) Sufficiently many edges are within A_i; (3) Byzantine penetration to A_1 is bounded; (4) Edges between harmless and byzantine identifiers are scarce; (5) Community growth in each step is bounded; (6) The conductance within $G|_{A_i}$ is sufficiently high; Then, the community may grow indefinitely with bounded byzantine penetration. Intuitively, it follows from the inability to add byzantines while maintaining sufficiently large conductance (Assumption 6), while maintaining a scarce cut between honest and byzantine identifiers (Assumption 4).

4 Outlook

Digital identity systems face the "Decentralized Identity Trilemma", of having to be (1) privacy-preserving, (2) sybil-resilient, and (3) self-sovereign, all at the same time [8]. It has been claimed that no existing identity system satisfies all three corners of the Trilemma [8]; our approach may be the first to do so.

While this paper provides a formal mathematical framework, we aimed the constructions to be readily amenable to implementation. Realizing the proposed solution entails developing additional components, notably sybil-resilient governance mechanisms, e.g. along the lines of [15]; a mechanism for encouraging honest behavior and discouraging corrupt behavior, e.g. along the lines of [14]; and a cryptocurrency to fuel such a mechanism and the system in general. Once all components have been designed, we aim to implement, simulate, test, deploy, and evaluate the proposed framework, hopefully realizing the potential of genuine personal identifiers.

Acknowledgements. Ehud Shapiro is the Incumbent of The Harry Weinrebe Professorial Chair of Computer Science and Biology. We thank the generous support of the Braginsky Center for the Interface between Science and the Humanities. Nimrod Talmon was supported by the Israel Science Foundation (ISF; Grant No. 630/19).

References

1. Abdul-Rahman, A.: The PGP trust model. EDI-Forum: J. Electron. Commer. **10**(3), 27–31 (1997)
2. Baird, R., et al.: Human tide: the real migration crisis (2007)
3. World Bank. Identification for development (ID4D) global dataset (2018)
4. Cardelli, L., Shahaf, G., Shapiro, E., Talmon, N.: Digital social contracts: a foundation for an egalitarian and just digital society (2020)
5. Fung, B., Garcia, A.: Facebook has shut down 5.4 billion fake accounts this year. CNN, November 2019
6. Geisler, C., Currens, B.: Impediments to inland resettlement under conditions of accelerated sea level rise. Land Use Policy **66**, 322–330 (2017)
7. Jacobovitz, O.: Blockchain for identity management. The Lynne and William Frankel Center for Computer Science Department of Computer Science. Ben-Gurion University, Beer Sheva Google Scholar, vol. 1, p. 9 (2016)
8. Laskus, M.: Decentralized identity trilemma (2018). http://maciek.blog/dit/?cookie-state-change=1574327093444
9. McCarthy, N.: Facebook deleted more than 2 billion fake accounts in the first quarter of the year. Forbes, May 2019
10. Mühle, A., Grüner, A., Gayvoronskaya, T., Meinel, C.: A survey on essential components of a self-sovereign identity. Comput. Sci. Rev. **30**, 80–86 (2018)
11. Reed, D., Sporny, M., Longley, D., Allen, C., Grant, R., Sabadello, M.: Decentralized identifiers (DIDs) v0. 12-data model and syntaxes for decentralized identifiers (DIDs). Draft Community Group Report, 29 (2018)
12. Rivest, R.L., Shamir, A., Adleman, L.: A method for obtaining digital signatures and public-key cryptosystems. Commun. ACM **21**(2), 120–126 (1978)
13. Searle, J.R., Willis, S., Vanderveken, D.: Foundations of Illocutionary Logic. CUP Archive (1985)
14. Seuken, S., Parkes, D.C.: Sybil-proof accounting mechanisms with transitive trust. In: Proceedings of the 2014 International Conference on Autonomous Agents and Multi-agent Systems, pp. 205–212. International Foundation for Autonomous Agents and Multiagent Systems (2014)
15. Shahaf, G., Shapiro, E., Talmon, N.: Sybil-resilient reality-aware social choice. In: IJCAI 2019, pp. 572–579 (2019)
16. Shahaf, G., Shapiro, E., Talmon, N.: Foundation for genuine global identities. arXiv preprint arXiv:1904.09630 (2019)
17. Shapiro, E.: Point: foundations of e-democracy. Commun. ACM **61**(8), 31–34 (2018)
18. W3C. Verifiable credentials data model 1.0 (2019)
19. Yasin, A., Liu, L.: An online identity and smart contract management system. In: 2016 IEEE 40th Annual Computer Software and Applications Conference (COMPSAC), vol. 2, pp. 192–198. IEEE (2016)

Political Framing: US COVID19 Blame Game

Chereen Shurafa[1(✉)], Kareem Darwish[2], and Wajdi Zaghouani[1]

[1] College of Humanities and Social Sciences, HBKU, Doha, Qatar
chereenalshurafa@gmail.com, wzaghouani@hbku.edu.qa
[2] Qatar Computing Research Institute, HBKU, Doha, Qatar
kdarwish@hbku.edu.qa

Abstract. Through the use of Twitter, framing has become a prominent presidential campaign tool for politically active users. Framing is used to influence thoughts by evoking a particular perspective on an event. In this paper, we show that the COVID19 pandemic rather than being viewed as a public health issue, political rhetoric surrounding it is mostly shaped through a blame frame (blame Trump, China, or conspiracies) and a support frame (support candidates) backing the agenda of Republican and Democratic users in the lead up to the 2020 presidential campaign. We elucidate the divergences between supporters of both parties on Twitter via the use of frames. Additionally, we show how framing is used to positively or negatively reinforce users' thoughts. We look at how Twitter can efficiently be used to identify frames for topics through a reproducible pipeline.

Keywords: Framing theory · COVID19 · Social media analysis

1 Introduction

The US has been the hardest hit country by the COVID19 pandemic with 100,000 deaths as of May 26, 2020. Though the pandemic is a public health problem, the response to the pandemic among politically active US Twitter users took a political turn due to diverging views between Republicans and Democrats. Both sides turned to actively placing blame on different entities for the spread of COVID19 through the use of carefully crafted phrases. For the Democrats, the effect is manifested through blaming Trump for his inability to act quickly and protect the nation from the rapidly spreading virus. Conversely, the Republicans rallied together to protect Trump and consistently shifted the blame to deep state conspiracies and to China for being the source of the pandemic. In this paper, we examine the divergence between supporters of the Republican (GOP) and Democratic (DNC) parties[1] in the context of framing theory, where frames shape how individuals construct their attitudes towards an issue [10]. Using

This paper was accepted at SocInfo2020. Please cite the SocInfo version.

[1] GOP: *Grand Old Party*; DNC: *Democratic National Committee*.

S. Aref et al. (Eds.): SocInfo 2020, LNCS 12467, pp. 333–351, 2020.
https://doi.org/10.1007/978-3-030-60975-7_25

an expectancy value model, an individual's overall attitude towards an issue is determined using a weighted average of different factors of an issue, where the weights correspond to the relative importance of these factors to the individual [30]. For the COVID19 pandemic, different factors may include perceived political responsibility, the state of the economy, the accessibility of health care, and effect of the pandemic on elections. In effect, a person's attitude is framed by the perceived importance of such factors. Framing works at three levels, namely the existing of factors in memory, meaning if the person is aware of the factor, whether this factor is mentally accessible when forming an attitude, and the perceived relevance of the factor to the situation [10]. We look at how framing is affecting the Twitter political discourse associated with the COVID19 pandemic. We base our analysis on the timeline tweets of 30 thousand Twitter users who discussed the pandemic in conjunction with politically suggestive words. We utilize unsupervised stance detection to automatically and accurately distinguish between supporters of the Democratic and Republican parties. We establish the salient background topics these users have discussed in the period between Jan. 1, 2020 and April 12, 2020, which includes the period leading to the pandemic and the period after the World Health Organization (WHO) declared COVID19 as a global pandemic. We contrast politically active users against a set of 20 thousand random US users. The contributions of this paper are as follows:

- We illustrate how framing can be used to shape the public discourse. Specifically, we show that supporters of the two main political parties in the US are framing the COVID19 pandemic as a political issue.
- We utilize multiple analysis tools ranging from stance detection to detect political leanings of users to conducting an analysis based hashtags, URLs, retweeted accounts, and rhetorical devices.
- We show that our analysis of different discourse elements can help identify underlying frames.

2 Related Work

Related research on political discourse framing is well established in the social science community as described in [10,16] and also by automatically analyzing political discourse and news articles using natural language processing techniques [4,8]. Other computational works focused on analyzing social media political discourse using opinion mining [1] and stance detection [15,22,24,35,38]. Stance detection is the task of identifying the political leaning or the position of a user towards a specific topic or entity. In the context of Twitter, User stances can be gleaned using a myriad of signals including content features (e.g. words and hashtags), profile information (e.g. profile description), interactions (e.g. retweets, likes, and mentions) [2]. There are multiple methods for performing stance detection including supervised, semi-supervised, and unsupervised methods [6,12,13]. In this work, we employ unsupervised stance detection due in most part to its ease and high accuracy [13].

Framing in public discourse such as in political speeches and news articles have been explored by [4,8,17]. While [9,19,33] discussed framing in the context of language bias and subjectivity [39]. On the other hand, [37] analyzed the effect of lexical choices in Twitter and how they can affect the propagation of the tweets. [3,31] studied sentiment analysis of the political discourse and [14,34] created a framework to measure and predict ideologies. [5,18] analyzed the voting patterns and polls based on political sentiment analysis. Furthermore, several social sciences researchers have discussed framing based on Twitter data and how it has been used to impact the public opinion [7,21,27]. For instance, [20] discussed the framing of Public sentiment in social media content during the 2012 U.S. presidential campaign, and [32] presented his research on social media and framing discourse in the digital public sphere based on riots and digital activism. Finally, [26] presented a model based on the tweet's linguistic features and several ideological phrase indicators to predict the general frame of political tweets.

3 Data Collection

Obtaining Tweets. Our target was to assemble a dataset composed of Twitter users who discuss COVID19. Specifically, we were interested in users who politically lean towards the Republican (GOP) or the Democratic (DNC) parties and, for reference, a random sample of US users. We collected tweets from the period of March 5 – 31, 2020 using the Twitter streaming API. We filtered the tweets by language, to retain English tweets, and by the following hashtags and keywords: #covid19, #CoronavirusOutbreak, #Coronavirus, #Corona, #CoronaAlert, #CoronaOutbreak, Corona, and COVID19. To filter by language, we used the language tag provided by Twitter for each tweet. In all, we obtained a set of 31.64 million tweets, which we henceforth refer to as the *base* dataset.

To identify users with interest in US politics, we further filtered tweets containing any of the following strings (or potentially sub-strings): Republican, Democrat, Trump, Biden, GOP, DNC, Sanders, and Bernie, which resulted in 2.48 million tweets. We sorted in descending order users by the number of tweets that mention COVID19 related words and politically indicative words and retained the top 30k users. These users had tweets that ranged in number between 1 and 1,302 tweets (average: 2.78 tweets and standard deviation: 6.72). We proceeded to crawl the timeline tweets for these 30k users. Twitter APIs allow the scrapping of the last 3,250 timeline tweets for a user. From timeline tweets, we retained tweets with dates ranging between January 1 and April 12, 2020, which were 92.02 million tweets. Of those, 18.55 million tweets contained the substrings *corona* or *covid*, and the users had tweets ranging in number between 1 and 3,250 tweets (average: 677.52 tweets and standard deviation: 292.88). We shall refer to this dataset as the *politicised* dataset.

To sample US users in general, we filtered all users with tweets in the *base* dataset by their locations to obtain US users. We deemed a user to be from the US if their specified location (from their Twitter profile) contained the tokens

"United States", "America", "USA", or the names of any state or its abbreviation (e.g. "Maryland" or "MD"). For USA and state abbreviations, they had to be capitalized. From the 117,408 matching users, we randomly sampled 20k users and obtained their timeline tweets. In the period of interest (Jan. 1 – April 12, 2020), these users had 14.12 million tweets, of which 773k tweets contained the substrings *corona* or *covid*. We refer to this as the *sampled* dataset.

One obvious difference between the *politicised* and the *sampled* datasets is that the users in the politicised dataset produced more than 4.3 times more tweets in general and 16 times more COVID19 related tweets. The difference may be an artifact of identifying the 30k most politically active users in the *base* dataset and could be the result of employing COVID19 for political messaging. However, this requires further investigation. Nonetheless the disparity implies that users in the *politicised* dataset seem to be actively spreading their message in light of their frames.

User Stance Detection. Given the *politicized* dataset filtered on COVID19 related terms, we wanted to determine which users politically lean towards the DNC or the GOP. To do so, we employed an unsupervised stance detection method, which attempts to discriminate between users based on the accounts that they retweet [13]. We elected to use this method because it was shown to produce nearly perfect user clusters. The method represents each user using a vector of the accounts that they retweeted. Then it computes the cosine similarity between users and projects them onto a two dimensional space using UMAP, which places the users in a manner where similar users are closer together and less similar users are further apart. Next, projected users are clustered using mean shift clustering. Of the 30k users, stance detection was able to assign 25,753 users to two main clusters and was unable to cluster the remaining users. By inspecting the two clusters, the first cluster of size 17,689 users was clearly composed of DNC leaning users and the second cluster of 8,064 users, who were

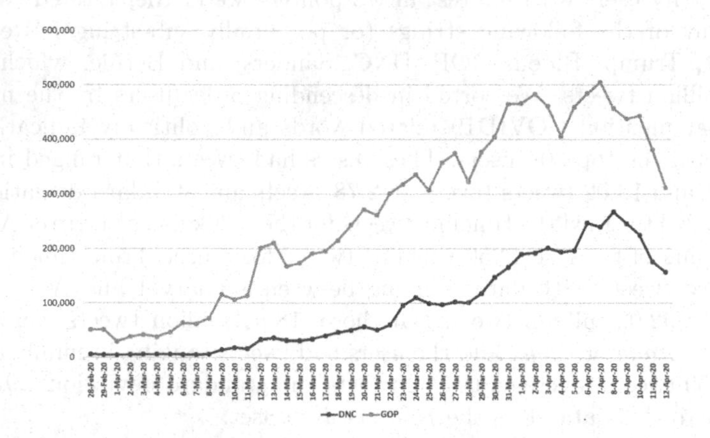

Fig. 1. The number of daily tweets of DNC and GOP learning users.

Table 1. Categorization of the 100 most frequent hashtags (excluding COVID19 hashtags) by pro-GOP users.

Category	Count	%	Examples
Conspiracy	231,132	17.8	#FilmYourHospital, #BillGates, #Qanon (deep state conspiracy)
Blame China	216,549	16.7	#ChineseVirus, #WuhanCoronaVirus
Trump support	135,602	10.4	#MAGA (Make America Great Again), #Trump2020, #KAG2020 (Keep America Great)
Anti liberal media	131,705	10.1	#FakeNews, #EnemyOfThePeople
Anti DNC	117,760	9.1	#DemocratsHateAmerica
Pro GOP	88,395	6.8	#walkaway, #TCOT (Top Conservatives on Twitter)
News	60,620	4.7	#BREAKING
Praising Trump's actions	60,378	4.7	#PaycheckProtectionProgram, #AmericaWorksTogether
COVID19 cure	56,420	4.3	#Hydroxychloroquine
Specific issues	51,629	4.0	#Iran, #FISA
Social solidarity	38,820	3.0	#InItTogether, #SocialDistancing
Holiday related	33,472	2.6	#Easter, #GoodFriday
Conservative media	29,144	2.2	#FoxNews
Voting conspiracy	19,401	1.5	#VoterId, #VoterFraud
Patriortic sentiment	19,374	1.5	#America

clearly leaning towards the GOP. Since less than 1% of the tweets were authored before Feb. 28, we restrict our analysis to the period from Feb. 28 to April 12, 2020. Figure 1 plots the number of daily tweets for DNC and GOP supporters.

4 Frames

Identifying Frames. We wanted to determine the frames for both the *sampled* and *politicized* dataset. For the *politicised* dataset, we analyzed the top 100 most frequently used hashtags for pro-DNC and pro-GOP users. Tables 1 and 2 categorize the 100 most frequent hashtags (excluding COVID19 hashtags) by pro-GOP and pro-DNC users respectively. In looking at these hashtags, we can better understand the frames of both groups. For pro-GOP users (Table 1), we can see that the key factors in framing are: COVID19 is a conspiracy, blaming China, Trump reelection, liberal media credibility, Trump's effectiveness, party loyalty, social solidarity, and patriotism. For the pro-DNC users (Table 2), the factors are: defeating Trump (anti-Trump/pro-DNC), Trump's ineptness, and social solidarity. As is apparent from these factors, there are 3 overarching frames for both groups, namely: 1. Assignment of blame: pro-GOP users blame conspiracies, China, and left-leaning media, while pro-DNC users blame Trump and the GOP; 2. Support for party candidate(s): pro-GOP users support Trump and GOP candidates, while pro-DNC users support DNC candidates; and 3. Social messaging concerning the COVID19 lockdown.

Pro-GOP users had an additional frame that "everything is OK" (e.g. possible cure, holiday greetings), implicitly supporting that status quo. Assignment of blame and support for party candidate(s) are associated with the Nov. 2020 election, which is both mentally accessible and relevant. They account for 77% and 87% of hashtag volume for pro-GOP and pro-DNC users respectively.

For the *sampled* dataset, we analyzed the top 200 most frequently hashtags, excluding COVID19 related hashtags. As shown in Table 3, the main themes were: entertainment (e.g. games, music, sports), following the news, politics (e.g. supporting/opposing Trump/GOP/DNC), science and technology (e.g. big data, artificial intelligence, machine learning), motivational, conspiracy theories (e.g. QAnon), and other issues. Unlike the *politicized* dataset, the volume of election related themes was significantly lower (27% of the total). This implies that politically active users are not necessarily representative of the entire US Twitter user population. However, their influence is visible in the general population.

Analyzing Frames. To further analyze the content of the tweets, we utilized multiple methods. The first method involved identifying the most distinguishing features for both camps. To do so, we employed a variation of the so-called valence score [11], which attempts to determine the distinctiveness of a given

Table 2. Categorization of the 100 most frequent hashtags (excluding COVID19 hashtags) by pro-DNC users.

Category	Count	%	Examples
Anti Trump	1,105,341	56.6	#TrumpVirus, #FireTrump
Pro DNC	214,975	11.0	Biden2020, VoteBlue2020
Blame Trump Administration	162,253	8.3	#PPE (Personal Protective Equipment), WhereAreTheTests
Anti GOP	144,175	7.4	#MoscowMitch, #GOPGenocide
Social solidarity	126,468	6.5	#StayHome, #FamiliesFirst
News	77,358	4.0	#BREAKING
Voting issues	43,550	2.2	#VoteByMail, #WisconsinPrimary
Conservative media	32,619	1.7	#BloodOnHisHandsHannity
Liberal media	32,233	1.7	#Madow, #MSNBC
Holocaust	14,285	0.7	#Auschwitz

Table 3. Categorization of the 200 most frequent hashtags (excluding COVID19 hashtags) by *sampled* group.

Category	Count	%	Examples
Entertainment	128,803	33.3	#AnimalCrossing, #BTS, #IHeartAwards, #NFLDraft, #Oscars
News	30,952	8.0	#Breaking, #News
Support trump/GOP	23,386	6.1	#MAGA, #Trump2020, #KAG
Social solidarity	21,464	5.6	#StayHome, #SocialDistancing
Science/Technology	21,138	5.5	#AI, #BigData, #ML
Business	19,837	5.1	#Business, #Marketing
Democrats	16,656	4.3	#DemDebate, #Biden
Attack trump/GOP	13,467	3.5	#TrumpVirus, #Resist
Support DNC	13,203	3.4	#YangGang, #Bernie2020
Inspiration	12,770	3.3	#Success, #Wisdom, #Leadership
Conspiracy theories	11,964	3.1	#Qanon, #wwg1wga
Republican	11,896	3.1	#TCOT, #GOP
Issues	10,258	2.7	#EarthDay, #BlackHistoryMonth
Generic	9,233	2.4	#Love, #Win, #Truth
Political	8,501	2.2	#Impeachment, #2020Census
Health care	7,090	1.8	#Health, #MedicareForAll, #MentalHealth
China	6,734	1.7	#China, #Wuhan, #CCP
Foreign issues	6,409	1.7	#WWIII, #Iran
Blame media	3,410	0.9	#FakeNews
Attack DNC	3,192	0.8	#MeToo (accusing Biden of harassment)

token, such as a hashtag or retweeted account, for a particular group. Valence for term t in tweets of group G_g is computed as follows:

$$V(t, G_g) = 2\frac{\frac{N(t,G_g)}{N(G_g)}}{\frac{N(t,G_g)}{N(G_g)} + \frac{N(t,G_{\neg g})}{N(G_{\neg g})}} - 1 \quad (1)$$

Where $N(t, G_g)$ is the frequency of term t in the tweets of group G_g (e.g. GOP tweets) and $N(t, G_{\neg g})$ is the frequency of the term in the other group (e.g. DNC), and $N(G_g)$ is the total number of all terms in group G_g. Valence values are bounded between -1 and 1, where -1 and 1 imply extreme disassociation or association respectively. We split the range into five equal intervals, and we computed the valence scores for all hashtags and retweeted accounts for the *politicized* dataset. Tables 4, 5, and 6 list the most polarizing hashtags, retweets, and URLs,

Table 4. Top hashtags from GOP and DNC supporters from *politicized* dataset

GOP Supporters		DNC Supporters	
Hashtag	Freq.	Hashtag	Freq.
China	29,993	TrumpVirus	37,229
WuhanCoronaVirus	20,151	TrumpLiesAboutCoronavirus	26,776
OANN	9,703	TrumpGenocide	20,502
QAnon	9,309	SmartNews	18,747
FakeNews	9,597	TrumpLiesAmericansDie	18,336
CCPVirus	6,643	TrumpLiesPeopleDie	17,185
ChineseVirus	7,914	TrumpOwnsEveryDeath	16,980
Democrats	7,748	TrumpPandemic	15,741
ChinaVirus	5,844	GOP	14,242
Trump2020	6,420	TrumpVirusCoverup	14,782
Hydroxycloroquine	5,582	StopAiringTrump	14,603
WuhanVirus	5,427	TrumpLiedPeopleDied	14,260
KAG2020	5,913	TrumpIsTheWORSTPresidentEVER	13,202
USNSMercy	4,853	Resist	12,162
WWG1WGA	4,591	FamiliesFirst	12,481
USNSComfort	4,596	TrumpPlague	11,622
WHO	5,912	MoscowMitch	10,133
NEW	4,600	PPE	8,245
KAG	6,133	WhereAreTheTests	8,099
Iran	5,296	25thAmendmentNow	7,749

where the absolute value of the valence score is ≥ 0.6, and they are ranked by the product the valence score and the log of the term frequency [36]. The most distinguishing hashtags seem to reinforce the aforementioned frames. On the GOP side, 6 out of 20 hashtags blame China (e.g. #ChinaVirus), 4 promote conspiracy theories (e.g. #QAnon), and 5 support Trump and his actions (e.g. #KAG2020, #USNSMercy). On the DNC side, 14 of the 20 hashtags in Table 4 are targeted directly at Trump (e.g. #TrumpVirus, #Resist, #25thAmendment-Now), 2 more hashtags implicitly attack Trump (#PPE – personal protective equipment, #WhereAreTheTests), and yet 2 more attack the GOP (#GOP, #MoscowMitch). The list of most prominent retweeted accounts is dominated by media, media personalities, politicians, and activists with conservative and liberal leanings respectively. Though all the tweets in the *politicized* dataset mention COVID19, the most distinguishing hashtags and retweeted accounts reinforce political framing. As for the most distinctive URLs, pro-GOP users cited multiple government websites (e.g. White House), articles attacking democrats,

Table 5. Top retweeted accounts by pro-GOP and pro-DNC users in *politicized* dataset

GOP Supporters			DNC Supporters		
Account	Count	Description	Account	Count	Description
Mitchellvii	100,680	Pro-Trump host	Funder	201,485	Scott Dworkin (liberal host)
TomFitton	79,787	President of conservative activist group	DrDenaGrayson	127,047	Medical doctor
JsolomonReports	73,168	Conservative pundit	tedlieu	126,776	Democratic congressman
RealCandaceO	64,828	conservative author	kylegriffin1	115,110	MSNBC producer
WhiteHouse	88,982	Official White House account	Yamiche	113,313	MSNBC contributor
RealJamesWoods	57,697	Actor/Trump supporter	JoeBiden	156,903	Biden's official account
TrumpWarRoom	57,258	pro-Trump account	MSNBC	116,555	Liberal media
IngrahamAngle	57,642	FoxNews host	TeaPainUSA	79,974	Liberal account
marklevinshow	40,487	Conservative radio host	joncoopertweets	75,994	Democratic politician
gatewaypundit	37,482	Conservative newsletter	SethAbramson	72,541	Newsweek columnist
BreitbartNews	37,504	Conservative media	JoyAnnReid	73,680	MSNBC corespondent
DonaldJTrumpJr	46,172	Trump's son	maddow	79,309	MSNBC host
SaraCarterDC	33,207	FoxNews host	atrupar	68,202	Vox journalist
catturd2	30,913	pro-Trump account	realTuckFrumper	57,377	liberal newsletter
charliekirk11	32,510	president of conservative activist group	mmpadellan	54,389	liberal blogger
seanhannity	36,724	FoxNews host	CREWcrew	53,963	liberal thinktank
DailyCaller	25,781	FoxNews host	MollyJongFast	52,534	Daily Beast editor
FLOTUS	30,819	First Lady's official account	washingtonpost	67,127	liberal media
TeamTrump	26,864	Official Trump campaign account	Amy_Siskind	49,278	liberal activist
JackPosobiec	24,586	alt-right activist/conspiracy theorist	eugenegu	49,639	medical doctor

China, and World Health Organization, and articles promoting the efficacy of hydroxychloroquine. Pro-DNC users conversely attacked Trump and the GOP.

The second method involved using DocuScope [23], which is a text visualization and analysis environment specifically designed to carry out rhetorical research with language and text. The DocuScope dictionary was developed based on David and Brian Butler's theoretical work on rhetoric [28] and their applied work in representational theories of language [29]. DocuScope is based on more than 60 million English linguistic patterns that map textual segments to fine-grained rhetorical effects which allows analysts to engage in deep cultural interpretation and extract sociocultural trends from text. We analyzed the tweets from both camps using DocuScope, and we used the output in two ways. First, we looked at the tweets that mention specific keywords, namely: China, Trump, Republican, Democrat, media, and conspiracy, which are strongly associated with the aforementioned frames. We identified the rhetorical devices and

Table 6. Top cited URLs from pro-GOP and pro-DNC users in *politicized* dataset.

Count	URL
GOP Supporters	
6,570	http://CoronaVirus.gov
2,803	https://nypost.com/2020/04/09/senate-dems-block-250-billion-for-coronavirus-small-business-loans/
2,606	http://WWW.GEORGE.NEWS (conservative news site)
2,511	https://www.wsj.com/articles/world-health-coronavirus-disinformation-11586122093
2,482	https://www.whitehouse.gov/briefings-statements/coronavirus-guidelines-america/
2,456	http://45.wh.gov/RtVRmD (White House Press Conference)
2,055	https://www.freep.com/story/news/local/michigan/detroit/2020/04/06/democrat-karen-whitsett-coronavirus-hydroxychloroquine-trump/2955430001/
2,048	https://www.foxnews.com/politics/after-mocking-trump-promoting-hydroxychloroquine-media-acknowleges-might-treat-coronavirus
2,027	https://www.usatoday.com/story/opinion/2020/04/07/time-put-china-lockdown-dishonesty-amid-coronavirus-pandemic-crisis-column/2954433001/
2,000	https://thegreggjarrett.com/coronavirus-crisis-americans-fear-the-lockdown-more-than-the-virus/
DNC Supporters	
5,252	https://www.latimes.com/politics/story/2020-04-07/hospitals-washington-seize-coronavirus-supplies
4,370	https://secure.actblue.com/donate/coronavirus-liar-video
4,062	https://www.nytimes.com/2020/04/11/us/politics/coronavirus-red-dawn-emails-trump.html
3,600	https://www.queerty.com/fox-news-officially-sued-peddling-coronavirus-misinformation-20200406
3,485	https://www.nytimes.com/interactive/2020/03/25/opinion/coronavirus-trump-reopen-america.html
3,392	https://www.npr.org/sections/coronavirus-live-updates/2020/04/08/829955099/federal-support-for-coronavirus-testing-sites-end-as-peak-nears
3,340	https://www.nytimes.com/2020/04/02/opinion/jared-kushner-coronavirus.html
3,417	https://www.thedailybeast.com/sen-kelly-loeffler-dumped-millions-in-stock-after-coronavirus-briefing
2,978	https://www.politico.com/news/2020/03/31/trump-obamacare-coronavirus-157788
3,201	https://www.state.gov/the-united-states-announces-assistance-to-combat-the-novel-coronavirus/

Table 7. Rhetorical devices used in conjunction with: China, Trump, Republicans, Democrats, Media, and Conspiracies

GOP Supporters		DNC Supporters	
China			
Negative/Bogus, Phony	distracted from China by **phony** impeachment	Narrative/Forgiveness	Trump should ask **forgiveness** for incompetence
Inquiry/Curiosity, Intrigue	**intriguing** if China had vaccine	Investigate/Detect	US gov failed despite **detecting** virus early
Description/Objects Gate	**Gates** wants to vaccine and track people	Negative/Incompetent	complaint that Trump is an **incompetent idiot**
Inquiry/Investigate	Complaint Congress has **been investigating** Trump	Strategic/Goals, Aim	Trump ended funds **aimed at** early warning
Reasoning/Ethics, Human	China lacks **human rights**	Negative/Trickery	Trump calling virus a **hoax**
Trump			
Information/Left	**the left** mad over Jared Kushner	Negative/Incompetence	Trump **incompetence**
Political Ideology/Democrat	**Democratic state** Rep says that Hydroxychloroquine saved her life	Negative/Firing	Trump **fired** the pandemic response team
Negative/Restrict	Trump **restricting** travel to China	Negative/Lapse, Failure	**failure of** Trump over COVID19
Negative/Trickery, Sham	calling the impeachment a **sham**	Negative/Bungle	Trump **bungled** testing
Political Ideology/Left wing	attacking **left-wing** media smear	Narrative/Turn Break	**major breaking**: Trumped blocked testing
Republicans			
Description/Drug	**hydroxychloroquine** is effective	Positive/Join our	**join our** fight against Republicans
Strategic/Goals, Target	COVID19 **targets** elderly (mostly republicans)	Interactive/Request help	**help** defeat Republicans
Negative/Burdens	Republicans calling models **used for projection** conflicted and cause undue **burden** on Americans	Negative/Suppress	GOP accused of voter **suppression**
Future/Projection		Information/Expunge, Remove	Republicans **could have removed** Trump but did not
Narrative/Used to		Information/Report cost	Accusing GOP **used** $500B to help rich donors
Democrats			
Description/Drug	Democratic representative praises Trump **touted** drug (**hydroxychloroquine**)	Positive/Join our	**help us/join our** effort to defeat **Republicans**
Strategic/Persuasion, Touting		Interactive/Request help	
Negative/llegal	Accusing DNC of demanding relief for **illegal immigrants**	Public/Political Ideology/Republican	
Description/Cards	DNC fighting for COVID19 **card** but against voted ID	Public/Legislature	Accusing WI **legislature** and **court** of conspiring to help GOP in election
Positive/Tremendous	relief bill **tremendous opportunity** that DNC missed	Public/Law/Courts	

(continued)

Table 7. (*continued*)

GOP Supporters		DNC Supporters	
Media			
Information/Left	the **left** leaning media refuses to air Trump's press conference	Narrative/Turn Break	**major breaking news:** Trump spreading misinformation
Character/Globalist	media is manipulated by **globalists**	Negative/Incompetence	media calls Trump's actions **incompetence**
Positive/Loyalty	Trump exposes media – follow **patriots** instead	Negative/Squander	Trump **squandered** his credibility
Negative/Trickery, Sham	media coverage is a **sham**	Negative/Bungle	Media exposes **bungled** COVID19 response
Negative/Bogus, Phony	media concentrated on **phony** impeachment	Negative/Act Aggressive, Fire	Trump & GOP spreading misinformation, while Trump **fired** response team
Conspiracy			
Negative/Fail to work	Fauci thinks **Hydroxychloroquine doesn't work. Beginning to think** he is either an **idiot** or **lying to** us.	Character/Personality	**Murdoch** (FoxNews) spreading conspiracy theories
FirstPerson/Anger		Updates/BreakingNews	**breaking news:** lawsuit against Trump, Murdoch, and FoxNews coming **soon**
Negative/Stupid person		Narrative/Time shift, soon	
Description/Drug		Description/Mask	Trump accuses nurses of stealing **masks**
Narrative/Embark		Public/Administer	Trump **administration** spreading conspiracy theories

ranked them using the the product of the valence score and the log of their frequency. Table 7 lists the top 5 most associated rhetorical devices that are used in conjunction with each target. Again, we can see that the rhetorical devices are used in a manner that is consistent with the frames. GOP supporters are blaming China (source of virus, no human rights), blaming Democrats (for "phony" impeachment, supporting illegal immigrants), attacking media (as tool of globalists and leftist, being a sham), touting a potential drug (hydroxychloroquine), and promoting conspiracy theories (e.g. COVID19 targets elderly GOP supporters). As for DNC supporters, they are used to blame Trump (as incompetent, failure), blame the GOP (for voter suppression, helping the rich), call for help to defeat republicans in elections, and accusing Trump and conservative media of promoting conspiracy theories.

The second way we used DocuScope output is that we inspected the top 5 positive and negative devices for both groups and ranked them using the product of the valence and log of the frequency. As shown in Table 8, the devices are consistent with the aforementioned frames. While positive devices were often used by GOP supporters to express support for Trump, the positive devices were used by DNC supporters to formulate attacks against Trump (e.g. going after public servants, seeking acquittal from crimes). Usage of negative devices were used in

Table 8. Top positive and negative rhetorical devices for GOP and DNC supporters

GOP Supporters		DNC Supporters	
Positive			
Positive/High quality	Food supply workers delivering **high quality** products	Positive/Relieve, Acquittal	Trump hid COVID19 to secure impeachment **acquittal**
Positive/Guardian, Prophylactic	India approved Hydroxycloroquine as **prophylactic**	Positive/Greater good	Trump not working for the **greater good**
Positive/Outstanding	Fauci defies Trump, calls WHO boss **'outstanding** person'	Positive/Empathy, Hear one out	**hear me out:** Trump failed
Positive/Tolerant	**"tolerant"** left cause violence	Positive/Speak truth to power	Trump pursuing public servants who **speak truth to power**
Information/Relieve, Vindicate	COVID19 **vindicates** Trump's immigration policy	Information/Relieve, Exculpatory	Trump's actions are not **exculpatory** (but damming)
Negative			
Negative/Illegal	DNC wants to help **illegal aliens**	Negative/Act aggressive, Fire	Trump **fired** pandemic team
Negative/Resist, Deny	**Unrelated** deaths blamed on COVID19	Negative/Incompetence	Trump is **incompetent**
Negative/Bogus, Phony	Complaint about **phony** impeachment	Negative/Unforgivable	Trump's inaction **unforgivable**
Negative/Trickery, Sham	Media is a **sham**	Negative/Mother fucker	Trump is ...
Negative/Fudge data	CDC **fudging the numbers** (for COVID19)	Negative/Fucked up	Trump ... COVID19 response

attacking media, CDC, and Democrats by GOP supporters, and for attacking Trump by the DNC supporters.

Discussion of Frames. Framing is autological. The purpose of frames is to "serve as bridges between elite discourse about a problem or issue and popular comprehension of that issue" [30]. Political groups dedicate a substantial amount of time towards regulating not just the information that is shared but also how it is presented [30]. Translated to Twitter, framing has become a vital tool for both the GOP and the DNC, in order for them to relay (or spin) information to the members of their parties and any users who display interest. Using Twitter for a political purpose, Democrats have reinforced a negative frame around Donald Trump, weaving the COVID19 pandemic into the picture for users that are avidly reading and absorbing the hashtags and the tweets. As for Republicans, there is a more positive form of framing that from the start has been built to evoke fellow party members to protect and support Trump. As a result, the COVID19 blame has been continuously shifted towards conspiracies theories (e.g. deep state, liberal media, etc.) and China as the source of the virus. Using both a blame frame and a support frame, COVID19 is framed as a political issue. These frames are somewhat predictable, because they reinforce the sentiment of each party, influencing users to continue to support what they already believe.

Table 9. Top tweets from GOP and DNC supporters from *politicized* dataset

Freq	Text
GOP Supporters	
5,008	RT @realDonaldTrump: .@OANN A key CoronaVirus Model is now predicting far fewer deaths than the number shown in earlier models. That's because the American people are doing a great job. Social Distancing etc. Keep going!
4,998	RT @realDonaldTrump: So now the Fake News @nytimes is tracing the CoronaVirus origins back to Europe, NOT China. This is a first! I wonder what the Failing New York Times got for this one? Are there any NAMED sources? They were recently thrown out of China like dogs, and obviously want back in. Sad!
3,795	RT @DonaldJTrumpJr: Take a look back and RT what the liberal media and Joe Biden said about President Trump's aggressive, early response to the #coronavirus. Thank God @realDonaldTrump is the one in charge during this scourge! https://t.co/d6OtmWrGmb
DNC Supporters	
6,435	RT @tribelaw: What if we were to learn that Trump suppressed scary information re COVID19 (and the needed federal response) in January to postpone the economic turndown until it could no longer endanger his Senate acquittal? Retweet if you wouldn't be surprised by his making that tradeoff
6,207	RT @RepMaxineWaters: Trump, you incompetent idiot! You sent 18 tons of PPE to China early but ignored warnings & called COVID19 concerns a hoax. You've endangered doctors, nurses, aids, orderlies, & janitors - all risking their lives to save ours. Pray 4 forgiveness for the harm that you're
6,114	RT @BrianKarem: Lies. On 2/28/20 on the South Lawn I ASKED you about the W.H.O. telling us the risk for COVID19 had increased. You blew off the question to tell us about Your rally that night. You blew it. Not the W.H.O. @realDonaldTrump https://t.co/DHhSgURGRt

According to framing theory, it is believed that "people draw their opinions from the set of available beliefs stored in memory. Only some beliefs become accessible at a given moment. Out of the set of accessible beliefs, only some are strong enough to be judged relevant or applicable to the subject at hand" [10]. In this case, the framing that is reinforced is relevant to the upcoming presidential campaign and has an emotional inclination due to factors such as COVID19. Both Democrats and Republicans have prominent communities on Twitter, and political framing can be observed in the top tweets by both party supporters. The purpose of these tweets and hashtags, as well as URLs and links, are to present to both audiences why Trump is fit or unfit to be reelected.

From inspecting the top retweeted tweets (Table 9), while the top DNC tweets are pretty self-explanatory, the emotions behind them are negatively geared against Trump. They are meant to demonstrate that Trump is not suitable to remain president, because he showed misfeasance or perhaps malfeasance in handling the pandemic. On the opposite end, the framing theory is seen in full effect for Republicans as a positive source of influence to show that Trump is a suitable president both now and in the future. For example, the tweet by @DonaldJTrumpJr (Table 9) is directly related to COVID19 and frames the perspective that Trump is a fit president that handled the pandemic properly from the beginning. The other two tweets by @realDonaldTrump are expressing hope that everything will be okay and attacking the media as fake news.

Aside from tweets that are direct, frames are regularly reinforced by both Democrats and Republicans through routine hashtags and URLS. The top hashtags from both parties in the period of study include Democratic hashtags such as #TrumpPlague #TrumpGenocide, and #TrumpLiesAboutCoronavirus. As for Republican influenced hashtags, these include hashtags such as #OANN and #WuhanCoronaVirus. Correlating with the framing theory, there is focus on the pandemic and it is framed to fit a certain understanding that is different for each party. For Democrats, it is directly blaming Trump for COVID19. As for Republicans, it is shifting the blame from Trump to conspiracies and China, which is where the virus originated. Adding emphasis to particular parts of issues allows for politicians and political parties to create an mental association and to put them in specific reference frames [25]. There is also usage of the hashtag #OANN, which refers to the conservative news channel (One America News Network). Since COVID19 updates are provided by the media, using the hashtag #OANN directly connects users to Republican influenced news. Through framing, OANN presents opinion and information from a Conservative perspective.

The top URLs from both Republican and Democratic Twitter accounts are primarily focused on two categories: they are COVID19 and the presidential election (Table 6). For Republican URLs, the top URLs include the official government website on COVID19 (CoronaVirus.gov), which displays pandemic updates, ways to protect yourself, and guidelines to re-open America. This influences the notion that Trump is acting accordingly as the president. There are also multiple URLs that link to websites such as USA Today and The Wall Street Journal, which focus on blaming China for not sharing early on how threatening COVID19 is. The usage of these URLs frames the idea that the blame and the emotions that come with the reaction to blame are to be focused on China.

From a Democratic perspective, the use of URLs as a source of political framing is used to influence users to blame Trump for his inability to contain the spread of COVID19. The URLs include a link to the Democratic Coalition's fundraiser to create ads aimed at exposing Trump's lies about COVID19. There are also links to articles in media such as Foreign Policy, Washington Post, and New York Times, which echo the sentiment of Trump's leadership failure both before COVID19 reached the United States and after it became a pandemic. Through political framing, the URLs are used to foreshadow President Trump's leadership efforts. Democrats directly set the agenda that Trump's inadequacy to act early, in order to prevent COVID19, renders him unfit for the upcoming presidency.

Through the usage of rhetorical devices, framing is reinforced to support positive or negative thoughts. The most popular Republican supported categories include, but are not limited to, conspiracy, blame China, praising Trump's actions, and Pro GOP. Hashtags that support President Trump evoke a positive rhetoric. For example, hashtags that directly relate to Trump are #AmericaWorksTogether and #KAG (Keep America Great). Examples of these hashtags strengthen the sentiment of America's togetherness and the continued need for it to remain great. Contrarily, the rhetoric used towards China displays a more negative connotation. This can be seen through hashtags such as #ChineseVirus and #WuhanCoronaVirus. These crafted phrases are used to frame the perception that China and COVID19 are synonymous with one another.

Aside from blame and support frames that are prominently used by both parties, a social solidarity frame is visible for both parties as reflected by the hashtags #SocialDistancing, #StayHome and #FamiliesFirst. As a rhetorical device, these hashtags illustrate that the GOP and DNC claim to prioritize the worth of people's health and families' financial well being.

5 Conclusion

In this paper, we illustrate how framing can serve to shape the public discourse on an issue. Specifically, we show how politically active supporters of the Democratic and Republican parties distinctly frame the COVID19 pandemic as a political issue – as opposed to it solely being a public health issue. We use unsupervised stance detection to effectively identify the supporters of each party, and we analyze their most distinctive hashtags, retweeted accounts, URLs, and rhetorical devices. In doing so, we are able to identify the frames that underlie their discourse. Supporters of both parties crafted two primary frames, namely: a blame frame, where GOP supporters blame China and conspiracies, and DNC supporters blame Trump and the GOP; and a support frame, where each supports the candidate(s) of their respective parties. These frames dominate the discourse of both groups, thus revealing that support and blame prominently influence the political sentiments attached to COVID19.

References

1. Abu-Jbara, A., King, B., Diab, M., Radev, D.: Identifying opinion subgroups in Arabic online discussions. In: Proceedings of the 51st Annual Meeting of the Association for Computational Linguistics (Volume 2: Short Papers), pp. 829–835 (2013)
2. Aldayel, A., Magdy, W.: Your stance is exposed! Analysing possible factors for stance detection on social media. In: Proceedings of the ACM on Human-Computer Interaction, vol. 3, no. CSCW, pp. 1–20 (2019)
3. Bakliwal, A., Foster, J., van der Puil, J., O'Brien, R., Tounsi, L., Hughes, M.: Sentiment analysis of political tweets: towards an accurate classifier. In: Proceedings of the Workshop on Language Analysis in Social Media, pp. 49–58. Association for Computational Linguistics, Atlanta, June 2013. https://www.aclweb.org/anthology/W13-1106

4. Baumer, E., Elovic, E., Qin, Y., Polletta, F., Gay, G.: Testing and comparing computational approaches for identifying the language of framing in political news. In: Proceedings of the 2015 Conference of the North American Chapter of the Association for Computational Linguistics: Human Language Technologies, pp. 1472–1482 (2015)
5. Bermingham, A., Smeaton, A.: On using Twitter to monitor political sentiment and predict election results. In: Proceedings of the Workshop on Sentiment Analysis where AI meets Psychology (SAAIP 2011), pp. 2–10. Asian Federation of Natural Language Processing, Chiang Mai, November 2011. https://www.aclweb.org/anthology/W11-3702
6. Borge-Holthoefer, J., Magdy, W., Darwish, K., Weber, I.: Content and network dynamics behind Egyptian political polarization on twitter. In: Proceedings of the 18th ACM Conference on Computer Supported Cooperative Work & Social Computing, pp. 700–711. ACM (2015)
7. Burch, L., Frederick, E., Pegoraro, A.: Kissing in the carnage: an examination of framing on twitter during the vancouver riots. J. Broadcast. Electron. Media **59**, 399–415 (2015). https://doi.org/10.1080/08838151.2015.1054999
8. Card, D., Boydstun, A., Gross, J.H., Resnik, P., Smith, N.A.: The media frames corpus: annotations of frames across issues. In: Proceedings of the 53rd Annual Meeting of the Association for Computational Linguistics and the 7th International Joint Conference on Natural Language Processing (Volume 2: Short Papers), pp. 438–444 (2015)
9. Choi, E., Tan, C., Lee, L., Danescu-Niculescu-Mizil, C., Spindel, J.: Hedge detection as a lens on framing in the GMO debates: a position paper. In: Proceedings of the Workshop on Extra-Propositional Aspects of Meaning in Computational Linguistics, pp. 70–79. Association for Computational Linguistics, Jeju, July 2012. https://www.aclweb.org/anthology/W12-3809
10. Chong, D., Druckman, J.N.: Framing theory. Annu. Rev. Polit. Sci. **10**, 103–126 (2007)
11. Conover, M., Ratkiewicz, J., Francisco, M.R., Gonçalves, B., Menczer, F., Flammini, A.: Political polarization on Twitter. ICWSM **133**, 89–96 (2011)
12. Darwish, K.: To kavanaugh or not to kavanaugh: that is the polarizing question. arXiv preprint arXiv:1810.06687 (2018)
13. Darwish, K., Stefanov, P., Aupetit, M.J., Nakov, P.: Unsupervised user stance detection on twitter. arXiv preprint arXiv:1904.02000 (2019)
14. Djemili, S., Longhi, J., Marinica, C., Kotzinos, D., Sarfati, G.E.: What does twitter have to say about ideology? In: NLP 4 CMC: Natural Language Processing for Computer-Mediated Communication/Social Media-Pre-conference Workshop at Konvens 2014, vol. 1, p. http-www. Universitätsverlag Hildesheim (2014)
15. Ebrahimi, J., Dou, D., Lowd, D.: A joint sentiment-target-stance model for stance classification in tweets. In: Proceedings of COLING 2016, the 26th International Conference on Computational Linguistics: Technical Papers, pp. 2656–2665 (2016)
16. Entman, R.M.: Framing: toward clarification of a fractured paradigm. J. Commun. **43**(4), 51–58 (1993)
17. Fulgoni, D., Carpenter, J., Ungar, L., Preoţiuc-Pietro, D.: An empirical exploration of moral foundations theory in partisan news sources. In: Proceedings of the Tenth International Conference on Language Resources and Evaluation (LREC 2016), pp. 3730–3736. European Language Resources Association (ELRA), Portorož, Slovenia, May 2016. https://www.aclweb.org/anthology/L16-1591

18. Gerrish, S., Blei, D.M.: How they vote: Issue-adjusted models of legislative behavior. In: Pereira, F., Burges, C.J.C., Bottou, L., Weinberger, K.Q. (eds.) Advances in Neural Information Processing Systems, vol. 25, pp. 2753–2761. Curran Associates, Inc. (2012). http://papers.nips.cc/paper/4715-how-they-vote-issue-adjusted-models-of-legislative-behavior.pdf
19. Greene, S., Resnik, P.: More than words: syntactic packaging and implicit sentiment. In: Proceedings of Human Language Technologies: The 2009 Annual Conference of the North American Chapter of the Association for Computational Linguistics, pp. 503–511. Association for Computational Linguistics, Boulder, June 2009. https://www.aclweb.org/anthology/N09-1057
20. Groshek, J., Al-Rawi, A.: Public sentiment and critical framing in social media content during the 2012 U.S. presidential campaign. Soc. Sci. Comput. Rev. **31**, 563–576 (2013). https://doi.org/10.1177/0894439313490401
21. Harlow, S., Johnson, T.: The arab spring—overthrowing the protest paradigm? How the new york times, global voices and twitter covered the Egyptian revolution. Int. J. Commun. **5** (2011). https://ijoc.org/index.php/ijoc/article/view/1239
22. Hasan, K.S., Ng, V.: Why are you taking this stance? Identifying and classifying reasons in ideological debates. In: Proceedings of the 2014 Conference on Empirical Methods in Natural Language Processing (EMNLP), pp. 751–762 (2014)
23. Ishizaki, S., Kaufer, D.: The docuscope text analysis and visualization environment. In: McCarthy, P., Boonthum, C. (eds.) Invited Chapter for Applied Natural Language Processing and Content Analysis: Identification, Investigation, and Resolution (2011)
24. Johnson, K., Goldwasser, D.: Identifying stance by analyzing political discourse on twitter. In: Proceedings of the First Workshop on NLP and Computational Social Science, pp. 66–75 (2016)
25. Johnson, K., Jin, D., Goldwasser, D.: Modeling of political discourse framing on Twitter. In: Eleventh International AAAI Conference on Web and Social Media (2017)
26. Johnson, K., Lee, I.T., Goldwasser, D.: Ideological phrase indicators for classification of political discourse framing on Twitter. In: Proceedings of the Second Workshop on NLP and Computational Social Science, pp. 90–99. Association for Computational Linguistics, Vancouver, August 2017. https://doi.org/10.18653/v1/W17-2913. https://www.aclweb.org/anthology/W17-2913
27. Jones Jang, M., Hart, P.: Polarized frames on "climate change" and "global warming" across countries and states: evidence from Twitter big data. Glob. Environ. Change **32**, 11–17 (2015). https://doi.org/10.1016/j.gloenvcha.2015.02.010
28. Kaufer, D., Butler, B.: Rhetoric and the Arts of Design. Routledge, Abingdon (1996)
29. Kaufer, D., Butler, B.: Designing Interactive Worlds with Words: Principles of Writing as Representational Composition. Routledge, Abingdon (2000)
30. Nelson, T.E., Oxley, Z.M., Clawson, R.A.: Toward a psychology of framing effects. Polit. Behav. **19**(3), 221–246 (1997)
31. Pla, F., Hurtado, L.F.: Political tendency identification in Twitter using sentiment analysis techniques. In: Proceedings of COLING 2014, the 25th International Conference on Computational Linguistics: Technical Papers, pp. 183–192. Dublin City University and Association for Computational Linguistics, Dublin, August 2014. https://www.aclweb.org/anthology/C14-1019
32. Pond, P., Lewis, J.: Riots and Twitter: connective politics, social media and framing discourses in the digital public sphere. Inf. Commun. Soc. **22**(2), 213–231 (2019). https://doi.org/10.1080/1369118X.2017.1366539

33. Recasens, M., Danescu-Niculescu-Mizil, C., Jurafsky, D.: Linguistic models for analyzing and detecting biased language. In: Proceedings of the 51st Annual Meeting of the Association for Computational Linguistics (Volume 1: Long Papers), pp. 1650–1659. Association for Computational Linguistics, Sofia, August 2013. https://www.aclweb.org/anthology/P13-1162
34. Sim, Y., Acree, B.D.L., Gross, J.H., Smith, N.A.: Measuring ideological proportions in political speeches. In: Proceedings of the 2013 Conference on Empirical Methods in Natural Language Processing, pp. 91–101. Association for Computational Linguistics, Seattle, Washington, October 2013. https://www.aclweb.org/anthology/D13-1010
35. Sridhar, D., Foulds, J., Huang, B., Getoor, L., Walker, M.: Joint models of disagreement and stance in online debate. In: Proceedings of the 53rd Annual Meeting of the Association for Computational Linguistics and the 7th International Joint Conference on Natural Language Processing (Volume 1: Long Papers), pp. 116–125 (2015)
36. Stefanov, P., Darwish, K., Atanasov, A., Nakov, P.: Predicting the topical stance and political leaning of media using tweets. In: ACL 2020 (2020)
37. Tan, C., Lee, L., Pang, B.: The effect of wording on message propagation: topic- and author-controlled natural experiments on Twitter. In: Proceedings of the 52nd Annual Meeting of the Association for Computational Linguistics (Volume 1: Long Papers), pp. 175–185. Association for Computational Linguistics, Baltimore, June 2014. https://doi.org/10.3115/v1/P14-1017. https://www.aclweb.org/anthology/P14-1017
38. Walker, M.A., Anand, P., Abbott, R., Grant, R.: Stance classification using dialogic properties of persuasion. In: Proceedings of the 2012 Conference of the North American Chapter of the Association for Computational Linguistics: Human Language Technologies, pp. 592–596. Association for Computational Linguistics (2012)
39. Wiebe, J., Wilson, T., Bruce, R., Bell, M., Martin, M.: Learning subjective language. Comput. Linguist. **30**(3), 277–308 (2004). https://doi.org/10.1162/0891201041850885

Measuring Adolescents' Well-Being: Correspondence of Naïve Digital Traces to Survey Data

Elizaveta Sivak and Ivan Smirnov(✉)

Institute of Education, National Research University Higher School of Economics,
Myasnitskaya ul., 20, Moscow 101000, Russia
{esivak,isbmirnov}@hse.ru

Abstract. Digital traces are often used as a substitute for survey data. However, it is unclear whether and how digital traces actually correspond to the survey-based traits they purport to measure. This paper examines correlations between self-reports and digital trace proxies of depression, anxiety, mood, social integration and sleep among high school students. The study is based on a small but rich multilayer data set ($N = 144$). The data set contains mood and sleep measures, assessed daily over a 4-month period, along with survey measures at two points in time and information about online activity from VK, the most popular social networking site in Russia. Our analysis indicates that 1) the sentiments expressed in social media posts are correlated with depression; namely, adolescents with more severe symptoms of depression write more negative posts, 2) late-night posting indicates less sleep and poorer sleep quality, and 3) students who were nominated less often as somebody's friend in the survey have fewer friends on VK and their posts receive fewer "likes." However, these correlations are generally weak. These results demonstrate that digital traces can serve as useful supplements to, rather than substitutes for, survey data in studies on adolescents' well-being. These estimates of correlations between survey and digital trace data could provide useful guidelines for future research on the topic.

Keywords: Adolescents · Depression · Psychological well-being · Digital traces · Validity · Sleep · Social networks · Social media

1 Introduction

Adolescents' everyday lives and well-being are a black box to researchers. It is not well-known how adolescents' behavioral patterns, moods, emotions, and other psychological states change over time. Youths are particularly difficult to study due to uncertainty about the accuracy and validity of adolescents' self-reports [1, 2]. Also, surveys are too expensive and time-consuming to be conducted frequently. Meanwhile, adolescents' well-being is attracting greater attention from researchers and policy-makers as mental health issues among adolescents increase [3, 4].

S. Aref et al. (Eds.): SocInfo 2020, LNCS 12467, pp. 352–363, 2020.
https://doi.org/10.1007/978-3-030-60975-7_26

Digital traces (data from social media and other digital platforms) present a promising new approach to studying adolescents' well-being that is fast, inexpensive, non-intrusive, and with high resolution. Previous literature has demonstrated that mental health conditions, such as depression and anxiety, can be predicted from mobile sensor data [5–8], social media engagement [9], language [9–12], and photos [13]. An increasing number of studies have analyzed emotive trends based on social media data [14, 15].

However, further research is needed to better understand how digital traces correspond to survey data and whether they can act as a substitute for surveys. It remains unclear whether and to what extent digital traces correspond to the characteristics they are assumed to reflect such as sleep patterns, social relations and integration, mood, and psychological well-being. High school students have been understudied as previous studies have typically focused on undergraduate students or adults [16–19]. Unfortunately, the results of these studies cannot necessarily be generalized to adolescents since high school students may use social media differently than undergraduates or adults. Thus, it is uncertain whether digital traces from social media can elucidate the psychological states or behavior of high school students.

Moreover, some online indicators of well-being have been studied much less than others. For instance, researchers have mostly used data from wearable devices and mobile sensors to predict sleep patterns [19–21], but these kinds of studies may be difficult to implement on a large scale. Using data that are more readily accessible, such as timestamps on posts or logins to a learning management system [22], could allow social scientists to study sleep patterns in a way that is both fine-grained and large-scale. However, there is still little evidence that such indicators reflect actual sleep patterns.

Another understudied measure of well-being is negative social ties. For adolescents, peer unpopularity and rejection are associated with depression and externalizing problems [23–25]. However, social media sites usually do not collect information about negative links, i.e., "dislikes" [26]. Therefore, it is important to study whether it is possible to infer negative links and unpopularity from the data available on social networking sites.

In this paper, we explore how naïve digital traces correspond to survey data on high school students' well-being, focusing on depression, anxiety, mood, social integration, and sleep. By naïve digital traces we mean digital trace data that seem to reflect behavior, attitudes, or states. The study is based on a unique data set regarding adolescents' social networks, social media use, psychological well-being, sleep, and demographics (N = 144). A public, anonymized version of the data set is available online[1] and could be used to further investigation of the validity of digital traces. Data were obtained from VK, Russia's most popular social networking site, and from adolescents' self-reports on depression, anxiety, and social ties. We also assessed participants' mood and sleep daily over a period of four months using a mobile app. This study analyzes whether the sentiments of adolescents' social media posts are correlated with depression and anxiety and how day-of-week sentiment dynamics is correlated with self-reported mood. We also investigate how the time of adolescents' posts corresponds to their sleep patterns

[1] Due to the sensitive nature of the data, the files are encrypted. The password to decrypt the files will be sent upon request. https://osf.io/b57rp/?view_only=f872ea8355cb4d818683e29967282a23.

and quality. Finally, we examine the relationships between (a) interaction-based metrics for online friendship and popularity on social media and (b) actual friendship ties and peer popularity and unpopularity.

2 Data and Methods

2.1 Participants

This study was conducted among students from one high school in Moscow, Russia. Participation was voluntary. All students between the ages of 16 and 17 were informed about the study and the opportunity to participate. The researchers held group meetings with all students who self-selected to participate where we explained the aims of the study and what data would be collected. Each participant who decided to participate, as well as one of his or her parents, signed an informed consent form. Students were informed that their participation was voluntary and that they could stop participating in the study at any time and request that any data already collected be deleted (to date no one has requested this). The baseline survey was filled out by 144 students, the endline by 78 students, and 118 participants answered everyday questions about mood and sleep at least once. Most of the participants (86%) were girls.

2.2 Procedures

The study took place over a 4-month period (November 2017–February 2018). Participants were asked to fill out a survey at the beginning and end of this period. Both surveys included items that measure depression and anxiety. In the baseline survey, we also asked about friendship and peer popularity and unpopularity; in the endline survey, we asked about sleep quality. Participants were also asked about their current mood three times per day via a mobile app, using the experience sampling method. Similarly, participants were asked once per day what time they had woken up that morning and gone to bed the previous night. Public data were gathered from profiles on VK with the participants' consent. All procedures used to obtain data are described in Table 1.

2.3 Measures

Depression

We used the Patient Health Questionnaire scale (PHQ-9)[2], which is used to calculate the severity of depressive symptoms. It includes nine items scored from 0 to 3 and generates a severity score of 0 to 27. Scores of 5, 10, 15, and 20 represent the cut-off points for mild, moderate, moderately severe, and severe depression, respectively [27]. The scale has been shown to be a valid tool in detecting depression among adolescents across various cultures [28–33].

We measured depressive symptoms twice. The test-retest reliability coefficient is 0.73 ($P < 10^{-11}$). In the analysis, we used the results of the first measurement as it has fewer missing values.

[2] https://www.phqscreeners.com/.

Table 1. Measures and data sources

Data sources	Measures
Surveys (baseline/endline)	• Sleep quality (PSQI scale), endline • Friendship ties (with whom are you friends?), baseline • Popularity (who do you consider popular/unpopular?), baseline • Depression (PHQ-9 scale), baseline and endline • Anxiety as a trait (trait anxiety subscale of the STAI), baseline and endline • Gender • Grade (10th or 11th)
Daily self-reports (mobile app RealLife Exp)	• Bedtime/wake up time (asked once per day in the morning) • Mood (assessed three times per day using a 5-point Likert scale, from 1 "very bad mood" to 5 "very good mood")
Public data gathered from profiles on social networking site (VK.com)	• For each public post: – Timestamp – Number of likes (from students who attend the same - school and overall) – Strength of negative and positive sentiment expressed in the post • Number of friends from the same school (based on the list of school students) and overall

According to the PHQ-9 scale, 10% of the sample had no symptoms of depression (scored 0–4), 41% exhibited mild symptoms (5–9), 26% moderate (10–14), 11.5% moderately severe (15–19), and 11.5% severe (20–27). These rates are unusually high: previous studies that used PHQ-9 to measure depression among school students reported the prevalence of moderately severe/severe depression to be 5–9% [28, 29, 34, 35]. In Russia, there is not enough data on the prevalence of depressive symptoms in adolescents to compare our results to. We attribute this high rate to the fact that the school in this study is selective and students may be under a great deal of academic pressure and stress. There is also a possibility of self-selection bias.

Anxiety

To assess anxiety, we used the subscale of Spielberger's State-Trait Anxiety Inventory (STAI) [36] measuring trait anxiety—that is, anxiety as a personal characteristic. The test-retest reliability coefficient for the two measures of anxiety is 0.82 ($P < 10^{-14}$).

Mood

We assessed mood daily over a period of 4 months via experience sampling using a 5-point Likert scale from 1 (very bad) to 5 (very good). Mood was assessed three times per day at random points within specific time periods: morning (8:00 a.m.–9:00 a.m. on school days, 10:00 a.m.–11:00 a.m. on days off), afternoon (3:00 p.m.–5:00 p.m.), and evening (8:00 p.m.–10:00 p.m.). We then computed the participants' average mood level for each day of the week.

Sleep

We assessed participants' sleep using a mobile app. Each morning, the app asked participants what time they had gone to bed the previous night and woke up that morning. We also assessed sleep quality using the Pittsburgh Sleep Quality Index (PSQI) [37].[3] The PSQI is a self-report questionnaire that assesses sleep quality over a 1-month interval. The measure consists of 19 individual items, creating seven components that produce one global score. Each item is weighted on a 0–3 interval scale. The global PSQI score is then calculated by totaling the seven component scores, providing an overall score ranging from 0 to 21, where lower scores denote healthier sleep quality.

Sentiments of Posts

We analyzed the sentiments of posts written within the 4-month study period ($N =$ 5,371) using the SentiStrength program [38].[4] SentiStrength estimates the strength of positive and negative sentiment in short texts, even for informal language. SentiStrength reports two sentiment strengths: -1 (not negative) to -5 (extremely negative) and 1 (not positive) to 5 (extremely positive). A team of linguists adjusted it for Russian language use.[5] We used the following as measures of the aggregate sentiment of users' posts: (a) average positive and negative sentiment across all posts and (b) proportion of strongly negative (scored 3, 4, or 5 on the scale of negative sentiment) and strongly positive (scored 3, 4, or 5 on the scale of positive sentiment) posts among all user's posts. We also separately computed the average positive sentiment of posts for each day of the week.

Night Posting

We computed the proportion of late-night posts (those written between 1:00 a.m. and 5:00 a.m.) to all the posts written over the 4-month study period and also the proportion of days when a participant wrote a late-night post to all the days when a participant wrote a post.

Online Friendship

We used several interaction-based measures of online friendship. For each participant, we measured (1) the number of "friends" they had on VK, including (a) those who studied at the same school and (b) the overall number of VK friends and (2) the average number of "likes" per post (a) from students from the same school and (b) overall. For each pair of participants A and B, we determined (1) whether A and B were friends on VK, 2) whether A and B had at least one reciprocal like (from A to B and from B to A), and 3) average intensity of likes per pair (sum of the proportion of B's posts liked by A and the proportion of A's posts liked by B, divided by two).

[3] https://www.sleep.pitt.edu/instruments/.

[4] http://sentistrength.wlv.ac.uk/.

[5] http://sentistrength.wlv.ac.uk/#Non-English.

Offline Friendship

In the survey, participants were asked to indicate up to 10 other students from the same school whom they considered friends. Since the sample included only 10% of all students at the school, who were almost all from different classes, we used unilateral friendship nominations as indicators of friendship, i.e., if at least one person named another as a friend, we considered them friends.

Popularity and Unpopularity

Participants were asked to name up to 10 of the most popular students, in their opinion, and up to 10 of the most unpopular students at their school. For each participant, we determined if she/he was mentioned as popular or unpopular.

3 Results

3.1 Psychological Well-Being: Depression, Anxiety, and Sentiments of Posts

We found that severity of depression is correlated with the proportion of strongly negative posts (Pearson's r = 0.24) and the average strength of negative sentiment expressed in posts (Pearson's r = 0.23). Pearson correlation coefficients are provided in Table 2. Negative emotions were, on average, more pronounced in the posts of participants who have symptoms of depression, with an average negative sentiment of 1.33, than in posts of students with no depressive symptoms, who had an average negative sentiment of 1.09 (P = 0.002). Students with moderately severe or severe depression wrote more posts, on average, that were strongly negative (22% of posts) than those with moderate or mild symptoms of depression (8.6% of posts) and wrote ten times more strongly negative posts than adolescents with no signs of depression (2% of posts). However, we found no significant correlation between anxiety and the sentiments of posts (see Table 2).

For sentiment analysis, we selected only participants who had written at least three posts within the study period. In previous studies, the higher threshold was used [39, 40]. However, it was not possible in our case due to the small sample size.

Table 2. Pearson correlation coefficients between depression, anxiety, and sentiments of posts

Variable	Pearson's r	P	90% CI
Depression			
Average strength of negative sentiment (participants with 3–200 posts, $n = 61$)	0.23	0.07	(−0.02, 0.42)
Proportion of strongly negative posts (participants with 3–200 posts, $n = 61$)	0.24	0.06	(0.03, 0.43)
Anxiety			
Average strength of negative sentiment (participants with 3–200 posts, $n = 61$)	0.08	0.54	(−0.13, 0.28)
Proportion of strongly negative posts (participants with 3–200 posts, $n = 61$)	0.14	0.28	(−0.07, 0.34)

3.2 Mood

According to the self-reports acquired via experience sampling, students' positive affect level was higher on weekends and Thursdays, when students leave school for a full day to attend classes of their choice at a partner university, than on school days (see Fig. 1[6] and Table 3). The sentiments of participants' posts on VK in general capture this pattern where the average positive sentiments of posts were higher on weekends and Thursdays than on school days. However, likely due to the small sample size and the lack of statistical power, these estimates did not differ significantly.

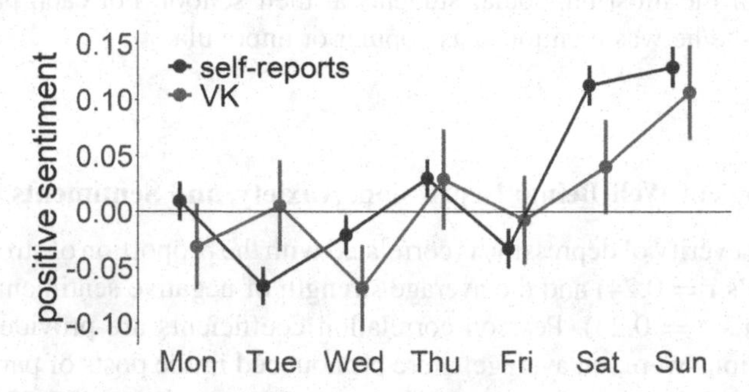

Fig. 1. Average level of positive sentiment (in SD units) over the course of the week based on self-reports and sentiment analysis of VK posts. Error bars indicate standard errors.

Table 3. Average mood and positive sentiment of VK posts on school days and days off

	Mean	90% CI
Average mood on school days	3.38	(3.37, 3.40)
Average mood on days off	3.50	(3.49, 3.52)
Average positive sentiment of VK posts on school days	1.43	(1.38, 1.49)
Average positive sentiment of VK posts on days off	1.48	(1.43, 1.53)

3.3 Online Activity and Sleep

We analyzed whether the times when participants published posts were related to sleep patterns and found that the proportion of late-night posts (written between 1 a.m. and 5 a.m.) is correlated with late bedtime (proportion of days when a participant went to bed after 1 a.m.; $r = 0.2$, $P = 0.024$). Also, see Fig. 2. The proportion of days when a participant wrote a late-night post to all the days with at least one post on VK is correlated

[6] We used a bootstrap approach for hierarchically structured data to compute the standard errors of the mean for positive sentiment and mood on each day of the week as well as 90% confidence intervals for the mean positive sentiment and mood on weekdays and weekends as a correction for repeated measures within the same participant. We implemented a bootstrap approach with replacement on the level of measurements http://biostat.mc.vanderbilt.edu/wiki/Main/HowToBootstrapCorrelatedData.

with a late bedtime (r = 0.25, P = 0.01). Among those who have at least one post written between 1 and 5 a.m., 99% on average go to sleep after midnight (71% after 1 a.m.). Those who don't post late are around 1.5 and 2 times less likely to have a late bedtime (69% go to sleep on average after midnight, 36% after 1 a.m.).

More importantly, we found that late-night posting indicated less sleep as well as poorer sleep quality. Students who posted late at night slept 30 min less on average than those who did not. On school days, students who posted late at night slept for 6 h and 40 min, in contrast to 7 h and 10 min for students who had no late-night posts (P = 0.0002, 95% CI for the difference of means [28, 30]). On days off, students who posted late at night slept for 7 h and 50 min, compared to 8 h and 20 min for students who had no late-night posts (P = 0.002, 95% CI = [28, 32]). Sleep quality was correlated with late bedtime (r = 0.26, P = 0.02, 95% CI [0.008, 0.44]) and on average was worse for those who had at least one late-night post (P = 0.02). The means are 0.87 standard deviations apart (6.2 and 7.7 points).

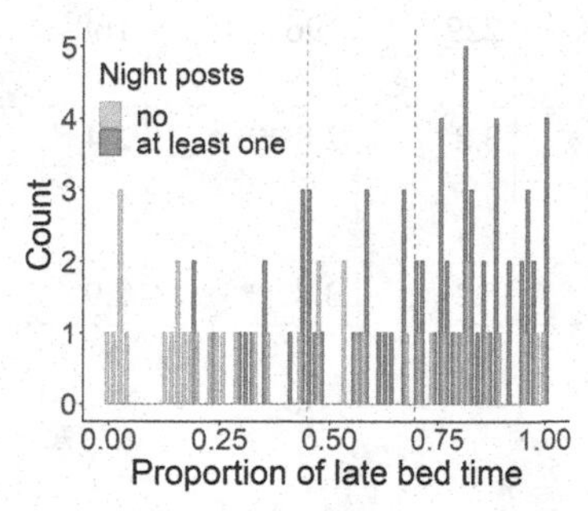

Fig. 2. The number of participants with different proportions of late bedtime (days when a participant went to bed after 1 a.m.) for those with at least one late-night post (written between 1 a.m. and 5 a.m.) and those without.

3.4 Offline Friendship, Social Status, and Online Behavior

Among all pairs of VK friends, 5.6% are also "offline" friends, according to self-reports. In contrast, the probability of friendship for students who are not friends on VK is only 0.02%.

If a pair of VK friends have at least one mutual "like" (i.e., at least one "like" from A to B and from B to A), the probability that they are "offline" friends is 11%; if they have no mutual likes, the probability is 3.6%. Even for pairs with a high intensity of likes (i.e., every second post is liked), the probability of an offline friendship is only 21%. This indicates that friendship nominations and the intensity of mutual likes reflect different kinds of relationships.

Students who were mentioned as popular had, on average, 1.5 times as many VK friends from the same school than those who were not mentioned as popular (see Table 4). In addition, their posts received more likes on average. Among students in the top 10% by the number of VK friends from the same school, 53% were also named by peers as among the most popular students (top 10% by mentions as popular). Those who were described by peers as unpopular had 1.5 times fewer friends. The same was not true for the overall number of friends on VK, which did not differ significantly for popular and unpopular students. These results suggest that, while online friendship and interactions could be a good indicator of offline social integration, it is important to account for the composition of an online social network, i.e., the overall number of VK friends is not necessarily related to one's status in a community of peers at school.

Table 4. Popularity, unpopularity, and online social ties

Variable	Named as popular		Named as unpopular	
	Yes	No	Yes	No
n friends on VK from the same school	35	24***	18	28***
n friends on VK overall	229	196*	168	212
n likes from students at the same school	3.8	2.8***	2.4	3.2***
n likes overall	5.2	3.8***	2.9	4.4***

* $P = 0.08$
*** $P < 10^{-4}$

4 Conclusion

Digital traces have the potential to greatly advance research on adolescent well-being. However, ready-made digital traces are framed by the affordances of the service providing the data, its code, cultural usage practices, and users' motives [41]. As such, we should not implicitly assume that digital trace measures accurately reflect adolescents' behavior, attitudes, relationships, or affective states, even if they seem to, as in the case of the sentiments of posts that are often assumed to reflect the author's mood and emotions. Instead, we should test the validity of such naïve digital trace measures among various populations.

In this paper, we analyzed the validity of digital trace data as an indicator of various aspects of high school students' well-being including depression, anxiety, mood, social integration, and sleep. We found that some features of online behavior are correlated with these indicators. For example, the temporal patterns of social media posts corresponded to sleep length and sleep quality. Also, students who were named as unpopular by their peers have, on average, fewer friends on social media at the same school than those who were not unpopular, though unpopular students do not have fewer online friends overall. However, these correlations are generally weak. Even if the absence of significant correlations can be explained by the small sample size, the upper bounds of the confidence intervals indicate no more than a moderate effect. Our results demonstrate that digital trace measures of well-being can be treated and used as complementary data rather than as close proxies.

The estimations of the correlations between digital trace measures and well-being can be useful for potential meta-analyses of the validity of digital trace data as a measure of adolescents' well-being. Unfortunately, the small sample size prevented us from performing a more detailed analysis. In particular, we do not compare different groups of students, e.g. boys vs girls or highly active VK users vs less active users, while it is possible that digital traces would correlate stronger with self-reports for some groups than for others. However, the data collected here yield novel insights into the relationship between digital markers, social networks, and the well-being of adolescents, considering that this kind of multilayer research on adolescents—combining social media, demographics, depression screeners, and social network data—is rare. The present study can facilitate future research on the validity of digital traces and on the impact of social interactions on adolescents' well-being.

Acknowledgements. This work was supported by a grant from the Russian Science Foundation (project №19-18-00271).

References

1. Fan, X., Miller, B.C., Park, K., Winward, B.W., Christensen, M., Grotevant, H.D.: An exploratory study about inaccuracy and invalidity in adolescent self-report surveys. Field Methods **18**(3), 223–244 (2006). https://doi.org/10.1177/152822X06289161
2. Robinson-Cimpian, J.P.: Inaccurate estimation of disparities due to mischievous responders: several suggestions to assess conclusions. Educ. Res. **43**(4), 171–185 (2014). https://doi.org/10.3102/0013189X14534297
3. Balazs, J., et al.: P-259-Prevalence of adolescent depression in Europe. Eur. Psychiatry **27**, 1 (2012). https://www.sciencedirect.com/science/article/abs/pii/S0924933812744267
4. Keyes, K.M., Gary, D., O'Malley, P.M., Hamilton, A., Schulenberg, J.: Recent increases in depressive symptoms among US adolescents: trends from 1991 to 2018. Soc. Psychiatry Psychiatric Epidemiol. **54**(8), 987–996 (2019). https://doi.org/10.1007/s00127-019-01697-8
5. Ghandeharioun, A., et al.: Objective assessment of depressive symptoms with machine learning and wearable sensors data. In: 2017 Seventh International Conference on Affective Computing and Intelligent Interaction (ACII), pp. 325–332 (2017). http://doi.org/10.1109/ACII.2017.8273620
6. Mohr, D.C., Zhang, M., Schueller, S.M.: Personal sensing: understanding mental health using ubiquitous sensors and machine learning. Ann. Rev. Clin. Psychol. **13**, 23–47 (2017). https://doi.org/10.1146/annurev-clinpsy-032816-044949
7. Place, S., et al.: Behavioral indicators on a mobile sensing platform predict clinically validated psychiatric symptoms of mood and anxiety disorders. J. Med. Internet Res. **19**(3), e75 (2017). https://doi.org/10.2196/jmir.6678. 1–9
8. Saeb, S., Lattie, E.G., Schueller, S.M., Kording, K.P., Mohr, D.C.: The relationship between mobile phone location sensor data and depressive symptom severity. PeerJ **4**, e2537 (2016). https://doi.org/10.7717/peerj.2537. 1–15
9. De Choudhury, M., Gamon, M., Counts, S., Horvitz, E.: Predicting depression via social media. In: Seventh International AAAI Conference on Weblogs and Social Media (2013). https://www.aaai.org/ocs/index.php/ICWSM/ICWSM13/paper/viewFile/6124/6351

10. Eichstaedt, J.C., et al.: Facebook language predicts depression in medical records. Proc. Natl. Acad. Sci. **115**(44), 11203–11208 (2018). https://doi.org/10.1073/pnas.1802331115
11. Tackman, A.M., et al.: Depression, negative emotionality, and self-referential language: a multi-lab, multi-measure, and multi-language-task research synthesis. J. Pers. Soc. Psychol. **116**(5), 817 (2019). https://doi.org/10.1037/pspp0000187
12. Resnik, P., Garron, A., Resnik, R.: Using topic modeling to improve prediction of neuroticism and depression in college students. In: Proceedings of the 2013 Conference on Empirical Methods in Natural Language Processing, pp. 1348–1353 (2013). https://www.aclweb.org/anthology/D13-1133.pdf
13. Reece, A.G., Danforth, C.M.: Instagram photos reveal predictive markers of depression. EPJ Data Sci. **6**, 1–12 (2017). https://doi.org/10.1140/epjds/s13688-017-0110-z
14. Bollen, J., Mao, H., Pepe, A.: Modeling public mood and emotion: Twitter sentiment and socio-economic phenomena. In: Fifth International AAAI Conference on Weblogs and Social Media (2011). https://www.aaai.org/ocs/index.php/ICWSM/ICWSM11/paper/viewFile/2826/3237
15. Garcia, D., Rimé, B.: Collective emotions and social resilience in the digital traces after a terrorist attack. Psychol. Sci. **30**(4), 617–628 (2019). https://doi.org/10.1177/0956797619831964
16. Eagle, N., Pentland, A.S.: Reality mining: sensing complex social systems. Pers. Ubiquit. Comput. **10**(4), 255–268 (2006). http://realitycommons.media.mit.edu/pdfs/realitymining_old.pdf
17. Aharony, N., Pan, W., Ip, C., Khayal, I., Pentland, A.: Social fMRI: investigating and shaping social mechanisms in the real world. Pervasive Mob. Comput. **7**(6), 643–659 (2011). https://doi.org/10.1016/j.pmcj.2011.09.004
18. Stopczynski, A., et al.: Measuring large-scale social networks with high resolution. PLoS One **9**(4), e95978 (2014). https://doi.org/10.1371/journal.pone.0095978
19. Wang, R., et al.: StudentLife: assessing mental health, academic performance and behavioral trends of college students using smartphones. In: Proceedings of the 2014 ACM International Joint Conference on Pervasive and Ubiquitous Computing, pp. 3–14 (2014). https://doi.org/10.1145/2632048.2632054
20. Bai, Y., Xu, B., Ma, Y., Sun, G., Zhao, Y.: Will you have a good sleep tonight?: sleep quality prediction with mobile phone. In: Proceedings of the 7th International Conference on Body Area Networks, pp. 124–130 (2012). https://dl.acm.org/doi/10.5555/2442691.2442720
21. Sathyanarayana, A., et al.: Sleep quality prediction from wearable data using deep learning. JMIR mHealth uHealth **4**(4), e125 (2016). https://doi.org/10.2196/mhealth.6562
22. Smarr, B.L., Schirmer, A.E.: 3.4 million real-world learning management system logins reveal the majority of students experience social jet lag correlated with decreased performance. Sci. Rep. **8**, 1–9 (2018). https://doi.org/10.1038/s41598-018-23044-8
23. Fergusson, D.M., Wanner, B., Vitaro, F., Horwood, L.J., Swain-Campbell, N.: Deviant peer affiliations and depression: confounding or causation? J. Abnorm. Child Psychol. **31**(6), 605–618 (2003). https://doi.org/10.1023/a:1026258106540
24. Kupersmidt, J.B., Coie, J.D.: Preadolescent peer status, aggression, and school adjustment as predictors of externalizing problems in adolescence. Child Dev. **61**(5), 1350–1362 (1990). https://doi.org/10.1111/j.1467-8624.1990.tb02866.x
25. Zimmer-Gembeck, M.J.: Peer rejection, victimization, and relational self-system processes in adolescence: toward a transactional model of stress, coping, and developing sensitivities. Child Dev. Perspect. **10**(2), 122–127 (2016). https://doi.org/10.1111/cdep.12174
26. Tang, J., Chang, S., Aggarwal, C., Liu, H.: Negative link prediction in social media. In: Proceedings of the Eighth ACM International Conference on Web Search and Data Mining, pp. 87–96 (2015). https://doi.org/10.1145/2684822.2685295

27. Kroenke, K., Spitzer, R.L., Williams, J.B., Löwe, B.: The patient health questionnaire somatic, anxiety, and depressive symptom scales: a systematic review. Gener. Hospit. Psychiatry **32**(4), 345–359 (2010). https://doi.org/10.1016/j.genhosppsych.2010.03.006
28. Andreas, J., Brunborg, G.S.: Depressive symptomatology among Norwegian adolescent boys and girls: the patient health Questionnaire-9 (PHQ-9) psychometric properties and correlates. Front. Psychol. **8**, 887 (2017). https://doi.org/10.3389/fpsyg.2017.00887
29. Tsai, F.J., Huang, Y.H., Liu, H.C., Huang, K.Y., Huang, Y.H., Liu, S.I.: Patient health questionnaire for school-based depression screening among Chinese adolescents. Pediatrics **133**, e402–e409 (2014). https://doi.org/10.1542/peds.2013-0204
30. Richardson, L.P., et al.: Evaluation of the Patient Health Questionnaire (PHQ-9) for detecting major depression among adolescents. Pediatrics **126**, 1117–1123 (2010). https://doi.org/10.1542/peds.2010-0852
31. Fatiregun, A.A., Kumapayi, T.E.: Prevalence and correlates of depressive symptoms among in-school adolescents in a rural district in southwest Nigeria. J. Adolescents **37**, 197–203 (2014). https://doi.org/10.1016/j.adolescence.2013.12.003
32. Ganguly, S., Samanta, M., Roy, P., Chatterjee, S., Kaplan, D.W., Basu, B.: Patient health questionnaire-9 as an effective tool for screening of depression among Indian adolescents. J. Adolescent Health **52**(5), 546–551 (2013). https://doi.org/10.1016/j.jadohealth.2012.09.012
33. Tafoya, S.A., Aldrete-Cortez, V.: The interactive effect of positive mental health and subjective sleep quality on depressive symptoms in high school students. Behavioral Sleep Medicine **17**(6), 818–826 (2019). https://doi.org/10.1080/15402002.2018.1518226
34. Tsehay, M., Necho, M., Mekonnen, W.: The role of adverse childhood experience on depression symptoms, prevalence, and severity among school going adolescents. Depress. Res. Treat. **2020**, 1–9 (2020). https://doi.org/10.1155/2020/5951792
35. Leung, D.Y., Mak, Y.W., Leung, S.F., Chiang, V.C., Loke, A.Y.: Measurement invariances of the PHQ-9 across gender and age groups in Chinese adolescents. Asia-Pac. Psychiatry, e12381 (2020). https://doi.org/10.1111/appy.12381
36. Spielberger, C.D., Sydeman, S.J., Owen, A.E., Marsh, B.J.: Measuring anxiety and anger with the State-Trait Anxiety Inventory (STAI) and the State-Trait Anger Expression Inventory (STAXI). In: Maruish, M.E. (ed.) The Use of Psychological Testing for Treatment Planning and Outcomes Assessment, pp. 993–1021. Lawrence Erlbaum Associates Publishers (1999)
37. Buysse, D.J., Reynolds, C.F., Monk, T.H., Berman, S.R., Kupfer, D.J.: The Pittsburgh Sleep Quality Index (PSQI): a new instrument for psychiatric research and practice. Psychiatry Res. **28**(2), 193–213 (1989). https://doi.org/10.1016/0165-1781(89)90047-4
38. Thelwall, M., Buckley, K., Paltoglou, G., Cai, D., Kappas, A.: Sentiment strength detection in short informal text. J. Am. Soc. Inf. Sci. Technol. **61**(12), 2544–2558 (2010). https://doi.org/10.1002/asi.21416
39. Kern, M.L., et al.: Gaining insights from social media language: methodologies and challenges. Psychol. Methods **21**(4), 507–525 (2016). https://doi.org/10.1037/met0000091
40. Jaidka, K., Guntuku, S.C., Buffone, A., Schwartz, H.A., Ungar, L.H.: Facebook vs. Twitter: cross-platform differences in self-disclosure and trait prediction. In: Proceedings of the Twelfth International AAAI Conference on Web and Social Media, pp. 141–150 (2018). https://aaai.org/ocs/index.php/ICWSM/ICWSM18/paper/view/17882
41. Jungherr, A.: Normalizing digital trace data. In: Stroud, N.J., McGregor, S. (eds.) Digital Discussions: How Big Data Informs Political Communication. Routledge (2018). https://doi.org/10.4324/9781351209434

A Computational Analysis of Polarization on Indian and Pakistani Social Media

Aman Tyagi(✉), Anjalie Field, Priyank Lathwal, Yulia Tsvetkov, and Kathleen M. Carley

Carnegie Mellon University, Pittsburgh, PA 15213, USA
{amant,anjalief,plathwal}@andrew.cmu.edu,
{ytsvetko,kathleen.carley}@cs.cmu.edu

Abstract. Between February 14, 2019 and March 4, 2019, a terrorist attack in Pulwama, Kashmir followed by retaliatory airstrikes led to rising tensions between India and Pakistan, two nuclear-armed countries. In this work, we examine polarizing messaging on Twitter during these events, particularly focusing on the positions of Indian and Pakistani politicians. We use a label propagation technique focused on hashtag co-occurrences to find polarizing tweets and users. Our analysis reveals that politicians in the ruling political party in India (BJP) used polarized hashtags and called for escalation of conflict more so than politicians from other parties. Our work offers the first analysis of how escalating tensions between India and Pakistan manifest on Twitter and provides a framework for studying polarizing messages.

Keywords: Polarization · Hashtags · Political communication strategies

1 Introduction

While social media platforms foster open communication and have the potential to offer more democratic information systems, they have simultaneously facilitated divisions in society by allowing the spread of polarizing and incendiary content [23,54]. Polarizing content can be beneficial by encouraging pride and solidarity, but it has also become a social cyber-security concern: foreign and domestic actors may employ polarizing social media content to sow divisions in a country, to demean other nations, or to promote political agendas [4,13,15,34].

Using automated methods to analyze social media offers a way to understand the type of content users are exposed to, the positions taken by various users, and the agendas pursued through coordinated messaging across entire platforms. Understanding the dynamics of this information landscape has become critical, because social media can strongly influence public opinion [13]. However, prior computational social science research on polarization has focused primarily on

A. Tyagi and A. Field—Equal Contribution.

S. Aref et al. (Eds.): SocInfo 2020, LNCS 12467, pp. 364–379, 2020.
https://doi.org/10.1007/978-3-030-60975-7_27

U.S. politics, and much attention has focused on the influence of Russian or Chinese state actors [4,5,23,30,34,38,52,54]. In contrast, we focus on polarizing social media content in India and Pakistan and how it can contribute to rising tensions between these two nations. Specifically, we examine communication patterns on Twitter following the terrorist attack in the Pulwama district, Jammu and Kashmir, India, on February 14, 2019.

We primarily investigate: *to what extent did entities on social media advocate for or against escalating tensions?*. India and Pakistan are both nuclear-armed countries and have a decades-long history involving multiple armed conflicts. The Pulwama attack in 2019 was followed by an escalation of tensions between these two nations that nearly approached full-fledged war [26,45,46]. Moreover, the relationship between these countries is an important agenda for political parties in both India and Pakistan. India has two primary political parties: the Indian National Congress (INC), which was dominant in the early 21st century, and the Bharatiya Janata Party (BJP), which rose to prominence on a populist and nationalist platform in 2014 and has been in power since [40]. Given this context, we first examine the tweets and communication patterns of general users in order to understand how polarizing the attack was and to what extent users with different viewpoints may have interacted with each other. We then examine the social media messaging of political party members and how it changed over the sequence of events in order to uncover possible political agendas.

Our core methodology uses a network-based label propagation algorithm to quantify the polarity of hashtags along specified dimensions: Pro-India vs. Pro-Pakistan and Pro-Aggression vs. Pro-Peace. We then aggregate the hashtag-level scores into tweet-level and user-level scores, e.g. the polarity of a given user on a given day. Unlike methodology that assumes users' opinions do not change [20,21,62], focuses on binary stances [12], or requires in-language annotations and feature-crafting [39], our methodology allows us to analyze degrees of polarization in a multilingual corpus and how they change over time.

We begin by providing an overview of the events between February 14 and March 1, 2019 (Sect. 2). Next, we describe the Twitter data collection (Sect. 3) and discuss methods (Sect. 4) and evaluation (Sect. 5). Our results (Sect. 6) suggest that more members of the BJP propagated a narrative of escalation than members of other political parties. This finding supports anecdotes reported by journalists [61] about these events. Through this research, we develop (1) the first analysis of escalating tensions between India and Pakistan on Twitter, (2) a data-driven investigation of social media messaging following the 2019 Pulwama attack, and (3) a novel and general methodology to examine polarization on multilingual social media.

2 Timeline of Events

We briefly provide background on relevant events, relying primarily on third party newspapers unaffiliated with either nation (The New York Times and BBC News) and noting where official accounts differ.

Feb. 14, 2019. A 22-year old native of Pulwama carried out a suicide attack against a convoy carrying approximately 2,500 security personnel in the Pulwama district in Kashmir, India. The attack resulted in the death of more than 40 Indian soldiers. Jaish-e-Mohammad (JeM) a militant group based in Pakistan (the group is formally banned in Pakistan) claimed responsibility [8,58].

Feb. 14–26, 2019. The Indian government responded to the attack with threats of retaliation against Pakistan, even though Pakistani officials denied any role [9]. Diplomatic ties deteriorated, e.g., India revoked Pakistan's most favored nation status, which had provided trade advantages. Pakistan threatened to retaliate if India pursued military action [10].

Feb. 26, 2019. The Indian Air Force (IAF) conducted a retaliatory airstrike against a JeM training camp inside Pakistan, which the Indian government termed "non-military, preemptive" [57]. According to Indian government officials, the JeM camp targeted by this airstrike was located 70km inside the Line of Control (LoC) – the military line dividing the Indian and Pakistani controlled parts of Jammu and Kashmir. Indian officials reported that the airstrike was "100% successful", went on "exactly as planned", and killed over 200 terrorists [44,57]. In contrast, Pakistani officials reported that the target of the attacks was located only 5–6km inside the LoC, that the Pakistani air force turned back the Indian fighters, and that the attacks landed in an empty area [22,55].

Feb. 27, 2019. The Pakistan Air Force (PAF) carried out retaliatory airstrikes along the LoC. Indian and Pakistani officials presented different details of the strikes, but both emphasized de-escalation: a Pakistani official reported that the PAF intentionally targeted open spaces, to demonstrate Pakistan's capabilities without inviting escalation, while an Indian official reported no deaths or civilian casualties [17,18,24]. However, in aerial combat following the strikes, an IAF pilot was captured by the Pakistani Army [7,56].

Mar. 1, 2019. Pakistan returned the IAF pilot to India on March 1 in what Pakistani Prime Minster Imran Khan called "a gesture of peace" [6].

3 Data

We collected tweets related to these events by first identifying a set of relevant hashtags. Our hashtag set is based on hashtags related to #pulwama found on best-hashtags.com.[1] We modified the hashtag set to ensure that it included both hashtags more likely to be used by Pro-India users (e.g., IndiaWantsRevenge) and hashtags more likely to be used by Pro-Pakistan users (e.g., PakistanZindabad).

[1] best-hashtags.com uses an algorithm to provide popular hashtags that are similar to the provided seed (#pulwama). Since our analysis, the website has stopped reporting Twitter hashtags.

We then collected all tweets using these terms, either as words or as hashtags during the events.[2]

Our final data set contains 2.5M unique tweets (including retweets) from 567K users that use 67K unique hashtags. All tweets occurred between February 14th and March 4th. The data contains a mix of languages including English, Urdu, and Hindi, and many users use multiple languages in the same tweet. While some tweets express neutral opinions, others contain incendiary language, such as: *"@PMOIndia @PMOIndia @narendramodi We r eagerly waiting for ur action of revenge...#PulwamaRevenge #IndiaWantsRevenge #PulwamaAttack"* and *"I feel time has come to give all support to #Balochistan activist. Let us #bleed Pakistan from all fronts. #NeverForget @PMOIndia @narendramodi #IndiaWantsRevenge"*.

4 Methodology

We develop a method to assign a polarity score to an aggregate group of tweets, and we analyze how polarities change over time for different groups of users. For instance, given pole A (e.g., Pro-Pakistan) and pole B (e.g., Pro-India), we aggregate all tweets by a given user and assign the user a polarity score between [a, b], where a score close to a indicates the user more likely supports A and a score close to b indicates the user more likely supports B. We could also aggregate only tweets by the user on one day and determine the user's Pro-A/Pro-B polarity on that day.

In the absence of annotated data, we use a weakly supervised approach. First, for pole A, we hand-select a small seed set of tokens S_A that are strongly associated with A, and we equivalently hand-select S_B. We assign each $s \in S_A$ a polarity score of a, and we assign each $s \in S_B$ a polarity score of b. Then, we use S_A and S_B to infer polarity scores over a larger lexicon of words or hashtags $\mathcal{V}$, where each $w \in \mathcal{V}$ is assigned a score in $[a, b]$. Finally, we estimate the polarity of an aggregated set of tweets by averaging the inferred polarity scores for all $w \in \mathcal{V}$ used in those tweets.

In order to propagate the hand-annotated labels in S_A and S_B to the larger lexicon $\mathcal{V}$, we use 3 variants of graph-based label propagation. In each variant, we construct a graph G, whose nodes consist of $w \in \mathcal{V}$ and whose edges and edge weights are defined based on similarity metrics between members of $\mathcal{V}$. We describe each variant in detail below.

Network-Based Hashtag Propagation. In the first variant, we define $\mathcal{V}$ to be the set of all hashtags used in our data set. Then, we construct G as a hashtag*hashtag co-occurrence network. Each node in G corresponds to a hashtag. Edges occur between hashtags that co-occur in the same tweet, and edge weights are proportional to how frequently the hashtags co-occur. Then, we use the label

[2] We provide further details, including the full list of keywords, data statistics, network densities and evidence that our data set is comprehensive in our project repository: https://github.com/amantyag/india_pakistan_polarization.

propagation algorithm detailed in Algorithm 1 to infer polarity scores for $w \in \mathcal{V}$ from S_A and S_B, where $a = -1$ and $b = 1$. The algorithm uses a greedy approach to assign labels to each node in G. If all nodes connected to a node n have been labeled, then node n is assigned a weighted average of all the adjacent nodes. This step is repeated until the maximum possible number of nodes are labeled. A low value of γ would label nodes neighboring unlabeled nodes, a high value would only label nodes neighboring unlabeled nodes after multiple iterations of the outer loop.

Our algorithm is similar to methods used to infer user-level polarities, in which a small seed of users is hand-annotated and a graph-based algorithm propagates labels to other users by assuming that users who retweet each other share the same views [21,62]. For example, [29] quantify polarity based on a graph structure by assuming that the controversial topics induce clusters of discussions, commonly referred to as "echo-chambers". However, we conduct propagation at a hashtag level, by assuming that hashtags that frequently occur in the same tweets indicate similar polarities. Also, our approach does not assume homophily in retweet network nor that user polarities are constant over time. Graph-based approaches have also been used to examine sentiment or for mixed tweet/hashtag/user-level analyses [19,48].

Network-Based Word Propagation. The second variant is similar to the first; however, instead of restricting $\mathcal{V}$ to be the set of hashtags in the corpus, we define $\mathcal{V}$ to be the set of all tokens, including words and hashtags. We then construct G as a token*token co-occurrence network, and as above, we infer labels using Algorithm 1 and obtain token-level polarity scores in the range $[-1, 1]$. Expanding $\mathcal{V}$ to all tokens instead of just hashtags allows our algorithm to incorporate more information, but also risks introducing noise, as we do not attempt to process nuances in language like negation.

Embedding-Based Word Propagation (SentProp). In the third variant, we define $\mathcal{V}$ to be the set of all tokens, as in the Network-based Word Propagation approach. Then, we train GloVe embeddings [47] over our entire corpus (limiting vocabulary size to 50K). We then use SentProp [31], a method for inferring domain-specific lexicons to infer labels over $\mathcal{V}$. In this method, as before, we construct a graph G where each $w \in \mathcal{V}$ is a node. However, rather than relying on raw co-occurrence scores, SentProp uses embedding similarity metrics to define edge weights and a random-walk method to propagate labels. We implement SentProp using the SocialSent package [31], where $a = 0$ and $b = 1$.

Once we have obtained hashtag-level or word-level polarity scores, we infer the polarity of a tweet or a group of tweets (e.g. all tweets by a given user) by averaging the polarity scores inferred by our algorithms for all the hashtags and words used in data subset. This approach is similar to the aggregation conducted in [12], but our label propagation allows for the incorporation of thousands of words and hashtags, rather than relying on only a small hand-annotated set. If the data subset does not contain any of the keywords labelled by our algorithm (e.g. in a hashtag-based approach, the tweet contains no hashtags), we consider it unclassified. In some cases, primarily for evaluation, we convert the polarity

Algorithm 1: Label Propagation Algorithm

Input: Graph G with nodes n and edges e with e_{ij} as the edge weight between $i \in n$ and $j \in n$
initialize $\gamma = 50/100$ and i=0;
for *each n* **do**
  define l = integer(i/γ); i+=1;
  for *each n* **do**
    if *n not labeled* **then**
      compute t = neighbors of n;
      compute t_l = labeled neighbors of n;
      if $|t_l| + l \geq t$ **then**
        initialize *score*, *c*
        for *each* $t_i \in t$ **do**
          score += label $t_i * e_{nt_i}$; c += e_{nt_i}
        update label n = *score/c*

scores into a ternary negative/neutral/positive position by using the cut-offs $\{<0, 0, >0\}$ for the $[-1, 1]$ scale and $\{<0.5, 0.5, >0.5\}$ for the $[0, 1]$ scale.

This methodology allows us to infer the polarity of any group of tweets along any dimensions, provided a small set of seed words or hashtags for each dimension. Thus, we can examine how polarities differed for different groups of users and how they changed over time. The two dimensions we focus on are Pro-India/Pro-Pakistan and Pro-Peace/Pro-Aggression. In practice we found that minor variations in the exact words in the seed set had no noticeable impact on our final results. For the network-based methods, we label Pro-India seeds as +1, Pro-Pakistan seeds as −1, Pro-Peace seeds as +1, and Pro-Aggression seeds as −1. For the embedding-based approach, we label Pro-India seeds as +1 Pro-Pakistan seeds as 0, Pro-Peace seeds as +1, and Pro-Aggression seeds as 0. For all word-based approaches, we limit the vocabulary size to 50K.[3]

5 Evaluation

Automated Evaluation. We first evaluate our methods by focusing on the Pro-India and Pro-Pakistan dimension and assuming that popular users in India are more likely to post Pro-India content and popular users in Pakistan are more likely to post Pro-Pakistan content. From the Socialbakers.com platform, we identified the 100 most followed Twitter accounts in India and in Pakistan. 16 of the Indian accounts and 15 of the Pakistani accounts do not occur in our data, leaving 84 Indian accounts with 2,199 tweets and 85 Pakistani accounts with 1,456 tweets for evaluation. For each account, we average word and hashtag

[3] We provide our manually defined seed sets and label propagation code on https://github.com/amantyag/india_pakistan_polarization/.

Table 1. Classification results for the 100 most followed Indian and Pakistani Twitter accounts, where Pro-India or Pro-Pakistan are treated as the dominant class, and the nationality of the account owner is treated as a gold label. %Unk denotes accounts that our algorithm was unable to classify and %Incorrect denotes accounts that received polar opposite labels (e.g. Indian accounts classified as Pro-Pakistan)

	Pro-India (84 accounts)				Pro-Pakistan (85 accounts)			
	Prec.	Recall	%Unk.	%Incorrect	Prec.	Recall	%Unk.	%Incorrect
Hashtag	0.91	0.25	0.68	0.07	0.90	0.61	0.36	0.02
Word	0.69	0.69	0.24	0.07	0.83	0.35	0.34	0.31
Sentprop	0.48	0.80	–	0.20	0.43	0.15	–	0.85

polarities over all tweets from the account, and binarize the resulting score into a Pro-Pakistan or Pro-India position.

Table 1 reports results. Both of the network-based approaches rely on hashtag or word co-occurrences to propagate labels. Thus, hashtags and words that do not have any co-occurrence links to the original seed list are unable to be labeled. For instance, in the hashtag propagation approach, our method labels 41,700 hashtags out of 67,059 total hashtags in the dataset. Any users who only use unlabeled words or hashtags are therefore unable to be classified by our algorithm, resulting in 88/169 unlabeled accounts for the hashtag approach and 49/169 unlabeled accounts for the word approach (%Unk in Table 1). In contrast, SentProp obtains polarity scores for all accounts, as it relies on embedding similarity and can propagate labels between words, even if they do not ever co-occur.

However, although SentProp labels more accounts, its precision is much lower than the network-based methods. The network-based hashtag propagation approach overall obtains the highest precision and the least explicit errors – lower recall scores occur because of accounts that it leaves unlabeled, rather than because of accounts that it labels incorrectly. Although the word-propagation approach labels more accounts and works well over the Indian accounts, its classification of the Pakistani accounts is close to random. We suspect that our method works well for hashtags, because they tend to be strongly polar and indicative of the overall sentiment of the tweet. A word-based approach likely requires more careful handling of subtle language cues like negation or sarcasm.

In our subsequent analysis, we use the network-based hashtag propagation method in order to infer polarities, thus favoring high precision and strong polarization, and choosing not to analyze data where we cannot infer polarity with high-confidence. Additionally, in examining the data set, we found that many of the top-followed accounts in India and Pakistan consisted of celebrities who avoided taking stances on politicized issues, which makes the high number of unclassified accounts in this subset of the data unsurprising.

Manual Evaluation. In order to further evaluate our methods, we compare the performance of the network-based hashtag model with a small sample of

Table 2. Inter-annotator agreement and classification accuracy over 100 manually annotated data points

	Krippendorff α	% Agree.	Hashtag Acc.	Soft Hashtag Acc.
India/Pakistan	0.77	88%	74%	89%
Aggression/Peace	0.60	74%	57%	76%

manually annotated tweets. We randomly sampled 100 users from our data set. For each user, we randomly sampled 1 day on which the user tweeted and aggregated all tweets from that day. Thus, we conduct this evaluation at a per-user-per-day level. Two annotators independently annotated each data sample as Pro-India/Pro-Pakistan/Neutral/Can't Determine and Pro-Peace/Pro-Aggression/Neutral/Can't Determine. For simplicity, we collapsed Neutral/Can't Determine and Unclassified into a single "Neutral" label. Notably, the Pro-Peace/Pro-Aggression and Pro-India/Pro-Pakistan dimensions are distinct. For example, users may write tweets that are Pro-Peace and Pro-Pakistan: *"Don't let people pull you into their War, pull them into your Peace... Peace for our World, Peace for our Children, Peace for our Future !! #PakSymbolOfPeace #SayNoToWar"* or that are Pro-Peace and Pro-India: *"Very mature conciliatory speech by #ImranKhan. We now urge him to walk the talk. Please return our #Abhinandan safely back to us. This will go a long way in correcting perceptions and restoring peace. #SayNoToWar"*.

Table 2 reports inter-annotator agreement, which is generally high. Additionally, most disagreements occurred when one annotator labeled Neutral/Can't Determine and the other did not, meaning polar opposite annotations were rare. If we only count polar opposite labels as disagreements, the percent agreement rises to 94% for both dimensions.

Then, the two annotators discussed any data points for which they initially disagreed and decided on a single gold label for each data point. We compare performance of the network-based hashtag propagation method against these gold annotations in Table 2. In this 3-way classification task, the accuracy of random guessing would be 33%, which our method easily outperforms. In particular, the "Soft" accuracy, in which we only consider the model output to be incorrect if it predicted the polar-opposite label, meaning neutral/unclassified predictions are not considered incorrect, is high for both dimensions.[4]

6 Results and Analysis

We investigate multiple aspects of our data set, including network structure, polarities of various entities, and changes over time. Based on prior work suggesting that political entities in India and Pakistan may use social media to

[4] We provide the manual annotations as well as additional metrics on https://github.com/amantyag/india_pakistan_polarization.

influence public opinion [2,3,36,51], we pay particular attention to the Twitter accounts of politicians as a method for uncovering political agendas.

Table 3. Overall polarities of users and tweets.

Position	Unique users	Total tweets	Position	Unique users	Total tweets
Pro-India	125K (23%)	1.16M (46%)	Pro-Aggression	78K (14%)	626K (25%)
Pro-Pakistan	117K (20%)	764K (30%)	Pro-Peace	252K (45%)	1.48M (59%)
Unclassified	325K (57%)	578K (23%)	Unclassified	237K (40%)	351K (16%)

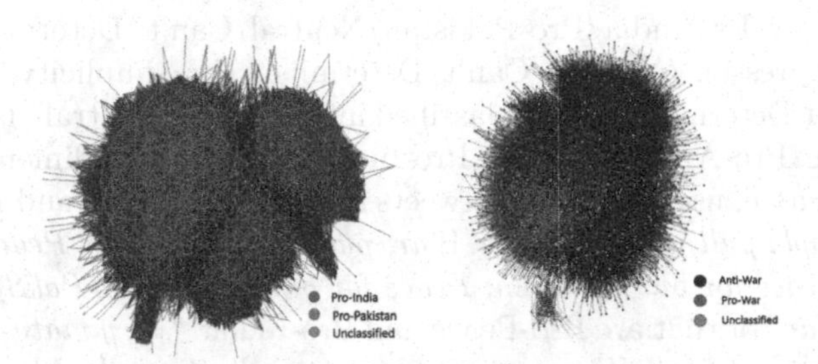

Fig. 1. 30-core all communication networks, colored by Pro-India/Pro-Pakistan polarity (left) and Pro-Peace/Pro-Aggression polarity (right). The Pro-India/Pro-Pakistan network displays more homophily than the Pro-Peace/Pro-Aggression network.

What Are the Overall Polarities of our Data Set? In Table 3, we obtain polarity scores for each user and tweet and then ternarize them into Pro-India/Pro-Pakistan/Unclassified and Pro-Peace/Pro-Aggression/Unclassified as in Sect. 5. At the user level, the classified accounts are approximately balanced between Pro-India and Pro-Pakistan. However, at the tweet level, the classified data contains a high percentage of Pro-India tweets, suggesting Pro-India users tweeted about this issue more prolifically. Further, there is a much higher percentage of Pro-Peace users than Pro-Aggression users. This pattern also holds at the tweet level, where only a small percentage of tweets are unclassified.

What Are Characteristics of the Communication Network? Next, we examine the communication network between users, particularly prevalence of echo chambers. Did users with opposite positions interact? Figure 1 shows a 30-core all communication network constructed using ORA-PRO [14]. Accounts are colored based on their Pro-India/Pro-Pakistan polarity (left) and Pro-Peace/Pro-Aggression polarity (right). An edge occurs between two users if one user retweeted, mentioned, or replied to the other and users with $\leq$30 links are not shown. Unsurprisingly, the Pro-India/Pro-Pakistan position is highly segregated, with little interaction between users with different positions. In contrast, the Pro-Peace/Pro-Aggression dimension is more mixed. Although there are some areas of high density for each position, there are interactions between users of different positions, which are potential avenues for users to influence each other's views.

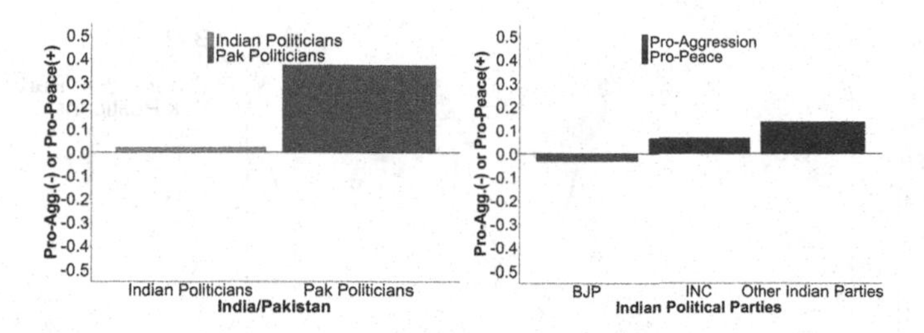

Fig. 2. Aggregate Pro-Peace and Pro-Aggression polarities of the most popular Indian (33/78) and Pakistani (36/66) politicians in our data set (left) and of members of Indian political parties (right).

How Polarized were Different Political Entities? We investigate the polarities projected by different political entities: specifically BJP politicians (currently in power in India), INC politicians (largest opposition party), other Indian politicians, and Pakistani politicians. We used the Socialbakers.com platform to obtain the Twitter handles of the 100 most followed politicians in India and Pakistan. Our data contained tweets from 66 Pakistani and 78 Indian politicians, and our hashtag model inferred scores for 36 Pakistani and 33 Indian politicians. Figure 2 (left) reports aggregate polarity scores over all tweets from these politicians. Pakistani politicians were predominantly Pro-Peace, while Indian politicians expressed mixed polarities, yielding a near neutral score.

We then examined a broader set of Indian politicians, subdivided by political party based on a list of members running for parliament elections in 2019 [37]. Out of the 1,360 Twitter handles in the list, our data set contained activity from 316 BJP accounts, 281 INC accounts, 204 other Indian party accounts.

Figure reffig:indiaspspakspspolarity (right) shows the overall polarities, aggregated from all tweets by verified members of each party. Strikingly, members of the BJP party are positioned as much more Pro-Aggression than the members of either the INC or other parties, and the party overall obtains a Pro-Aggression polarity score. This score is not dominated by 1–2 strongly polarized members of the party: if we aggregate the polarity scores by individuals instead of by party, 15% of BJP members had net Pro-Aggression scores and 13% had net Pro-Peace scores, in comparison to 10% Pro-Aggression/25% Pro-Peace for INC, and 6% Pro-Aggression/29% Pro-Peace for other parties. The language used by BJP politicians was often openly Pro-Aggression: *#IndiaWantsRevenge We need to give a befitting reply to Pakistan, we will strike back....*

These results support observations made by journalists and community members about the role of the BJP party in these events. BJP is well-known for promoting nationalism, and several journalists have speculated that conflict with

Pakistan would increase Prime Minister Modi's chances of winning the upcoming elections in April and have accused the BJP of war-mongering [28,41,59].

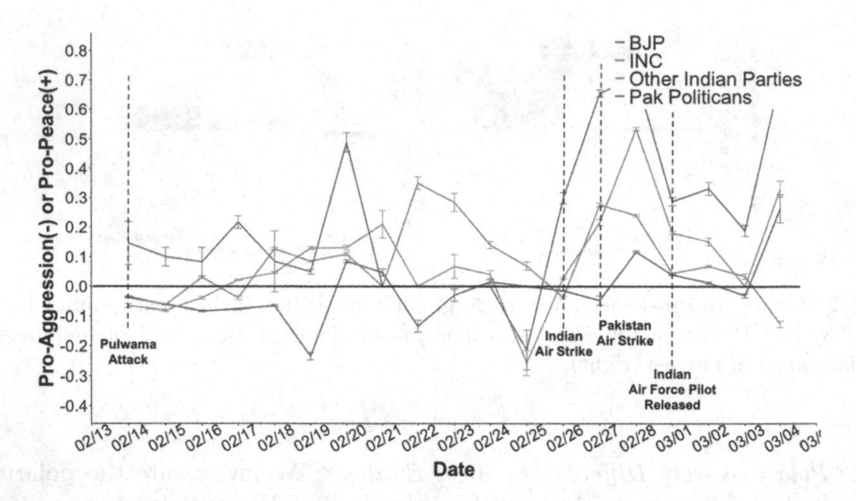

Fig. 3. Daily Pro-Peace/Pro-Aggression positions of political entities. Negative values denote net Pro-Aggression polarity and positive values denote net Pro-Peace. The error bars represent ± 1 standard deviation.

How did Polarization Change Over Time? Figure 3 shows how this polarity changed over the two-week period of events: we infer a Pro-Peace/Pro-Aggression polarity score for all tweets posted by members of the specified political subgroup, and we plot the average score across tweets posted each day.

Immediately following the initial attack on 2/14, the tweets from all Indian political party members are inclined towards Pro-Aggression, suggesting initial outrage. However, over the next few days, while tweets from INC and other Indian political party members switch towards Pro-Peace, tweets from BJP politicians remain consistently Pro-Aggression. There is high volatility between 2/20 and 2/26. However, there was a much lower volume of tweets about the Pulwama incidents during this time period,[5] and we do not believe these fluctuations are meaningful. The volume of tweets increases once again following the Indian (2/26) and Pakistani (2/27) airstrikes. Tweets by Pakistani politicians generally fall on the Pro-Peace side, but they become more polarized after the Indian airstrike and reach a peak following the Pakistani airstrike. This is consistent with reported quotes by Pakistani officials (Sect. 2), saying that the airstrike was designed to avoid escalation. Similarly, tweets by Indian politicians from the INC and other parties become strongly Pro-Peace directly following the Indian airstrike, with polarity increasing after the Pakistani airstrike. In contrast, on the day of the Pakistani airstrike, tweets by BJP politicians remain

[5] Tweet volume is provided in our project repository.

Pro-Aggression, possibly focusing either on praise for the Indian airstrike or condemnation of the Pakistani airstrike. The polarity of the BJP tweets belatedly switches to Pro-Peace on the following day (2/28), though the strength of the Pro-Peace polarity still remains weaker for BJP tweets than for tweets by other politicians.

7 Discussion and Related Work

The potential that social media platforms have for manipulating public opinion has led to growing interest in information operations and the development of social cyber security as a field of research [15,52]. While we do not claim that social media coverage of the Pulwama incident constituted an information operation, e.g. coordinated efforts to manipulate public opinion and change how people perceive events [52], we do find similarities between our observations and other work in this area. Notably, as described in Sect. refsec:events, the Indian and Pakistani governments maintain starkly different accounts about the events that occurred, particularly whether or not the 2/26 airstrikes resulted in 200 casualties. Similarly, Russian and Ukranian governments circulated conflicting narratives about the cause of the crash of Malaysian Airlines Flight MH17 in 2014, which prompted analyses of information operations about this incident. In a work similar to ours, [30] examine social media coverage of the incident by using a set of hashtags to collect all relevant tweets during a set time frame. Other work has examined the media influence of Chinese and Russian state actors in various domains, including US and UK elections and the Syrian War [27,34,35,50,53]. [4] examine Russian influence in polarizing movements on Twitter, particularly the #BlackLivesMatterMovement, and observe how Russian actors attempted to increase tensions between involved parties. Furthermore, the polarization that we observe in our data align with the "Excite" and "Dismay" strategies, which are tools of public opinion manipulation described in the BEND forms of maneuver [11].

Almost all of these works are focused on U.S. social media, possibly involving Chinese and Russian actors. In general, most work on polarization and public opinion change has focused on U.S. politics [16,21,33], with a few exceptions focusing on Germany [20], Egypt [12,62], and Venezuela [42]. Work on social media in India and Pakistan has focused on healthcare [1], natural disasters [43], self-promotion (e.g. "brand marketing") primarily in relation to elections [2,3,36], or on election forecasting [32,51], though [60] does argue that the Pakistan Army uses social media to subvert democracy. While these works only focus on intra-country analysis, our work also examines tensions between India and Pakistan. A small selection of work has also looked at the incidents in Pulwama and the implications of rising tensions. [26] and [46] discuss the sociopolitical context and implications of events from a non computational perspective. [45] additionally conduct a social media analysis, but they use YouTube data and focus on identifying deescalating language. Their timeline of escalation and deescalation is generally consistent with our findings.

Our primary methodology involves using label propagation to infer aggregated polarity scores. In language corpora, label propagation has typically relied on embedding similarity [31,49]. Instead, our approach takes advantage of the short-text nature of Twitter through co-occurrences networks, as well as the strong semantic signals provided by hashtags [25]. Prior methods for analyzing polarization focus on inferring user-level scores [20,21] or require in-language annotations and feature-crafting [39], whereas our method facilitates analyzing how user polarities can change over time in a multilingual corpus.

Conclusions. Polarizing language on social media can have long-lasting sociopolitical impacts. Our analysis shows how Twitter users in India and Pakistan used polarizing language during a period of escalating tensions between the two nations, and our methodology offers tools for future work in this area.

Acknowledgements. We thank anonymous reviews and colleagues who provided feedback on this work. The authors would like to acknowledge the support of center for Computational Analysis of Social and Organizational Systems (CASOS), Carnegie Mellon University. This research was also supported in part by Public Interest Technology University Network Grant No. NVF-PITU-Carnegie Mellon University-Subgrant-009246-2019-10-01. The second author of this work is supported by NSF-GRFP under Grant No. DGE1745016. Any opinions, findings, and conclusions or recommendations expressed in this material are those of the authors and do not necessarily reflect the views of the NSF.

References

1. Abbasi, R.A., et al.: Saving lives using social media: analysis of the role of twitter for personal blood donation requests and dissemination. Telematics Inform. **35**(4), 892–912 (2018). https://doi.org/10.1016/j.tele.2017.01.010
2. Ahmed, S., Jaidka, K., Cho, J.: The 2014 Indian elections on Twitter: a comparison of campaign strategies of political parties. Telematics Inform. **33**(4), 1071–1087 (2016). https://doi.org/10.1016/j.tele.2016.03.002
3. Antil, A., Verma, H.V.: Rahul Gandhi on Twitter: an analysis of brand building through Twitter by the leader of the main opposition party in India. Global Busi. Rev. **0**(0) (2019). https://doi.org/10.1177/0972150919833514
4. Arif, A., Stewart, L.G., Starbird, K.: Acting the part: examining information operations within# blacklivesmatter discourse. In: Proceedings of CSCW, p. 20 (2018)
5. Badawy, A., Ferrara, E., Lerman, K.: Analyzing the digital traces of political manipulation: the 2016 Russian interference Twitter campaign. In: Proceedings of ASONAM, pp. 258–265 (2018)
6. BBC News: Abhinandan: Captured Indian pilot handed back by Pakistan (2019). https://www.bbc.com/news/world-asia-47412884
7. BBC News: India Pakistan: Kashmir fighting sees Indian aircraft downed (2019). https://www.bbc.com/news/world-asia-47383634
8. BBC News: Kashmir attack: Tracing the path that led to Pulwama (2019). https://www.bbc.com/news/world-asia-india-47302467
9. BBC News: Pulwama attack: India will 'completely isolate' Pakistan (2019). https://www.bbc.com/news/world-asia-india-47249133

10. BBC News: Pulwama attack: Pakistan warns India against military action (2019). https://www.bbc.com/news/world-asia-india-47290107
11. Beskow, D.M., Carley, K.M.: Social cybersecurity: an emerging national security requirement. Milit. Rev. **99**(2), 117 (2019)
12. Borge-Holthoefer, J., Magdy, W., Darwish, K., Weber, I.: Content and network dynamics behind Egyptian political polarization on Twitter. In: Proceedings of CSCW, p. 700–711 (2015). https://doi.org/10.1145/2675133.2675163
13. Bradshaw, S., Howard, P.N.: Challenging truth and trust: A global inventory of organized social media manipulation. The Computational Propaganda Project (2018)
14. Carley, K.M.: ORA: a toolkit for dynamic network analysis and visualization. In: Encyclopedia of Social Network Analysis and Mining (2017). https://doi.org/10.1007/978-1-4614-7163-9_309-1
15. Carley, K.M., Cervone, G., Agarwal, N., Liu, H.: Social cyber-security. In: Thomson, R., Dancy, C., Hyder, A., Bisgin, H. (eds.) SBP-BRiMS 2018. LNCS, vol. 10899, pp. 389–394. Springer, Cham (2018). https://doi.org/10.1007/978-3-319-93372-6_42
16. Chen, S., Khashabi, D., Yin, W., Callison-Burch, C., Roth, D.: Seeing things from a different angle: discovering diverse perspectives about claims. In: Proceedings of NAACL, pp. 542–557 (2019)
17. CNBC: India and Pakistan say they've launched airstrikes against each other. Here's what you need to know (2019). https://www.cnbc.com/2019/02/27/india-pakistan-air-strike-claims-what-you-need-to-know.html
18. CNBC: Pakistan says it shot down Indian jets, carried out air strikes in Kashmir (2019). https://www.cnbc.com/2019/02/27/indian-air-force-plane-crashes-in-kashmir-says-indian-police-official.html
19. Coletto, M.,et al.: Sentiment-enhanced multidimensional analysis of online social networks: Perception of the Mediterranean refugees crisis. In: Proceedings of ASONAM, pp. 1270–1277 (2016)
20. Darius, P., Stephany, F.: Twitter "Hashjacked": online polarisation strategies of Germany's political far-right. In: Proceedings Socio Informatics, pp. 188–201 (2019)
21. Darwish, K.: Quantifying polarization on twitter: the Kavanaugh nomination. In: Proceedings of Socio Informatics, pp. 188–201 (2019)
22. Dawn: Indian aircraft violate LoC, scramble back after PAF's timely response: ISPR (2019). https://www.dawn.com/news/1466038/indian-aircraft-violate-loc-scramble-back-after-pafs-timely-response-ispr
23. Demszky, D., et al.: Analyzing polarization in social media: Method and application to tweets on 21 mass shootings. In: Proceedings of NAACL, pp. 2970–3005 (2019). https://doi.org/10.18653/v1/N19-1304
24. Economic Times: How Pakistan failed to do a Balakot-type strike on India on February 27 (2019). https://economictimes.indiatimes.com/news/defence/how-pakistan-failed-to-do-a-balakot-type-strike-on-india-on-february-27/articleshow/68592269.cms
25. Ferragina, P., Piccinno, F., Santoro, R.: On analyzing hashtags in Twitter. In: Proceedings of ICWSM (2015)
26. Feyyaz, M.: Contextualizing the Pulwama attack in Kashmir-a perspective from Pakistan. Perspect.Terrorism **13**(2), 69–74 (2019)
27. Field, A., Kliger, D., Wintner, S., Pan, J., Jurafsky, D., Tsvetkov, Y.: Framing and agenda-setting in Russian news: a computational analysis of intricate political strategies. In: Proceedings of EMNLP, pp. 3570–3580 (2018)

28. Forbes: India's fight with Pakistan seen lifting Modi's election chances (2019). https://www.forbes.com/sites/kenrapoza/2019/02/27/indias-fight-with-pakistan-seen-lifting-modis-election-chances/#1df2795a397c
29. Garimella, K., Morales, G.D.F., Gionis, A., Mathioudakis, M.: Quantifying controversy on social media. Trans. Soc. Comput. **1**(1) (2018). https://doi.org/10.1145/3140565
30. Golovchenko, Y., Hartmann, M., Adler-Nissen, R.: State, media and civil society in the information warfare over Ukraine: citizen curators of digital disinformation. Int. Affairs **94**(5), 975–994 (2018)
31. Hamilton, W.L., Clark, K., Leskovec, J., Jurafsky, D.: Inducing domain-specific sentiment lexicons from unlabeled corpora. In: Proceedings of EMNLP, vol. 2016, p. 595 (2016)
32. Kagan, V., Stevens, A., Subrahmanian, V.S.: Using Twitter sentiment to forecast the 2013 Pakistani election and the 2014 Indian election. IEEE Intell. Syst. **30**(1), 2–5 (2015)
33. Khosla, S., Chhaya, N., Jindal, S., Saha, O., Srivastava, M.: Do events change opinions on social media? Studying the 2016 US Presidential debates. In: Proceedings of Socio Informatics, pp. 287–297 (2019)
34. King, G., Pan, J., Roberts, M.E.: How the Chinese government fabricates social media posts for strategic distraction, not engaged argument. Am. Polit. Sci. Rev. **111**(3), 484–501 (2017)
35. Kriel, C., Pavliuc, A.: Reverse engineering Russian Internet Research Agency tactics through network analysis. Defence Strategic Communication, pp. 199–227 (2019)
36. Kumar, A., Dhamija, S., Dhamija, A.: Political marketing: the horizon of present era politics. SCMS J. Indian Manag. **13**(4), 116–125 (2016)
37. Kumaraguru, P., et al.: Social media to win elections: Analysis of #LokSabhaElections2019 in India. Precog Technical report (2019)
38. Le, H., Boynton, G., Shafiq, Z., Srinivasan, P.: A postmortem of suspended Twitter accounts in the 2016 US presidential election. In: Proceedings of ASONAM, pp. 258–265 (2019)
39. Magdy, W., Darwish, K., Abokhodair, N., Rahimi, A., Baldwin, T.: #isisisnotislam or#deportallmuslims? Predicting unspoken views. In: Proceedings of WebScience, pp. 95–106 (2016)
40. McDonnell, D., Cabrera, L.: The right-wing populism of India's Bharatiya Janata Party (and why comparativists should care). Democratization **26**(3), 484–501 (2019). https://doi.org/10.1080/13510347.2018.1551885
41. Mishra, S.: Emerging electoral dynamics after Pulwama tragedy. Observer Research Foundation (2019)
42. Morales, A., Borondo, J., Losada, J.C., Benito, R.M.: Measuring political polarization: Twitter shows the two sides of Venezuela. Chaos Interdisc. J. Nonlinear Sci. **25**(3), 033114 (2015)
43. Murthy, D., Longwell, S.A.: Twitter and disasters. Inf. Commun. Soc. **16**(6), 837–855 (2013). https://doi.org/10.1080/1369118X.2012.696123
44. NDTV: India strikes after Pulwama terror attack, hits biggest Jaish-e-Mohammed camp in Balakot (2019). https://www.ndtv.com/india-news/india-struck-biggest-training-camp-of-jaish-in-balakot-large-number-of-terrorists-eliminated-governm-1999390
45. Palakodety, S., KhudaBukhsh, A.R., Carbonell, J.G.: Hope speech detection: a computational analysis of the voice of peace. In: Proceedings of ECAI (2020)

46. Pandya, A.: The future of Indo-Pak relations after the Pulwama attack. Perspect. Terrorism **13**(2), 65–68 (2019)
47. Pennington, J., Socher, R., Manning, C.: GloVe: global vectors for word representation. In: Proceedings of EMNLP, pp. 1532–1543 (2014)
48. Pollacci, L., Sîrbu, A., Giannotti, F., Pedreschi, D., Lucchese, C., Muntean, C.I.: Sentiment spreading: An epidemic model for lexicon-based sentiment analysis on Twitter. In: Esposito, F., Basili, R., Ferilli, S., Lisi, F.A. (eds.) Proceedings of AI*IA, pp. 114–127 (2017)
49. Rothe, S., Ebert, S., Schütze, H.: Ultradense word embeddings by orthogonal transformation. In: Proceedings of NAACL, pp. 767–777 (2016)
50. Rozenas, A., Stukal, D.: How autocrats manipulate economic news: evidence from Russia's state-controlled television. J. Polit. **81**(3), 982–996 (2019)
51. Singh, P., Kumar, K., Kahlon, K.S., Sawhney, R.S.: Can tweets predict election results? Insights from Twitter analytics. In: Luhach, A.K., Jat, D.S., Hawari, K.B.G., Gao, X.Z., Lingras, P. (eds.) Proceedings of ICAICR, pp. 271–281 (2019)
52. Starbird, K., Arif, A., Wilson, T.: Disinformation as collaborative work: surfacing the participatory nature of strategic information operations. In: Proceedings of CSCW, pp. 1–26 (2019)
53. Starbird, K., Arif, A., Wilson, T., Van Koevering, K., Yefimova, K., Scarnecchia, D.: Ecosystem or echo-system? exploring content sharing across alternative media domains. In: Proceedings of ICWSM (2018)
54. Stewart, L.G., Arif, A., Nied, A.C., Spiro, E.S., Starbird, K.: Drawing the lines of contention: Networked frame contests within #BlackLivesMatter discourse. In: Proceedings of CSCW, pp. 1–23 (2017)
55. The Express Tribune: No blood. no bodies. no debris. no tragedy (2019). https://tribune.com.pk/story/1919080/1-no-blood-no-bodies-no-debris-no-tragedy/
56. The Hindu: IAF plane shot down, pilot taken captive by Pak. army (2019). https://www.thehindu.com/news/national/iaf-plane-shot-down-pilot-taken-captive-by-pak-army/article26390980.ece
57. The Hindu: India bombs Jaish camp in Pakistan's Balakot (2019). https://www.thehindu.com/news/national/air-strikes-hit-balakot-in-pakistan-initial-assessment-100-hit-sources/article26373318.ece
58. The New York Times: Kashmir suffers from the worst attack there in 30 years (2019). https://www.nytimes.com/2019/02/14/world/asia/pulwama-attack-kashmir.html
59. The Week: Imran's party slams Modi's 'warmongering' for escalating Indo-Pak tension (2019). https://www.theweek.in/news/india/2019/02/28/imran-s-party-slams-modi-on-escalating-indo-pak-tension.html
60. Upadhyay, A.: Decimating democracy in 140 characters or less: Pakistan army's subjugation of state institutions through Twitter. Strat. Anal. **43**(2), 101–113 (2019). https://doi.org/10.1080/09700161.2019.1600823
61. Washington Post: After Pulwama, the Indian media proves it is the BJP's propaganda machine (2019). https://www.washingtonpost.com/opinions/2019/03/04/after-pulwama-indian-media-proves-it-is-bjps-propaganda-machine/
62. Weber, I., Garimella, V.R.K., Batayneh, A.: Secular vs. Islamist polarization in Egypt on Twitter. In: Proceedings of ASONAM, pp. 290–297 (2013). https://doi.org/10.1145/2492517.2492557

Young Adult Unemployment Through the Lens of Social Media: Italy as a Case Study

Alessandra Urbinati[1](✉), Kyriaki Kalimeri[2](✉), Andrea Bonanomi[3], Alessandro Rosina[3], Ciro Cattuto[2], and Daniela Paolotti[2]

[1] Università di Torino, Turin, Italy
alessandra.urbinati@unito.it
[2] ISI Foundation, Turin, Italy
kyriaki.kalimeri@isi.it
[3] Università Cattolica del Sacro Cuore (UNICATT), Milan, Italy

Abstract. Youth unemployment rates are still in alerting levels for many countries, among which Italy. Direct consequences include poverty, social exclusion, and criminal behaviours, while negative impact on the future employability and wage cannot be obscured. In this study, we employ survey data together with social media data, and in particular likes on Facebook Pages, to analyse personality, moral values, but also cultural elements of the young unemployed population in Italy. Our findings show that there are small but significant differences in personality and moral values, with the unemployed males to be less agreeable while females more open to new experiences. At the same time, unemployed have a more collectivist point of view, valuing more in-group loyalty, authority, and purity foundations. Interestingly, topic modelling analysis did not reveal major differences in interests and cultural elements of the unemployed. Utilisation patterns emerged though; the employed seem to use Facebook to connect with local activities, while the unemployed use it mostly as for entertainment purposes and as a source of news, making them susceptible to mis/disinformation. We believe these findings can help policymakers get a deeper understanding of this population and initiatives that improve both the hard and the soft skills of this fragile population.

Keywords: Personality traits · Moral foundations · Human values · Inequalities · Social media · Facebook · Unemployment · NLP · Topic modeling

1 Introduction

Youth unemployment is one of the most significant challenges that modern societies face, with more than 3.3 million unemployed young people (aged 15–24 years)

A. Urbinati and K. Kalimeri—These authors contributed equally to this work.

S. Aref et al. (Eds.): SocInfo 2020, LNCS 12467, pp. 380–396, 2020.
https://doi.org/10.1007/978-3-030-60975-7_28

in 2019 in the EU alone. In 2014, youth unemployment in Italy reached 46% [36]. Today, this rate is at 37% [36], substantially lower than five years ago, but still in alerting levels and well above the average European trend[1]. The "dejuvenation" of the Italian population exacerbates this phenomenon [12], urging for a better understanding of this vulnerable segment of the society and devising better policies.

Drivers of youth unemployment vary over place and time: inexperience in connecting with the job market, limited job offers, and lack of qualifications are some of the most common determinants [5,46,48]. The social and economical impact of high unemployment rates among the younger age groups cannot be overstated. High inactivity rates create a vicious circle of poverty, homelessness, social exclusion [25], marginalisation [33], and criminal behaviours [11,22]. At an individual level, youths are in a fragile phase where the hardships of getting a job can cause pronounced, long-term feedback effects leaving a negative impact on later employability [24,29,50] and the future wage [30]. Early unemployment also has a potential impact on physical and mental health [35,55], as well as psychological distress, depression, and suicides [1,34].

Traditionally, scientists gained insights into this phenomenon via survey administration to representative population groups [15]. In this work, we adopt a mixed approach employing both traditional surveys administered on Facebook, and large scale observational digital data. We engaged a large cohort in Italy via a Facebook-hosted app, whose main functionality was to administer a series of questionnaires regarding personality traits, moral values, general interests, and political views, while at the same time, gauged the participants' Likes on Facebook Pages. Observing these digital behaviours, we aim to shed light on facets of youth unemployment that are difficult, or even impossible, to assess via traditional means.

We contribute to the current state of the art by exploring associations between moral values, political views and employability, accounting at the same time for demographic, behavioural and geography bias. Our findings show that the employed appear to cherish individualistic values. On the other hand, the unemployed tend to trust more the EU and the national government. At the same time, they are less open to the introduction and assimilation of migrants in society, and they are particularly worried about protecting their traditional values. Thoroughly analysing the digital patterns of interests, as expressed through their Page "Likes", we found that the macro interests of both communities largely overlap. However, the way employed and unemployed utilise the Facebook platform presents subtle differences.

These results push back against conventional ideas about professional success; instead, they show proof of the fact that unemployment is not driven by a lack of interests, rather than a considerable lack of opportunities. Finally, we believe that these findings can inform policymakers to support initiatives that do not solely improve the hard but also the soft skills of this fragile population.

[1] https://ec.europa.eu/social/main.jsp?catId=1036.

2 Background and Related Work

The role of psychological attributes in the labour market has received increasing interest in economics and organisation psychology during the past years. Extended literature across various scientific communities highlights for instance the relationship between personality traits and employability skills, e.g. [20,40,58] (see [2] for a review). Getting a job that is in line with one's personality traits was shown to lead to higher career satisfaction [44], higher productivity [19,41], and consequently, well-being and health quality [23]. However, evidence on the relationship between personality traits and unemployment is still scarce [9,57,58]. Limited studies also exist for other psychological constructs such as human values [53] and culture [10], which were recently shown to be related to individuals' ability to connect to the job market.

New data sources such as search engine queries, mobile phone records, and social media data, allow for unobtrusive observation of both psychological and behavioural patterns [38,42]. Pioneering work in predicting unemployment rates from search engine query data [17] paved the way for other nationally-wide studies which proved the feasibility of observing and predicting unemployment rates from query data [26], also in small and emerging economies [49].

More recently, social media platforms were exploited to create useful indicators of the socio-economical status of population groups that are very hard to reach with more traditional means of study [51]. Analysing Twitter data [3] created a social-media based index of job loss, which tracks the initial unemployment insurance claims at medium and high frequencies. Llorente et al. [43] with geotagged tweets collected from Spain, found a strong positive correlation between unemployment and the Twitter penetration rate, due to diversity in mobility fluxes, daily rhythms, and grammatical styles of Twitter users. Bokanyi et al. [8] studied how employment statistics of counties in the US are encoded in the daily rhythm of people. At the same time, Proserpio et al. [50] explored the relations between psychological well-being and the macroeconomic shock of employment instability, showing that SM can capture and track changes in psychological variables over time. Employing Facebook data [9], predicted youth unemployment, and in particular the NEET status (Not in Employment, Education, or Training population), leveraging on people's "Likes" on Facebook Pages.

In this study, we are not interested in predicting the unemployment status or rate; rather we focus on the association between political views and moral values to employability, which, in contrast to other psychological traits, is largely unexplored. Additionally, we employ Facebook as a social sensor, focusing on the understanding of interests and cultural elements from the digital patterns of Facebook Page "Likes". We provide, in a data-driven approach, the important topics of interest of the employed and unemployed, but also insights on how the communities utilise the Facebook Platform.

3 Data Collection

The data were collected employing a Facebook-hosted application, designed to act as a surveying tool since a recent study showed that Facebook is a valid scientific tool to administer demographic and psychometric surveys [37]. The app was disseminated through a traditional media campaign. Upon providing their informed consent, participants agreed to provide us with their "Likes" on Facebook Pages. After entering the app, the participants were requested to provide basic demographic information, namely gender, employment status and province of residence. Then, they proposed a series of questionnaires regarding (i) psychometric attributes, (ii) general cultural interests, and (iii) political opinions. Inserting their demographic information as well as completing the questionnaires and surveys was on a strict volunteering basis. The application was mainly deployed in Italy; it was initially launched in March 2016, while the data used here were downloaded in September 2019.

Demographic Information. Table 1 reports the demographic information of the participants. Data are not complete, given that the form was not mandatory. We report the percentages for gender, age, employment status and location of the general population as obtained from the official national statistics for 2019 [36] (Census). Next, we report the total of individuals that have accessed the app (Facebook App) and the individuals that have declared their qualification and are under 44 years of age (Complete Dataset). The Balanced Dataset consists of a sample of the Complete Dataset aimed to avoid gender, employment, location, as well as digital activity bias. More details are reported in the "Data processing" section.

Psychometric Surveys. The Big Five Personality Dimensions [27] and the Moral Foundations Questionnaire [28,31,32] were administered. Both are scientifically validated and extensively employed in the literature to assess personality and morality, respectively. The Big Five Personality Dimensions (BIG5) personality traits model characterises personality based on five dimensions, (i) Openness to experience, (ii) Conscientiousness, (iii) Extraversion, (iv) Agreeableness, (v) Neuroticism. MFT focuses on the psychological basis of morality, identifying the individualistic group consisting of the(i) Care/Harm, and (ii) Fairness/Cheating foundations, and the social binding group consisting of the (iii) Loyalty/Betrayal, (iv) Authority/Subversion, and (v) Purity/Degradation foundations.

Additional Surveys. The app also proposed a series of surveys aimed at addressing broader political topics but also the general opinions of participants. Inspired by the 41-item inventory proposed by Schwartz et al. [52] and the updated version of Barnea et al. [6] adapted to the Italian context, the Political Opinions (PO) survey explores essential societal issues. The survey consists of the following 8-items from the original inventory: *Q1* - It is extremely important to defend our traditional religious and moral values, *Q2* - People who come to live here from other countries generally make our country a better place to live,

Q3 - I trust the President of the Republic, *Q4* - I believe every person in the world should be treated in the same way. I believe that everyone should have the same opportunities in life, *Q5* - I trust the national government, *Q6* - The less the government gets involved with the business and the economy, the better off this country will be, *Q7* - I trust the European Union, *Q8* - I strongly believe that the state needs to be always aware of the threads, both internal and external. In the Culture and Interests (CI) survey, participants were asked to rate their interest in several categories in a 5-point Likert scale: Travel, Sport, Science, Food, Culture, Nature, Society and Politics, Education, Health, Hobbies, Business, Shopping.

Table 1. Demographic breakdown of our data according to gender, age, qualification, and geographic location. The first column, "Census", reports the percentages per attribute according to the statistics provided by the official census bureau [36]. The "Facebook App" column reports the percentages of the total number of participants. We use "Others", in the qualification section, as an umbrella term that includes students, or missing responses. The "Complete Dataset" consists only of the people that have declared their qualification and are under 44 years of age. Finally, the "Balanced Dataset" consists of a random sample of the "Complete Dataset" taking into consideration gender, qualification, and location biases.

		Census	Facebook App	Complete Dataset	Balanced Dataset
			$n = 63{,}980$	$n = 3079$	$n = 842$
Gender	Female	51.6%	31.7%	39.6%	50%
	Male	48.4%	68.3%	60.4%	50%
Age	16–17	4.8%	1.6%	0.1%	0.1%
	18–24	4.9%	38.3%	12.8%	16.7%
	25–34	11.0%	28.5%	56.0%	55.9%
	35–44	13.8%	13.0%	24.2%	27.2%
	45+	53.9%	18.7%	–	–
Qualification	Employed	54.2%	42.4%	80.2%	50%
	Unemployed	8.6%	7.2%	19.8%	50%
	Others(*)	37.2%	50.4%	–	–
Location	North	44.1%	53.7%	50.8%	39.4%
	Center	19.5%	18.5%	19.5%	26.5%
	South	36.4%	29.2%	29.2%	34.1%

Facebook Data. For the participants who entered the app, we gathered information on the public Facebook Pages which they "liked", as well as some necessary metadata about the page, such as the name, description, category[2], and the popularity of the page in terms of followers. Importantly, all the linguistic information available to us is the content by the page creator. No information about the posts or the comments of the page is available for this analysis. Table 2 reports the total number of Facebook Pages, Likes, and categories collected per qualification status.

Table 2. Total number of unique Facebook Pages and total number of Likes per qualification type in the "Complete Dataset" and the "Balanced Dataset". We report the number of pages liked only by the employed or unemployed, respectively. We also report the total number of unique Facebook categories in the two datasets.

		Complete Dataset	Balanced Dataset
		$n = 3079$	$n = 842$
#Pages	Employed	730,237	189,199 (just E 41,151)
	Unemployed	307,916	229,548 (just U 58,275)
#Likes	Employed	2,063,944	332,610
	Unemployed	780,693	456,971
#Category		1,396	1,214

3.1 Data Processing

Demographic Information. From the initial cohort of the app, we derived the "Complete Dataset", a subset of individuals for which we have complete information records about their gender, employment status, and geographic location, and who are younger than 44 years old. Table 1 reports the demographic breakdown of the population according to the official census as well as the datasets employed in this study. Lombardy region is over-represented, a phenomenon explained by the fact that the project was initially launched in that region. At the same time, we notice that the area of Marche is under-represented, while all other regions in our cohort follow the distribution of the census closely[3]. Our cohort deviates from the census distribution for gender, with males to be over-represented. We also have an over-representation of the 18–24 and 25–34 age groups. In terms of qualification, we notice that we have an over-representation of employed individuals.

To avoid gender and qualification biases (see Table 1) which affect the linguistic analysis and the topic modelling, we created a subset of the Complete Dataset, namely "Balanced Dataset". The Balanced Dataset consists of randomly selected

[2] Facebook Pages, are assigned to one of the predefined Facebook categories. Link to the full list of Facebook Categories: https://www.facebook.com/pages/category/.

[3] Figure 4 in SI provides a clear picture of the geographical distribution per region as compared to the expected values with respect to the census.

participants who are equally distributed for gender, qualification, and geography. To obtain this subset, firstly, we identified N_{min}, the population of the smallest demographic subgroup in our dataset, in terms of gender, qualification, and location breakdown. Then, we divided our participants into four quadrants according to their "likes per user" activity. From each quadrant, we randomly sampled N_{min} participants, resulting with a total population that consists of $4 \star N_{min}$ participants uniformly distributed with respect to age, gender, and geography. The "Balanced Dataset" is much smaller, however, less impacted by demographic and activity biases. It is employed only in the language and topic modelling analysis of the following section.

Facebook Data. Among the Pages liked by participants in the sample we retain for the analysis, 65% of them are in Italian, 28% in English, 2% in French, and 5% in other languages. For this study, we analyse only the pages in Italian and English. We dropped the generic Page category "community", since it may contain pages with diverse content, and also the page of the Italian newspaper "Repubblica" due to a communication campaign carried out on that journal.

4 Methods and Results

Questionnaire Analyses. We assessed the relationship of unemployment to the demographic and psychological attributes by means of logistic regression. A strong statistical significant effect of gender emerged (results reported in Table 5 in SI), with women to be more often jobless as confirmed by the literature [4]. Geography too emerges as a significant factor, with the southern Italian regions to be more heavily affected as supported by the literature [45].

To avoid the known gender biases of the psychological attributes [18,56], we divided the analysis per gender. We assess differences in the psychological attributes, interests, and political views between the employed (E) and unemployed (U) populations using the Mann-Whitney U test [47], since our data are ordinal and not necessarily follow a normal distribution.

Personality Traits. We assess differences in the personality traits between the employed and unemployed populations, employing the Mann-Whitney U test. As reported in Table 3, employed males are more agreeable while unemployed females are more open to new experiences.

Moral Foundations. Unemployed females value significantly more the protection of others, they value more the in-group loyalty, and respect to authority. Both males and females have a higher purity score, and hence exhibiting a more social binding worldview (see Table 3). The relationship of moral values with unemployment has received much less attention by the scientific community, despite playing a critical motivational role in job selection and successful school-to-work transition [53,54], and the disengagement from the labour market [7].

Political Opinions. Turning to the political views of the two populations, we find that unemployed males tend to trust less the EU as well as the national government and president. Along the same lines, female unemployed trust less the presidency and national government, while they see the privatisation of the

Table 3. Mann-Whitney U test between the employed and unemployed populations per gender. Averages and confidence intervals of the comparison between the Male Employed (ME) versus Male Unemployed (MU) and the Female Employed (FE) versus the Female Unemployed (FU) with the respective p-values. Only statistical significant findings are reported here. Please check the SI section for the complete Table. Statistical significance is reported as follows: p-value - "****": $p < 0.0001$, "***": $p < 0.001$, "**": $p < 0.01$, "*": $p < 0.05$, otherwise – for non-significant results.

Big five	EM	UM	pvalue	EF	UF	pvalue
Extraversion	8.83 ± 0.21	9.02 ± 0.52	–	8.94 ± 0.17	9.04 ± 0.26	–
Agreeableness	9.41 ± 0.24	8.83 ± 0.54	*	9.55 ± 0.21	9.66 ± 0.26	–
Conscientiousness	9.16 ± 0.22	9.47 ± 0.45	–	9.27 ± 0.19	9.61 ± 0.27	–
Openness	8.74 ± 0.24	8.98 ± 0.52	–	9.26 ± 0.22	9.59 ± 0.33	*
Neurotisism	9.91 ± 0.22	10.02 ± 0.51	–	9.93 ± 0.20	10.20 ± 0.27	–
Moral foundation						
Care	18.18 ± 0.34	18.32 ± 0.81	–	18.77 ± 0.33	19.21 ± 0.45	*
Loyalty	15.33 ± 0.40	15.96 ± 0.89	–	15.94 ± 0.41	16.82 ± 0.52	***
Authority	14.48 ± 0.42	14.84 ± 0.96	–	15.43 ± 0.39	16.32 ± 0.54	***
Purity	14.56 ± 0.46	15.84 ± 1.05	**	16.40 ± 0.41	17.33 ± 0.57	**
Individualism	38.64 ± 0.54	38.01 ± 1.49	–	39.13 ± 0.57	39.63 ± 0.82	*
SocialBinding	44.37 ± 1.09	46.65 ± 2.56	*	47.77 ± 1.07	50.47 ± 1.44	***
Political opinion						
Q1	2.89 ± 0.07	3.13 ± 0.19	**	2.90 ± 0.13	2.97 ± 0.23	–
Q2	3.27 ± 0.06	3.03 ± 0.16	**	3.34 ± 0.11	3.19 ± 0.25	–
Q3	3.26 ± 0.07	2.90 ± 0.18	****	3.30 ± 0.12	3.03 ± 0.25	*
Q4	4.43 ± 0.05	4.27 ± 0.14	*	4.56 ± 0.08	4.43 ± 0.18	–
Q5	2.74 ± 0.06	2.51 ± 0.17	**	2.73 ± 0.12	2.46 ± 0.24	*
Q6	2.81 ± 0.06	2.86 ± 0.17	–	2.77 ± 0.12	2.98 ± 0.21	*
Q7	3.47 ± 0.07	3.28 ± 0.19	*	3.66 ± 0.12	3.42 ± 0.27	–
Q8	3.79 ± 0.06	3.96 ± 0.15	**	3.76 ± 0.11	3.69 ± 0.24	–

public enterprises and the minimisation of the governmental involvement in the economy [59] positively. Employed males feel less the need to protect their traditional values and believe in the fair treatment of individuals believe the state should always be aware of national and international threats. The current literature [21] also supports several of these such findings reported in Table 3.

Culture and Interests. As reported in Table 1, in the Complete Dataset, we have very different representation in the participant's demographics. To overcome this issue, we normalised the category frequencies for each gender and qualification subgroup. Merging the ten most frequently liked categories by each subgroup, i.e. unemployed female, employed male and so on, we obtained the 15 most frequently liked categories by females (a) and males (b). Figure 1 depicts this distribution. Marked differences are observed among the female population, with unemployed women to be more interested in health and beauty pages and

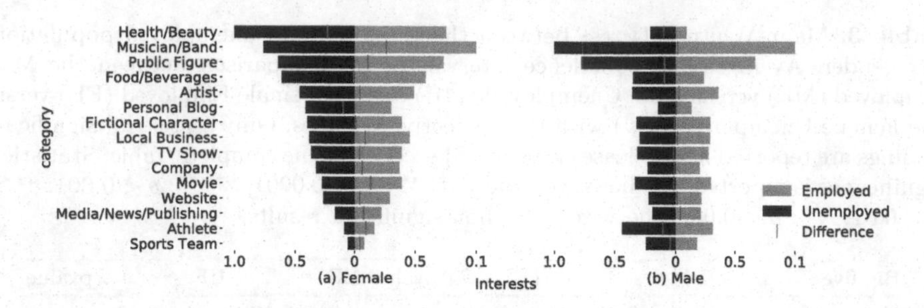

Fig. 1. Top 15 Facebook Page categories liked by the Employed (light-coloured bars) and Unemployed (dark-coloured bars) participants in the Complete Dataset. The category frequencies are normalised per gender and qualification. The red vertical lines indicate the direction of the difference between the two populations. (Color figure online)

personal blogs to the employed ones. After visual inspection, these blogs refer to beauty advice and recipes. Their employed counterparts are slightly more interested in musicians and bands, movies, news magazines, and public figures like athletes. In the male cohort, there are less pronounced differences. Unemployed are interested in health and beauty, public figures, athletes, food and beverages, while employed are more fond of music bands, artists, and news magazines. Health and beauty pages are the only common interest both for male and female unemployed, probably since these Pages often promote discounts.

Mann-Whitney U test on the CI survey's self-assessments, shows that the employed population declares to be more interested in topics that regard the society and politics and hobbies ((see Fig. 2 (a) and (c)). Mapping the original page categories (1,396 in total) to the twelve more generic interest categories of the CI survey, we assessed the extent to which the self-reporting patterns of interests correspond to the actual digital ones. The mapping is primarily based on Facebook's original hierarchical structure, where for instance, the various medical specialities are under the generic term "Medical and Health". Then, we manually classified a limited amount of categories that were not part of the twelve generic categories of the questionnaire, for instance, "Vitamins and Supplements" under "Health", "Non-Profit Organization" under "Society", "Elementary School" under "Education". An obvious limitation of this approach is that the frequency of visit to the Liked Page is not available to us, still, liking a page is a clear signal of interest. Importantly, and despite the limitations, Fig. 2 (b, d) shows that the responses in the online questionnaires reflect the actual online activity of the users, with only exceptions shopping and hobbies that emerge to be significantly more of interest to the female unemployed population, and health and sports for the unemployed male population.

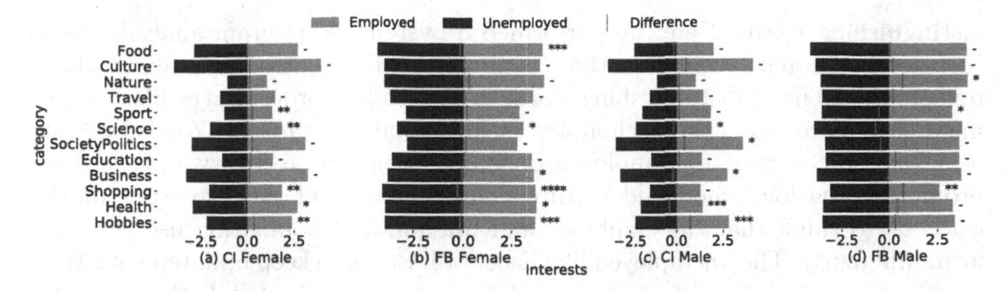

Fig. 2. Mann-Whitney U test between the actual (FB) and the self-reported interests (CI) by the Employed (light-coloured bars) and the Unemployed (dark-coloured bars) participants. The Pages categories are manually classified in the twelve macro-categories of the CI questionnaire. The bar amplitude corresponds to the median values for the Employed and Unemployed respectively, while the red vertical lines indicate the direction of difference between the two means. Statistical significance is reported as follows: p-value - "****": $p < 0.0001$, "***": $p < 0.001$, "**": $p < 0.01$, "*": $p < 0.05$, otherwise – for non-significant results. (Color figure online)

Table 4. Top 20 distinguishing words for the employed and unemployed communities for pages in Italian (IT) and English (EN), respectively.

Unemployed	IT	Roma, series, film, INTER, welcome, Napoli, love, fashion, creation, advertisement, football, direction, good, photo, passion, best, channel, jewellery, director, beauty
	EN	Music, fashion, love, film, game, time, series, band, news, years, album, team, products, released, house, games, Italy, life, based, people
Employed	IT	Association, activity, project, territory, Calabria, culture, cultural, Salerno, centre, Bologna, Genova, cuisine, restaurant, municipal, Reggio, wine, local, social, trapped, research
	EN	Nail, polish, health, nutrition, architecture, design, Salento, BMW, food, healthy, diet, Versailles, steampunk, cancer, science, gothic, lacquer, SAT, disease, medical

4.1 Language and Topic Modelling

Often, category information is too generic; for instance, a business page could refer to various types of business, like restaurants, or beauty centres. To overcome this issue, we focus on the linguistic content of pages' metadata, which includes the page about, and the page description. To avoid demographic and behavioural biases, we analysed the Pages included in the "Balanced Dataset" (see Table 1 and 2). We employed statistical language modelling to distinguish the two communities. Considering the differences between the probability distribution of words occurring in pages liked by the two populations and the overall probability distribution of words in the entire dataset, we can identify the most distinguishing words. Here, we performed a systematic bi-gram analysis which surfaces common sub-sentences that tend to occur frequently in the text. Table 4 reports the 20 most distinguishing words in descending order. Pages liked by the unemployed are talking about hobbies; fashion, photography and YouTube channels; on the contrary, the employed ones are about the territory, and projects promoting the local food and culture. This snapshot of the two communities leads us to think that the employed and unemployed population use the platform differently. The unemployed like Facebook Pages to keep updated with their hobbies in a relaxed and fun way but also as a news source, while the employed use the platform to stay in touch with the local communities.

Moving to higher level constructs, we aim to uncover latent behavioural patterns in the interest and culture of the two populations. We apply a topic modelling approach based on the NMF algorithm [13], to analyse the linguistic information contained in the Page title, about information, and the full description. We included all pages liked by the population of interest, excluding ones liked by both groups (see Table 2). We create a model per language and qualification for $k = 10$ components. Figures 3 depict the most important words; both the topics and the words within each topic are raked in decreasing weight order[4].

Noteworthy is the fact that the communities exhibit several differences in the first two topics (T1 and T2) while from the third topic (T3) onward (see Footnote 4), the differences are negligible. Interestingly, the primary topic of the unemployed community regards entertainment TV shows and news, while the respective one for the employed is about non-profit organisations. In contrast, the news appears only in the fourth topic. This is of significant importance since it indicates that the unemployed use the Facebook platform to get news

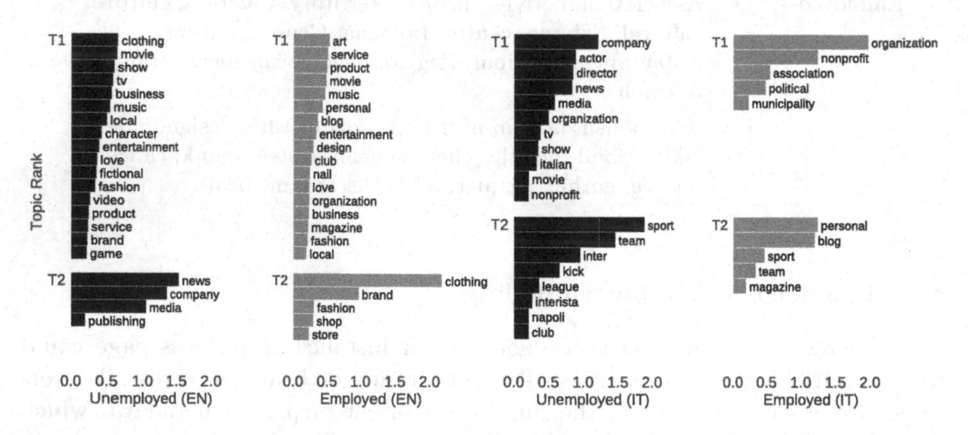

Fig. 3. The ten topics according to the NMF algorithm with the most contributing words per topic, for the Italian and English language, respectively. See the Appendix for the complete topic presentation.

[4] Due to space limitations the full topic descriptions are presented in the Appendix, please refer to Fig. 5 for the Italian and English languages respectively.

information. Observing the emergent topics, we notice that there are no striking differences between the two communities. Another important note lies in the fact that our cohort consists of its vast majority by Italians. This influences the content differences between the topics in Italian and English language; in English, prevalent topics are entertainment, art, and TV shows. Interestingly, both employed and unemployed exhibit the same behavioural pattern.

5 Discussion and Conclusions

Youth unemployment in Italy is still in alerting levels, with numerous implications both for the individual and the society. Engaging a large cohort of young Italians on the Facebook Platform, we provide a more in-depth understanding of psychological, cultural, but also differences in digital behaviours between the employed and unemployed communities.

Are there associations between psychological constructs and employability? Accounting for the significant demographic, behavioural, and geography biases, our analysis shows that employed males are more agreeable. Unemployed females exhibit high levels of openness. Moving to higher level constructs, we find that unemployed individuals tend to have a more social binding worldview, prioritising purity values. In particular, unemployed females are strongly valuing in-group loyalty and authority. Moral values reflect our political viewpoints and personal narratives. Coherent with our previous findings, employed males are not really concerned with protecting their traditional values, while for them is fundamental the fair treatment of everyone in the society.

Are there divides between the generic interests and cultural attributes with respect to the occupational status? Self-reported scores did not show any striking differences to the general interests of employed and unemployed. Employed declared to be more interested in societal issues and hobbies, while unemployed in business and food pages. The two communities have broadly the same interests; this finding of extreme importance since it shows that the employability status is not related with the interests or cultural differences but rather with the psychological aspects and life opportunities.

Assessing the linguistic content of the liked pages allows for a more in-depth understanding of peoples' interest. When comparing the linguistic content of pages likes by the two communities, the probability distributions (language models) show that the most distinguishing words for the unemployed are about hobbies and entertainment while for the employed about cultural projects and associations promoting the wine, food and in general the territory. This finding leads us to the conclusion that the two communities may have similar interests but make different use of the platform. Excluding the commonly liked content, the NMF topic models confirm that the general interests of the two communities are along the same lines; however, in the prevalent topic of the unemployed community, news and media play a major role followed by sports. On the other side, the employed community focuses more on non-profit associations and personal blogs. This finding indicates that the unemployed population use Facebook as a source of information. Such finding is alerting given the susceptibility of social media platforms to mis/disinformation [14,16,39].

Concluding, our findings show small but significant differences in the psychological and moral values of the two communities. Importantly, it emerged that the two communities utilise the Facebook platform differently, with the unemployed to experience it as a news resource and the employed to connect with their local activities. We believe that these findings can inform policymakers to devise better policies that do not solely improve the hard but also the soft skills of this fragile population.

A Appendix

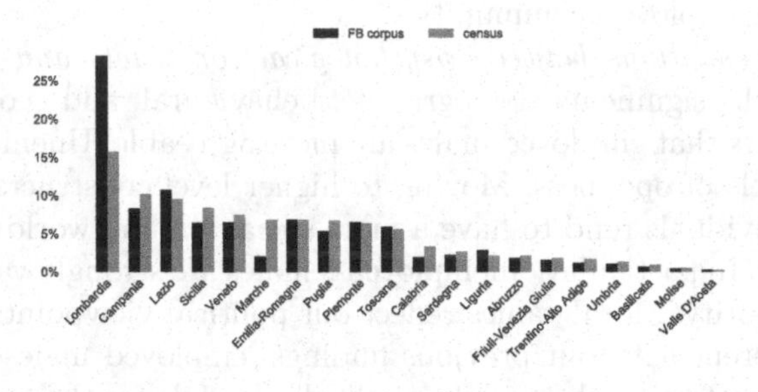

Fig. 4. Geographic distribution of the population in the Complete Dataset (dark-coloured bars) as compared to the expected population distribution according to the official census data per region (light-coloured bars).

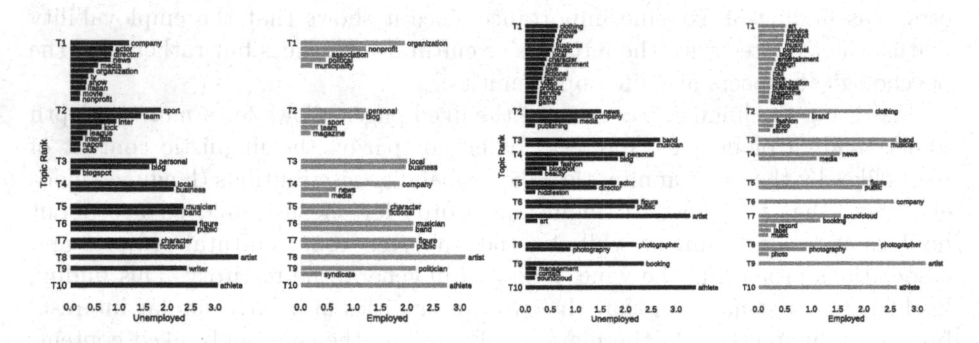

Fig. 5. The ten topics according to the NMF algorithm with the most contributing words per topic, for the Italian (first two columns) and the English language (last two columns).

Table 5. Logistic regression models predicting qualification using demographic (D), Big Five (B5), Moral Foundation (MFT), Political Opinion (PO). The Table shows alongside their coefficient estimate and their corresponding p-values (Bonferroni-adjusted). Confidence levels: $p < 0.001$ ***, $p < 0.01$ **, $p < 0.05$ *.

	D	MFT	D+MFT	B5	D+B5	PO	D ± PO
n	3079	786	786	687	687	1114	1114
R^2_{MD}	0.075	0.026	0.09	0.042	0.084	0.029	0.050
Intercept	−1.287(***)	−1.298	−1.123	−2.056(*)	−1.777	−1.857	−0.870
Gender (M)	−1.007(***)		−0.726(***)		−0.686(***)		−0.740(***)
Center	0.189(***)		0.080		0.168		−0.104
South	1,052(***)		0.199(***)		1.217(***)		0.385
Care		0.052	0.033				
Fairness		−0.084(**)	−0.070				
Loyalty		0.010	0.016				
Authority		−0.008	−0.004				
Purity		0.066(**)	0.046				
Extraversion				0.002	−0.015		
Agreeableness				0.050	−0.060		
Conscientiousness				0.094	0.103		
Openness				0.056	0.032		
Neurotisism				0.011	0.003		
Q1						0.230	0.244
Q2						−0.052	−0.044
Q3						−0.026	−0.254
Q4						−0.108	−0.132
Q5						−0.080	−0.068
Q6						0.071	0.071
Q7						0.019	0.161
Q8						−0.050	−0.036

References

1. Adler, N., et al.: How search engine data enhance the understanding of determinants of suicide in India and inform prevention: observational study. J. Med. Internet Res. **21**(1), e10179 (2019)
2. Almlund, M., Duckworth, A.L., Heckman, J., Kautz, T.: Personality psychology and economics. In: Duckworth, A., Almlund, M., Kautz, T. (eds.) Handbook of the Economics of Education, vol. 4, pp. 1–181. Elsevier, Amsterdam (2011)
3. Antenucci, D., Cafarella, M., Levenstein, M., Ré, C., Shapiro, M.D.: Using social media to measure labor market flows. Technical report, National Bureau of Economic Research (2014)
4. Azmat, G., Güell, M., Manning, A.: Gender gaps in unemployment rates in OECD countries. J. Labor Econ. **24**(1), 1–37 (2006)
5. Baah-Boateng, W.: The youth unemployment challenge in Africa: what are the drivers? Econ. Labour Relat. Rev. **27**(4), 413–431 (2016)
6. Barnea, M.F., Schwartz, S.H.: Values and voting. Polit. Psychol. **19**(1), 17 40 (1998)
7. Barr, A., Miller, L., Ubeda, P.: Moral consequences of becoming unemployed. Proc. Natl. Acad. Sci. **113**(17), 4676–4681 (2016)

8. Bokányi, E., Lábszki, Z., Vattay, G.: Prediction of employment and unemployment rates from Twitter daily rhythms in the US. EPJ Data Sci. **6**(1), 14 (2017)
9. Bonanomi, A., Rosina, A., Cattuto, C., Kalimeri, K.: Understanding youth unemployment in Italy via social media data. In: 28th IUSSP International Population Conference (2017)
10. Brügger, B., Lalive, R., Zweimüller, J.: Does culture affect unemployment? evidence from the röstigraben (2009)
11. Buonanno, P.: Crime and labour market opportunities in Italy (1993–2002). Labour **20**(4), 601–624 (2006)
12. Caltabiano, M., Rosina, A.: The dejuvenation of the Italian population. J. Modern Italian Stud. **23**(1), 24–40 (2018)
13. Cichocki, A., Zdunek, R., Amari, S.I.: New algorithms for non-negative matrix factorization in applications to blind source separation. In: 2006 IEEE International Conference on Acoustics, Speech and Signal Processing, ICASSP 2006 Proceedings, vol. 5, p. V. IEEE (2006)
14. Cinelli, M., Morales, G.D.F., Galeazzi, A., Quattrociocchi, W., Starnini, M.: Echo chambers on social media: A comparative analysis (2020)
15. Clark, K.B., Summers, L.H.: The dynamics of youth unemployment. In: Freeman, R.B., WiseT, D.A. (eds.) The Youth Labor Market Problem: Its Nature, Causes, and Consequences, pp. 199–234. University of Chicago Press, Chicago (1982)
16. Cossard, A., Morales, G.D.F., Kalimeri, K., Mejova, Y., Paolotti, D., Starnini, M.: Falling into the echo chamber: the Italian vaccination debate on Twitter. arXiv preprint arXiv:2003.11906 (2020)
17. Ettredge, M., Gerdes, J., Karuga, G.: Using web-based search data to predict macroeconomic statistics. Commun. ACM **48**(11), 87–92 (2005)
18. Feingold, A.: Gender differences in personality: a meta-analysis. Psychol. Bull. **116**(3), 429 (1994)
19. Finnerty, A.N., Kalimeri, K., Pianesi, F.: Towards happier organisations: understanding the relationship between communication and productivity. In: Aiello, L.M., McFarland, D. (eds.) SocInfo 2014. LNCS, vol. 8851, pp. 462–477. Springer, Cham (2014). https://doi.org/10.1007/978-3-319-13734-6_33
20. Finnerty, A.N., Muralidhar, S., Nguyen, L.S., Pianesi, F., Gatica-Perez, D.: Stressful first impressions in job interviews. In: Proceedings of the 18th ACM International Conference on Multimodal Interaction, pp. 325–332 (2016)
21. Finseraas, H., Pedersen, A.W., Bay, A.H.: When the going gets tough: the differential impact of national unemployment on the perceived threats of immigration. Polit. Stud. **64**(1), 60–73 (2016)
22. Freeman, R.B.: Crime and the job market. Technical report, National Bureau of Economic Research (1994)
23. Friedman, H.S., Kern, M.L.: Personality, well-being, and health. Ann. Rev. Psychol. **65**, 719–742 (2014)
24. Furnham, A.: 2. unemployment in the digital age. The Multi-generational and Aging Workforce: Challenges and Opportunities, p. 39 (2015)
25. Gallie, D., Paugam, S., Jacobs, S.: Unemployment, poverty and social isolation: is there a vicious circle of social exclusion? Eur. Soc. **5**(1), 1–32 (2003)
26. Gao, J., Zhang, Y.C., Zhou, T.: Computational socioeconomics. Phys. Rep. **817**, 1–104 (2019)
27. Gosling, S.D., Rentfrow, P.J., Swann, W.B.: A very brief measure of the big-five personality domains. J. Res. Pers. **37**(6), 504–528 (2003)
28. Graham, J., Haidt, J., Nosek, B.A.: Liberals and conservatives rely on different sets of moral foundations. J. Pers. Soc. Psychol. **96**(5), 1029 (2009)

29. Gregg, P.: The impact of youth unemployment on adult unemployment in the NCDS. Econ. J. **111**(475), F626–F653 (2001)
30. Gregg, P., Tominey, E.: The wage scar from male youth unemployment. Labour Econ. **12**(4), 487–509 (2005)
31. Haidt, J., Graham, J.: When morality opposes justice: conservatives have moral intuitions that liberals may not recognize. Soc. Justice Res. **20**(1), 98–116 (2007)
32. Haidt, J., Joseph, C.: Intuitive ethics: how innately prepared intuitions generate culturally variable virtues. Daedalus **133**(4), 55–66 (2004)
33. Hällsten, M., Edling, C., Rydgren, J.: Social capital, friendship networks, and youth unemployment. Soc. Sci. Res. **61**, 234–250 (2017)
34. Hammarström, A., Janlert, U.: Early unemployment can contribute to adult health problems: results from a longitudinal study of school leavers. J. Epidemiol. Commun. Health **56**(8), 624–630 (2002)
35. Hammer, T.: Unemployment and mental health among young people: a longitudinal study. J. Adolesc. **16**(4), 407–420 (1993)
36. I.StatDatabase: Istat database. data on unemployed rate. http://dati.istat.it. Accessed 28 Oct 2018
37. Kalimeri, K., Beiró, M.G., Bonanomi, A., Rosina, A., Cattuto, C.: Traditional versus Facebook-based surveys: evaluation of biases in self-reported demographic and psychometric information. Demographic Res. **42**, 133–148 (2020)
38. Kalimeri, K., Beiró, M.G., Delfino, M., Raleigh, R., Cattuto, C.: Predicting demographics, moral foundations, and human values from digital behaviours. Comput. Hum. Behav. **92**, 428–445 (2019)
39. Kalimeri, K., Beiró, G., Urbinati, M., Bonanomi, A., Rosina, A., Cattuto, C.: Human values and attitudes towards vaccination in social media. In: Companion Proceedings of The 2019 World Wide Web Conference, pp. 248–254 (2019)
40. Kalimeri, K., Lepri, B., Pianesi, F.: Causal-modelling of personality traits: extraversion and locus of control. In: Proceedings of the 2nd International Workshop on Social Signal Processing, pp. 41–46 (2010)
41. Kalimeri, K., Lepri, B., Pianesi, F.: Going beyond traits: multimodal classification of personality states in the wild. In: Proceedings of the 15th ACM on International Conference on Multimodal Interaction, pp. 27–34 (2013)
42. Kosinski, M., Stillwell, D., Kohli, P., Bachrach, Y., Graepel, T.: Personality and website choice. In: ACM Web Sciences 2012. ACM Conference on Web Sciences, January 2012. https://www.microsoft.com/en-us/research/publication/personality-and-website-choice/
43. Llorente, A., Garcia-Herranz, M., Cebrian, M., Moro, E.: Social media fingerprints of unemployment. PLOS ONE **10**(5), 1–13 (2015). https://doi.org/10.1371/journal.pone.0128692
44. Lounsbury, J.W., Steel, R.P., Gibson, L.W., Drost, A.W.: Personality traits and career satisfaction of human resource professionals. Hum. Resour. Dev. Int. **11**(4), 351–366 (2008)
45. Manacorda, M., Petrongolo, B.: Regional mismatch and unemployment: theory and evidence from Italy, 1977–1998. J. Popul. Econ. **19**(1), 137–162 (2006)
46. Mayer, T., Moorti, S., McCallum, J.K.: The Crisis of Global Youth Unemployment. Routledge, London (2018)
47. Nachar, N., et al.: The Mann-Whitney U: a test for assessing whether two independent samples come from the same distribution. Tutor Quant. Methods Psychol. **4**(1), 13–20 (2008)

48. O'Reilly, J., et al.: Five characteristics of youth unemployment in Europe: flexibility, education, migration, family legacies, and EU policy. Sage Open **5**(1), 2158244015574962 (2015)
49. Pavlicek, J., Kristoufek, L.: Nowcasting unemployment rates with google searches: evidence from the Visegrad group countries. PLoS ONE **10**(5), e0127084 (2015)
50. Proserpio, D., Counts, S., Jain, A.: The psychology of job loss: using social media data to characterize and predict unemployment. In: Proceedings of the 8th ACM Conference on Web Science, pp. 223–232. ACM (2016)
51. Rama, D., Mejova, Y., Tizzoni, M., Kalimeri, K., Weber, I.: Facebook ads as a demographic tool to measure the urban-rural divide. In: Proceedings of The Web Conference 2020, pp. 327–338 (2020)
52. Schwartz, S.H.: Universals in the content and structure of values: theoretical advances and empirical tests in 20 countries. In: Advances in Experimental Social Psychology, vol. 25, pp. 1–65. Elsevier (1992)
53. Sortheix, F.M., Chow, A., Salmela-Aro, K.: Work values and the transition to work life: a longitudinal study. J. Vocat. Behav. **89**, 162–171 (2015)
54. Sortheix, F.M., Dietrich, J., Chow, A., Salmela-Aro, K.: The role of career values for work engagement during the transition to working life. J. Vocat. Behav. **83**(3), 466–475 (2013)
55. Strandh, M., Winefield, A., Nilsson, K., Hammarström, A.: Unemployment and mental health scarring during the life course. Eur. J. Pub. Health **24**(3), 440–445 (2014). https://doi.org/10.1093/eurpub/cku005
56. Tangney, J.P., Dearing, R.L.: Gender differences in morality (2002)
57. Uysal, S.D., Pohlmeier, W.: Unemployment duration and personality. J. Econ. Psychol. **32**(6), 980–992 (2011)
58. Viinikainen, J., Kokko, K.: Personality traits and unemployment: evidence from longitudinal data. J. Econ. Psychol. **33**(6), 1204–1222 (2012)
59. Westholm, A., Niemi, R.G.: Youth unemployment and political alienation. Youth Soc. **18**(1), 58–80 (1986)

Women Worry About Family, Men About the Economy: Gender Differences in Emotional Responses to COVID-19

Isabelle van der Vegt[1] and Bennett Kleinberg[1,2](✉)

[1] Department of Security and Crime Science, University College London, London, UK
{isabelle.vandervegt,bennett.kleinberg}@ucl.ac.uk
[2] Dawes Centre for Future Crime, University College London, London, UK

Abstract. Among the critical challenges around the COVID-19 pandemic is dealing with the potentially detrimental effects on people's mental health. Designing appropriate interventions and identifying the concerns of those most at risk requires methods that can extract worries, concerns and emotional responses from text data. We examine gender differences and the effect of document length on worries about the ongoing COVID-19 situation. Our findings suggest that *i)* short texts do not offer as adequate insights into psychological processes as longer texts. We further find *ii)* marked gender differences in topics concerning emotional responses. Women worried more about their loved ones and severe health concerns while men were more occupied with effects on the economy and society. This paper adds to the understanding of general gender differences in language found elsewhere, and shows that the current unique circumstances likely amplified these effects. We close this paper with a call for more high-quality datasets due to the limitations of Tweet-sized data.

Keywords: Gender differences · COVID-19 · Emotions · Language

1 Introduction

The COVID-19 pandemic is having an enormous effect on the world, with alarming death tolls, strict social distancing measures, and far-reaching consequences for the global economy. In order to mitigate the impact that the virus may have on mental health, it is crucial to gain an understanding of the emotions, worries and concerns that the situation has brought about in people worldwide. Text data are rich sources to help with that task and computational methods potentially allow us to extract information about people's worries on a large scale. In previous work, the *COVID-19 Real World Worry Dataset* was introduced, consisting of 5,000 texts (2,500 short + 2,500 long) asking participants to express their emotions regarding the virus in written form [1]. In the current study, we delve deeper into that dataset and explore potential gender differences with regards to real-world worries about COVID-19. Building on a substantial evidence base for gender differences in language, we examine whether linguistic information can reveal whether and how men and women differ in how they respond to the crisis. Importantly, we also

S. Aref et al. (Eds.): SocInfo 2020, LNCS 12467, pp. 397–409, 2020.
https://doi.org/10.1007/978-3-030-60975-7_29

examine whether men or women are potentially more affected than the other, which may hold implications for developing mitigation strategies for those who need them the most.

1.1 Real World Worry Dataset

The Real World Worry Dataset (RWWD) is a text dataset collected from 2,500 people, where each participant was asked to report their emotions and worries regarding COVID-19 [1]. All participants were from the UK and completed the study through crowdsourcing platform Prolific Academic. The participants selected the most appropriate emotion describing what they were experiencing, choosing from anger, anxiety, desire, disgust, fear, happiness, relaxation, or sadness. They then rated the extent to which they worried about the COVID-19 situation and scored each of the eight emotions on a 9-point scale. Participants – all social media users – then wrote both a long text (avg. 128 tokens) and a short, Tweet-sized text (avg. 28 tokens) about said emotions and worries. The instructions read: "write in a few sentences how you feel about the Corona situation at this very moment. This text should express your feelings at this moment."

1.2 Gender Differences in Language

A large body of research has studied gender differences in language. Researchers have adopted both closed- and open vocabulary approaches. A closed vocabulary refers to approaches, where gender differences are measured through predefined word lists and categories. The LIWC (Linguistic Inquiry and Word Count) is a prominent example that measures linguistic categories (e.g., pronouns and verbs), psychological processes (e.g., anger and certainty), and personal concerns (e.g., family and money) [2]. The LIWC outputs the percentage of a document that belongs to each category. An open vocabulary approach is data-driven, in that gender differences are assessed without the use of predefined concepts or word lists. For example, *n*-grams or topics may be used to study gender differences.

In a closed-vocabulary study of 14,324 text samples from different sources (e.g. stream-of-consciousness essays, emotional writing), gender differences for LIWC categories were examined [3]. It was found that women used more pronouns (Cohen's $d = -0.36$)[1] words referring to emotions ($d = -0.11$), including anxiety ($d = -0.16$) and sadness ($d = -0.10$), as well as social words referring to friends ($d = -0.09$) and family ($d = -0.12$), and past ($d = -0.12$) and present-tense ($d = -0.18$) verbs. On the other hand, men used more articles ($d = 0.24$), numbers ($d = 0.15$), swear words ($d = 0.22$), and words referring to occupations ($d = 0.12$) and money ($d = 0.10$)[2]. Another approach partially replicated these results, showing that gender differences also emerged for language on social media (Facebook status updates from 75,000 individuals) [6]. In addition to examining differences in LIWC categories, the authors extended

[1] The effect size Cohen's *d* expresses the magnitude of the difference after correcting for sample size and, therefore, offers a more nuanced metric than p-values [4]. A *d* of 0.20, 0.50 and 0.80 can be interpreted as a small, moderate and large effect, respectively [5].

[2] Throughout this paper, positive values show that the respective concept is used more by men and negative values show that it is used more by women.

their approach to a data-driven, open-vocabulary approach, using both *n*-grams and topics. For example, unigrams such as 'excited', 'love', and 'wonderful' were used more frequently by women, whereas unigrams such as 'xbox', 'fuck', and 'government' were used more by men. In terms of topics, women more often mentioned family and friends and wished others a happy birthday, whereas men spoke more often about gaming and governmental/economic affairs [6].

1.3 Contribution of This Paper

With the current paper, we aim to add to the timely need to understand emotional reactions to the "Corona crisis". A step towards finding intervention strategies for individuals in need of mental health support is gaining an understanding of how different groups are affected by the situation. We examine the role of gender in the experienced emotions as well as potential manifestations of the emotional responses in texts. Specifically, we look at gender differences *i)* in self-reported emotions, *ii)* in the prevalence of topics using a correlated topic model approach [7], *iii)* between features derived from an open-vocabulary approach, and *iv)* in psycholinguistic constructs. We use the COVID-19 Real World Worry Dataset and thereby also test whether the differences emerge similarly in long compared to Tweet-sized texts.

2 Method

2.1 Data

The RWWD contains 5,000 texts of 2,500 individuals - each of whom wrote a long and short text expressing their emotions about COVID-19. We applied the same base exclusion criteria as [1] (i.e. nine participants who padded their statements with punctuation), and further excluded those participants who did not report their gender ($n = 55$). The sample used in this paper consisted of $n = 2{,}436$ participants, 65.15% female, 34.85% male.

2.2 Gender Differences in Emotion

We examined whether there were gender differences in the self-reported emotion scores (i.e. how people felt about COVID-19) using Bayesian hypothesis testing [8, 9]. In short, we report the Bayes factor, BF_{10} which expresses the degree to which the data are more likely to occur under the hypothesis that there is a gender difference, compared to the null hypothesis (i.e. no gender difference). For example, $BF_{10} = 20$ means that the data are 20 times more likely under the hypothesis that there is a gender difference. Importantly, Bayesian hypothesis testing allows for a quantification of support for the null hypothesis, too. While a $BF_{10} = 1$ implies that the data are equally likely under both hypotheses, a BF_{10} smaller than 1 indicates the support for the null, since $BF_{10} = \frac{1}{BF_{01}}$. In addition, we report Bayesian credible intervals. The Bayesian credible interval is that interval in which the parameter of interest (here: the raw mean delta between female and male values) lies with a probability of 95% (i.e. it includes the 95% most credible values) [12, 13]. We calculated the equal-tailed credible interval with 10,000 sampling iterations.

2.3 Topic Differences

As preprocessing steps, prior to topic modelling, all text data were lower-cased, stemmed, and all punctuation, stopwords, and numbers removed. First, we assess whether there are differences in topics between long and short texts. We construct a topic model[3] for all texts (long + short) and select the number of topics by examining values for semantic coherence and exclusivity [10, 11]. That approach assigns a probability for each document belonging to each topic. Here, we assign each document to its most likely topic (i.e., the highest topic probability for the document). We use a Chi-square test to examine whether there is an association between document type (long vs short) and topic occurrence. Standardised residuals (z-scores) are used to assess what drives a potential association.

Gender differences in topic occurrences are assessed in the same way for long and short texts separately. A Chi-square test is applied to test for an association between gender (male vs. female) and topic occurrence.

2.4 Open-Vocabulary Differences

In addition, we also look at differences in *n*-grams (unigrams, bigrams, trigrams) without the assumption of topic membership. Specifically, we calculate Bayes factors for male vs female comparisons on all *n*-grams to explore how both genders differ. We conduct that analysis for the short and long texts separately.

2.5 Closed-Vocabulary Differences

Since *n*-grams might not capture higher-order constructs (e.g., analytical thinking, anxiety) and psychological processes, we run the same analysis as in **2.4** using the LIWC 2015 [2].

3 Results

3.1 Gender Differences in Reported Emotions

We compared the self-reported emotions (ranging from 1 = very low to 9 = very high) calculating Bayes factors and Bayesian credible intervals. Table 1 suggests that there was extreme evidence (based on the Bayes factor [8]) that women were more worried, anxious, afraid and sad than men. There was strong evidence that women were angrier than men. Conversely, men reported considerably more desire and more relaxation than women. We also assessed whether gender was associated with the "best fitting" chosen emotion.

A Chi-square test, $X^2(7) = 43.83$, $p < .001$, indicated an association between the chosen emotion and gender. Standardised residuals showed that this effect was driven by disparities between females choosing fear significantly more often than males ($z = -2.62$) and males choosing relaxation significantly more often ($z = 4.40$) than females. Thus, while anxiety was overall the most chosen emotion (55.36%, see [1]), gender did play a role for the selection of fear and relaxation.

[3] All topic models were constructed with the *stm* R package [10].

Table 1. Means (SD) for reported emotions about COVID-19 and statistical test results for gender differences.

	Female	Male	BF_{10}	Mean delta [95% Bayesian CI]
Worry	6.84 (1.62)	6.01 (1.90)	$>10^{25}$	−0.82 [−0.96;−0.68]
Anger	4.04 (2.22)	3.68 (2.26)	54.97	−0.36 [−0.54; −0.17]
Anxiety	6.78 (2.14)	5.95 (2.43)	$>10^{14}$	−0.82 [−1.01; −0.64]
Desire	2.72 (1.96)	3.48 (2.09)	$>10^{15}$	0.76 [0.60; 0.94]
Disgust	3.23 (2.13)	3.21 (2.15)	0.05	−0.01 [−0.19; 0.17]
Fear	6.00 (2.17)	5.04 (2.34)	$>10^{20}$	−0.95 [−1.14; −0.76]
Happiness	3.57 (1.87)	3.74 (1.93)	0.41	0.17 [0.01; 0.32]
Relaxation	3.73 (2.10)	4.36 (2.13)	$>10^{9}$	0.62 [0.45; 0.79]
Sadness	5.91 (2.21)	4.96 (2.36)	$>10^{19}$	−0.95 [−1.13; −0.76]

Note. Positive delta values indicate that the variable was scored higher by men than women, and vice versa.

3.2 Differences in Topic Occurrence

Long vs Short Texts
For the topic model with long and short texts, 15 topics were selected based on semantic coherence and exclusivity of topic words. A significant association was found between text length (long vs. short) and topic occurrence, $X^2(14) = 1776.6$, $p < .001$. The six topics that differed most (i.e., highest standardised residuals) between long and short texts are depicted in Table 2[4]. Long texts were more likely to concern worries about both family and work, as well as the societal impact of the virus. Short texts were more likely to concern lockdown rules, staying home, and negative emotions.

Gender Differences
The topic model for long texts contained 20 topics[4]. For this model, we observed a significant association between gender and topic occurrence, $X^2(19) = 140.02$, $p < .001$. Table 3 show the topics with the largest gender difference and suggests that men were more likely to write about the national impact of the virus, worries about health, and the future outlook regarding COVID-19. Women spoke more about family and friends, as well as sickness and death.

We selected a model with 15 topics to represent the short texts. Here, we also observed a significant association between gender and topic occurrence, $X^2(14) = 101.47$ $p < .001$. Women spoke more about missing family and friends, and included more calls for solidarity. In contrast, men spoke more about the international impact of the virus, governmental policy, and returning to normal.

[4] For a full list of topics and terms, see: https://osf.io/ktwhp/.

Table 2. Topic differences (z-scores) long and short texts

Topic	Terms	Std. residual
Lockdown rules	stay, home, follow, live, rule, peopl, nhs, help, wish, other, please, protect, listen, save, selfish	−22.53
Family worries	worri, work, family, health, children, risk, catch, also, worker, job, food, elder, virus, shop, high	19.00
Staying home	can, keep, safe, stay, get, home, try, right, best, time, now, possible, position, make, love	−17.32
Work worries	feel, also, lot, sad, work, quit, howev, family, find, job, make, anxious, other, difficult, anxiety	13.93
Negative emotions	feel, anxious, situat, scare, fear, sad, end, like, anxiety, come, whole, moment, corona, job, stress	−12.49
Societal impact	concern, situat, health, impact, current, effect, economy, person, term, social, lockdown, mental, affect, may, society	10.35

Note. A positive standardised residual indicates that the topic was more likely to occur in long texts.

3.3 Open-Vocabulary Differences

Before extracting unigrams, bigrams and trigrams, the corpus was lower-cased and stop-words and punctuation were removed. Table 4 indicates that, in long texts, women use "anxious" (both in unigram and in bigram/trigram form) markedly more often than men, and mention "family" and "children" more often. The findings are, in part, corroborated for short texts, which, in addition, include the unigrams "scared" and "scary". Interestingly, unigrams which were more frequently used by men were "hopefully" and "calm". In broad lines, these findings reflect the differences found using a topic-based approach, where women expressed more fear and men were more likely to write about a (hopeful) return to normal. We also observe that *n*-gram-based differences are more pronounced in the longer texts than in shorter ones.

3.4 Closed-Vocabulary Differences

To capture potential differences in higher-order constructs, we also looked at gender differences for the LIWC variables. Table 5 suggests that men had a higher score on analytical thinking, used more articles and more "big" words (more than six letters). Women, on the other hand, used more pronouns (all, personal pronouns, and first-person pronouns), more verbs, and expressed more anxiety and references to their family. We also observe that women had a substantially higher focus on the present than men.

For short texts, we see that the differences are less pronounced (BFs smaller than ten only constitute moderate evidence [8] and are ignored here). The data show that men scored higher on references to power and used more articles, while women used more conjunctions and scored higher on authenticity.

Table 3. Topic proportions with significant gender differences

Topic	Terms	Std. residuals
Long texts		
National impact	countri, govern, way, can, taken, get, feel, like, hope, world, also, response, much, better, china	5.34
Family and friends	feel, will, anxious, also, know, sad, famili, friend, worri, like, end, dont, see, day, situation	−4.45
Sickness/death	worri, get, one, feel, also, sick, love, scare, anxious, die, husband, mum, fear, day, alone	−4.34
Health worries	health, famili, worri, member, also, situat, concern, mental, able, risk, time, current, condition, older, however	3.92
Outlook	will, virus, year, people, world, death, get, also, know, catch, many, end, ill, feel, human	3.43
Short texts		
Missing family & friends	famili, life, see, friend, like, day, really, miss, cant, enjoy, wait, week, even, can't, fed	−4.91
International impact	world, covid-, concern, death, mani, covid, news, economy, real, never, test, must, infect, term	4.82
Policy	govern, scare, countri, social, distance, angry, proud, public, crisis, line, response, boris, frontline, fron, nhs	4.18
Solidarity	stay, home, everyone, live, safe, pleas, insid, save, protect, listen, nhs, indoor, advice, quicker, want	−3.79
Return to normal	get, can, thing, normal, back, better, well, bit, possible, start, worse, hand, sick, told, tire	3.38

Note. Positive standardised residuals indicate that the topic occurrence was higher for men than women, and vice versa.

3.5 Follow-up Exploration: Concerns, Drives and Time Focus

To understand gender differences in emotional responses to COVID-19 on the psycholinguistic level better, we zoom in on three LIWC clusters (Table 6). We look at the clusters "personal concerns", "drives", and "time orientation" - each of which consists of sub-categories (e.g., concerns: work, death, drives: risk, achievement, time orientation: future, present). Men scored higher on the variables risk (e.g., cautious, dangerous), work, and money, whereas women had higher values for affiliation (e.g., friend, party, relatives), home (e.g., neighbour, pet) and a focus on the present. Again, these bottom-up findings seem to align with the topic models from a psycholinguistic angle.

Table 4. Top 10 ngrams with the largest gender differences for long and short texts.

n-gram	BF_{10}	Mean delta [95% Bayesian CI]
Long texts		
anxious	$>10^{10}$	−0.20 [−0.26; −0.15]
Sad	$>10^{9}$	−0.16 [−0.20; −0.11]
Feel	$>10^{8}$	−0.40 [−0.52; −0.29]
Family	$>10^{6}$	−0.20 [−0.26; −0.13]
i_feel	$>10^{6}$	−0.29 [−0.38; −0.19]
Makes	$>10^{5}$	−0.09 [−0.12; −0.06]
Feel_Very	$>10^{4}$	−0.08 [−0.10; −0.05]
Children	$>10^{4}$	−0.09 [−0.12; −0.06]
Feel_anxious	$>10^{4}$	−0.06 [−0.09; −0.04]
i_feel_anxious	$>10^{4}$	−0.05 [−0.08; −0.03]
Short texts		
Family	308.00	−0.05 [−0.08; −0.03]
Sad	201.16	−0.04 [−0.06; −0.02]
Scared	66.41	−0.04 [−0.06; −0.02]
Friends	14.19	−0.03 [−0.05; −0.01]
Scary	8.03	−0.02 [−0.03 −0.01]
Hopefully	7.47	0.02 [0.01; 0.03]
Loved_ones	7.41	−0.02 [−0.04; −0.01]
Calm	7.19	0.01 [0.01; 0.02]
Stay	4.96	−0.07 [−0.12; −0.03]
Stay_home	4.73	−0.05 [−0.05 −0.01]

Note. Negative delta values indicate that the variable was scored higher by women than men, and vice versa.

4 Discussion

This study elucidated gender differences in emotional responses to COVID-19 in several (linguistic) domains. Analyses were performed for both long and short, Tweet-sized texts. We first discuss the implications of using the respective text lengths for understanding emotions and worries. Then, we review the observed gender differences and relate them to previous research on this topic. Lastly, some potential future avenues for understanding worries from text data are identified.

4.1 Short vs Long Texts

In previous work using the same dataset, it was suggested that important differences emerge when participants are asked to write about their worries in long versus short,

Table 5. Top 10 LIWC variables with the largest gender differences for long and short texts.

LIWC variable	Meaning	BF_{10}	Mean delta [95% Bayesian CI]
Long texts			
Analytic	Analytical thinking	$>10^{24}$	10.59 [8.67; 12.51]
Pronoun	Pronouns (all)	$>10^{22}$	−1.81 [-2.14; −1.48]
Ppron	Personal pronouns	$>10^{19}$	−1.49 [−1.79; −1.20]
Article	Articles	$>10^{17}$	0.92 [0.73; 1.12]
Verb	Verbs	$>10^{14}$	−1.36 [−1.67; −1.05]
Anx	Emotional process: anxiety	$>10^{13}$	−0.63 [−0.78; −0.48]
i	First person singular	$>10^{13}$	−1.28 [−1.57; -0.98]
Focuspresent	Time orientation: present	$>10^{12}$	−1.11 [−1.39; −0.84]
Sixltr	Words with more than 6 letters	$>10^{10}$	1.44 [1.06; 1.82]
Family	Social processes: family	$>10^{8}$	−0.33 [−0.42; −0.24]
Short texts			
Power	Drives: power	28.06	0.64 [0.28; 0.98]
Conj	Conjunctions	24.66	−0.84 [−1.30; −0.37]
Authentic	Authenticity	23.01	−5.48 [−8.69; −2.46]
Article	Articles (all)	12.22	0.72 [0.29; 1.13]
Money	Personal concerns: money	8.28	0.21 [0.08; 0.33]
Work	Personal concerns: work	3.91	0.37 [0.13; 0.62]
Analytic	Analytical thinking	3.32	4.08 [1.36; 6.79]
Home	Personal concerns: home	3.27	−0.44 [-0.73; −0.14]
i	First person singular	2.61	−0.64 [-1.08; −0.19]
insight	Cognitive processes: insight	1.73	−0.39 [−0.68; −0.11]

Note. Positive delta values indicate that the variable was scored higher by men than women, and vice versa.

Tweet-size texts: in topic models that were constructed for long and short texts separately, longer texts seemed to shed more light on the worries and concerns of participants. In contrast, shorter texts seemed to function as a call for solidarity [1]. In the current study, we were able to statistically test that idea by constructing a topic model for both text types together. Indeed, when testing for an association between text length and topic occurrence, we found that topics significantly differed between text types. Our results confirmed that short Tweet-like texts more frequently referred to calls to 'stay at home, protect the NHS, save lives' (a UK government slogan during the crisis at the time of data collection). Longer texts more effectively related to the specific worries participants had, including those about their family, work, and society.

Table 6. Means (SD) of drives, concerns and time orientation LIWC clusters and test results for gender differences.

Variable	Females	Males	BF_{10}	Mean delta [95% Bayesian CI]
Drives: affiliation	2.99 (2.04)	2.61 (2.09)	594.41	−0.38 [−0.55; −0.21]
Drives: achievement	1.73 (1.32)	1.74 (1.32)	0.05	0.01 [−0.10; 0.12]
Drives: power	2.30 (1.61)	2.35 (1.75)	0.06	0.05 [−0.09; 0.18]
Drives: reward	1.33 (1.17)	1.28 (1.14)	0.07	0.05 [−0.05; 0.14]
Drives: risk	0.97 (0.98)	1.22 (1.18)	$>10^3$	0.25 [0.16; 0.34]
Concerns: work	2.24 (1.76)	2.68 (2.02)	$>10^3$	0.44 [0.28; 0.59]
Concerns: leisure	1.18 (1.15)	1.14 (1.22)	0.07	−0.04 [−0.14; 0.06]
Concerns: home	1.24 (1.11)	0.98 (1.07)	$>10^3$	−0.26 [−0.35; −0.17]
Concerns: money	0.73 (0.98)	0.96 (1.17)	$>10^3$	0.22 [0.13; 0.31]
Concerns: work	0.07 (0.31)	0.07 (0.26)	0.05	0.00 [−0.03; 0.02]
Concerns: death	0.31 (0.56)	0.29 (0.61)	0.06	−0.02 [−0.06; 0.03]
Time orientation: past	1.80 (1.68)	1.85 (1.73)	0.07	0.06 [−0.08; 0.20]
Time orientation: present	15.10 (3.23)	13.98 (3.50)	$>10^{11}$	−1.11 [−1.40; −0.84]
Time orientation: future	2.22 (1.68)	2.32 (1.83)	0.12	0.10 [−0.04; 0.24]

Note. Positive delta values indicate that the variable was scored higher by men than women, and vice versa.

4.2 A Cautionary Note on Using Short Text Data

The apparent differences between long and Tweet-sized texts suggest that researchers need to exercise caution in relying largely on Twitter datasets to study the COVID-19 crisis, and other more general social phenomena. Indeed, several Twitter datasets have been released [14–18] for the research community to study responses to the Corona crisis. However, the current study shows that such data may be less useful if we are interested in gaining a deep understanding of emotional responses and worries. The exact reasons for that difference (e.g., possibly different functions that each text form serves) merit further research attention. One possibility is that instructing participants to write texts of a specific length also influences emotion expression. However, the observation that Tweet-sized texts in this study are less able than long texts to convey emotions and worries about COVID-19, on top of the classical limitations of social media data (e.g., demographic biases [19], participation biases [20], or data collection biases [21]), are reasons to be more cautious for issues as timely and urgent as mental health during a global pandemic. Ultimately, making use of the convenience of readily-available social media data might come at the expense of capturing people's concerns and could lead to skewed perceptions of these worries. We urge us as a research community to (re-) consider the importance of gathering high-quality, rich text data.

4.3 Emotional Responses to COVID-19: Males vs Females

Gender differences emerged throughout this paper. Reported emotions showed that women were more worried, anxious, afraid, and sadder then men and these results were supported by linguistic differences. For instance, topic models suggested that women discussed sickness and death more often than men. *N*-gram differences showed that women used 'anxious', 'sad', and 'scared' more than men and LIWC differences showed that women used more words related to anxiety. Importantly, this is not to say that men did not worry about the crisis, as reported negative emotions were still relatively high (e.g., the average score of 6 out of 9 for worrying). Furthermore, topic models showed that men wrote more frequently about worries related to their health than women.

The results further illustrated differences in what men and women worry about with regards to the crisis. Women's focus on family and friends emerged consistently in the topic models, *n*-gram differences, and LIWC differences. On the other hand, men more frequently worried about national and international impact (topic models) and wrote more frequently about money and work than women (LIWC). All in all, these results seem to follow previous work on gender differences in language more generally. For example, similar to our results, previous studies have found that women, in general, score higher than men on the LIWC categories for anxiety, family, and home [3]. Our results also seemed to have further replicated that men use more articles, and use the LIWC categories money and work more often than women [3]. In light of these findings, it is of key importance to discern whether the gender differences that emerged in this study are specific to COVID-19 worries, or are simply a reflection of gender differences that emerge regardless of the issue in question.

There are some indications in our data to imply that the differences are in line with general gender differences but in a more pronounced way. Although there is no agreed upon (meta-analytic) benchmark of gender differences in language to which we can compare our results, there is some previous work with which comparisons can be made [3]. For example, general linguistic differences in anxiety ($d = -0.16$, here: $d = -0.36$), family ($d = -0.12$, here: $d = -0.34$), present focus ($d = -0.18$, here: $d = -0.33$), article use ($d = 0.24$, here: $d = 0.38$), references to work ($d = 0.12$, here: $d = 0.23$) and money ($d = 0.10$, here: $d = 0.20$) have been found previously [3]. All of these are present in our data as well but often with an effect twice the size (note: the d values are calculated for comparison purpose but are not included in the tables). Thus, it is possible that the COVID-19 situation and the accompanying toll on mental health exacerbated the language differences between men and women further. If we follow the line of previous work [3], the intensified linguistic gender differences can be interpreted as serving different functions for men and women. It has been suggested that women predominantly discuss other people and internal, emotional processes, while men focus more on external events and objects [3, 6]. Similar patterns are discoverable in our data, where women were more likely to discuss loved ones and their own negative emotions, whereas men were more likely to write about the external effects of the virus on society. It is important to note that the findings here (and elsewhere) do not imply that these patterns are all-encompassing or exclusive to men or women: men do also discuss their family and women also worry about the economy, but on average the reported differences emerged. All in all, the current results seem to fall in line with previous empirical work as well as persisting gender stereotypes.

4.4 Next Steps in Understanding Worries from Text Data

For the social and behavioural sciences during and after the Corona crisis, a principal research question revolves around the mental health implications of COVID-19. The current study leveraged text data to gain an understanding of what men and women worry about. At the same time, it is of vital interest to develop mitigation strategies for those who need them the most. While this paper attempted to shed light on gender differences, other fundamental questions still need answering.

First, relatively little is known about the manifestation of emotions in language and the subsequent communication of it in the form of text data (e.g., how good are people in expressing their emotions and, importantly, which emotions are better captured computationally than others?). Second, the type of a text (e.g., stream-of-consciousness essay vs pointed topical text vs Tweet) seems to determine the findings to a great deal (i.e. different topics, different effect sizes). Ideally, future research can illuminate how the language and aims of an individual change depending on the type of text. Third, to map out the worries on a large scale and use measurement studies to understand the concerns people have, further data collection in different (offline) contexts and participant samples beyond the UK alone will be needed. Sample demographics (e.g. socio-economic status, ethnicity) in this study could have influenced text writing or emotion expression, thus replications in different samples will be needed to understand these potential effects. Further large-scale data collection will also allow for increased attention to prediction models constructed on high-quality ground truth data.

5 Conclusion

The current paper contributes to our understanding of gender differences in worries related to the COVID-19 pandemic. Gender was related to the reported emotions, topics, *n*-grams, and psycholinguistic constructs. Women worried mainly about loved ones and expressed negative emotions such as anxiety and fear, whereas men were more concerned about the broader societal impacts of the virus. The results showed that long texts differed from short Tweet-size texts in reflecting emotions and worries.

References

1. Kleinberg, B., van der Vegt, I., Mozes, M.: Measuring emotions in the COVID-19 real world worry dataset. arXiv:2004.04225 [cs]. http://arxiv.org/abs/2004.04225 (2020)
2. Pennebaker, J.W., Boyd, R.L., Jordan, K., Blackburn, K.: The development and psychometric properties of LIWC2015. The University of Texas at Austin (2015). https://repositories.lib.utexas.edu/handle/2152/31333
3. Newman, M.L., Groom, C.J., Handelman, L.D., Pennebaker, J.W.: Gender differences in language use: an analysis of 14,000 text samples. Discourse Process. **45**(3), 211–236 (2008). https://doi.org/10.1080/01638530802073712

4. Lakens, D.: Calculating and reporting effect sizes to facilitate cumulative science: a practical primer for t-tests and ANOVAs. Front. Psychol. **4**, 863 (2013). https://doi.org/10.3389/fpsyg.2013.00863
5. Cohen, J.: Statistical Power Analysis for the Behavioral Sciences. Academic Press, Cambridge (1988)
6. Schwartz, H.A., et al.: Personality, gender, and age in the language of social media: the open-vocabulary approach. PLoS ONE **8**(9), e73791 (2013). https://doi.org/10.1371/journal.pone.0073791
7. Blei, D.M., Lafferty, J.D.: A correlated topic model of science. Ann. Appl. Stat. **1**, 17–35 (2007). https://projecteuclid.org/euclid.aoas/1183143727
8. Ortega, A., Navarrete, G.: Bayesian hypothesis testing: an alternative to null hypothesis significance testing (NHST) in psychology and social sciences. Bayesian Infer. (2017). https://doi.org/10.5772/intechopen.70230
9. Wagenmakers, E.-J., Lodewyckx, T., Kuriyal, H., Grasman, R.: Bayesian hypothesis testing for psychologists: a tutorial on the Savage-Dickey method. Cogn. Psychol. **60**(3), 158–189 (2010). https://doi.org/10.1016/j.cogpsych.2009.12.001
10. Roberts, M.E., Stewart, B.M., Tingley, D.: stm: R package for structural topic models. J. Stat. Softw. **41**, 1–40 (2014)
11. Mimno, D., Wallach, H., Talley, E., Leenders, M., McCallum, A.: Optimizing semantic coherence in topic models, vol. 11 (2011)
12. Kruschke, J.K.: Bayesian estimation supersedes the T test. J. Exp. Psychol. Gen. **142**(2), 573–603 (2013). https://doi.org/10.1037/a0029146
13. Kruschke, J.K., Liddell, T.M.: The Bayesian new statistics: hypothesis testing, estimation, meta-analysis, and power analysis from a bayesian perspective. Psychon. Bull. Rev. **25**(1), 178–206 (2017). https://doi.org/10.3758/s13423-016-1221-4
14. Banda, J.M., et al.: A Twitter dataset of 150 + million tweets related to COVID-19 for open research. Zenodo, 5 April 2020. https://doi.org/10.5281/zenodo.3738018
15. Chen, E., Lerman, K., Ferrara, E.: #COVID-19: the first public coronavirus Twitter dataset. Python. (2020). https://github.com/echen102/COVID-19-TweetIDs
16. Lamsal, R.: Corona virus (COVID-19) tweets dataset. IEEE, 13 March 2020. https://ieee-dataport.org/open-access/corona-virus-covid-19-tweets-dataset
17. Jacobs, C.: Coronada: Tweets about COVID-19. Python. (2020). https://github.com/BayesForDays/coronada
18. Basile, V., Caselli, T.: TWITA - long-term social media collection at the university of Turin, 17 April 2020. http://twita.di.unito.it/dataset/40wita
19. Morstatter, F., Pfeffer, J., Liu, H., Carley, K.M.: Is the sample good enough? comparing data from Twitter's streaming API with Twitter's FIrehose. arXiv:1306.5204 [physics]. http://arxiv.org/abs/1306.5204 (2013)
20. Solymosi, R., Bowers, K.J., Fujiyama, T.: Crowdsourcing subjective perceptions of neighbourhood disorder: interpreting bias in open data. Br. J. Criminol. **58**(4), 944–967 (2018). https://doi.org/10.1093/bjc/azx048
21. Pfeffer, J., Mayer, K., Morstatter, F.: Tampering with Twitter's sample API. EPJ Data Science **7**(1), 1–21 (2018). https://doi.org/10.1140/epjds/s13688-018-0178-0

Jump on the Bandwagon? – Characterizing Bandwagon Phenomenon in Online NBA Fan Communities

Yichen Wang(✉), Jason Shuo Zhang, Xu Han, and Qin Lv

Department of Computer Science, University of Colorado Boulder,
Boulder, CO 80309, USA
{yiwa6864,JasonZhang,xuha2442,Qin.Lv}@colorado.edu

Abstract. Understanding user dynamics in online communities has become an active research topic and can provide valuable insights for human behavior analysis and community management. In this work, we investigate the "bandwagon fan" phenomenon, a special case of user dynamics, to provide a large-scale characterization of online fan loyalty in the context of professional sports teams. We leverage the existing structure of NBA-related discussion forums on Reddit, investigate the general bandwagon patterns, and trace the behavior of bandwagon fans to capture latent behavioral characteristics. We observe that better teams attract more bandwagon fans, but they do not necessarily come from weak teams. Our analysis of bandwagon fan flow also shows different trends for different teams, as the playoff season progresses. Furthermore, we compare bandwagon users with non-bandwagon users in terms of their activity and language usage. We find that bandwagon users write shorter comments but receive better feedback, and use words that show less attachment to their affiliated teams. Our observations allow for more effective identification of bandwagon users and prediction of users' future bandwagon behavior in a season, as demonstrated by the significant improvement over the baseline method in our evaluation results.

Keywords: Reddit · Bandwagon · Loyalty · Online community · Sports fan behavior

1 Introduction

The proliferation of online social networks has led to the increasing popularity of online communities, which allow like-minded people to connect and discuss their shared interests, values, and goals without geographic constraints. For instance, in Reddit, a popular online community platform, users can participate in a large number of communities [10,25], which are referred to as subreddits and denoted by "r/" followed by the community's topic, such as r/politics. The plethora of online communities provides a great opportunity to understand human interactions across multiple communities. Some "loyal" users may commit to one

S. Aref et al. (Eds.): SocInfo 2020, LNCS 12467, pp. 410–426, 2020.
https://doi.org/10.1007/978-3-030-60975-7_30

particular community and maintain a stable engagement [8], some may jump across several communities [25], while others may change their community affiliation over time. Such phenomena are fundamental examples of user dynamics in online communities, and understanding how users migrate across communities is an important problem. By analyzing and characterizing these dynamics in terms of users' behavior patterns can provide useful insights for designing better communities, guide community managers to provide better services to improve user engagement, and help community-related stakeholders (e.g., sports teams, celebrities, advertisers) to better promote their business.

The bandwagon phenomenon is a widespread phenomenon in online sports communities. By definition, "bandwagon fans" refer to sports fans who start following a sports team only because of its recent success, and this group of fans will be gone immediately after the team performs poorly. Reddit officially introduced the bandwagon mechanism in some sports-related discussion groups (e.g., r/NBA and r/NFL) several years ago, which allows users to change their team affiliation and self-identify themselves as bandwagon fans during playoffs. For instance, during playoffs in NBA season 2016–17, 17.9% of Cavaliers' fans in r/NBA were bandwagon fans. Our work focuses on examining this specific phenomenon, "bandwagon fan", in online NBA fan communities. We leverage the existing structure of NBA-related discussion forums on Reddit to study users' bandwagon behavior in the context of professional sports, a domain that is understudied yet closely connected to people's daily life. We choose online fan groups of professional sports teams as a testbed for the following reasons. First, professional sports play a significant role in modern life and a large population is actively engaged [7,18]. Second, professional sports teams are unambiguously competitive in nature and users affiliated with different teams have clearly different preferences [30].

Present Work. Specifically, using users' posts and comments in r/NBA and individual team subreddits across three recent NBA seasons, we aim to answer three research questions: (1) What is the general pattern of bandwagon phenomenon and its relationship with team performance? (2) Are there behavioral features that differentiate bandwagon users from non-bandwagon users? and (3) How effective can we identify bandwagon users and predict future bandwagon behavior using these features? Our large-scale study reveals that better teams attract more bandwagon fans and bandwagon fans typically switch to better-performing teams, but not all bandwagon fans come from weak teams. Also, as the playoffs progresses, bandwagon fans from different teams show different team change flow patterns. We also identify clear behavioral differences between bandwagon and non-bandwagon users in terms of their activity and language usage after applying a matching technique, e.g., bandwagon users tend to leave shorter comments but receive better feedback; and they are less attached to their affiliated teams. Using the features we identify, we are able to improve the bandwagon fan classification and prediction results over the bag-of-words baseline method, with 18.9% and 47.6% relative improvement, respectively.

Our work contributes to the following aspects: First, to the best of our knowledge, this is the first large-scale analysis of bandwagon behavior in sports community, which reveals clear behavioral characteristic differences of bandwagon fans compared with non-bandwagon fans. Second, using the observed behavioral characteristics, we can better distinguish bandwagon users and predict future bandwagon behaviors. Third, our work offers new insights for user loyalty research and online community management.

2 Related Work

User Engagement and Loyalty in Online Communities. Online community engagement has been a topic of active research [1,4,13,22,25,27,28]. Danescu-Niculescu-Mizil et al. [4] build a framework to track users' linguistic change in an intra-community scenario and find they follow a determined two-stage lifecycle: an innovative learning phase and a conservative phase. Tan et al. [25] study users' multi-community engagement in Reddit through community trajectory, language usage and feedback received. They find that over time, users span more communities, "jump" more and concentrate less; users' language seems to continuously grow closer to the community's; frequent posters' feedback is continually growing more positive; and departing users can be predicted from their initial behavioral features. Loyalty is a fundamental measure of users maintaining or changing affiliation with single or multiple communities. Users' loyalty with a single community is studied via churning in question-answering platform [5,20], where gratitude, popularity related features, and temporal features of posts are shown to be predictive. Multi-community loyalty in both community and user level is studied by Hamilton et al. [8], where loyalty is defined as users making the majority or minority of their posts to a certain forum in a multi-forum site at a specific time. They find loyal communities tend to be smaller, less assertive, less clustered and have denser interaction networks, and user loyalty can be predicted from their first three comments to the community. Different from prior work, our study focuses on the self-defined bandwagon status in sub-communities (different teams) within a large, single community (r/NBA).

Bandwagon Phenomenon. The bandwagon phenomenon is found in many fields such as politics [15], information diffusion [17], sports [29], and business applications [24]. In Sundar et al.'s work [24], by conducting an experiment using fake products with different star ratings, number of customer reviews, and sales rank on an e-commerce site, they provide the preliminary support for bandwagon effect on users' intention and attitude toward products. Zhu et al. [31] also find that other people's opinions significantly sway people's own choices via an experiment in an online recommender system. In Wang et al.'s work [26], they predict article life cycles in Facebook discussion groups using the bandwagon effect. Voting is another behavior that is related to this phenomenon, where voters may or may not follow their own opinion to make a voting decision but follow the majority [12,16]. Most of these research efforts conclude at observing the bandwagon phenomenon at the application level. Further study is needed to analyze the specific characteristics, especially in the context of online communities.

3 Dataset

We focus on the professional sports context derived from NBA-related discussion forums (r/NBA and individual team subreddits) on Reddit, an active community-driven platform where users can submit posts and make comments. The r/NBA and team subreddits provide an ideal testbed for observing and understanding bandwagon fans, because a user's team affiliation can be acquired directly through a mechanism known as "flair". Flair appears as a short text next to the username in posts and comments (e.g., Lakers *username*). In r/NBA, users can use flairs to indicate support and each user cannot have multiple flairs at the same time. After a specific date (referred to as the *bandwagon date* in the rest of this paper) which is usually shortly before playoffs begin in each season, each user is given the option to change his/her flair to a bandwagon flair of a different team (e.g., Warriors Bandwagon *username*). A user can change the flair as many times as he/she wants. We obtain 0.6M posts and 30M comments as well as their received feedback in NBA-related subreddits from https://pushshift.io [2], where flair is used to determine the user's bandwagon status. As pointed out by Zhang et al. [29], offline NBA seasons are reflected in users' behavior in these NBA-related subreddits. As such, we organize our data according to the timeline of NBA seasons and focus on three seasons: 2015–16, 2016–17, and 2017–18.

4 General Patterns of Bandwagon Phenomenon

Our first research question aims to identify the general pattern of bandwagon behavior. We extract all flair changes from team A to team B bandwagon, where A $\neq$ B. A general user flow network in is shown in Fig. 3.

Observation 1: Better Teams Attract More Bandwagon Fans, but Not All of Them Come from Weak Teams. We first investigate which teams bandwagon fans switch to (i.e., target team). Intuitively, we expect the bandwagon fans move to better teams and abandoning the weak ones. Here, we consider the number of bandwagon fans who switched to team B (the target team) and its correlation with either team B's standing (i.e., rank) or the difference between team B' standing and team A (source team)'s standing. Table 1 shows the correlations computed for each scenario and each season. We can see strong correlations in all but one case. The correlations are negative because lower standing means better performance (e.g., the best team ranks 1), and if team B is better than team A then the standing difference (B's standing - A's standing) should be negative. Based on the results, we can conclude that better teams tend to attract more bandwagon fans.

Table 1. Correlation results for team standing and standing difference with user count. **Throughout this paper, the number of stars indicate p-values:** $^{***}: p < 0.001$, $^{**}: p < 0.01$, $^{*}: p < 0.05$

Season	Correlation between team B's standing and user count	Correlation between standing difference and user count
2015-16	-0.675^{*}	-0.617^{**}
2016-17	-0.673^{**}	-0.489^{**}
2017-18	-0.789^{*}	0.188 (not significant)

We further study where the bandwagon fans switch from (i.e., the source team A). We examine the correlation of team A's standing with user count. To our surprise, we do not observe any significant correlation for all the three seasons. This indicates that many bandwagon fans also come from strong teams. To better understand this, we calculate each team's bandwagon fan ratio and select the teams with a ratio above the median of all teams. We find that in the three seasons, there are 8, 7, 8 teams respectively which are in the playoffs but have above-median ratio of fans leaving team. For instance, Raptors and Spurs are top 3 in their conference in all three seasons but still have a high percentage of fans leaving. Although bandwagon is only one aspect of loyalty, this result supplements the finding in Zhang et al.'s work on fan loyalty [29]: top teams tend to have lower fan loyalty, in terms of user retention. As can be seen in Fig. 3, not all bandwagon fans come from weak teams and strong teams can also lose a large number of bandwagon fans.

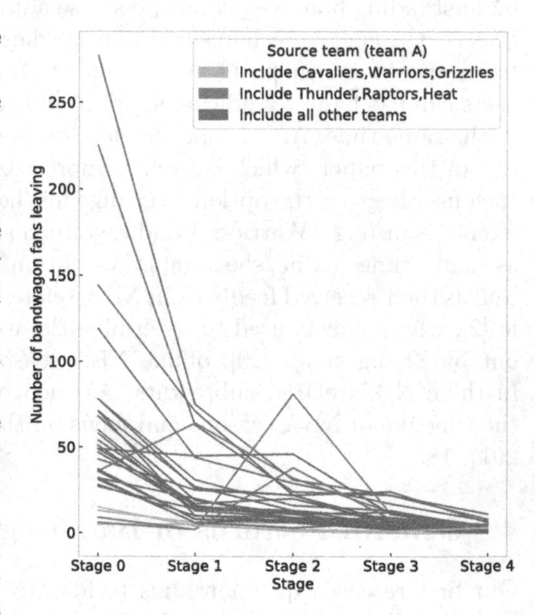

Fig. 1. Bandwagon fan flair change trend in season 2015–16.

Observation 2: Bandwagon Flair Changes Exhibit Different Stage-Wise Trends Across Teams. NBA playoffs are elimination tournaments which include 4 rounds: conference first round, conference semi-final, conference final and overall final. To keep our analysis consistent with the temporal structure of NBA playoffs, we divide the period after the *bandwagon date* into 5 stages: the pre-playoffs stage (stage 0) and the other 4 stages corresponding to the 4 rounds in playoffs, to examine the temporal dynamics of user's bandwagon behaviors. As shown in Fig. 1, we examine the number of bandwagon fans in the 5 stages. We find three types of representative trends: (1) Most teams' bandwagon fans abandon their original teams in stage 0 and there are still fans leaving in later stages but in decreasing numbers. For instance, bottom teams such as Lakers and Knicks have a large number of vagrant fans, and better teams such as Clippers and Heat also fall into this category due to some vagrant fans. (2) Some teams have very few bandwagon fans in the whole period. Such teams include top ones (e.g., Cavaliers and Warriors) and bottoms ones (e.g., Grizzles). Those teams have a strong fan base. (3) Some teams have a spike during the playoffs, which means most of their bandwagon fans abandon the team in a specific stage and very few leave before or after that stage. Three teams fall into this category in season 2015–16: Thunder, Raptors, and Celtics. These teams have a relatively strong fan base compared with the teams in (1), but some fans lose interest in the middle (e.g., team is eliminated) and choose to support a stronger team. Please note that the teams falling into these three trend categories can vary from season to season depending on each team's performance in a season.

5 Behavioral Features of Bandwagon Users

Our second research question aims to identify behavioral features that help differentiate bandwagon fans from non-bandwagon fans. To do this, we apply a matching technique to make the two groups comparable. We first give a formal definition of *bandwagon users* and identify all these users. Then we match each *bandwagon user* with a *non-bandwagon user* who has a similar activity level. This allows us to directly compare the behavioral features of bandwagon and non-bandwagon fans.

5.1 Identify Bandwagon Users for Behavioral Comparison

To identify bandwagon users within one season, we first define *active* users in r/NBA as those who have at least five activities before the *bandwagon date* and one activity after, where an *activity* refers to either submitting a post or making a comment.

We view posting/commenting with a team's flair in r/NBA as an indication of support towards that team. We define an active user as a *fan* of a team during a specific period of time if the user indicates support (flair) only for that team and such support sustains over all activities during that time period. Note that this flair does not contain the word "bandwagon". If the user only uses flair with "bandwagon" for a team, we call that user a *bandwagon fan* of that team.

We further define bandwagon user and non-bandwagon user based on whether a *fan* changes to a *bandwagon fan* of another team. To summarize, we consider the following two groups of users:

- *Non-bandwagon user*: *Fan* of a team A throughout the whole season.

- *Bandwagon user*: *Fan* of a team A till a time point, and after that point, becomes *bandwagon fan* of team B (B $\neq$ A) for a period of time, regardless of any bandwagon changes thereafter.

The terms "fan", "bandwagon fan", "bandwagon users" and "non-bandwagon users" in this section all refer to the definition above.

There are also a number of users who are not in the aforementioned two groups. For instance, one user can be a fan for different teams during the whole season, while others may not be fans of any team. We do not consider those cases in our analysis since they do not link directly to the bandwagon phenomenon that our work focuses on. Statistics of the bandwagon and non-bandwagon users for our analysis are shown in Table 2.

Table 2. Number of bandwagon and Non-bandwagon Users

Season	#Bandwagon users	#Non-bandwagon users
2015-16	2,562	23,165
2016-17	1,526	29,955
2017-18	1,163	36,053

5.2 User Matching

To make a fair comparison between the two user groups, we need to rule out the influence of activity level. Specifically, for each bandwagon user who is a fan of team A at the beginning, we find a matching non-bandwagon user who has a similar activity level and supports the same team. As a result, we have 2562, 1526, and 1163 user pairs after user matching in the three seasons, respectively.

To evaluate the result of our matching procedure, we check distributional differences in terms of the number of activities between the treatment group (bandwagon users) and the control group (non-bandwagon users). We compare their empirical cumulative distributions before and after matching, using the Mann-Whitney U test [14]. Prior to matching, the p-value for this feature is very close to 0.0, indicating significant difference. After matching, we find no difference between the treatment group and the matched control group at the 5% significance level in all three seasons ($p = 0.485$, 0.489, 0.49 for the three seasons), indicating that the data is balanced in terms of activity level after user matching. Figure 4 in the Appendix shows the CDF plots before and after matching. Using the matched user groups, we then characterize how bandwagon users behave differently in terms of their posting/commenting activities and language usage. For consistency, we only consider user activities that occur before the *bandwagon date*. Since users cannot change their flairs to bandwagon flairs prior to that date, the differential features we observe can also be used for predicting future bandwagon behavior in playoffs.

5.3 Activity Features

Observation 1: Bandwagon Users Are Less Active Than Non-bandwagon Users in Individual Team Subreddit. We compare users' activity level in the same individual team subreddit after matching in r/NBA, and find that bandwagon users have fewer activities (i.e., more silent) than non-bandwagon users in terms of posting/commenting count: (19.91 vs. 29.40*, 23.55 vs. 35.63*, 23.51 vs. 32.4* for the three seasons, respectively). This is reasonable since individual team subreddits tend to attract fans who are more loyal/dedicated to their teams and participate more actively in their subcommunities.

Observation 2: Bandwagon Users Write Shorter Comments but Receive Better Feedback in R/NBA. Here, we compute a score for each comment (#upvotes - #downvotes) as a measure of received feedback. The higher the score, the better the feedback it receives. Our results show that bandwagon users write significantly shorter comments (18.71 vs. 20.75**, 18.34 vs. 20.30**, 18.26 vs. 19.55** for the three seasons, respectively), but receive better feedback (11.08 vs. 9.88*, 13.11 vs 11.80*, 16.64 vs 16.32* for the three seasons, respectively). This observation is consistent with the general pattern that users who "wander around" across diverse communities are more likely to receive better feedback [25].

5.4 Language Usage Features

Although we use both posting and commenting as indicators of user activity level, we focus more on comments in language usage analysis since posts are more about news and game reports, and less reflective of users' personal characteristics.

Observation 3: Bandwagon Users Talk Less About Specific Players and Teams. To analyze the content of users' comments, we first conduct preprocessing steps including lowercasing words and removing stop words and hyperlinks from comments. After that, Latent Dirichlet Allocation (LDA) [3], a widely-used topic modeling method, is applied to extract keywords and topics. In our case the perplexity score drops significantly when the number of topics increases from 5 to 10, but remains stable afterwards (from 10 to 30). Therefore, we use 10 as our topic number in this analysis.

We compare the topics of bandwagon comments and that of non-bandwagon ones. While they share many similar topics (e.g., game strategy related words: defense, offense; emotional expressions related words: shit, lol), there still exist some differences. Non-bandwagon users talk more about specific teams/players (e.g., Harden, spursgame), even when they are talking about similar topics as bandwagon users. It shows that bandwagon users appear to be less concerned about the details of teams/players, indicating a relatively indifferent attitude towards their affiliated teams.

Observation 4: Bandwagon Users Are Less Dedicated to Discussions in Terms of Word Usage. Inspired by Hamilton et al. [8] and previous observations, we examine the two groups' word usage. To capture the esoteric content and users' attachment to the teams, we calculate the proportion or summary variable of different types of words in their comments using LIWC word categories [19], a well known set of word categories that were created to capture people's social and psychological states. We find significantly less word use of bandwagon comments in five word categories: clout [11] (high clout value suggests that the author is texting from the perspective of high expertise and is confident), social process (words that suggest human interaction), cognitive process (words that suggest cognitive activity), drives (an overarching dimension that captures the words that represent the needs, motives and drives including achievement, power, reward, etc.), and future focus (future tense verbs and references to future events/times). The results are shown in Table 3. Please note that the results show the average value of bandwagon users versus that of non-bandwagon users. All these five lexicon categories show that non-bandwagon users have a closer attachment to their affiliated teams and a more proactive attitude towards discussions.

Table 3. LIWC Word Categories Analysis Results

Word category	Season 2015-16	Season 2016-17	Season 2017-18
Clout	52.36 vs. 53.76***	52.24 vs. 53.07**	52.13 vs. 53.39***
Social process	9.26 vs. 9.42*	9.45 vs. 9.70*	9.48 vs. 9.74*
Cognitive process	10.52 vs. 10.86***	10.52 vs. 10.90**	10.40 vs. 10.64 ($p < 0.1$)
Drives	7.44 vs. 7.58*	7.59 vs. 7.77*	7.41 vs. 7.59 ($p < 0.1$)
Future focus	1.09 vs. 1.13*	1.10 vs. 1.18**	1.12 vs. 1.19*

6 Bandwagon User Classification and Prediction

To demonstrate the differential power of the behavioral features we have identified in the previous section, we formulate a classification task and a prediction task to investigate how the activity and language usage features can be used for identifying bandwagon users and inferring future bandwagon behavior.

6.1 Experimental Setup

Tasks. Our first task is a classification task that aims to distinguish between bandwagon and non-bandwagon users. We take all users (after matching) across the three seasons, and randomly select 80% of the users' data as the training set and the remaining 20% as the testing set.

Our second task is to predict whether a user will become a bandwagon user (i.e., change his/her flair to a bandwagon flair) during a season, based on his/her behavior before the bandwagon date in that season. We take users' data in season 2015–16 and 2016–17 to train our model, and apply it to the data in season 2017-18 to predict if a user will jump on the bandwagon in that season.

Features. Based on previous observations, we extract two types of features to train our classification and prediction models.

- Activity features: This set of features includes average comment length and average comment feedback score as discussed in the previous section.
- Language features: This set of features includes the average summary variable of clout [11], average word proportion in terms of social process, cognitive process, drives, and future focus, as discussed in the previous section.

We use Bag-of-words (BOW) features as a strong baseline since BOW not only effectively captures the content of users' comments [9], but also requires no pre-observational study. Please note that all the aforementioned features are extracted from the period between season beginning and *bandwagon date*.

Evaluation Procedure. As the analyses before, we conduct the same user matching process in the first place. We label bandwagon users as positive examples. To evaluate the effectiveness of the classification and prediction tasks using the behavioral features we have identified, we deploy a standard $l2$-regularized logistic regression classifier, and use grid search to find the best parameters. All the results in the classification task are derived after 5-fold cross-validation. We consider both precision and recall as the evaluation metrics.

6.2 Results

Figure 2 summarizes the classification and prediction performance when using different feature sets. As shown in the figure, the two sets of behavioral features that we identify (activity and language) improve precision for both tasks. When combined with BOW features, there is further improvement (18.9% for classification and 47.6% on prediction, as compared with the baseline result). These results indicate that the activity and language features we identify are good complements for BOW text representations to reduce false positive.

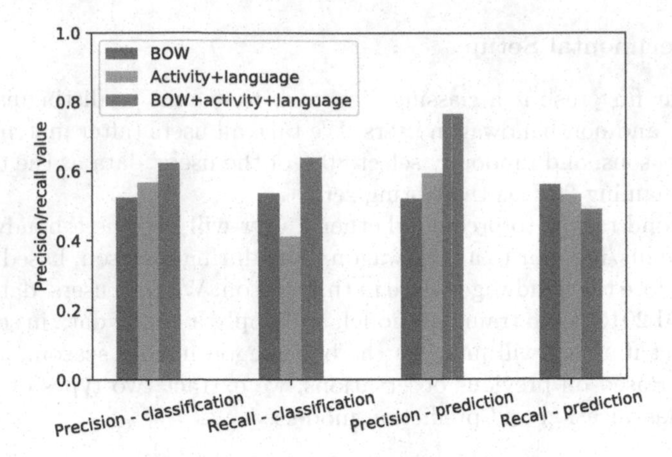

Fig. 2. Classification and prediction performance using different feature sets.

However, we do notice that our activity and language features do not work as well on improving recall in both tasks. Although they improve the classification recall when combined with BOW features, the recall is lower than the BOW only scenario. One possible reason is that some non-bandwagon users actually behave as bandwagon ones but fail to report their bandwagon flairs on Reddit. We find some non-bandwagon users who start to follow another stronger team's games and news after their original affiliated teams are eliminated, but do not change their flair, especially for users in season 2017–18. These "fake" non-bandwagon users can confuse our model, resulting in missed bandwagon detection. The recall performance of using all features for the classification task has been improved because we include all three seasons' users in our training set and the combined features can catch some good indicators of "fake" non-bandwagon users for each season, while for the prediction task, the training data does not contain any users' information in the 3rd season, 2017–18.

7 Concluding Discussion

In this work, we have analyzed the bandwagon phenomenon (a common case of user dynamics) using NBA-related subreddits data from Reddit. We find that better teams attract more bandwagoners, but bandwagoners do not necessarily come from weak teams. Most teams' vagrant fans leave their teams and jump on another team's bandwagon at the beginning, while some teams have a relatively stronger fan base and their bandwagon fans leave when the teams are eliminated. In the comparison after user matching, we find that bandwagon users write shorter comments but receive better feedback, and use words that show less attachment to their affiliated teams. These features can effectively help classify bandwagon users and predict users' future bandwagon behavior.

Implications for User Loyalty. Our results show that loyalty plays an important role in online communities. Bandwagoners have clear behavioral differences from non-bandwagoners. It is crucial for community managers to identify loyal and vagrant fans with the goals of maintaining and growing their user base. To this end, our classification and prediction models show the feasibility of automating these identification processes, and demonstrate the great potential of incorporating such capabilities as a more standard pipeline.

Our work also complements research on user and multi-community engagement, and offers insights on how users behave across sub-communities. The bandwagon phenomenon in our study is about user's preference change in sub-communities, where users share the common interest (basketball), but have different preferences towards the teams.

In addition, we find that bandwagoning in r/NBA is different from general loyalty. We notice that around 80% of the bandwagoners in seasons 2015-16 and 2016-17 change their flairs back to their original teams in the following season, which means that bandwagoning is a "temporary" non-loyal behavior for most users. Thus, one possible future direction is to investigate what factors account for their choice of bandwagoning and willingness to stay with the new team.

Implications for Sports Community Management. Our findings on bandwagon users' characteristics can be useful for sports team management. Firstly, our observations reveal that bandwagon fans are not necessarily from weak teams, which suggest that some higher-ranked teams also need to pay close attention to maintaining their fan base. As mentioned earlier, Spurs is a good example. Furthermore, since bandwagon users tend to move "up" to higher-ranked teams, a strategy to gain some temporary support for the strong-but-not-top teams is to attract more "travelers" during the playoffs. For example, during the 2016 western semi-final between Thunder and Warriors, Thunder acted as a challenger and highlighted their two star players Kevin Durant and Russell Westbrook. These actions brought them lots of fans from other teams, especially from the teams that were defeated by Warriors.

Secondly, it is important to keep fans engaged in online discussion during off-season, especially when certain fans' affiliated teams are eliminated. Prior work has shown that incorporating group identity can help strengthen member attachment in online communities [21]. Our results show that this bandwagon mechanism, i.e., allowing users to switch team affiliation, does have some effective impact on not only encouraging some weak teams' fans to change their flairs and participate in other teams' discussion, but also encouraging certain strong teams' fans to go "up" to a better team when their teams are eliminated during playoffs.

Limitations and Future Work. One key limitation of our work is the representativeness of our dataset. Although Goldschein [6] suggests that /r/NBA is now playing an important role among fans, the NBA fan communities on Reddit may not be representative of the whole communities. Another limitation is that the bandwagon identity requires users' self-identification. As discussed earlier, the recall is low because there are some "fake" loyal users who do not use bandwagon flairs but act the same as *bandwagon users*. We also notice that the number of *bandwagon users* is decreasing, which means that fewer users are "serious" about using this bandwagon flair mechanism. Future directions to address this include designing better strategies in online communities to promote user behavior diversity, and designing better metrics and algorithms to identify real vagrant fans.

Another important question to ask is why users choose to bandwagon. In our analysis we find some fans jump on the bandwagon because their original teams are eliminated and they turn to another team just to have something to watch. Another finding is that some fans jump to teams which are opponents of their "enemy" team, i.e., "Enemy of my enemy is my friend". Answering this question will help provide a fundamental explanation to the bandwagon behaviors.

In addition to online community, the bandwagon effect also plays an important role in information diffusion [17], and impacts the propagation of fake news [23], where the popularity of news allows users to bypass the responsibility of verifying information. One future direction is to investigate how bandwagoning affects users and helps fake news spread, and how to identify the impacted users.

A Appendix

Fig. 3 shows the general bandwagon fan flow of seasons 2015–16 and 2016–17, which validate Observation 1 in Sect. 4. Figure 5 is a stage-by-stage bandwagon fan flow. As shown in these flow graphs, bandwagon users keep leaving their original teams and going to stronger ones. There are fewer bandwagon users in later stages than that of the early stages. In stages 3 and 4, only three teams remain as bandwagon target teams. Figure 4 shows the results of Mann-Whitney U test before and after user matching.

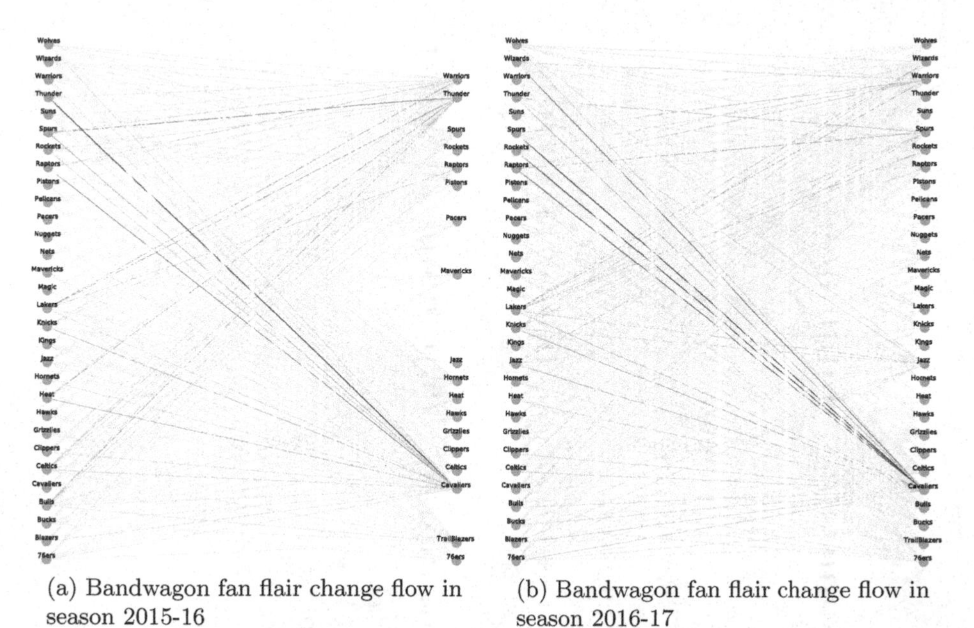

(a) Bandwagon fan flair change flow in season 2015-16

(b) Bandwagon fan flair change flow in season 2016-17

Fig. 3. General bandwagon fan flair change flow in different stages in seasons 2015–16 and 2016–17: from source team (team A, left nodes) to target team (team B, right nodes).

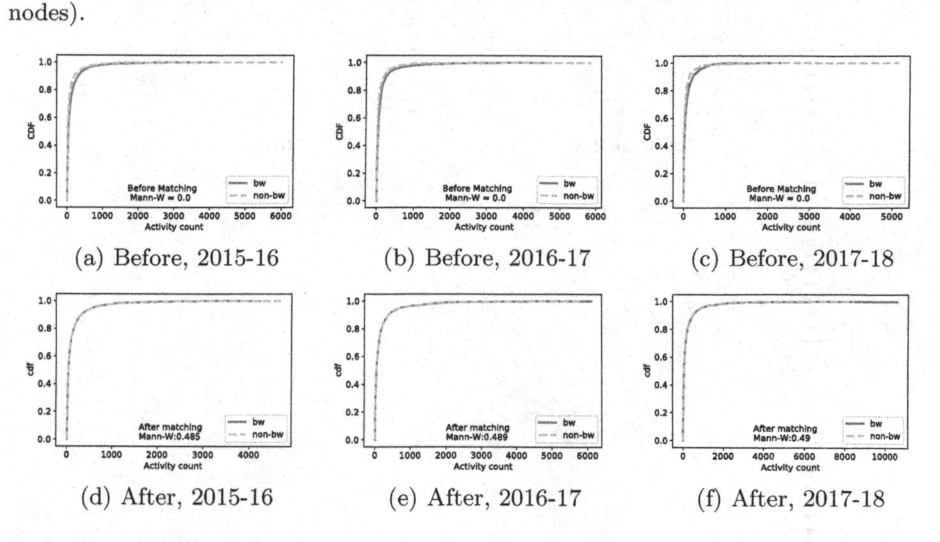

(a) Before, 2015-16 (b) Before, 2016-17 (c) Before, 2017-18

(d) After, 2015-16 (e) After, 2016-17 (f) After, 2017-18

Fig. 4. Mann-whitney U test results before and after matching. Before matching: randomly select the same number of non-bandwagon users as the number of bandwagon users and run the test. After matching: run the test on the two matched user groups.

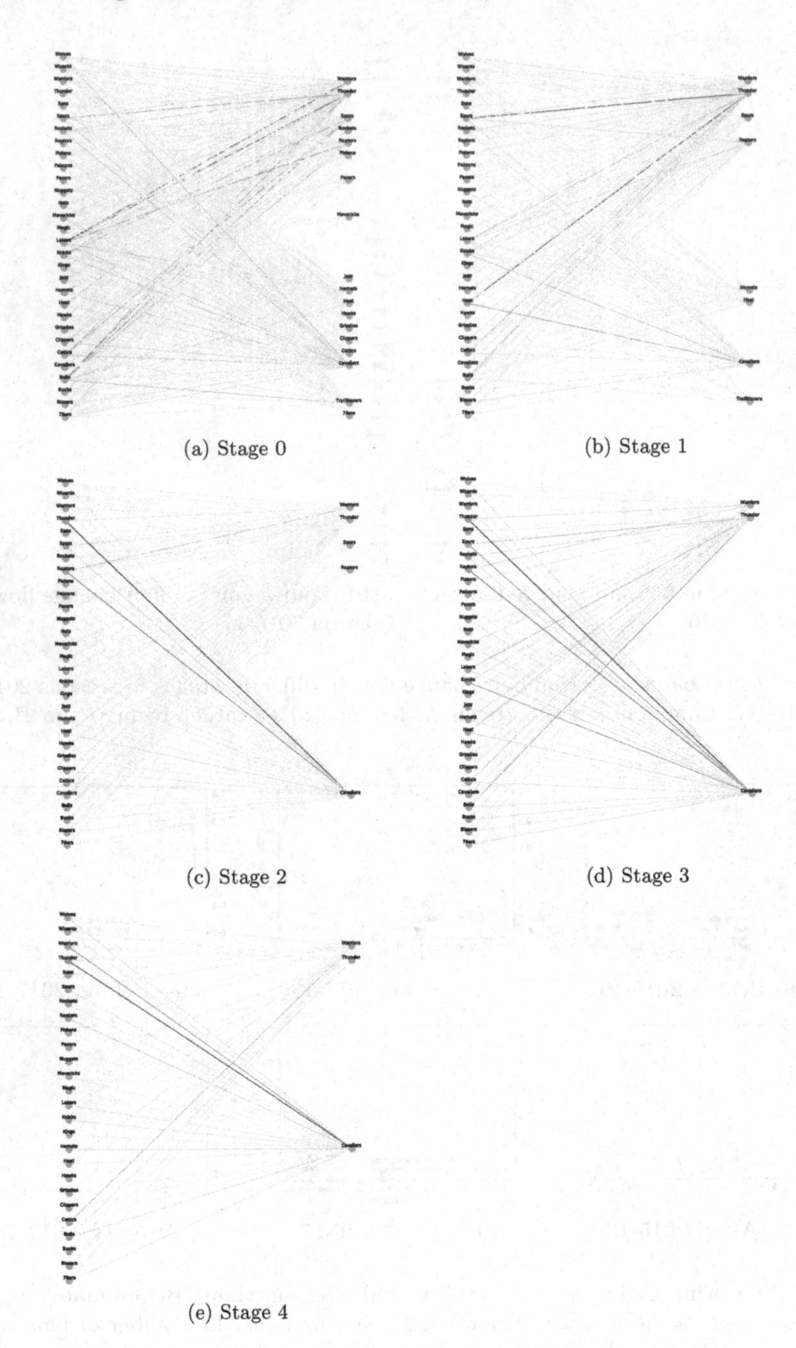

(a) Stage 0

(b) Stage 1

(c) Stage 2

(d) Stage 3

(e) Stage 4

Fig. 5. Bandwagon fan flow in different stages in season 2015-16: From source team (team A, left nodes) to target team (team B, right nodes).

References

1. Aldous, K.K., An, J., Jansen, B.J.: Predicting audience engagement across social media platforms in the news domain. In: Weber, I., et al. (eds.) SocInfo 2019. LNCS, vol. 11864, pp. 173–187. Springer, Cham (2019). https://doi.org/10.1007/978-3-030-34971-4_12
2. Baumgartner, J.: Reddit dataset (2018). https://files.pushshift.io/reddit/
3. Blei, D.M., Ng, A.Y., Jordan, M.I.: Latent dirichlet allocation. JMLR **3**, 993–1022 (2003)
4. Danescu-Niculescu-Mizil, C., West, R., Jurafsky, D., Leskovec, J., Potts, C.: No country for old members: user lifecycle and linguistic change in online communities. In: WWW 2013, pp. 307–318. ACM, New York, May 2013
5. Dror, G., Pelleg, D., Rokhlenko, O., Szpektor, I.: Churn prediction in new users of yahoo! answers. In: WWW 2012, pp. 829–834. Companion, Association for Computing Machinery, New York, April 2012
6. Goldschein, E.: It's time to give /r/nba the respect it deserves (2015). https://www.sportsgrid.com/as-seen-on-tv/media/its-time-to-give-rnba-the-respect-it-deserves
7. Guttmann, A.: From Ritual to Record: The Nature of Modern Sports. Columbia University Press, New York (2004)
8. Hamilton, W.L., Zhang, J., Danescu-Niculescu-Mizil, C., Jurafsky, D., Leskovec, J.: Loyalty in online communities. In: ICWSM (2017)
9. Harris, Z.S.: Distributional structure. Word **10**(2–3), 146–162 (1954)
10. Hessel, J., Tan, C., Lee, L.: Science, askscience, and badscience: on the coexistence of highly related communities. In: ICWSM, March 2016
11. Kacewicz, E., Pennebaker, J.W., Davis, M., Jeon, M., Graesser, A.C.: Pronoun use reflects standings in social hierarchies. J. Lang. Soc. Psychol. **33**(2), 125–143 (2014)
12. Kiss, Á., Simonovits, G.: Identifying the bandwagon effect in two-round elections. Public Choice **160**(3), 327–344 (2013). https://doi.org/10.1007/s11127-013-0146-y
13. Mahmud, J., Chen, J., Nichols, J.: Why are you more engaged? Predicting social engagement from word use. arXiv preprint arXiv:1402.6690 (2014)
14. Mann, H.B., Whitney, D.R.: On a test of whether one of two random variables is stochastically larger than the other. Ann. Math. Stat. **18**, 50–60 (1947)
15. McAllister, I., Studlar, D.T.: Bandwagon, underdog, or projection? opinion polls and electoral choice in britain, 1979–1987. J. Polit. **53**(3), 720–741 (1991)
16. van der Meer, T.W.G., Hakhverdian, A., Aaldering, L.: Off the fence, onto the bandwagon? A Large-scale survey experiment on effect of Real-Life poll outcomes on subsequent vote intentions. Int. J. Public Opin. Res. **28**(1), 46–72 (2016)
17. Nadeau, R., Cloutier, E., Guay, J.H.: New evidence about the existence of a bandwagon effect in the opinion formation process. Int. Polit. Sci. Rev. **14**(2), 203–213 (1993)
18. Nielsen.com: The year in sports media report: 2015 (2016). https://www.nielsen.com/us/en/insights/reports/2016/the-year-in-sports-media-report-2015.html
19. Pennebaker, J.W., Boyd, R.L., Jordan, K., Blackburn, K.: The development and psychometric properties of LIWC2015. Technical report (2015)
20. Pudipeddi, J.S., Akoglu, L., Tong, H.: User churn in focused question answering sites: characterizations and prediction. In: WWW 2014, pp. 469–474. Companion, Association for Computing Machinery, New York, April 2014

21. Ren, Y., et al.: Building member attachment in online communities: applying theories of group identity and interpersonal bonds. MIS Q. **36**, 841–864 (2012)
22. Rowe, M.: *Changing with Time*: modelling and detecting user lifecycle periods in online community platforms. In: Jatowt, A., et al. (eds.) SocInfo 2013. LNCS, vol. 8238, pp. 30–39. Springer, Cham (2013). https://doi.org/10.1007/978-3-319-03260-3_3
23. Shao, C., et al.: Anatomy of an online misinformation network. PLoS ONE **13**(4), e0196087 (2018)
24. Sundar, S.S., Oeldorf-Hirsch, A., Xu, Q.: The bandwagon effect of collaborative filtering technology. In: CHI 2008 extended abstracts on Human Factors in Computing Systems, pp. 3453–3458. ACM (2008)
25. Tan, C., Lee, L.: All who wander: On the prevalence and characteristics of multi-community engagement. In: WWW, International World Wide Web Conferences Steering Committee, pp. 1056–1066 (2015)
26. Wang, K.C., Lai, C.M., Wang, T., Wu, S.F.: Bandwagon effect in Facebook discussion groups. In: Proceedings of the ASE BigData & Social Informatics 2015, p. 17. ACM (2015)
27. Zhang, J.S., Keegan, B.C., Lv, Q., Tan, C.: A tale of two communities: characterizing reddit response to covid-19 through/r/china_flu and/r/coronavirus. arXiv preprint arXiv:2006.04816 (2020)
28. Zhang, J.S., Lv, Q.: Event organization 101: Understanding latent factors of event popularity. In: ICWSM (2017)
29. Zhang, J.S., Tan, C., Lv, Q.: This is why we play: characterizing online fan communities of the NBA teams. In: Proceedings of the ACM on Human-Computer Interaction, vol. 2, no. CSCW (2018). Article no. 197
30. Zhang, J.S., Tan, C., Lv, Q.: Intergroup contact in the wild: characterizing language differences between intergroup and single-group members in NBA-related discussion forums. In: Proceedings of the ACM on Human-Computer Interaction, vol. 3, no. CSCW, pp. 1–35 (2019)
31. Zhu, H., Huberman, B., Luon, Y.: To switch or not to switch: understanding social influence in online choices. In: SIGCHI. CHI 2012, pp. 2257–2266. ACM, New York, May 2012

ALONE: A Dataset for Toxic Behavior Among Adolescents on Twitter

Thilini Wijesiriwardene[1], Hale Inan[4], Ugur Kursuncu[1(✉)], Manas Gaur[1], Valerie L. Shalin[2], Krishnaprasad Thirunarayan[3], Amit Sheth[1], and I. Budak Arpinar[4]

[1] AI Institute, University of South Carolina, Columbia, USA
{thilini,amit}@sc.edu, kursuncu@mailbox.sc.edu, mgaur@email.sc.edu
[2] Department of Psychology, Wright State University, Dayton, USA
valerie.shalin@wright.edu
[3] Department of Computer Science and Engineering, Wright State University, Dayton, USA
t.k.prasad@wright.edu
[4] Department of Computer Science, University of Georgia, Athens, USA
hale.inan25@uga.edu, budak@cs.uga.edu

Abstract. The convenience of social media has also enabled its misuse, potentially resulting in toxic behavior. Nearly 66% of internet users have observed online harassment, and 41% claim personal experience, with 18% facing severe forms of online harassment. This toxic communication has a significant impact on the well-being of young individuals, affecting mental health and, in some cases, resulting in suicide. These communications exhibit complex linguistic and contextual characteristics, making recognition of such narratives challenging. In this paper, we provide a multimodal dataset of toxic social media interactions between confirmed high school students, called ALONE (AdoLescents ON twitTEr), along with descriptive explanation. Each instance of interaction includes tweets, images, emoji and related metadata. Our observations show that individual tweets do not provide sufficient evidence for toxic behavior, and meaningful use of context in interactions can enable highlighting or exonerating tweets with purported toxicity.

Keywords: Toxicity · Harassment · Social media · Resource · Dataset

1 Introduction

The language of social media is a socio-cultural product, reflecting issues of relevance to the sample population and evolving norms in the exchange of coarse language and acceptable sarcasm, employing toxic, questionable language, and sometimes constituting actual harassment. According to a 2017 Pew Research Center survey, 41% of U.S. adults claim to have experienced some type of online

T. Wijesiriwardene and H. Inan—Equally contributed.

S. Aref et al. (Eds.): SocInfo 2020, LNCS 12467, pp. 427–439, 2020.
https://doi.org/10.1007/978-3-030-60975-7_31

harassment, offensive name-calling, purposeful embarrassment, physical threats, harassment over a sustained period of time, sexual harassment or stalking[1].

Toxic behavior is prevalent among adolescents, sometimes leading to aggression [26,27]. Adolescents exemplify a population that is particularly vulnerable to disturbing social media interactions[2] [47], and this behavior is observable in a network of high school students [5]. Further, a toxic online environment may cause mental health problems for this population[3] [2,20,40,48]. While a victim may experience a negative reaction from a toxic environment of offensive language, this differs from *targeted* toxicity which is usually directed whose content collected and confirmed with a unique method towards one individual. The analysis of single tweets or individual users is potentially misleading as the context of interactions between the two people (e.g., source and target) dictates the determination of toxicity. In other words, two individuals who are friends may use coarse keywords or language that is seemingly toxic, but it may be sarcastic, exonerating them from toxicity.

In this paper, we provide a dataset and its details, specific to toxic behavior in social media communications. This dataset has two particular contributions: (i) the population is *high school students* whose content was collected and confirmed with a unique method, and (ii) it was designed based on the *interactions* between participants. The detection of true toxic behavior against a persisting background of coarse language poses a challenging task. Moreover, the scope of the original crawl has great bearing on the prevalence of toxicity features and the criteria for toxic behavior itself. To address these issues, we have assembled a social media corpus from Twitter for a sample of midwestern American High School Students. We assert a dyadic, directed interaction, between a source and a target. Existing related datasets (see Related Work section) focus mainly on the user or tweet level for the task of detecting toxic content. Such datasets fail to capture adequately the fundamental and contextual nuances in the language of these conversations. Thus, our corpus preserves and aggregates the social media interaction history between participants. This enables the determination of existing friendship and hence possible sarcasm. Because individuals can communicate with multiple partners, we have the potential of detecting unique toxic person-victim pairings that would be otherwise undetectable in the raw original crawl.

Each entry in our dataset consists of 12 fields: *Interaction Id, Count, Source User Id, Target User Id, Emoji, Emoji Keywords, Tweets, Image Keywords, created at, favorite count, in reply to screenname* and *label* where the *Tweets* field contains an aggregation of the tweets between a specific pair of source and target. For preliminary analysis, we define a single dimension of *toxic language*, pegged

[1] https://www.pewresearch.org/fact-tank/2017/07/11/key-takeaways-online-harassment/.
[2] https://www.cim.co.uk/newsroom/release-half-of-teens-exposed-to-harmful-social-media/.
[3] https://www.cnn.com/2016/12/14/health/teen-suicide-cyberbullying-continues-trnd/index.html.

at one end by benign content and the other by harassment. This dimension can be partitioned into several, partially overlapping classes, determined by a decision rule. We have identified and experimented with three levels of toxic interactions between source and target: *Toxic (T), Non-Toxic (N), or Unclear (U)*. However, the boundaries between levels are discretionary, accommodating construct definitions that are, at best, debatable.

We include examples across the continuum of toxic language, with sufficient context to determine the nature of toxicity. We detect true toxicity on Twitter by analyzing interactions among a collection of tweets, in contrast with prior approaches where the main focus is performing user or tweet level analysis. Further, we assert that detecting a user as a toxic person with respect to one victim does not provide evidence of being a universal toxic person because they can be friendly to a majority of others.

2 Related Work

We reviewed prior work for the variety of overlapping constructs related to toxic exchanges. The social media literature related to toxic behavior lacks crisp distinctions between: offensive language [14,19,37], hate speech [4,10,14,50], abusive language [14,31,33] and cyberbullying [8,11,18]. For example, the following definition of offensive language substantially overlaps with the subsequent definition of hate speech. According to [14], *offensive language is profanity, strongly impolite, rude or vulgar language expressed with fighting or hurtful words in order to insult a targeted individual or group. Hate speech is language used to express hatred towards a targeted individual or group, or is intended to be derogatory, to humiliate, or to insult the members of a group, on the basis of attributes such as race, religion, ethnic origin, sexual orientation, disability, or gender.* [42] classifies swearing, aggressive comments, or mentioning the past political or ethnic conflicts in a non-constructive and harmful way as hateful: @user_name *nope you just a stupid hoe who wouldn't know their place* 😂😂 comprises both offensive and hate speech. Specifically, the challenge lies in operationalizing the contextual differences between offensiveness, hate speech and harassment. As the existing work on offensive content, harassment and hate speech fails to take into account the nature of the relationship between participants, we focus our attention on the context-aware analyses of targeted exchanges.

Offensive–[4] annotated 16 K tweets from [52] with the labels, racist, sexist or neither. 3383 and 1972 tweets were sexist and racist respectively, and others were labeled as neither. In [31], their aim was to detect abusive language on online user comments posted on Yahoo. 56,280 comments were labeled as "Abusive" and 895,546 comments as "Clean".

Hate Speech–[44] developed a dataset to identify the main targets of online hate speech including the nine categories such as race, behavior, physical, sexual orientation, class, gender, ethnicity, disability, religion, and other for non-classified hate targets. 178 most popular targets from Whisper and Twitter were manually labeled, unveiling new forms of online hate that can be harmful to people. [10] focused on distinguishing hate speech from other forms of offensive language. They extracted 85.4 million tweets from 33,458 users, and randomly sampled 25 K tweets containing words from a hate speech lexicon. Individual tweets were labeled as hate speech, offensive or neither. [43] presented an annotated corpus of tweets classified by different levels of hate to provide an ontological classification model to identify harmful speech. They randomly sampled 14,906 tweets and developed a supervised system used for detection of the class of harmful speech. In [52], tweets were sampled from the 130 K tweets, and in addition to "racism", "sexism", and "neither", the label "both" was added. A character n-gram based approach provided better performance for hate speech detection. [51] examined the influence of annotators' knowledge for hate speech on classification models, labeling individual tweets. Considering only cases of full agreement among amateur annotators, they found that amateur annotators can produce relatively good annotations as compared to expert annotators.

Harassment–A number of researchers have attempted to identify dimensions or factors underpinning harassment. [29] drew on the model [6] that conceptualized aggression on four dimensions: *verbal, physical, direct-indirect,* and *active-passive.* [38] analyses the linguistics aspects of harassment based on different harassment types. Consistent with our interest in interaction history between participants, cyberbullying emphasizes the repetitiveness of aggressive acts [35]. The harasser may target a victim over a period of time, or a group of harassers may target a victim about the same demeaning characteristic or incident. Apart from repetitiveness, the difference of power between the harasser and victim suggests cyberbullying. However, this work [35] is not computationally oriented. Golbeck [16] introduced a large, human labeled corpus of online harassment data including 35,000 tweets with 5495 non-harassing and 29505 harassing examples.

In contrast to this literature, our approach to the problem is to focus on interactions between participants to capture the context of the relationship rather than solely tweets or users. As online toxic behavior is a complex issue that involves different contexts and dimensions [1,21,22], tweet-level or user-level approaches do not adequately capture the context with important nuances due to the fluidity in the language. Our interaction-based dataset will enable researchers to uncover critical patterns for gaining a better understanding of toxic behavior on social media. Additionally, our dataset is unique in its focus on high school student demographic.

3 Dataset

For the dataset ALONE, we retrieved 469,786 tweets from our raw Twitter data, and used a harassment lexicon provided by [39] to filter tweets that are likely to contain toxic behavior, obtaining a collection of 688 interactions with aggregated 16,901 tweets.

3.1 Data Collection

We focused on tweets as the source for our dataset because of its public access. Besides text, tweets can contain images, emoji and URLs as additional content.

To create a ground truth dataset, we reviewed public lists of students, such as the list of National Merit Scholars published in newspapers, identifying 143 names of the attendees of a high school. Using the list of identified individuals, we searched Twitter for the profiles associated with these students using Twitter APIs. Then, with the guidance of our cognitive scientist co-author, we confirmed that the users that we retrieved were high school students, through their profiles and tweets conversing on their school mascot, clubs or faculty members. The 143 user profiles with their tweets constituted the seed corpus.

Dataset Expansion: As a typical network of high school students is larger than 143 users, we expanded the network using the friend and follower relationships. We followed the following procedure:

- Collect friends and followers lists for each seed profile.
- Exclude non-student accounts: We identified the accounts following each other considering them as candidate students, and removed accounts that are not both following and being followed by the accounts in the friends and followers lists of seed accounts (not common profiles). As the adults, such as teachers, would notice any toxic behavior, such as harassment, bullying or aggression, which may have consequences, students with potentially toxic behavior would avoid following their social media accounts [30,46] to sequester social network behavior [28,30]. We obtained 8805 accounts that follow and are being followed by at least one seed account, as candidates for student accounts in the high school. We removed 80 accounts as they were suspended or deleted or otherwise protected by account owners.
- Retain only the peer profiles that follow and are being followed by more than 10% of the seed profiles, yielding 320 likely peers. To confirm the absence of false positives, 50 accounts out of the 320 likely peers were randomly selected and manually validated that all the 50 were confirmed student accounts. When tweets of the newly added 320 accounts were crawled, seven accounts were deleted or restricted. Hence, we removed them from the dataset, resulting in 456 accounts (143 seed and 313 added).

After we finalized the 456 accounts, tweets (up to 3200 if available) were collected for each, starting from the most recent (May 2018), along with their account metadata, using the Twitter API.

Interaction-Based Dataset: As our toxic behavior construct requires interactions between participants, we pruned the tweet corpus to retain a dataset that consists of interactions. We define an interaction as a collection of tweets exchanged between the two participants (e.g., source and target) in one direction, and on Twitter, we consider mentions (including replies) and retweets as interactions. For instance, one user may mention another user in a tweet for harassing, bullying or insulting. Moreover, retweeting a harassing tweet potentially boosts popularity, which creates the role of bystander for the source, suggesting that the retweeting user (source) is actually supporting or helping the harasser (target). We have left retweet indicators (e.g., RT @username:) in the data. Further, some tweets are included in multiple interactions; hence, these communications are a part of a group communication that is not dyadic. For some instances, source and target are the same users, and we left these conversations in the dataset as they may be likely a part of group aggression.

We aggregated tweets that qualify as interactions between users, potentially reducing the false alarm rate of an analysis solely based on the presence of characteristics of offensive language [3]. This allows for the detection of a particularly intriguing combination of positive and negative sentiment lexical items, suggestive of sarcasm, e.g., *happy birthday @user_name love you but hate your feet* 💖💗💝💕💖 and *Happy birthday ugly!!* 💗. The presence of "Happy Birthday" or positive emoji (see above) alters the interpretation of content that would otherwise be regarded as potentially suggestive of toxic behavior and the phenomenon of conflicting valence *exoneration* content, assuming that the toxic content is sarcastic, e.g., the source does not really believe the recipient has unattractive feet or is generally ugly. Moreover, contextual analysis reveals that some of these are not truly toxic. Prior tweets in an interaction provides exonerating context, by indicating the presence of friendship, thus correcting the false positive. Designing the dataset based on interactions captures the context of the relationship between the two user; thus, enabling one to employ computational techniques to retrieve meaningful information concerning true toxicity.

A portion of the tweets does not include any interaction indicator, but they refer to a person indirectly without mentioning or writing the name with malicious intent, to avoid the authority figures. This is called *Subtweeting*[4,5] [9,13,36]. Adolescents have specifically developed such practice due to their own privacy concerns and parental intrusion. For each user, we aggregated the tweets that do not mention the target explicitly, and indicated the target as *"None"*.

Then, a harassment lexicon [39] was utilized to filter the interactions that potentially contain toxic content. For online harassment, source and target dyads can be considered as *harasser-victim* or *bystander-harasser*. Further, as capturing context to determine the toxicity in the content is critical, an interaction should include a sufficient number of tweets. Therefore, we set an empirical threshold for one interaction as having at least three tweets, to capture context.

We have fully *de-identified* the interactions by replacing; (i) Twitter usernames and mentions in tweets with a numeric user id, (ii) URLs with the token of *<url>*, and (iii) person names with the token of *<name>*. We have also included the following metadata for each tweet in the interactions: timestamp, favorite counts, and the de-identified user id of the replied user (if the tweets is a reply). Thus, researchers will have the ability to study a variety of aspects of this problem such as time series analysis. The finalized dataset includes 688 interactions with 16,901 tweets. The fields in an instance are as follows: Interaction Id,

[4] https://www.theguardian.com/technology/blog/2014/jul/23/subtweeting-what-is-it-and-how-to-do-it-well.

[5] http://bolobhi.org/abuse-subtweeting-tweet-school-cyber-bullying/.

Count, Source User Id, Target User Id, Label, Emoji, Emoji Keywords, Tweets, Image Keywords, Timestamp, in reply to and favorite count. "Count" field holds information for the number of tweets in an interaction. "Source and Target User Id" fields hold numeric identification *(after de-identification)* information. A "Label" field holds the assigned label (T,N,U) for the interaction. While the "Emoji" field holds the emoji being used in the tweets, "Emoji Keywords" field provides the keywords that explain the meaning of the emoji, retrieved from EmojiNet [53]. The "Tweets" field has the tweets, and the following fields holds the metadata for each tweet: (i) Timestamp: time information of a tweet, (ii) in reply to: (non-real) user id of the target if the tweet is a reply, (iii) favorite count: number of favorites. See Table 1 for example interactions from the dataset with four fields.

Table 1. Examples from the dataset with labels Toxic (T), Non-Toxic (N) and Unclear (U). The expletives were replaced with the first letter followed by as many *dots* as there are remaining letters.

Label	Tweets
T	if you gon say n.... this much, the LEAST you could do is hit the tanning bed < *url* > *** you're f...... the most hideous and racist piece of s... *** YOU ARE LITERALLY F...... RACIST SHUT THE F... UP *** yeah 😂😂 you're not racist at all !!!!!!!! *** are you in f...... politics no, you're like 17 s... the f... up and stop putting your factsön...
T	ight f... you again *** nah f... all of you frfr bunch of f...... f...... *** f... you < *url* > you have no room to be talking s... shut your bum a.. up frfr 😂 ** you're halarious, f... you and everyone that favorited that and retweeted that
N	"Kix is the handjob of cereals"- John Doe 😂😂 < *imageurl* > *** Explain to that i..... that doing it spreads the word and the chance of someone donating XD fedora wearing as... *** get the f... off my twitter b.... BOI *** guys follow bc he's an i.... and forgot his password.
U	This tweet was dumb I agree with u this time *** hahaha I'm so dumb *** that's my mom f.... *** boob *** never seen a bigger lie on the Internet then this one right here

Multimodality: As it will be described in Section Descriptive Statistics, different modalities of data, such as text, image, emoji, appear in Toxic and Non-Toxic interactions with different proportions. Therefore, we provided explanations of potentially valuable emoji and images. Each image name was created by combining "source user id", "target user id", and "tweet number" in an interaction that each image pertains to. For example: the image 0023.0230.5.jpg is from a tweet between "user 0023" and "user 0230" and the 5th tweet in their interaction. We processed these images utilizing a state-of-the-art image recognition tool, ResNet[6] [17], providing the objects recognized in images with their probabilities (top-5 accuracy= 0.921). We kept the top 20 (empirically set) recognized object names. For example, an image has the following set of recognized objects: "television", "cash machine", "screen", "monitor", "neck brace", "toyshop", "medicine chest", "library", "home theater", "wardrobe", "scoreboard", "moving van", "entertainment center", "barbershop", "desk", "web site". We utilized EmojiNet[7] [53] to retrieve the meanings of the emoji in the interactions, and provided in the dataset. For instance, for the emoji 😂, EmojiNet provides the following set of keywords: "face", "tear", "joy", "laugh", "happy", "cute", "funny", "joyful", "hilarious", "teary", "laughing", "person", "smiley", "lol", "emoji", "wtf", "cry", "crying", "tears", "lmao". Specifically, the significant difference in the use of image, video and emoji between the content of Toxic and Non-Toxic interactions, suggests that the contribution of multimodal elements would likely be critical.

Privacy and Ethics Disclosure: We use only public Twitter data, and our study does not involve any direct interaction with any individuals or their personally identifiable private data. This study was reviewed by the host institution's IRB and received an exemption determination. As noted above, we follow standard practices for anonymization during data collection and processing by removing any identifiable information including names, usernames, URLs. We do not provide any Twitter user or tweet id, or geolocation information. Due to privacy concerns and terms of use by Twitter, we make this dataset available upon request to the authors, and researchers will be required to sign an agreement to use it only for research purposes and without public dissemination.

Table 2. For **three** and **two** labels, agreement scores between the three annotators using **Krippendorff's alpha**.

Three label	Two label
0.63	0.65

Table 3. Pairwise agreement for the **three** label scheme, agreement scores between the three annotators (A,B,C) using Cohen Kappa

Kappa	A	B
B	0.77	–
C	0.52	0.62

3.2 Annotation

Capturing truly toxic content on social media for humans requires reliable annotation guidelines for training annotators. Our annotators have completed a rigorous training process including literature reviews and discussions on online toxic behavior and its socio-cultural context among adolescents. Three annotators labeled the interactions using three labels:

Table 4. Pairwise agreement for **two** labels, agreement scores between the three annotators (A, B, C) using Cohen Kappa

Kappa	A	B
B	0.82	–
C	0.49	0.63

[6] https://github.com/onnx/models/tree/master/vision/classification/resnet.
[7] http://wiki.aiisc.ai/index.php/EmojiNet.

Toxic (T), Non-Toxic (N) and *Unclear (U)*. The annotators were trained by our co-author cognitive scientist to consider the context of the interaction rather than individual tweets while determining the label of an interaction. We developed a guideline for annotators to follow that comprises intent-oriented criteria for labeling interactions as Toxic (T). That is, a tweet is toxic if the interactions contain: (i) Threat to harm a person, (ii) Effort to degrade or belittle a person, (iii) Express dislike towards a person or a group of people, (iv) Promote hate/violence/offensive language towards a person or a group of people, (v) Negatively stereotype a person or a minority, (vi) Support and defend xenophobia, sexism or racism.

Table 5. (a) Descriptive statistics of tweets per interaction. (b) Descriptive statistics of emoji per interaction. (c) Descriptive statistics of URLs per interaction. (d) Descriptive statistics of images per interaction. There were 140 images showing Toxic Behavior and 471 images showing Non-Toxic Behavior.

Number of tweets	Mean	Min	Max
Toxic	13.28	3.0	304.0
Non-Toxic	7.15	3.0	99.0

(a)

Number of emoji	Mean	Min	Max
Toxic	6.72	0.0	290.0
Non-Toxic	3.51	0.0	60.0

(b)

Number of URLs	Mean	Min	Max
Toxic	2.70	0.0	73.0
Non-Toxic	1.63	0.0	26.0

(c)

Number of images	Mean	Min	Max
Toxic	1.18	0.0	20.0
Non-Toxic	0.86	0.0	12.0

(d)

If an annotator could not arrive at a conclusion after assessing the interaction following this guideline, it was labeled as *Unclear*. After the annotations were completed by *the three annotators*, the labels were finalized by majority vote. Then, agreement scores were computed utilizing Krippendorff's alpha (α) and Cohen's Kappa (κ). Note that the instances labelled Unclear (U) can be included in the training to exercise the robustness of a learned model, or they can be removed as they add noise (as per the consensus of the annotators). To accommodate both scenarios, we create two schemes: (i) three label (T, N, U), (ii) two label (T, N) removing Unclear (U) instances [15]. We perform two annotation analysis for both schemes: (i) A group-wise annotator agreement to find the robustness of the annotation by the three annotators using Krippendorff's alpha (α) [45], (ii) A pair-wise annotator agreement using Cohen's Kappa (κ) to identify the annotator with highest agreement with others. In the three-label scheme, α was computed as 0.63, and for the two label scheme, (α) was 0.65.

The agreement scores reported in Table 2 imply substantial agreement[8] [7]. We also computed the agreement between annotators using κ and provided in Table 3 and Table 4, for three label and the two label, respectively. While the annotators A and B have substantial and near perfect agreement, C has moderate and substantial agreement with A and B, both for the three and two label schemes respectively [7].

3.3 Descriptive Statistics

In this section, we provide descriptive statistics of the dataset concerning the distribution of tweets, images, emoji and URLs with respect to labels. Table 6 shows the overall distribution of the instances as Toxic interactions constitute the 17.15% of the dataset, while 79.51% remains as Non-Toxic. A minority group of interactions with 3.34% comprises the Unclear instances where annotators agreed that no conclusion could be derived. While the imbalance in the dataset provides challenges in the modeling of toxic behavior, it is reflective of the nature of occurrence in real life. On the other hand, although the number of toxic interactions is smaller, they are richer in content as well as multimodal elements, compared to non-toxic interactions [23] (see Tables 5a, 5b, 5c, 5d, and 7). Prior research shows that appropriate incorporation of multimodal elements in modeling with social media data would improve performance [12,23,24,32]. In Table 5a, we see mean and maximum number of tweets per interaction for Toxic ones being significantly higher than Non-toxic ones, suggesting the intensity of the toxic content. Further, according to Tables 5a, 5b, 5c, 5d, and 7, in the Toxic content, the use of multimodal elements such as image, video, and emoji, is clearly higher, suggesting that the incorporation of these different modalities in the analysis of this dataset will be critical for a reliable outcome [12,23,24,32].

Table 6. Overall distribution of the data instances over the three labels.

Toxic	Non-toxic	Unclear
118 (17.15%)	547 (79.51%)	23 (3.34%)

Table 7. Different types of URLs in **toxic** interactions.

Type of URLs	Number of URLs
Image URLs	140 (43.88%)
Video URLs	44 (13.79%)
Text URLs	48 (15.04%)

4 Discussion and Conclusion

We created and examined the multimodal ALONE dataset for adolescent participants utilizing a lexicon [39] that divides offensive language into different types concerning appearance, intellectual, political, race, religion, and sexual preference. Given its unique characteristics concerning (i) adolescent population and (ii) interaction-based design, this dataset is an important contribution to

[8] http://homepages.inf.ed.ac.uk/jeanc/maptask-coding-html/node23.html.

the research community, as ground truth to provide a better understanding of online toxic behavior as well as training machine learning models [25,42] and performing time-series analysis. Specifically, quantitative as well as qualitative analysis of this dataset will reveal patterns with respect to social, cultural and behavioral dimensions [34,41,49] and shed light on etiology of toxicity in relationships. Further, researchers can develop guidelines for different kinds of toxic behavior such as harassment and hate speech, and annotate the dataset accordingly. Lastly, we reiterate that the ALONE dataset will be available upon request to the authors, and the researchers will be required to sign an agreement to use it only for research purposes and without public dissemination.

Acknowledgement. We acknowledge partial support from the National Science Foundation (NSF) award CNS-1513721: "Context-Aware Harassment Detection on Social Media". Any opinions, conclusions or recommendations expressed in this material are those of the authors and do not necessarily reflect the views of the NSF.

References

1. Arpinar, I.B., Kursuncu, U., Achilov, D.: Social media analytics to identify and counter islamist extremism: systematic detection, evaluation, and challenging of extremist narratives online. In: 2016 International Conference on Collaboration Technologies and Systems (CTS), pp. 611–612. IEEE (2016)
2. Arseneault, L., Bowes, L., Shakoor, S.: Bullying victimization in youths and mental health problems: "much ado about nothing"? Psychol. Med. **40**, 717 (2010)
3. Badjatiya, P., Gupta, M., Varma, V.: Stereotypical bias removal for hate speech detection task using knowledge-based generalizations. In: The World Wide Web Conference, pp. 49–59 (2019)
4. Badjatiya, P., Gupta, S., Gupta, M., Varma, V.: Deep learning for hate speech detection in tweets. In: WWW (2017)
5. Brener, N.D., Simon, T.R., Krug, E.G., Lowry, R.: Recent trends inviolence-related behaviors among high school students in the United States. JAMA **282**, 440–446 (1999)
6. Buss, A.H.: The psychology of aggression (1961)
7. Carletta, J., Isard, A., Isard, S., Kowtko, J.C., Doherty-Sneddon, G., Anderson, A.H.: The reliability of a dialogue structure coding scheme (1997)
8. Chatzakou, D., Kourtellis, N., Blackburn, J., De Cristofaro, E., Stringhini, G., Vakali, A.: Mean birds: detecting aggression and bullying on twitter. In: ACM Web Science (2017)
9. Crumback, D.: Subtweets: the new online harassment (2017)
10. Davidson, T., Warmsley, D., Macy, M., Weber, I.: Automated hate speech detection and the problem of offensive language. In: AAAI-ICWSM (2017)
11. Dinakar, K., Reichart, R., Lieberman, H.: Modeling the detection of textual cyberbullying. In: AAAI-ICWSM (2011)
12. Duong, C.T., Lebret, R., Aberer, K.: Multimodal classification for analysing social media. arXiv preprint arXiv:1708.02099 (2017)
13. Edwards, A., Harris, C.J.: To tweet or "subtweet"?: impacts of social networking post directness and valence on interpersonal impressions. Comput. Hum. Behav. **63**, 304–310 (2016)

14. Founta, A., et al.: Large scale crowdsourcing and characterization of Twitter abusive behavior (2018)
15. Gaur, M., et al.: Knowledge-aware assessment of severity of suicide risk for early intervention. In: The World Wide Web Conference, pp. 514–525. ACM (2019)
16. Golbeck, J., et al.: A large labeled corpus for online harassment research. In: ACM Web Science (2017)
17. He, K., Zhang, X., Ren, S., Sun, J.: Deep residual learning for image recognition. In: CVPR (2016)
18. Hosseinmardi, H., Mattson, S.A., Rafiq, R.I., Han, R., Lv, Q., Mishra, S.: Analyzing labeled cyberbullying incidents on the Instagram social network. In: SocInfo (2015)
19. Jay, T., Janschewitz, K.: The pragmatics of swearing. J. Polit. Res. Lang. Behav. Cult. **4**, 267–288 (2008)
20. Kumpulainen, K., Räsänen, E., Puura, K.: Psychiatric disorders and the use of mental health services among children involved in bullying. Aggressive Behav. J. **27**, 102–110 (2001)
21. Kursuncu, U.: Modeling the persona in persuasive discourse on social media using context-aware and knowledge-driven learning. Ph.D. thesis, University of Georgia (2018)
22. Kursuncu, U., et al.: Modeling islamist extremist communications on social media using contextual dimensions: religion, ideology, and hate. In: Proceedings of the ACM on Human-Computer Interaction, vol. 3,no. CSCW, pp. 1–22 (2019)
23. Kursuncu, U., et al.: What's ur type? Contextualized classification of user types in marijuana-related communications using compositional multiview embedding. In: 2018 IEEE/WIC/ACM International Conference on Web Intelligence (WI), pp. 474–479. IEEE (2018)
24. Kursuncu, U., Gaur, M., Lokala, U., Thirunarayan, K., Sheth, A., Arpinar, I.B.: Predictive analysis on Twitter: techniques and applications. In: Agarwal, N., Dokoohaki, N., Tokdemir, S. (eds.) Emerging Research Challenges and Opportunities in Computational Social Network Analysis and Mining. LNSN, pp. 67–104. Springer, Cham (2019). https://doi.org/10.1007/978-3-319-94105-9_4
25. Kursuncu, U., Gaur, M., Sheth, A.: Knowledge infused learning (K-IL): towards deep incorporation of knowledge in deep learning. In: Proceedings of the AAAI 2020 Spring Symposium on Combining Machine Learning and Knowledge Engineering in Practice. Stanford University, Palo Alto, California, USA. AAAI-MAKE (2020)
26. Liu, J., Lewis, G., Evans, L.: Understanding aggressive behaviour across the lifespan. J. Psychiatric Ment. Health Nurs. **20**, 156–168 (2013)
27. Lowry, R., Powell, K.E., Kann, L., Collins, J.L., Kolbe, L.J.: Weapon-carrying, physical fighting, and fight-related injury among us adolescents. Am. J. Prevent. Med. **14**, 122–129 (1998)
28. Mishna, F., Schwan, K.J., Lefebvre, R., Bhole, P., Johnston, D.: Students in distress: unanticipated findings in a cyber bullying study. Child. Youth Serv. Rev. **44**, 341–348 (2014)
29. Namie, G., Namie, R.: Bully at work: what you can do to stop the hurt and reclaim your dignity on the job (2009)
30. Nilan, P., Burgess, H., Hobbs, M., Threadgold, S., Alexander, W.: Youth, social media, and cyberbullying among australian youth: "sick friend". Soc. Media + Soc. **1**, 2056305115604848 (2015)
31. Nobata, C., Tetreault, J., Thomas, A., Mehdad, Y., Chang, Y.: Abusive language detection in online user content. In: WWW (2016)
32. O'Halloran, K., Chua, A., Podlasov, A.: The role of images in social media analytics: a multimodal digital humanities approach. In: Visual communication (2014)

33. Papegnies, E., Labatut, V., Dufour, R., Linarès, G.: Detection of abusive messages in an on-line community. In: CORIA (2017)
34. Parent, M.C., Gobble, T.D., Rochlen, A.: Social media behavior, toxic masculinity, and depression. Psychol. Men Masculinities **20**(3), 277 (2019)
35. Patchin, J.W., Hinduja, S.: Bullies move beyond the schoolyard: a preliminary look at cyberbullying. Youth Violence Juvenile Justice **4**, 148–169 (2006)
36. Rafla, M., Carson, N.J., DeJong, S.M.: Adolescents and the internet: what mental health clinicians need to know. Curr. Psychiatry Rep. **16**(9), 472 (2014)
37. Razavi, A.H., Inkpen, D., Uritsky, S., Matwin, S.: Offensive language detection using multi-level classification. In: Farzindar, A., Kešelj, V. (eds.) AI 2010. LNCS (LNAI), vol. 6085, pp. 16–27. Springer, Heidelberg (2010). https://doi.org/10.1007/978-3-642-13059-5_5
38. Rezvan, M., Shekarpour, S., Alshargi, F., Thirunarayan, K., Shalin, V.L., Sheth, A.: Analyzing and learning the language for different types of harassment. PLoS One **15**(3), e0227330 (2020)
39. Rezvan, M., Shekarpour, S., Balasuriya, L., Thirunarayan, K., Shalin, V.L., Sheth, A.: A quality type-aware annotated corpus and lexicon for harassment research. In: ACM Web Science (2018)
40. Rivers, I., Poteat, V.P., Noret, N., Ashurst, N.: Observing bullying at school: the mental health implications of witness status. School Psychol. Quart. **24**, 211 (2009)
41. Safadi, H., et al.: Curtailing fake news propagation with psychographics. Available atSSRN 3558236 (2020)
42. Salminen, J., et al.: Anatomy of online hate: developing a taxonomy and machine learning models for identifying and classifying hate in online news media. In: ICWSM, pp. 330–339 (2018)
43. Sharma, S., Agrawal, S., Shrivastava, M.: Degree based classification of harmful speech using Twitter data. arXiv preprint arXiv:1806.04197 (2018)
44. Silva, L., Mondal, M., Correa, D., Benevenuto, F., Weber, I.: Analyzing the targets of hate in online social media. In: AAAI-ICWSM (2016)
45. Soberón, G., Aroyo, L., Welty, C., Inel, O., Lin, H., Overmeen, M.: Measuring crowd truth: disagreement metrics combined with worker behavior filters. In: CrowdSem 2013 Workshop (2013)
46. Søndergaard, D.M.: Bullying and social exclusion anxiety in schools. Br. J. Sociol. Educ. **33**, 55–372 (2012)
47. Unicef, et al.: An everyday lesson: end violence in schools (2018)
48. Viner, R.M., et al.: Roles of cyberbullying, sleep, and physical activity in mediating the effects of social media use on mental health and wellbeing among young people in England: a secondary analysis of longitudinal data. Lancet Child Adolescent Health **3**, 685–696 (2019)
49. Wandersman, A., Nation, M.: Urban neighborhoods and mental health: psychological contributions to understanding toxicity, resilience, and interventions. Am. Psychol. **53**(6), 647 (1998)
50. Warner, W., Hirschberg, J.: Detecting hate speech on the world wide web. In: ACL (2012)
51. Waseem, Z.: Are you a racist or am i seeing things? Annotator influence on hate speech detection on twitter. In: NLP-CSS (2016)
52. Waseem, Z., Hovy, D.: Hateful symbols or hateful people? Predictive features for hate speech detection on Twitter. In: NAACL (2016)
53. Wijeratne, S., Balasuriya, L., Sheth, A., Doran, D.: EmojiNet: an open service and API for emoji sense discovery. In: AAAI-ICWSM (2017)

Cross-Domain Classification of Facial Appearance of Leaders

Jeewoo Yoon[1], Jungseock Joo[2], Eunil Park[1], and Jinyoung Han[1(✉)]

[1] Sungkyunkwan University, Seoul, South Korea
{yoonjeewoo,eunilpark,jinyounghan}@skku.edu
[2] University of California, Los Angeles, United States
jjoo@comm.ucla.edu

Abstract. People often rely on visual appearance of leaders when evaluating their traits and qualifications. Prior research has demonstrated various effects of thin-slicing inference based on facial appearance in specific events such as elections. By using a machine learning approach, we examine whether the pattern of face-based leadership inference differs in different domains or some facial features are universally preferred across domains. To test the hypothesis, we choose four different domains (business, military, politics, and sports) and analyze facial images of 272 CEOs, 144 4-star generals of U.S. army, 276 U.S. politicians, and 81 head coaches of professional sports teams. By extracting and analyzing facial features, we reveal that facial appearances of leaders are statistically different across the different leadership domains. Based on the identified facial attribute features, we develop a model that can classify the leadership domain, which achieves a high accuracy. The method and model in this paper provide useful resources toward scalable and computational analyses for the studies in social perception.

Keywords: Face · Facial display · Leadership · Machine learning

1 Introduction

Selecting a leader who can well represent his/her organization, e.g., a general of the army or a politician of a party, is important. Hence, people seek to judge and verify the candidates of their leaders with elaborate efforts. Studies have shown that people often rely on facial appearance of others while inferring their personality traits and making judgements [2,5,8,12,15]. For instance, Todorov *et al.* [15] showed that voters assess the competence of politicians from their faces and the estimated visual competence can predict actual election outcomes.

This in turn has attracted researchers to study the visual perception on leaders in diverse domains. Olivola *et al.* [12] showed that right-leaning voters tend to prefer 'republican-looking' politicians. Rule *et al.* [13] showed that the visual perception on the CEO's facial appearance is significantly related to the financial success of the company. Harms *et al.* [6] also showed that the visual

S. Aref et al. (Eds.): SocInfo 2020, LNCS 12467, pp. 440–446, 2020.
https://doi.org/10.1007/978-3-030-60975-7_32

perception on the CEO's personality and effectiveness is associated with the organizational performance. The visual appearance of the leaders in military has been also investigated; e.g., West Point cadets with more dominant faces are more likely to go higher ranks such as captain or lieutenant [10].

This paper studies whether such the visual appearance of leaders in different domains (e.g., politics, business) show distinctive patterns. If the leaders in a specific domain share distinctive facial appearance, the domain-specific facial stereotypes by people may also affect the leader selection process and this would add a new layer to the role of the face-based leadership selection [11]. Olivola *et al.* conducted an experiment, where participants were presented with two facial images from two different domains (e.g.., politics-business, military-sports) and selected one of them to which they assume it belongs, and showed that leaders from different domains (except politics) are distinguishable by their facial images [11]. Our work goes one step further by applying a computational approach using machine learning techniques. In particular, we seek to answer the following questions: *"Are there distinctive features in facial appearance of leaders?" "If so, can leadership domains be inferred by such facial visual features?"*

To answer the questions, we first collect facial images of leaders in four leadership domains: politics, business, military, and sports. We crawled facial images of 272 CEOs, 144 4-star generals of U.S. army, 276 U.S. politicians (house representatives), and 81 head coaches of professional sports teams. By extracting the facial features using the face recognition API[1], we analyze whether there exist distinct facial patterns in different leadership domains. Base on lessons learned, we develop a machine learning model that classifies leadership domains, which achieves a high accuracy.

2 Methodology

In this section, we first describe the collected face data in four leadership domains. We also describe the facial feature extraction process used in our work. We then propose a model that classifies leadership domains with facial images.

2.1 Leader Face Data Collection

We collected facial images of 272 CEOs, 144 4-star generals of U.S. army, 276 U.S. politicians (i.e., house representatives), and 81 head coaches of professional sports teams (i.e., in NBA, NFL, and MLB). Note that we only considered Caucasian male leaders to control gender and ethnic variables like [11]. The facial images of CEOs were collected from the official website of Forbes (https://forbes.com), a popular American business magazine. The 4-star general faces were collected from Wikipedia (https://en.wikipedia.org) and the images of politicians were collected from GovTrack (https://www.govtrack.us/). Note that we collected the images of politicians who were active members after the U.S. 2018 mid election and 4-star generals who were ranked from 1977 to 2019. We exclude the gray-scale images among them. For the sports domain, we collected the face

[1] https://azure.microsoft.com/en-us/services/cognitive-services/face.

images of head coaches from their sports teams' official websites such as https://www.mlb.com.

2.2 Facial Feature Extraction

We next extract the facial features from the collected images using the Microsoft Face API that performs the face-related computer vision tasks such as face detection, face grouping, etc. We use the facial detection function of the API, which takes an image as input and returns the location of the detected face, its attributes (e.g., age, gender, hair color), and its landmarks. By controlling the factors related to non-facial attributes such as hair color or glasses like [11], we finally use the following nine facial attributes in our analysis: Age, Happiness, Neutral, Anger, Contempt, Disgust, Fear, Sadness, and Surprise. Note that the age feature is the value of the estimated age of a given face by the face recognition model. The facial expression features (i.e., happiness, sadness, neutral, anger, contempt, disgust, surprise, and fear) are provided with the normalized confidence scores.

2.3 The Model

To classify different leadership domains based on facial features, we apply the eXtreme Gradient Boosting (XGBoost), a popular gradient boosting algorithm [4]. Given a leader face data $\mathcal{D} = \{(\boldsymbol{x}_i, y_i)\}_{i=1}^{n} (\boldsymbol{x}_i \in \mathbb{R}^m, y_i \in \{1...c\})$ with n face images, m facial attributes, and c leadership domain classes, our ensemble model for classification tasks can be generalized as follows:

$$\hat{y}_i = \phi(\boldsymbol{x}_i) = \sum_{k=1}^{K} f_k(\boldsymbol{x}_i) \tag{1}$$

where $\hat{y}_i$ is the predicted class for the i-th facial image, f_k is the k-th independent tree, K is the number of trees, and $f_k(\boldsymbol{x}_i)$ is the prediction score given by the k-th independent tree on the facial features extracted from the i-th facial image.

Then, the objective function of our model, $\mathcal{L}(\phi)$, can be calculated as follows:

$$\mathcal{L}(\phi) = \sum_{i} l(y_i, \hat{y}_i) + \sum_{k} \Omega(f_k) \tag{2}$$

where l is the squared error loss function between the predicted class $\hat{y}_i$ and the target class y_i, and $\sum_k \Omega(f_k)$ is the regularization term that penalizes the complexity of the model, respectively. Note that $\Omega(f) = \gamma T + \frac{1}{2}\lambda||w||^2$, where λ and γ controls the penalty for the number of leaves T and magnitude of the leaf weights w, respectively.

Since our model is trained in an additive manner, for each iteration, we add the f_k and minimize the following objective function for the k-th iteration:

$$\mathcal{L}^k(\phi) = \sum_{i=1}^{n} l(y_i, \hat{y}_i^{(k-1)} + f_k(\boldsymbol{x}_i)) + \Omega(f_k) \tag{3}$$

We then use the second-order approximation to optimize the Eq. 3:

$$\mathcal{L}^k(\phi) = \sum_{i=1}^{n}[g_i f_k(\boldsymbol{x}_i) + \frac{1}{2}h_i f_k^2(\boldsymbol{x}_i)] + \Omega(f_k) \tag{4}$$

where $g_i = \partial_{\hat{y}_i^{(k-1)}} l(y_i, \hat{y}_i^{(k-1)})$ and $h_i = \partial^2_{\hat{y}_i^{(k-1)}} l(y_i, \hat{y}_i^{(k-1)})$.

To measure the gain of splitting a leaf node in a tree, we calculate the score of the leaf node after the split as follows:

$$\mathcal{L}_{split} = \frac{1}{2}\left[\frac{(\sum_{i\in I_L} g_i)^2}{\sum_{i\in I_L} h_i + \lambda} + \frac{(\sum_{i\in I_R} g_i)^2}{\sum_{i\in I_R} h_i + \lambda} - \frac{(\sum_{i\in I} g_i)^2}{\sum_{i\in I} h_i + \lambda}\right] - \gamma \tag{5}$$

where I_L and I_R are the sets of left and right nodes after the split, and $I = I_L \cup I_R$. Note that the first, second, and third term in Eq. 5 represents the score of the left, right, and original (parent) leaf, respectively.

Table 1. Statistics of the age, happiness, and neutral features of the facial images.

	Age		Happiness		Neutral	
Domain	Mean	SD	Mean	SD	Mean	SD
Business	56.4	10.1	0.634	0.424	0.339	0.400
Military	54.2	4.77	0.348	0.380	0.645	0.376
Politics	49.3	8.36	0.952	0.165	0.047	0.163
Sports	44.2	8.63	0.796	0.370	0.198	0.360

3 Facial Attribute Analysis

We analyze the estimated age and facial expression features of the given leader face images for four different leadership domains. Table 1 shows the descriptive statistics of the age, happiness, and neutral features for four leadership domains. Note that we only report the happiness and neutral features among the eight facial expression features because other facial expressions were barely detected. In order to quantify the differences between leadership domains, we conduct the Kruskal-Wallis (KW) test followed by the Dwass-Steel-Critchlow-Flinger (DSCF) pairwise comparison.

We first analyze the estimated ages of facial images for different leadership domains. It has been reported that 'age' often plays an important role in selecting leaders [14]. Although the estimated age may not be the actual leader's biological age, the estimated age can inform how much does the given face look young or old, compared to others. The result of the KW test followed by the DSCF pairwise comparison shows that there is a significant difference in mean age values among four leadership domains ($\epsilon^2 = 0.185, p < 0.001$). The mean age of the business leaders ($M = 56.4, SD = 10.1$) is higher than the other domain leaders, i.e., military ($M = 54.2, SD = 4.77$, $p = 0.001$), politics ($M = 49.3, SD = 8.36$, $p < 0.001$), and sports ($M = 44.2, SD = 8.63$, $p < 0.001$). The estimated mean age of the leaders in sports is smaller than the other domains ($p < 0.001$). Note that most of sports leaders (i.e., team manager) are likely to start their career at early 40s, hence resulting in a relatively low mean age.

We next investigate the facial expression of leaders in different leadership domains. The facial expression has been known as a great tool for leaders to present themselves as they intended [7,9,16]. The result shows that there are significant differences among leadership domains in terms of the happiness ($p < 0.001, \epsilon^2 = 0.316$) and neutral features ($p < 0.001, \epsilon^2 = 0.325$). As shown in Table 1, politicians tend to display more happiness (represented by smile) than leaders in other domains ($p < 0.001$). This may be because politicians are likely to be guided smile when they take official photos to be seen as friendly to their voters. In the case of neutral facial expression, military leaders are likely to display more neutral faces than others ($p < 0.001$). Note that the neutral faces of military leaders can be linked to a prior work that found cadets with less smiling (revealed in their portraits) tend to go higher military ranks [10].

4 Leadership Domain Classification

We propose a model for classifying leadership domains based on the facial attribute features. The model learns nine facial features identified in the previous section, and classifies the face into one of the given domain pair. For example, the model classifies the leader face into one of the business-military leadership pair. Note that we use all nine features, since it shows the best performance in our feature selection process. We used other well-known classifiers including SVM or neural network, but we only report the results of the XGBoost classifier as it performs better than others.

Table 2. Performance results of the classification model as well as human recognition.

	Model				Human	
Classification	ACC	TPR	TNR	AUC	ACC	SD
Military-Politics	0.92	0.92	0.92	0.95	0.61	0.14
Business-Military	0.82	0.85	0.80	0.85	0.65	0.17
Military-Sports	0.83	0.67	0.92	0.86	0.69	0.15
Business-Politics	0.75	0.85	0.64	0.78	0.45	0.12
Politics-Sports	0.70	0.30	0.82	0.66	0.63	0.15
Business-Sports	0.73	0.63	0.76	0.74	0.51	0.19

We report the following performance metrics: (i) True Positive Rate (TPR), (ii) True Negative Rate (TNR), (iii) Area Under the Curve (AUC), and (iv) Accuracy (ACC). Since the numbers of face images for different domains are different, there exists the class imbalance problem. To remedy this issue, we apply the Synthetic Minority Over-sampling TEchnique (SMOTE) [3], which allows us to learn with over-sampled instances from the minority class. Note that we apply this technique only in the learning set.

Table 2 shows the performance results of the leadership domain classification model. As shown in Table 2, our model accurately distinguishes leader faces in (i) politics and military, (ii) sports and military, and (iii) military and business, whose accuracies are higher than 80%. In particular, the model shows the highest accuracy of 92% in politics and military classification, which may be due to the big difference in facial expression between politicians and military generals; smiling politicians vs. less smiling military generals. The relatively lower accuracy (i.e., 70%) on leader faces in sports and politics implies that these domains share similar facial features (i.e., age and happiness).

In order to compare our model's performance with human recognition, we recruited 180 human participants (30 participants for each classification) via Amazon Mechanical Turk. Note that the participants were limited to U.S. citizens. We provide 20 randomly selected faces (from the same test set) to the participants and ask them to select the leadership domain for the given face. If the participants already know the given face, the participants are able to select "I know this person", and we exclude those faces in calculating the accuracy.

Overall, our proposed machine-learning-based model can show higher accuracy in classifying leadership domains with facial attributes than human recognition as shown in Table 2; human performance: 45%–69% whereas our model: 70%–92%. The possible reason of the lower human performance may be due to human bias or stereotypes on leader faces [11]; on the contrary, the machine-learning model statistically analyzes the distribution of the facial features and makes a precise decision boundary that could make a higher accuracy. The computational model also reveals that the facial stereotypes on leadership domains (by people) may not exactly match with the actual facial appearance. While it was reported that human participants were not able to distinguish political leaders from other domain leaders [11], we demonstrate that machine learning can accurately distinguish leadership domains including politics by facial features.

5 Concluding Remarks

This paper applied a computational approach using the machine learning techniques to investigate facial attributes of four different leadership domains – business, military, politics, and sports. We demonstrated that the proposed machine-learning model achieves a higher accuracy than human recognition. This implies that facial attributes extracted by a computational approach are key predictors in inferring leadership domains. We believe such a model is useful in understanding visual perception on leaders in different leadership domains. Furthermore, the model can be used as a research tool to study human psychology and social perception, e.g., as a baseline model for measuring facial stereotypes [1].

Acknowledgements. This research was supported in part by Basic Science Research Program through the National ResearchFoundation of Korea (NRF) funded by the Ministry of Education (NRF-2018R1D1A1A02085647) and the MSIT (Ministry of Science and ICT), Korea, under the ICAN (ICT Challenge and Advanced Network of HRD) program (2020-0-01816) supervised by the IITP (Institute of Information & Communications Technology Planning & Evaluation).

References

1. Araújo, C.S., Meira, W., Almeida, V.: Identifying stereotypes in the online perception of physical attractiveness. In: Spiro, E., Ahn, Y.-Y. (eds.) SocInfo 2016. LNCS, vol. 10046, pp. 419–437. Springer, Cham (2016). https://doi.org/10.1007/978-3-319-47880-7_26
2. Bakhshi, S., Shamma, D.A., Gilbert, E.: Faces engage us: Photos with faces attract more likes and comments on instagram. In: Proceedings of the SIGCHI Conference on Human Factors in Computing Systems. pp. 965–974. ACM (2014)
3. Chawla, N.V., Bowyer, K.W., Hall, L.O., Kegelmeyer, W.P.: Smote: synthetic minority over-sampling technique. Int. J. Artif. Intell. Res. **16**, 321–357 (2002)
4. Chen, T., Guestrin, C.: Xgboost: A scalable tree boosting system. In: Proceedings of the SIGKDD International Conference on Knowledge Discovery and Data Mining. pp. 785–794. ACM (2016)
5. Fiore, A.T., Taylor, L.S., Mendelsohn, G.A., Hearst, M.: Assessing attractiveness in online dating profiles. In: Proceedings of the SIGCHI Conference on Human Factors in Computing Systems. pp. 797–806. ACM (2008)
6. Harms, P.D., Han, G., Chen, H.: recognizing leadership at a distance: a study of leader effectiveness across cultures. J. Leadersh. Organ. Stud. **19**(2), 164–172 (2012)
7. Humphrey, R.H.: The many faces of emotional leadership. Leadersh. Q. **13**(5), 493–504 (2002)
8. Joo, J., Steen, F.F., Zhu, S.C.: Automated facial trait judgment and election outcome prediction: Social dimensions of face. In: Proceedings of the IEEE International Conference on Computer Vision. pp. 3712–3720. IEEE (2015)
9. Masters, R.D., Sullivan, D.G.: Facial displays and political leadership in france. Behav. Processes **19**(1–3), 1–30 (1989)
10. Mueller, U., Mazur, A.: Facial dominance of west point cadets as a predictor of later military rank. Soc. Forces **74**(3), 823–850 (1996)
11. Olivola, C.Y., Eubanks, D.L., Lovelace, J.B.: The many (distinctive) faces of leadership: inferring leadership domain from facial appearance. Leadersh. Q. **25**(5), 817–834 (2014)
12. Olivola, C.Y., Sussman, A.B., Tsetsos, K., Kang, O.E., Todorov, A.: Republicans prefer republican-looking leaders: political facial stereotypes predict candidate electoral success among right-leaning voters. Soc. Psychol. Personal. Sci. **3**(5), 605–613 (2012)
13. Rule, N.O., Ambady, N.: The face of success: inferences from chief executive officers' appearance predict company profits. Psychol. Sci. **19**(2), 109–111 (2008)
14. Spisak, B.R.: The general age of leadership: older-looking presidential candidates win elections during war. PLoS ONE **7**(5), e36945 (2012)
15. Todorov, A., Mandisodza, A.N., Goren, A., Hall, C.C.: Inferences of competence from faces predict election outcomes. Science **308**(5728), 1623–1626 (2005)
16. Trichas, S., Schyns, B.: The face of leadership: perceiving leaders from facial expression. Leadersh. Q. **23**(3), 545–566 (2012)

The Effect of Structural Affinity on the Diffusion of a Transnational Online Movement: The Case of #MeToo

Xinchen Yu[1], Shashidhar Reddy Daida[1], Jeremy Boy[2,3](✉), and Lingzi Hong[1](✉)

[1] College of Information, University of North Texas, Denton, TX 76207, USA
{XinchenYu,ShashidharReddyDaida}@my.unt.edu, lingzi.hong@unt.edu
[2] United Nations Global Pulse, NewYork, USA
[3] UNDP Accelerator Labs Network, NewYork, USA
jeremy.boy@undp.org

Abstract. Social media platforms intrinsically afford the diffusion of social movements like #MeToo across countries. However, little is known about how extrinsic, country-level factors affect the uptake of such a transnational movement. In this paper, we present a macro, comparative study of the transnational #MeToo movement on Twitter across 33 countries. Our aim is to identify how socio-economic and cultural variables might have influenced the in-country scale of participation, as well as the timings of in-country peak surges of messages related to #MeToo. Our results show that total in-country participation over a three-year timeframe is highly related to the scale of participation in peak surges; the scale of in-country participation is related to the population size and income level of the country; and the timings of peak surges are related to the country's population size, gender equality score, and the language used in messages. We believe these findings resonate with theoretical frameworks that describe the formation of transnational social movements, and provide a quantitative perspective that complements much of the qualitative work in the literature.

Keywords: #MeToo · Transnational movement · Structural affinity · Diffusion.

1 Introduction

We present the results of a macro, comparative study of the transnational #MeToo movement on Twitter across 33 countries. Our goal is to understand how country-level socio-economic and cultural differences might have influenced the diffusion of the movement.

Social activism is increasingly brought online, where social media platforms like Twitter and Facebook offer a space for people to gather and advocate

S. Aref et al. (Eds.): SocInfo 2020, LNCS 12467, pp. 447–460, 2020.
https://doi.org/10.1007/978-3-030-60975-7_33

for the change they desire [18]. These platforms facilitate the dissemination of information across large, decentralized networks of individuals, and intrinsically afford the cross-country diffusion of online-, digital-, or cyber-activism movements [8] like #MeToo. However, other extrinsic factors are likely to influence the in-country reception of such movements. Anthropological research and gender studies have investigated the motivations and cognitive processes behind individual engagement in #MeToo [25], but only few studies have sought to compare broader social characteristics across several countries to understand their relation to the scale or timing of in-country engagement in the movement.

In this paper, we describe a quantitative analysis of 3 years of Twitter messages related to the #MeToo movement, including associated hashtags in various languages, posted from 33 different countries. The "me too" movement was launched in 2006 by American activist Tarana Burke with the intent to help women suffering from sexual harassment or violence stand up for themselves. On October 15, 2017, actress Alissa Milano popularized the movement, bringing it to Twitter and effectively making it viral, quickly turning the hashtag MeToo (*#MeToo*) into a symbol of the exposure and condemnation of gender based violence. Today, the online movement reaches across borders, through platforms like Twitter, Reddit, and Facebook, with #MeToo being translated into a variety of different languages, or associated with other, in-country movements.

Inspired by the theoretical framework in [13], we investigate the effects of country-level *population size*, *income level*, *language*, and, since #MeToo is a gender-related movement, *gender equality score* [16] on 1) the *scale of participation* in the movement, defined by the volume of Twitter messages posted from each country, which we separate into *total in-country participation* over the 3 years, and *participation in-country peak surges of online messaging*; and 2) the *timings of the first surge* and *highest surge* in each country, relative to the very first transnational peak of #MeToo messages posted on Twitter—which occurred in the United States. Our results indicate that the scale of participation is related to a country's population size and income level and that the timings of surges are related to a country's population size, gender equality score, and the language used in messages. We believe these findings provide early evidence to support existing theoretical frameworks that describe the diffusion of transnational social movements, while also shedding light on certain characteristics specific to online activism.

2 Related Work

2.1 The Study of Transnational Movements

Several theoretical frameworks attempt to describe the emergence of transnational social movements. Giugni [13] proposes a general model of "crossnational" similarities among social movements that integrates three broad concepts: globalization, structural affinity, and diffusion. *Globalization* relates to the increasing interconnectedness of economies, cultures, and ideologies that can stimulate movements simultaneously across countries. *Structural affinity* relates to the similarity of structures in different countries that can foster similar movement activity. *Diffusion* relates to the direct and indirect flows of information across countries that diffuse ideas from one place to another. Porta and Tarrow [10] also emphasize the accelerating effect of globalization on diffusion. International travels, languages that are commonly spoken, and the Internet all contribute to transnational movements. Bennett [6] further identifies *political ideology*, *organization models*, and *social technologies* as three important factors that can transform local movements into transnational ones.

Some studies attempt to compare the uptake of transnational social movements in countries with different cultural backgrounds. Batliwala [5] finds that one common characteristic in transnational movements is the cognitive changes of the grassroots people who seek their rights of accountability. Walgrave et al. [28] propose a theoretical typology of motivations for participating in protests across issues and countries, where motivations are highly contextual, and partially dependent on individual characteristics, as well as on the issue at stake and the type of movement. Thomas et al. [27] develop a model to describe the formation of a transnational movement, and of group consciousness of the plight of Syrian refugees. The model accounts for the world-views and exposure to iconic events through social media of individuals from six countries. Finally, Merry [20] analyzes the means and entities through which transnational ideas like a human rights approach to gender based violence translate down and up between global and local arenas. The author finds that community leaders, non-governmental organizations, and social activists play a critical role, and advocates for a deeper anthropological analysis of how these translators contribute to the circulation of human rights ideas and interventions.

While insightful for our work, these studies mainly focus on movements and protests that have taken place in 'conventional', offline settings. Our interest is in the diffusion of an online movement across a multitude of different countries.

2.2 The MeToo Movement

Several studies focus on the #MeToo movement as a whole, independent of country borders. Xiong et al. [29] and Bogen et al. [7] analyze the semantics and themes discussed in online messages related to the movement; Suk et al. [26] look at the temporal dynamics of case reporting in MeToo; and Clark-Parsons [9] explores how "hashtag feminism", through the lense of #MeToo, has a potentially contentious tendency to bring the personal into the political realm, by making it visible online.

Other studies focus on the #MeToo movement within specific countries. Pegu [22] conducts in-depth interviews with women in India to understand their views of the significance and impact of the movement. Results reveal a perception of failure of due process in addressing gender based violence, and the author advocates for politics of solidarity and mutual empowerment, rather than of patriarchal, individualistic capitalism. Alvinius and Homlberg [3] study the call made by 1768 Swedish women, in the context of #Metoo, to end gender based violence and harassment in the military. They find that #MeToo is challenging the norms of hyper-masculine military organizations. Kaufman et al. [15] show how #MeToo has changed people's information-seeking behaviors about sexual harassment online in the United States. Rottenberg et al. [24] find that newspapers and the press have heightened the visibility of #MeToo in the United Kingdom. In contrast, Hassan et al. [14] focus on the non-participation of South Asian women in the movement. They conduct an online survey and in-depth interviews with Bangladeshi women to understand whether they have reasons and a motivation to participate in the movement, and what might suppress them. The authors find that despite being supportive of the movement, cultural differences, social resistance, and the lack of actionable hope generally hold these women back.

Further comparative studies investigate differences in the uptake of the #MeToo movement between countries and languages. Lopez et al. [17] compare the content of online messages containing either #MeToo or #BalanceTonPorc, the French counterpart to #MeToo. They find messages using #BalanceTonPorc contain more aggressive and accusatory content, while a subset of messages using the #MeToo hashtag from India focus on religion-related issues. Kunst et al. [16] conduct cross-cultural surveys investigating gender differences in attitudes towards the #MeToo movement in the United States and Norway. Their results suggest that differences are rather ideological than related to the groups themselves. Finally, Askanius and Hartley [4] examine the similarities and differences between #MeToo in Sweden and Denmark. In both countries, the issue of sexual harassment is framed as an individual problem, rather than a collective, social problem. Further, more voices are critical of the movement in Denmark.

We build on these latter, comparative studies. We are interested in establishing a macro view of the diffusion of the #MeToo movement, and its associated local campaigns, in a multitude of different countries. We investigate the influence of country-level socio-economic and cultural variables on the *scale of participation* and the *timings of peak surges* in online activity related to the movement—a quantitative perspective that we believe complements the more qualitative work described above.

3 Data, Variables, and Transformations

3.1 Data Collection

We focus on #MeToo on Twitter, as it is where the movement first went viral. To gather a comprehensive data set, we worked with policy specialists at UN Women, the United Nations entity for gender equality and the empowerment of women. They provided us with the following list of hashtags used in different countries and languages, all related to #MeToo:

#MeToo, #YoTambien, #SendeAnlat, #WithYou, #WeToo, #QuellaVoltaChe, #AnaKaman, #RiceBunny, #BalanceTonPorc, #Cuéntalo, #TimesUp, #TimeisNow, #MeQueer #немолчи, #СегаКажувам, #米兔, and #미투 .

We acknowledge this list may not be globally exhaustive. However, it captures the major campaigns associated with #MeToo movement in languages like English, Spanish, French, Chinese, and Russian. We consider this sufficient for a first comparative study of the potential country-level factors that may have influenced #MeToo's diffusion across countries with different socio-economic and cultural backgrounds. Further, we understand that building a corpus and derived data sets based only on a search for hashtags might not be representative of the full extent of online conversations related to a topic. Messages can be related to gender based violence without containing any of these specific hashtags—or any hashtag at all. However, we chose to favor precision over recall here.

We used the platform Crimson Hexagon (now Brandwatch) to collect the data [1]. Crimson Hexagon provides access to statistics derived from the full Twitter firehose (*e.g.*, the daily volume of messages for a given query), and random samples of 10,000 unique tweets for given timeframes. We retrieved statistics on the daily volume of messages posted from each of the 33 countries between October 1, 2016, and October 1, 2019 (1095 days in total) to measure *total in-country participation*, and *participation in-country peak surges of online messaging*. We used a three-year range to ensure we captured the possible delayed diffusion of the movement in different countries. We also retrieved sampled tweets content (up to 10,000 per month for 36 months) to measure the use of languages in each country. For countries with fewer than 360,000 tweets, all posts are retrieved.

Crimson Hexagon also provides provenance information at country level for a subset of retrieved messages—essentially either those messages that are geolocated, or for which the author has provided some location information in their profile. While we acknowledge this is imperfect, we did not investigate further methods for locating messages based on language, as *e.g.*, posts in English can easily come from the United States, England, Canada, India, or Malaysia. Further, as we could only download random samples of the actual messages using Crimson Hexagon, and since we could not ensure geographic representativity of those samples, we chose not to investigate other crawling methods for identifying language-specific communities (as done in [23]) to infer location. In the end, we limited our study to the 33 most active countries, from each of which more than 45,000 messages containing at least one of the hashtags in our list were posted in the three-year timeframe, according to the statistics provided by Crimson Hexagon. We chose this threshold, because the number of located messages sent from the 34th country plummeted to 10,000, indicating either low usage of Twitter, or simply non-participation in #MeToo.

3.2 Summary Information

Figure 1 shows the proportion of different languages used, as well as the total number of located messages per country in our data set. We see that English is most common, even in messages from countries where it is not the primary spoken language, like Finland, India, Indonesia, Pakistan, or Malaysia. This seems to indicate that people who take part in the #MeToo movement on Twitter in

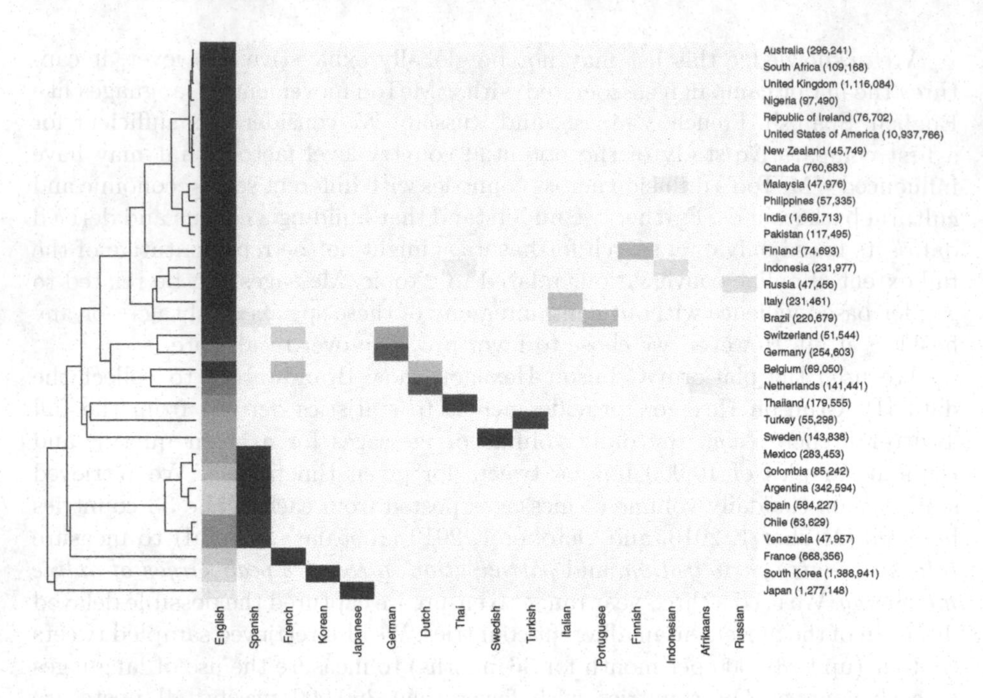

Fig. 1. Number of tweets in different languages across nations.

these countries belong to specific demographics that are likely not representative of the entire national population—these demographics might in fact simply correlate with those of Twitter users in these countries. However, we do not consider this lack of representativity a prohibitive issue, as social activists are never representative of an entire population, and our focus is specifically on #MeToo on Twitter. Messages from other countries like Switzerland, Germany, the Netherlands, Thailand, and Turkey are also posted in English, but to a lesser extent. Conversely, messages from France, South Korea, and Japan are hardly ever posted in English. Finally, we see that Spanish is also commonly used in Latin American countries—where it generally is the primary spoken language—like Mexico, Colombia, Chile, and Venezuela.

Figure 2 shows the Empirical Cumulative Distribution Functions (ECDF) of the daily volumes of messages in all 33 countries. We see that the trends are quite similar across countries: activity is low for 90%–95% of all days, and very high for less than 1% of all days. Note the United States differs slightly, with more days having high activity. This general trend is further illustrated in Fig. 3, which shows the daily volume of messages posted from the United States, Mexico, and South Korea. We see that volumes are overall low, but there are peak surges that last roughly three to five days during which activity is very high.

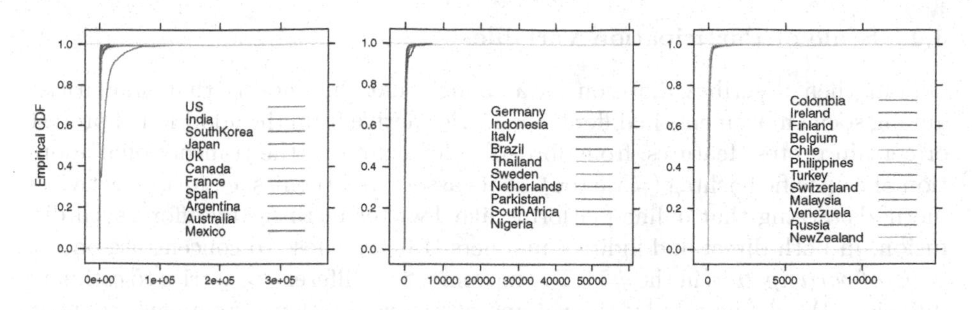

Fig. 2. Empirical cumulative distribution of daily volume of tweets across nations.

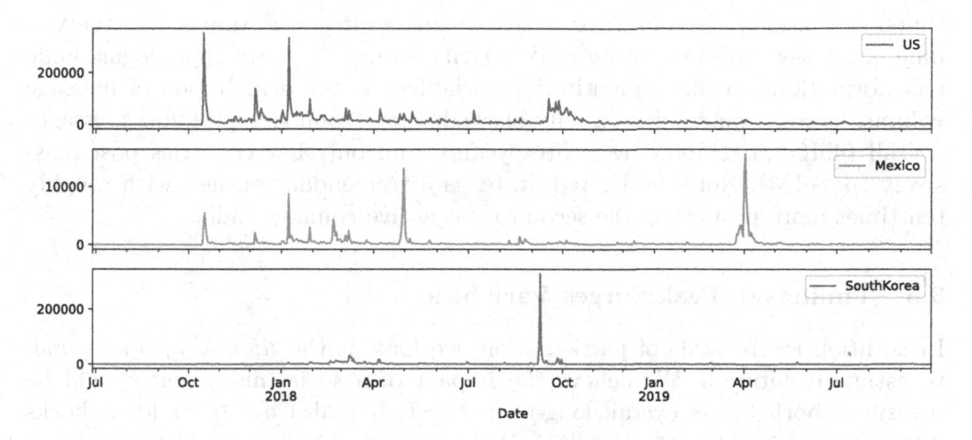

Fig. 3. Number of tweets of the US, Mexico and South Korea in daily granularity.

3.3 Socio-Economic and Cultural Variables

Using the concepts in Guigni's general model [13], we look at the effects of *structural affinity* (independent variable) on *diffusion* (dependent variable) of the transnational #MeToo movement. Guigni describes structural affinity as "the presence of similar structures among different nations [...] while the globalization model refers to structural or contingent events on the transnational level, the structural affinity model focuses upon structural similarities on the national level." We interpret these structures as country-level socio-economic and cultural variables, and investigate the effect of *population size* [21], *income level* [12]—which we bin into *low-*, *middle-*, and *high-income* levels—, *language*, and *gender equality score* [12] on diffusion.

3.4 Scale of Participation Variables

Guigni then describes diffusion as a model that "maintains that similarities among social movements in different countries derive from the adoption of protest or certain protest features, from abroad"—for example, the transnational adoption of a specific hashtag (#MeToo), or its associated themes, claims, or activism methods. Seeing that online platforms like Twitter intrinsically afford such diffusion, in both direct and indirect manners [13], we chose to concentrate on the *scale of participation* in the #MeToo movement in different countries to estimate diffusion. We define this by the volume of messages posted from each country, which we separate into: 1) the *total in-country participation*, *i.e.*, the total volume of messages posted from each country, accumulated over our three-year timeframe of analysis; and 2) *participation in-country peak surges*, *i.e.*, the volume of messages posted during peak activity surges. Note we apply logarithmic transformations to these quantitative variables, as the distribution of message volumes across countries is heavily skewed: most countries post few messages overall (45K$< n <$ 200K over three years), and only few countries post massively ($n >$ 1M). Note the United States is a tremendous outlier, with roughly ten times more posts than the second most active country, India.

3.5 Timings of Peak Surges Variables

In addition to the scale of participation, we look at the *timings of peak surges* to estimate diffusion. We believe the impact of a social movement should be measured both by its overall long-term effect, but also by the sudden shocks it provokes on public attention [15]. We apply a continuous wavelet transform-based peak detection algorithm to identify surges of activity in each country [11]. We set the minimum distance between two peaks to 5, as a manual inspection revealed that surges usually last three to five days. We also set a lower threshold for peaks at 1/3 of the highest peak to filter out less influential surges. We then extract the *first surge* and *highest surge* for each country. Finally, to measure the offset of diffusion in different countries, we set the very first transnational peak as a benchmark, and compute the relative number of days between that benchmark and each in-country surge. We find that first surges occur up to 595 days after the benchmark, and highest surges occur up to 622 days after the benchmark. These various transformations enable us in the next section to evaluate the effects of the structural affinities we have identified on our diffusion variables.

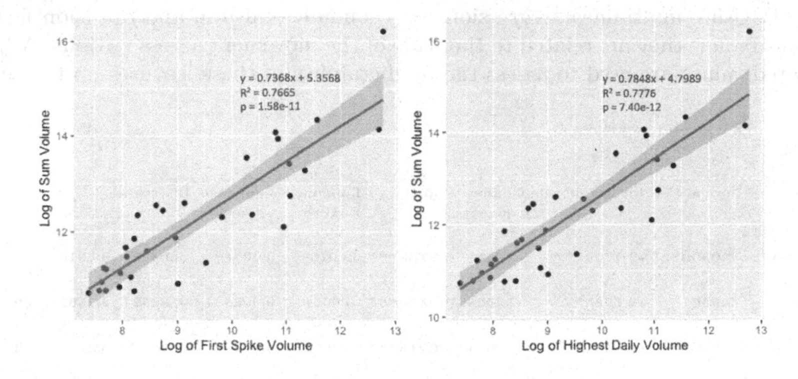

Fig. 4. Correlations of log of sum volume with log of first surge volume and log of highest surge volume.

4 Results

4.1 Comparing Peak Surges and Total In-Country Participation in the #MeToo Movement Across Countries

We first examine the interactions between our diffusion variables to establish whether the scales of the first and highest surges are indicative of the total in-country participation over the three-year timeframe of analysis. Figure 4 shows the linear relations between the log volumes of messages. We see a positive correlation between both the amount of messages posted during the first surge and the total in-country volume, and the amount of messages posted during the highest surge and the total in-country volume. A Pearson's correlation test further confirms this, with a correlation coefficient of **0.8797** ($p < 0.01$) for the first comparison; and a correlation coefficient of **0.8857** ($p < 0.01$) for the second comparison. These results indicate that the overall, in-country scale of participation in #MeToo is largely determined by the scale of the first and highest surges.

4.2 Relating Socio-Economic and Cultural Variables to the Scale of Participation in #MeToo

We next run a multivariate regression on the different scales of participation using our socio-economic and cultural variables. We aim to identify whether these variables are predictive of the scale of the movement, and to what extent; and which variables are statistically significant, and how these affect our diffusion variables. We regress the features separately on the three dependent, scale of participation variables. Table 2 shows the results of the three regressions (one per line). As mentioned above, we bin income into *low-*, *middle-*, and *high-income* levels, and show the coefficients for each bin independently. We also limit the

Table 1. Using multi-linear regressions on economic, cultural and diffusion features, we quantify how they are related to the scale of the movement across nations. Adjusted R-squared values are used to assess the predictability of these features on the scale.

Scale of participation	Explanatory variables							Adjusted R^2	p-value
	Income low	Income middle	Gender equality index	Pop (log)	Language English	Language Spanish	Intercept		
log(Total Vol)	**-3.0300*** **	**-2.1703*** **	-1.9411	**0.9738*** **	0.0070	0.0091	-3.0436	0.5916	2.974e-5
log(1st Surge Vol)	**-2.5361***	**-1.5935***	-2.4598	**0.9374*** **	-0.0071	0.0058	-4.3352	0.3562	0.0061
log(Highest Vol)	**-2.3537** **	**-1.6197** **	-3.6873	**0.9013*** **	-0.0052	0.0124	-2.8808	0.4209	0.0019

Signicance levels: 0 '***'; 0.001 '**'; 0.01 '*'; 0.05 '.'; 0.1 ' '

language variable to English and Spanish, the two most used in our data, and use percentages of messages posted in these languages for the regression.

We see that all regressions are statistically significant ($p < 0.01$; $0.35 <$ adj. $R^2 < 0.59$). The regression for the total in-country participation has the highest adjusted R^2, which means that this dependent variable is best predicted by the socio-economic and cultural variables we focus on. Further, income level and the log(population) are the most significant variables to explain the variation of participation across countries. The log(population) is positively correlated with participation, while both low and middle income levels are negatively correlated with participation. This indicates that there is more participation in #MeToo in countries with larger populations, and less participation in poorer countries.

To assess whether these results are influenced by the amount of Twitter users in each country—hypothesizing that larger, wealthier populations are likely to include more Twitter users who might be exposed to, and engage in the #MeToo movement than smaller, poorer populations—, we conduct a post-hoc evaluation, including data on Twitter user populations in 19 of the 33 countries of interest in our regression [2]. We find that results are not statistically significant ($p = 0.26$), which suggests that the amount of in-country Twitter users likely has no influence on the amount of in-country participation.

Meanwhile, the gender equality score and the language used seem irrelevant here. These results somewhat differ from previous studies on people's motivations for participating in more conventional, offline movements, *e.g.*, protests and demonstrations. For example, Walgrave et al. [28] find that people's participation in anti-war demonstrations is partially determined by instrumental and expressive motives.

4.3 Relating Socio-Economic and Cultural Variables to the Relative Timing of Peak Surges

Finally, we analyze the effect of our socio-economic and cultural variables on the timings of peak surges. Table 2 shows the results for two multivariate regressions applied to the timings of the first and highest surges. Here we use population instead of log(population) as the models fit better.

Table 2. Using multi-linear regressions on economic, cultural and diffusion features, we quantify how they are related to the time points of outbursts. Adjusted R-squared values are used to assess the predictability of time delayed for outbursts.

Time of Movement	Explanatory Variables							Adjusted R^2	p-value
	Income low	Income middle	Gender equality index	Pop	Language English	Language Spanish	Intercept		
1st Surge Delay	14.03	67.07	-587.6	**2.083e-7***	**-1.874***	-0.3183	570.6	0.3463	0.0073
Highest Surge Delay	148.2	72.94	**-1072***	6.364e-8	-0.8479	1.735	885.8	0.3948	0.0031

Signicance levels: 0 '***'; 0.001 '**'; 0.01 '*'; 0.05 '.'; 0.1 ' '

Both regressions are statistically significant ($p < 0.01$; $0.34 <$ adj. $R^2 < 0.39$). However, compared to the scale of participation variables, the timings of surges seem relatively poorly predicted by our socio-economic and cultural variables. For the first surge, the population and percentage of tweets in English are statistically significant ($p < 0.01$). The use of English is also negatively correlated with the timing of the surge, meaning that countries from which a higher percentage of messages are posted in English tend to see their first surge in activity earlier. This is likely simply explained by the fact that #MeToo first went viral in English, and that English is an international language that is easily diffused. Meanwhile, the gender equality score is statistically significant ($p < 0.01$), and negatively correlated with the timing of the highest surge. This indicates that countries with lower gender equality tend to see their highest surge delayed more. All other socio-economic and cultural variables investigated here do not seem relevant for predicting the timing of the highest surge. This resonates with the findings of Hassan et al. [14], which indicate that people are likely suppress their feelings in countries with low gender equality, and not voice their claims as much.

5 Discussion

Our results indicate that:

1. the total in-country participation in #MeToo is largely determined by the scale of participation in the *first* and *highest* in-country surges;
2. the in-country scale of participation in #MeToo is related to the population size an income level of the country; and

3. the timings of the first and highest in-country surges are related to the country's population size, gender equality score, and the language used in messages.

While an increasing amount of academic attention is brought to transnational movements, most comparative studies limit themselves to only a handful of countries. Scholars tend to focus on the similarities across countries, like the collective themes and mobilization strategies, or on individuals' motivations, like their perceptions and psychological processes. To our knowledge, our study is the first longitudinal investigation of the diffusion of #MeToo across 33 countries.

That said, we acknowledge our work currently has several limitations. First, Twitter is our only source of data for characterizing the #MeToo movement. #MeToo is spread all over the Internet, with people participating on different social media platforms, like Facebook, Reddit, or Instagram. Previous work [19] has shown that users tend to have slightly different participation behaviors on Twitter and Reddit: they tend to share more personal stories of sexual harassment or assault on Reddit, while they post more messages of sympathy and encouragement on Twitter. While studying the content shared on these other platforms will be necessary, we stand by our decision to start with Twitter, as it is where the movement first went viral, and it is fairly widely used across different countries (as opposed to *e.g.*, Reddit). Second, we understand that by exclusively using located message, we only capture the participation of people who specify their locations, or have geolocation enabled on their account. While this is still imperfect, since people might declare being in one location while they are in another, we believe it remains the most robust way to identify where messages come from, in order to establish a comparison between countries. Finally, we understand the socio-economic and cultural variables we selected are arbitrary and not exhaustive. More will need to be investigated.

6 Conclusion

In this paper, we have presented a macro, comparative study of the transnational #MeToo movement on Twitter across 33 countries. We have performed multivariate regressions to examine the influence of socio-economic and cultural variables on the in-country scale of participation, and the timings of peak surges of messages related to #MeToo. In the future, we intend to focus more on the up and down translation of transnational ideas between local and global arenas, as called for by Mery [20]. Here, we have focused on the effect of what Guigni [13] refers to as the *structural affinity* of countries on *diffusion*, but we have not looked into the thematic and rhetoric similarities in messages from different countries. This would require analyzing the structure and content of messages in different languages, which is challenging. However, it should help shed light on how the general idea of gender-based violence moves back and forth between national and transnational scales.

In a time when online movements are increasingly influential, we hope this study will inspire more theoretical work on transnational movements, especially in online environments. We also hope our early insights can help international organizations like UN Women identify opportunities for diffusing transnational campaigns.

Acknowledgement. This work has been done in collaboration with the joint fellowship program between UN Global Pulse and UN Women.

References

1. Crimson hexagon (2019). https://forsight.crimsonhexagon.com/. Accessed 27 Dec 2019
2. Leading countries based on number of twitter users as of April 2020 (2020). https://www.statista.com/statistics/242606/number-of-active-twitter-users-in-selected-countries/. Accessed 28 May 2020
3. Alvinius, A., Holmberg, A.: Silence-breaking butterfly effect: Resistance towards the military within# metoo. Work & Organization, Gender (2019)
4. Askanius, T., Hartley, J.M.: Framing gender justice: a comparative analysis of the media coverage of# metoo in denmark and sweden. Nordicom Review **40**(2), 19–36 (2019)
5. Batliwala, S.: Grassroots movements as transnational actors: implications for global civil society. Voluntas **13**(4), 393–409 (2002)
6. Bennett, W.L.: Social movements beyond borders: understanding two eras of transnational activism. Trans. Protest Global Activism **203**, 26 (2005)
7. Bogen, K.W., Bleiweiss, K.K., Leach, N.R., Orchowski, L.M.: #MeToo: disclosure and response to sexual victimization on Twitter. J. Interpers. Violence (2019). https://doi.org/10.1177/0886260519851211. Epub ahead of print. PMID: 31117851
8. Breuer, A., Landman, T., Farquhar, D.: Social media and protest mobilization: evidence from the tunisian revolution. Democratization **22**(4), 764–792 (2015)
9. Clark-Parsons, R.: "i see you, i believe you, i stand with you": #MeToo and the performance of networked feminist visibility. Feminist Media Stud. (2019). https://doi.org/10.1080/14680777.2019.1628797
10. Della Porta, D., Tarrow, S.: Transnational processes and social activism: an introduction. Trans. Protest Global Activism **1**, 175–202 (2005)
11. Du, P., Kibbe, W.A., Lin, S.M.: Improved peak detection in mass spectrum by incorporating continuous wavelet transform-based pattern matching. Bioinformatics **22**(17), 2059–2065 (2006)
12. Forum, W.E.: The global gender gap report (2018). http://reports.weforum.org/global-gender-gap-report-2018/. Accessed 16 Dec 2019
13. Giugni, M.: The other side of the coin: explaining crossnational similarities between social movements. Mobilization: Int. Q. **3**(1), 89–109 (1998)
14. Hassan, N., Mandal, M.K., Bhuiyan, M., Moitra, A., Ahmed, S.I.: Nonparticipation of bangladeshi women in# metoo movement. In: Proceedings of the Tenth International Conference on Information and Communication Technologies and Development. p. 29. ACM (2019)
15. Kaufman, M.R., Dey, D., Crainiceanu, C., Dredze, M.: #MeToo and google inquiries into sexual violence: a hashtag campaign can sustain information seeking. J. Interpers. Violence (2019). https://doi.org/10.1177/0886260519868197. Epub ahead of print. PMID: 31441695

16. Kunst, J.R., Bailey, A., Prendergast, C., Gundersen, A.: Sexism, rape myths and feminist identification explain gender differences in attitudes toward the# metoo social media campaign in two countries. Media Psychol. **22**(5), 818–843 (2019)
17. Lopez, I., Quillivic, R., Evans, H., Arriaga, R.I.: Denouncing sexual violence: a cross-language and cross-cultural analysis of #MeToo and #BalanceTonPorc. In: Lamas, D., Loizides, F., Nacke, L., Petrie, H., Winckler, M., Zaphiris, P. (eds.) INTERACT 2019. LNCS, vol. 11747, pp. 733–743. Springer, Cham (2019). https://doi.org/10.1007/978-3-030-29384-0_44
18. Luo, X.R., Zhang, J., Marquis, C.: Mobilization in the internet age: internet activism and corporate response. Acad. Manag. **59**(6), 2045–2068 (2016)
19. Manikonda, L., Beigi, G., Liu, H., Kambhampati, S.: Twitter for sparking a movement, reddit for sharing the moment:# metoo through the lens of social media. arXiv preprint arXiv:1803.08022 (2018)
20. Merry, S.E.: Transnational human rights and local activism: mapping the middle. Am. Anthropol. **108**(1), 38–51 (2006)
21. Nations, U.: World population prospects: The 2019 revision (2019). https://population.un.org/wpp/Download/Standard/Population/. Accessed 12 Jan 2020
22. Pegu, S.: MeToo in India: building revolutions from solidarities. DECISION **46**(2), 151–168 (2019). https://doi.org/10.1007/s40622-019-00212-x
23. Pratikakis, P.: twawler: A lightweight twitter crawler. CoRR abs/1804.07748 (2018), http://arxiv.org/abs/1804.07748
24. Rottenberg, C., Orgad, S., De Benedictis, S.: # metoo, popular feminism and the news: A content analysis of uk newspaper coverage. Eur. J. Cult. Stud. **22**(5–6), 718–738 (2019)
25. Schneider, K.T., Carpenter, N.J.: Sharing# metoo on twitter: incidents, coping responses, and social reactions. An International Journal, Equality, Diversity and Inclusion (2019)
26. Suk, J., Abhishek, A., Zhang, Y., Ahn, S.Y., Correa, T., Garlough, C., Shah, D.V.: #MeToo, networked acknowledgment, and connective action: How "empowerment through empathy" launched a social movement. Soc. Sci. Comput. Rev. (2019). https://doi.org/10.1177/0894439319864882
27. Thomas, E.F., Smith, L.G., McGarty, C., Reese, G., Kende, A., Bliuc, A.M., Curtin, N., Spears, R.: When and how social movements mobilize action within and across nations to promote solidarity with refugees. Eur. J. Soc. Psychol. **49**(2), 213–229 (2019)
28. Walgrave, S., Van Laer, J., Verhulst, J., Wouters, R.: Why do people protest? comparing demonstrators' motives across issues and nations. Media, movements and politics (2013)
29. Xiong, Y., Cho, M., Boatwright, B.: Hashtag activism and message frames among social movement organizations: semantic network analysis and thematic analysis of twitter during the# metoo movement. Public Relat. Rev. **45**(1), 10–23 (2019)

Author Index

Abdelali, Ahmed 237
Abeliuk, Andrés 206
Alani, Harith 28
Almeida, Jussara Marques 252
Arpinar, I. Budak 427

Bailey, Michael 1
Barbouch, Mohamed 15
Boldrini, Chiara 267, 283
Bonanomi, Andrea 380
Boy, Jeremy 447
Burel, Grégoire 28

Capozzi, Arthur 43
Carley, Kathleen M. 364
Carrington, Martin 168
Casino, Fran 181
Cattuto, Ciro 380
Chen, Lucia Lushi 58
Conti, Marco 267, 283
Cummings, Patrick 206

Daida, Shashidhar Reddy 447
Darwish, Kareem 237, 333
De Francisci Morales, Gianmarco 43
Delgado, Myriam 312
Deshmukh, Jayati 137

Farrell, Tracie 28
Ferreira, Carlos Henrique Gomes 252
Field, Anjalie 364
Frizzo, Gabriel 312

Gaur, Manas 427
Goel, Rahul 67

Han, Jinyoung 124, 440
Han, Sumin 81
Han, Xu 410
Hassan, Sabit 237
Helic, Denis 152
Hisano, Ryohei 95
Hong, Lingzi 447

Inan, Hale 427
Iyetomi, Hiroshi 95

Johnston, Drew 1
Joo, Jungseock 440

Kalimeri, Kyriaki 380
Khandelwal, Saransh 108
Khare, Prashant 28
Kim, Seungbae 124
Kleinberg, Bennett 397
Koley, Gaurav 137
Koncar, Philipp 152
Kuchler, Theresa 1
Kursuncu, Ugur 427

Lathwal, Priyank 364
Lee, Dongman 81
Lee, Kyumin 220
Lerman, Kristina 206
Li, Boxuan 168
Lüders, Ricardo 312
Lv, Qin 410
Lykousas, Nikolaos 181

Magdy, Walid 58
Marbach, Peter 168
Mejova, Yelena 43, 192
Mensio, Martino 28
Mizuno, Takayuki 95
Mokhberian, Negar 206
Monti, Corrado 43
Mou, Guanyi 220
Mubarak, Hamdy 237

Nobre, Gabriel Peres 252

Ollivier, Kilian 267

Panisson, André 43
Paolotti, Daniela 43, 380
Paraskevopoulos, Pavlos 283
Park, Eunil 440

Park, Kinam 81
Pasi, Gabriella 297
Passarella, Andrea 267, 283
Patsakis, Constantinos 181
Putri, Divi Galih Prasetyo 297

Rosina, Alessandro 380
Routray, Aurobinda 108
Russel, Dominic 1

Senefonte, Helen 312
Shahaf, Gal 320
Shalin, Valerie L. 427
Shapiro, Ehud 320
Sharma, Rajesh 67
Sheth, Amit 427
Shurafa, Chereen 333
Silva, Thiago 312
Silver, Daniel 312
Sivak, Elizaveta 352
Smirnov, Ivan 352
Srinivasa, Srinath 137
State, Bogdan 1
Stroebel, Johannes 1
Suarez-Lledó, Víctor 192

Takes, Frank W. 15
Talmon, Nimrod 320
Thirunarayan, Krishnaprasad 427
Tsvetkov, Yulia 364
Tyagi, Aman 364

Urbinati, Alessandra 380

van der Vegt, Isabelle 397
Verberne, Suzan 15
Viviani, Marco 297

Wang, Yichen 410
Whalley, Heather 58
Wijesiriwardene, Thilini 427
Wolters, Maria 58

Yoon, Jeewoo 440
Yu, Xinchen 447

Zaghouani, Wajdi 333
Zhang, Jason Shuo 410

Printed in the United States
By Bookmasters